LEAN PROJECT DELIVERY AND INTEGRATED PRACTICES IN MODERN CONSTRUCTION

Lean Project Delivery and Integrated Practices in Modern Construction is the new and enhanced edition of the pioneering book *Modern Construction* by Lincoln H. Forbes and Syed M. Ahmed.

This book provides a multi-faceted approach for applying lean methodologies to improve design and construction processes. Recognizing the wide diversity in the landscape of projects, and encompassing private and public sector activity, buildings and infrastructure, the book expands upon the detailed coverage of integrated project delivery and new lean tools and techniques to include:

- Greater emphasis on the importance of creating a lean culture and the initiatives required to transform the industry;
- Expanded discussions of the foundational writings in lean construction theory;
- Exploration of the synergies between "lean" and "green" initiatives;
- Specific procedures for modifying planning and scheduling activities to improve the performance of the project team;
- Expanded sections on quality, and topics that have become a part of the lean lexicon, such as Choosing by Advantages, "line of balance"/location-based scheduling, virtual design teams, takt time planning and set-based design;
- Discussion questions for beginners and advanced lean practitioners; and
- Improved cross-referencing within the text to help the reader navigate the frameworks, techniques and tools to support the application of lean principles.

The techniques described here enhance the use of resources, reducing waste, minimizing delays, increasing quality and reducing overall costs. They enable practitioners to improve the quality of the built environment, secure higher levels of customer/owner satisfaction, and simultaneously improve their profitability. This book is essential reading for all those wanting to be at the forefront of construction management and lean thinking.

Lincoln H. Forbes, PhD, PE, CM-Lean, specializes in Lean Project Delivery and lean management systems. A professional engineer, he has over 35 years' experience in quality/performance improvement as well as facilities design and construction management support. Dr. Forbes is a member of the Lean Construction Institute (LCI) and the International Group for Lean Construction (IGLC). He holds the CM-Lean credential from the Associated General Contractors of America (AGC). He is a fellow of the Institute of Industrial and Systems Engineers (IISE), and is the founder and past president of their Construction Division. A senior member of the American Society for Quality (ASQ), he also holds the LEED AP designation. He held senior facilities-related positions with Miami-Dade County Public Schools earlier in his career. Dr. Forbes has also served in adjunct positions in construction management programs at East Carolina University, Florida International University, and Drexel University. He is a principal consultant for Harding Associates, Inc. in Miami, Florida. He holds PhD, MBA, and MSIE degrees from the University of Miami. He has authored several papers, columns, book chapters and books; he has served as a reviewer for the ASCE *Journal of Construction Engineering and Management*, and the *Journal of the Built Environment Project and Asset Management* (BEPAM).

Syed M. Ahmed, PhD, is a full Professor and Chairperson of the Department of Construction Management at East Carolina University (ECU) in Greenville, North Carolina. He was Associate Professor and Graduate Program Director in the Department of Construction Management of the College of Engineering & Computing at Florida International University (FIU) in Miami. He was also an affiliated faculty member of the Department of Civil & Environmental Engineering at FIU. Dr. Ahmed has over 35 years of international experience in industry, teaching, research, and consulting in the United States, Hong Kong, Pakistan, Mexico, and Jamaica. He is the author of several books and has published extensively with over 160 peer-reviewed publications in international journals and conferences in his areas of expertise. Dr. Ahmed is currently on the editorial board of more than six international journals and serves as the associate editor of the American Society of Civil Engineers (ASCE) *Journal of Construction Engineering and Management* and the *Built Environment Project and Asset Management* (BEPAM) journal.

LEAN PROJECT DELIVERY AND INTEGRATED PRACTICES IN MODERN CONSTRUCTION

Second Edition

Lincoln H. Forbes and Syed M. Ahmed

Routledge
Taylor & Francis Group
LONDON AND NEW YORK

Second edition published 2020
by Routledge
2 Park Square, Milton Park, Abingdon, Oxon, OX14 4RN

and by Routledge
52 Vanderbilt Avenue, New York, NY 10017

Routledge is an imprint of the Taylor & Francis Group, an informa business

First edition published by Routledge 2010

British Library Cataloguing-in-Publication Data
A catalogue record for this book is available from the British Library

Library of Congress Cataloging-in-Publication Data
Names: Forbes, Lincoln H., author. | Ahmed, Syed M. (Syed Mahmood), 1960– author.
Title: Lean project delivery and integrated practices in modern construction/
Lincoln H. Forbes & Syed M. Ahmed.
Other titles: Modern construction
Description: Second edition. | Abingdon, Oxon ; New York, NY : Routledge, 2020. |
Originally published: Modern construction. Boca Raton, FL : CRC Press, c2011. |
Includes bibliographical references and index.
Identifiers: LCCN 2019033716 (print) | LCCN 2019033717 (ebook) |
ISBN 9781138311244 (hbk) | ISBN 9780429458989 (ebk)
Subjects: LCSH: Building. | Lean manufacturing. | Construction industry—Management. |
Construction industry—Cost control. | Project management.
Classification: LCC TH145 .F67 2020 (print) | LCC TH145 (ebook) | DDC 690.068/4—dc23
LC record available at https://lccn.loc.gov/2019033716
LC ebook record available at https://lccn.loc.gov/2019033717

ISBN: 978-1-138-31124-4 (hbk)
ISBN: 978-0-429-45898-9 (ebk)

Typeset in Bembo
by Apex CoVantage, LLC

To members of my family:
My wife Laurel, son Richard, daughter-in-law Anet,
and grandson Alex, for their encouragement and support
From Lincoln H. Forbes

CONTENTS

ACKNOWLEDGMENTS

This book has materialized through the encouragement, ideas, and contributions of many colleagues and associates in several sectors of the design and construction community as well as the academic community. The authors would like to extend great appreciation to several individuals for their kind support as follows: Dr. Tariq Abdelhamid of Michigan State University who provided many ideas on the content for an update; Prof. Bolivar Senior of Colorado State University for suggestions on helpful topics for both practitioners and students; Dr. Glenn Ballard, LCI Co-founder, for his thoughts on the triple-bottom line, lean leadership, and research ideas; Dr. Zofia Rybkowski and Dr. Cynthia Tsao for ideas on several lean concepts. The authors thank Samir Emdanat, Director of VPlanner and Digby Christian, Director of Integrated Lean Project Delivery for Sutter Healthcare for documenting their experiences with integrated project delivery (IPD) based on many projects with Sutter, a pioneer owner in the lean movement, as well as Ms. Amanda Barnes of Southland Industries for pictures of lean projects. We thank Chris Dierks, Project Executive and East Coast Director of Lean with DPR Construction for valuable guidance on engaging project teams and creating a lean mindset; Rebecca Snelling, VP and National Lean Director of JE Dunn Construction Company, Inc. for demonstrating how high commitment to lean enables lean successes. Special thanks are extended to Boulder Associates Architects, especially principals Romano Nickerson and Todd Henderson for sharing their successes with lean initiatives in their design practice.

We extend our appreciation to Tom Plumb, CEO of Kinetic Construction Limited in Canada for inspiring words on implementing lean; to Trevor Coyle, Lean Integration Manager of Kinetic for a highly informative chronology of Kinetic's lean journey, and their successes; Prof. Olli Seppanen of Aalto University in Finland, who provided a valuable service to readers in explaining location-based management systems and takt planning; Prof. Farook Hamzeh and his co-contributors for their research findings on lean metrics; Prof. Salman Azhar for his guidance on building information modeling; and Gilles Belanger, engineer and consultant, for examples of lean and Six Sigma tools and techniques; and Lachmi Khemlani, Ph.D. Editor of AECbytes for information on BIM 360. Thanks to Murray Guy, Commissioning Authority of Eco-Smart (Canada) for information on lean commissioning and to other colleagues who provided important input, such as Ed Anderson of Lean Implementation Services of Florida, Ted Angelo, Executive VP of Grunau Company; Dr. Paz Arroyo of DPR Construction; Dr. Al Attah of MACVAL Associates LLC; and Tom Polyzois and Bill Miller of Limbach Holdings, Inc. for sharing their successes with prefabrication. Our kindest regards go to Henry Nutt III, Preconstruction/Business Development

Manager of Southland Industries and Klas Berghede, Director of Production for The Boldt Company for providing information on efforts to update the Last Planner System, with guidance from Dr. Glenn Ballard.

The authors would like to express their special appreciation to Ms. Janie Poncelet, BSW, Ms. Amelia Saul, MS, LRT/CTRS, and Ms. Nihal Erian, Graduate Research Assistant, for their support. Nihal's input on the safety and sustainability chapters was very helpful.

PREFACE

This book was written to support the emergence of lean design and lean construction as a strategy for improving the construction industry that continues to fall short of expectations. Construction is a large sector of many countries' economies, hence its performance directly influences their competitiveness. The book is intended to provide strategies that relate primarily to the management of people and processes, and that can be implemented worldwide. Many of these tools and strategies have been developed on a foundation of principles and methodologies that have long served the manufacturing and service industries and have brought them to levels of world class performance.

Over several decades, these industries have greatly increased their levels of productivity, quality and profitability through the application of process improvement techniques and information technology. A major contributor to the competitiveness of these sectors is their embrace of the Lean production philosophy that evolved from the "Just-in Time" system that was the foundation of Toyota's success in automotive manufacture. By comparison, the construction industry worldwide still lags behind the manufacturing and service sectors with regard to the application of performance improvement and optimization techniques. According to the Bureau of Labor Statistics (BLS), in the past 50 years the productivity of non-farm industries has increased by over 100%, while construction industry productivity has experienced rates of growth that have alternated around an essentially flat level of performance.

Several research studies have pointed to high levels of waste, i.e., wasted resources, in the construction industry in the order of 30% or more because of entrenched attitudes, practices and behaviors. Waste in design and construction occurs especially in the interaction between project participants, and in the hand-offs of work from one discipline or specialist to another. Studies by the Construction Industry Institute (CII) have identified that 10% of project cost is typically spent on field rework. As global competitiveness increases, so will the expectation of higher levels of productivity in the delivery of constructed facilities. Increasingly, customers in other industries expect their suppliers to be flexible, communicative, and responsive to their needs in today's fast-paced business environment. Construction organizations that fail to utilize the foregoing approaches will find themselves at a competitive disadvantage

Construction industry practitioners have been continually seeking to apply better technologies and processes to improve project delivery, but change has come very slowly because of the lack of a unified strategy, and because there is little incentive to change.

A significant turning point began in the early 1990s with the emergence of "lean construction", precipitated by the research of Lauri Koskela, and followed by the tireless work of many other researchers, including Glenn Ballard, Greg Howell and Iris Tommelein. These efforts led to the creation of the

International Group for Lean Construction (IGLC) and The Lean Construction Institute in the USA. The advocacy of these organizations has led to a growing awareness of Lean Project Delivery in many countries. In the USA, the Lean Construction Institute (LCI) was created in 1997 and has seen significant growth in the attendance at its annual conferences, and a gradually increasing number of projects have been mobilized based on lean construction practices or forms of contract. Nevertheless, a change from long established practices takes time. Many members of the OAEC (Owners, Architects, Engineers, and Contractors) community are still not aware of the benefits of lean project delivery, or are reluctant to try a new approach. Beyond awareness, even more time is needed before that awareness translates into deployment in actual projects. Although the parties to a traditional construction project are interdependent, they tend to make decisions that further their own specific interests. Industry specialization has separated designers from the management of the production process, and this is perpetuated by the potential legal ramifications of apportioning project risk.

This book is intended to share knowledge that will, hopefully, encourage the adoption of lean project delivery by the OAEC community. Owners, especially need to have confidence in lean project delivery because design and construction projects essentially begin with them. Lean practices enhance the utilization of resources, reducing waste and overall costs as a result. They also enable practitioners to improve the quality of the built environment, secure higher levels of customer/owner satisfaction, and simultaneously improve their profitability.

Students are an important part of the target audience, as they represent the future of the industry as they move on to become a part of a project team, or may assume decision-making roles in project delivery.

What's in this book

This book builds on the mission of the first edition, which was to provide an understanding of the lean methodology and its application in the form of techniques to enhance the project management process. The book organizes the treatment of lean methods to improve the availability and understanding of specific procedures for planning and scheduling design and construction activities; this approach improves the interaction between the parties, reduces waste, increases stakeholder value, and improves overall design and construction performance.

In order to accommodate a wide variety of readers the book explains the principles behind a number of design and construction practices. It is assumed for example that some construction practitioners may not have a detailed knowledge of project definition and lean design. By the same token, non-practitioners of construction are provided with an overview of construction delivery methods as a foundation for applying process improvement techniques that may already be familiar to them.

Chapter 1 provides background information on the construction industry and the relationships among its participants – designers, general contractors, construction managers, contractors, subcontractors, owners, and material suppliers, etc. It explains the recent history of the construction sector as a significant portion of a nation's economy, pointing to the consequences of comparatively low productivity. The chapter provides a brief overview of productivity concepts, and how current design and construction practices relate to them.

Chapter 2, entitled "Foundations of lean construction," explains lean theory and provides a very important foundation for the book by explaining how lean concepts apply to the construction environment. It provides a historical background of the Toyota System and the production management concepts that grew from it. It presents the principles of reducing waste and adding value, as well as meeting the requirements for a lean environment. It describes the foundational work by Lauri Koskela, Glenn Ballard, Greg Howell, Sven Bertelsen, Iris Tommelein, and other researchers.

Chapter 3 provides an overview of the lean project delivery system (LPDS) as a proven framework for delivering projects on time, within budgets, with high product quality and minimal safety incidents. This treatment emphasizes the programming and design phase of projects. The application of the Last Planner®

System is explained from the perspective of design as well. The chapter also describes the integration of critical activities in project design and formation such as target value design (TVD), building information modeling (BIM), and designing for sustainability.

Chapter 4A presents lean in the construction phase of the LPDS, including lean supply, lean assembly, and use. The Last Planner® System is described extensively with regard to the joint activities of different specialists in the project team as they plan project execution typically from week to week; using reverse phase scheduling or pull planning to improve work flow predictability and reliability. It also includes the integration of prefabrication with construction processes to leverage the economic advantage of making building sub-assemblies in a controlled manufacturing environment.

Chapter 4B presents the topics of location based management systems (LBMS) and takt time planning with regard to improving workflow. Professor Olli Seppanen explains the functioning of both methods and describes their pros and cons.

Chapter 5 introduces building information modeling (BIM)/Information and Communication Technology (ICT) with an overview to describe how these systems can significantly improve the design of facilities to meet the needs of owners and users. It also explains the important factors to consider such as legal responsibilities in the selection and use of BIM systems.

Chapter 6 addresses the measurement of performance in the construction environment. It explains the techniques in two sections – Part A and Part B. Part A explains the use of traditional measures such as Earned Value and Performance Ability Ratio (PAR) values. It also includes frequently used lean metrics such as PPC (percent planned complete) and Analysis of reasons for non-completion (RNC).

Part B covers more complex metrics such as Tasks Anticipated (TA), Planned Work Ready (PWR), Delta 1, Delta 2 and Delta 3, as well as Percent Required Complete and Ongoing (PRCO).

Chapter 7 describes important cultural factors in construction that are essential for aligning the attitudes and behaviors of design and construction stakeholders to secure successful lean projects. It outlines prerequisites for lean projects, underscores the importance of respect for people, and suggests approaches for organizing lean projects. It describes effective use of Big Room activities, stresses the importance of Respect for People and the role of clear Conditions of Satisfaction (CoS) to ensure that stakeholders' expectations are met effectively.

Chapter 8 (management and worker factors) presents several topics that relate to human performance in the work environment. These topics can have a significant impact on a construction team's ability to maximize its performance. Part A includes: Management and motivation, Theory X and Theory Y, Maslow's Hierarchy of needs, and the criteria of the Malcolm Baldrige National Quality Award. The benefits of diversity are also discussed. Part B includes approaches for improving workforce effectiveness, such as temperature and wind effects, and the effects of overtime. Part C presents work sampling, time studies, and the learning curve. Part D provides selected readings on ergonomics.

Chapter 9 – Integrated Project Delivery (IPD): discusses the management of IPD projects to integrate people, systems, business structures and practices to accomplish high levels of construction performance. It is based on several highly successful projects conducted for Sutter Healthcare in California. The integrated form of agreement IFOA and the financial practices of risk/reward sharing are described. The original IPD project by Westbrook Air Conditioning and Plumbing is also described for its foundational practices. Samir Emdanat of VPlanner Inc. and Digby Christian of Sutter Healthcare share valuable lessons and first-hand experiences that will greatly benefit IPD stakeholders.

Chapter 10 provides several examples of projects that represent lean applications, but at an informal level without the benefit of a lean contract or an Integrated Form of Agreement. Some of the companies involved took the initiative to introduce lean in projects that were otherwise traditional, such as design-build and CM at Risk. By using this approach they were able to derive many of the benefits of lean projects. The examples presented include initiatives of general contractor organizations such as Kinetic Construction Limited of Canada and JE Dunn Construction of the USA.

Chapter 11 – quality management in construction and Six Sigma: A complement to lean construction has two sections: Part A provides important background on the history of quality in construction, including the contributions of several Quality Gurus. These include Deming, Juran, Crosby, and others. It describes Total Quality as an approach to doing business that attempts to maximize the competitiveness of an organization through the continual improvement of the quality of its products, services, people, processes and environments. Part B describes the application of Six Sigma techniques in construction projects.

Chapter 12 ("sustainable construction: environmental sustainability and commissioning") has two sections: It first presents sustainability as an important foundation for lean construction, based on the principles advanced by the US Green Building Council and the Department of Energy. The chapter also discusses synergies between "lean and green" in construction projects.

The second section describes Commissioning as serving two important roles – ensuring sustainable building performance through "Green" certification, and supporting project-wide quality control.

Chapter 13: safety management, addresses construction site safety practices from a lean perspective. It is an area of concern for employers of construction workers, this concern for safety is extended to the employees of subcontractors as well. Successful project delivery requires that safety incidents be minimized. The chapter describes several strategies for improving safety, beginning with the design process, especially as a foundation for lean construction.

Chapter 14: general lean tools presents a wide range of problem-solving approaches in one convenient location. Examples of the tools are the PDCA cycle, Pareto analysis, the Ishikawa diagram, Kaizen, 5 Whys, and A3 Diagrams.

Chapter 15 is titled "learning from projects and enhancing: lean project delivery and IPD". It describes the use of retrospectives to learn from project experiences both during a project and at its completion. Post Occupancy Evaluation (POE) is also presented, and is linked to the Deming Cycle for continuous improvement. It promotes POE as an important extension of the weekly review that is conducted on lean-based projects, as learning from project outcomes is essential for improving design and construction processes.

Chapter 16 discusses Advanced Lean Metrics. It takes a deep dive into more complex metrics and some new ones from recent research (by Dr. Farook Hamzeh and his team) such as CFI – Construction Flow Index and Capacity to load ratio.

Glossary of lean terms provides useful definitions that will help users to better understand lean construction concepts.

1
OVERVIEW OF THE CONSTRUCTION INDUSTRY

This chapter provides background information on the construction industry as a major component of the US economy, and its lagging productivity that calls for new approaches. The chapter also introduces the topic of lean design and construction as an antidote to the shortcomings related to construction underperformance. The subsequent chapters provide approaches, tools and techniques for applying the lean methodology to improve the performance of the construction industry. The term "construction" is used in a general sense that includes both design and construction processes, unless specified otherwise.

Background on industry performance

The construction industry has traditionally been one of the largest industries in the United States. As reported by the Bureau of Labor Statistics (BLS), US Department of Labor, the value of construction in 2018 was $1,636.5 billion, representing 4% of the gross domestic product (GDP) (Figure 1.1). [Bureau of Economic Analysis. 2019. Construction Gross Output. Online at www.bea.gov/. Accessed on May 1, 2019].

The industry employed approximately 7.4 million people in 2018 (Figure 1.2). By its very nature construction activity in the United States has not been subjected to the trend toward offshoring that has plagued both the manufacturing and service industries. The BLS report titled "Industry employment and output projections to 2024" forecasts that "the construction industry is projected to have the largest industry increase in employment, but construction employment is not expected to reach pre-recession levels before 2024." Although other industries have blazed a trail to higher levels of quality and performance, the majority of construction work is still based on long-established techniques.

Real output in the construction sector is projected to grow 2.8% annually from 2014 to 2024 to reach just under $1.4 billion. This growth in output almost makes up for the annual decline of 3.0% that occurred during the 2004–14 period. Construction is projected to be the fastest growing of all of the USA's goods-producing sectors in terms of output and will grow faster than the economy as a whole. Construction is expected to be the fourth fastest growing sector.

Construction productivity has been lower than that of other industries. The BLS does not maintain an official productivity index for the construction industry, the only major industry that is not tracked consistently. Productivity in construction is usually expressed in qualitative terms and cannot be analyzed scientifically. Several independent studies estimated that construction productivity increased between 1966 and 2003, at a rate of 33%, or 0.78% per year.

Table 1.1 Examples of problematic construction projects

Name of Project	*Budget $Millions*	*Final Cost $Millions*	*Percent Cost Growth %*
The Boston Big Dig (2005)	2600	14,600	460
Kennedy Center Parking Lot	28	88	210
Denver Airport (1995)	1700	4800	180
Capitol Hill Visitor Center (2008)	265	621	134
Hanford Nuclear Facility (2001)	715	1600	120
National Ignition Laser Facility	2100	3300	57

This productivity growth is less than one-half that of US nonagricultural productivity gains during the same period, which averaged 1.75% annually. An estimate by the National Institute of Science and Technology (NIST) points to a decline in US construction productivity over a 40-year period to 2007 that was –0.6% per year, versus a positive nonfarm productivity growth of 1.8% per year.

The comparatively poor performance of the construction industry is borne out by a small sample of well-known projects in Table 1.1 below. They experienced very significant cost and schedule growth. But they were probably able to sustain these impacts because they were government-sponsored projects. Other studies have shown that cost and schedule overruns are also very typical of smaller private sector projects. Undoubtedly, many of us are familiar with private sector projects that resulted in lawsuits and bankruptcies that resulted from cost and schedule over-runs that left many stakeholders greatly dissatisfied. Such experiences have led to the perception of construction as a litigious environment in which everyone is focused on self-interest.

There is clearly a need for change in the construction industry as many projects that have been delivered through traditional forms of contract have had unsatisfactory outcomes in cost and schedule. They often do not meet clients' quality expectations (Alarcon and Mesa 2012; Lichtig 2006). These shortcomings have been linked to a lack of communication, coordination, and integration as well as other factors. There is extensive waste in the industry, in the range of 25 to 50% in coordinating labor and in managing, moving, and installing materials; a broad range of values for "wasteful activities" (1.6 to 93.1%, with an average of 49.6% (Horman and Kenley 2005). The US-based Building SMART Alliance at the National Institute for Building Sciences estimates that more than 50% of the cost of a building is waste; compared to other industrial sectors, the construction sector has experienced consistently rising costs over many years. Construction industry practitioners have been continually seeking to apply better technologies and processes to improve project delivery, but the rate of change has been hampered by the lack of a unified motivation to change the organizational culture.

The advent of lean construction

Lean construction is now transforming the construction industry and is the main subject of this book. It came into existence in 1993, propelled by a thesis by Lauri Koskela, a researcher, who was later joined by several other thought leaders including Greg Howell and Glenn Ballard. Lean construction has been gaining strength during the past two decades and beyond, through continuous improvement as it has been gradually adopted by the OAEC community (owners, architects, engineers, and contractors). The lean methodology has undoubtedly proven its validity as it has yielded measurable improvements in the form of lower project costs, shorter schedules, better quality and improved safety. The McGraw Hill Company's Smart Market Report (ENR 2013) took a survey of construction providers. It cited many lean benefits reported by survey respondents. The percentages of respondents reporting medium and high levels of achievement were: Higher Quality Construction 84%, Greater Customer Satisfaction 80%, Greater

Productivity 77%, and Improved Safety, 77%. Glenn Ballard and Yong-Woo Kim (2007) conducted a major study for the Construction Industry Institute (CII Report 234-11) that saw significant positive outcomes with lean projects – both tangible and intangible. They documented savings of 10% or more and high levels of satisfaction for owners as well as project team members. In some cases both costs and schedules were reduced by 10% or more.

Lean construction (LC) is an innovative project delivery approach that addresses many of the shortcomings of traditional project management that have been mentioned above as depressing construction productivity. It has several interpretations, including Lean Project Delivery, Collaborative Project Delivery, and Integrated Project Delivery. It came into existence as a response to industry concerns over low construction productivity, commonly observed errors, delays, cost overruns, and safety incidents. It was developed to address deep-seated systemic inefficiencies in construction delivery, in the way increasingly specialized stakeholders interact. Lean construction aims to address these interactions in a comprehensive manner.

Lean construction is based on the "lean" manufacturing principles that are a foundation of the Toyota Production System (TPS) (Howell 1999). Lean construction relates to both design and construction and seeks to maximize value for project owners and minimize wastes; these include seven wastes. (Womack and Jones 1996). Lean *production* pursues the ideal to: "(1) do what the customer wants, (2) in no time, and (3) with nothing in stores" (Tommelein 2015). Lean construction is also defined as "a holistic facility design and delivery philosophy with an overarching aim of maximizing value to all stakeholders through systematic, synergistic, and continuous improvements in the contractual arrangements, the product design, the construction process design and methods selection, the supply chain, and the workflow reliability of site operations" (Abdelhamid 2013). IPD is defined as a delivery system that seeks to align interests, objectives and practices, even in a single business, through a team-based approach.

Koskela (1992) addressed the application of manufacturing techniques to construction. His study, titled "Applications of The New Production Philosophy to Construction", characterized construction as a form of production. Modeling the Toyota Production System (TPS), Koskela sought to reduce the seven wastes, i.e., non-value-adding activities as identified by Shingo and Ohno (Ohno 1988). Howell *et al.* (1993) addressed the combined impact of work flow variability and dependence, and their implication for the design of operations. Glenn Ballard, through collaboration with Koskela, Howell, and others, pioneered the development of The Last Planner® System (LPS). The LPS was based on the concept of reducing the hierarchical layers of construction management and empowering field-based actors in the construction process to optimize the allocation of available resources in the weekly planning, scheduling, and execution of work. Ballard further refined the LPS, transitioning from weekly work planning to look ahead planning, and to phase scheduling. This refinement focused on managing flows in the construction process, reducing flow variation from the plan, and on using buffers to limit the impacts of any remaining variability in these flows. (Ballard 2000a). He subsequently developed the Lean Project Delivery System LPDS (Ballard 2000b, 2008).

The principles and theories involved in the genesis and development of lean construction will be described in detail in Chapter 2.

The challenge of stagnant productivity will be described further before exploring the benefits and ramifications of lean construction.

The productivity challenge

The construction industry has traditionally been one of the largest industries in the United States. As reported by the Bureau of Labor Statistics (BLS), US Department of Labor, the value of construction in 2018 was $1.636 trillion, representing 4% of the gross domestic product (GDP) (Figure 1.1). The industry employed approximately 7.4 million people in 2018 (Figure 1.2).

There was a significant downturn in the US economy starting in 2008 because of poor practices such as over-leveraged assets that led to a loss of confidence and massive failures in the banking industry. The

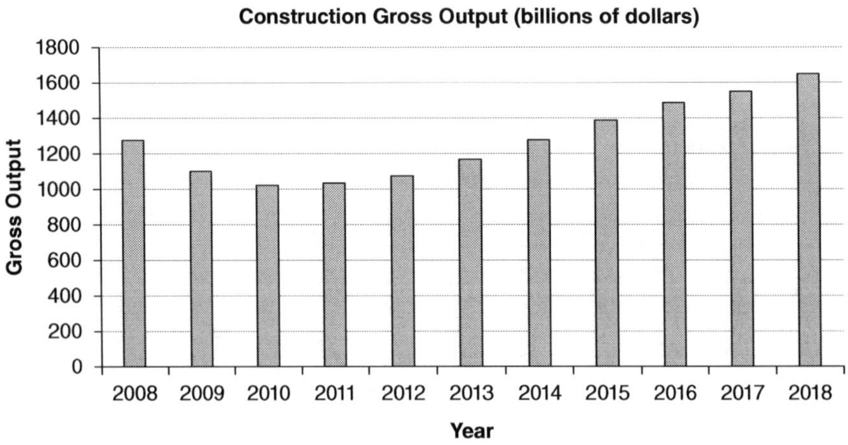

Figure 1.1 Construction gross output 2008–2018

Source: Courtesy of Bureau of Economic Analysis, US Department of Commerce

Figure 1.2 Construction employment: 2009–2019

Source: Courtesy of the US Department of Labor, Bureau of Labor Statistics

downturn had a negative effect on the entire economy; massive job losses reduced housing demand; as consumer spending shrank commercial construction was reduced, and government spending for infrastructure was similarly affected. Because of the dependence of construction activity on the availability of financing by the banking industry, the reductions in available financing led to immediate reductions in virtually all sectors of construction activity. Various initiatives such as an overhaul of the banking industry with tightened regulations gradually improved consumer confidence and a cautious improvement in construction activity. By its very nature construction activity in the United States has not been subjected to the trend toward offshoring that has plagued both the manufacturing and service industries, although foreign builders and designers have increased their presence as competitors in the US marketplace.

Although other industries have blazed a trail to higher levels of quality and performance, the majority of construction work is still based on long-established techniques.

Despite the improvements in the graphs above since 2010, the fundamentals have not changed significantly, as construction productivity has been lower than that of other industries. Much of the improvements represent a rebound from recessionary conditions in 2006 to 2008. In fact, construction employment is not expected to reach pre-recession levels until 2024.The BLS does not maintain an official productivity index for the construction industry, the only major industry that is not tracked consistently. Productivity in construction is usually expressed in qualitative terms and cannot be analyzed scientifically.A few independent studies estimated that construction productivity increased between 1966 and 2003, at a rate of 33%, or 0.78% per year.

This productivity growth is less than one-half that of US nonagricultural productivity gains during the same period, which averaged 1.75% annually. An estimate by the National Institute of Science and Technology (NIST) points to a decline in US construction productivity over a 40-year period to 2007 that was –0.6% per year, versus a positive nonfarm productivity growth of 1.8% per year.The increased productivity may be attributable to mechanization, automation, prefabrication, less costly and easier to use materials, and a reduced real cost of labor.

As reported by the McKinsey Global Institute, the MGI Construction Productivity Survey was sent out in August and September 2016 to construction-industry CEOs representing asset owners, engineering and construction firms, suppliers, other institutions s to rank the relative importance of root causes of low productivity and indicate what their companies were doing to address them.The survey results identified ten areas that have the greatest impact on construction productivity:

1. Increasing project and site complexities
2. Extensive regulation, land fragmentation, and the cyclical nature of public investment
3. Informality and potential for corruption distort the market
4. Construction is opaque and highly fragmented
5. Contractual structures and incentives are misaligned
6. Bespoke or suboptimal owner requirements
7. Design processes and investment are inadequate
8. Poor project management and execution basics
9. Insufficiently skilled labor at frontline and supervisory levels
10. The industry underinvests in digitization, innovation, and capital.

According to the MGI study, the construction industry is a growing and dynamic sector from a worldwide perspective. Around $10 trillion a year is spent on the buildings, infrastructure, and industrial installations that are the backbone of the global economy, and that amount is projected to increase to $14 trillion in 2025. However, low labor productivity is identified as an obstacle. It causes a major loss of value, a shortcoming that has impacted the industry in many geographic regions.

The US Bureau of Economic Analysis indicates that the construction industry has significant economic potential as an estimated additional $0.86 of economic activity was generated by every $1 of construction sector GDP in 2012, making it one of the industries with the largest economic spillover effects.

Worldwide, labor-productivity growth in the construction industry has averaged only 1% a year over the past two decades, compared with a rate of 2.8% in the case of the total economy and 3.6% in manufacturing. This is very similar to the situation experienced in the USA.

Most of the improvement in construction productivity has been the result of research and development (R&D) work in the manufacturing industry related to construction machinery. Earthmoving equipment has become larger and faster; power saws have replaced handsaws. However, as the use of power tools increases the consumption of energy, their contribution to productivity improvement may be eroded.

Tilt-up construction is one example of a technology that has improved construction productivity. According to the Tilt-up Concrete Association, tilt-up construction has proven to be a fast and economical construction methodology for small buildings of one to four stories but is far from new. It became popular in California in the late 1950s, initially for warehouses, then its use subsequently extended to shopping centers, churches, townhouses, and so on. Successful introduction of the technique is credited to one Robert Aiken who constructed several buildings using "lifting jacks" to raise building panels in Illinois and Ohio. It is believed by some that a Roman architect discovered – 2000 years earlier – that it would be easier to "cast a slab of concrete on the ground, then hoist (tilt) it up into position, than to build two wood forms. Place concrete between them, and then remove the forms."

However, digitization seems to have been inadequately addressed. The MGI study identified underinvestment in digitization and innovation as a major root cause of low productivity. While technology and digital tools would enable significant productivity gains, the construction industry is one of the most underdigitized sectors. In the US construction sector, only 1.5% of gross value added was invested compared with 5.7% in financial intermediation, 3.3% in manufacturing, and the all-sector average of 3.6%. There is a robust correlation between the level of digitization in a sector and its productivity growth over the past ten years. As the MGI study points out, the construction industry has yet to adopt an integrated platform that spans project planning, design, construction, operations, and maintenance.

The MGI study noted that the construction industry's adoption of technology is lower than in other sectors, except for agriculture. A 2016 survey by the Associated General Contractors of America noted that new software solutions (including building information modeling, BIM), productivity apps, augmented reality design, radio frequency identification, sensors for material management, drones, and virtual reality devices are increasingly adopted by contractors, 21% of respondents said that they were investing in labor-saving equipment, 13% in off-site prefabrication, and 7% in BIM.

The post-downturn recovery of the construction sector has led to worker shortages. Nearly half of the survey respondents said that they had raised base pay and invested in in-house training to cope with these shortages.

Productivity and construction

Productivity is the measure of how well resources are brought together in organizations and utilized for accomplishing a set of goals. It is measured as the ratio of outputs to inputs, so productivity is greatest when the highest level of performance is achieved with the least expenditure of resources. Productivity. In the construction environment may be represented as the constant-in-place value divided by inputs such as the dollar value of material and labor. Productivity measurements may be used to evaluate the effectiveness of using supervision, labor, equipment, materials, and so on to produce a building or structure at the lowest feasible cost.

The Bureau of Labor Statistics (BLS) defines productivity as a ratio relating output (goods and services) to one of the inputs (labor, capital, energy, etc.), that are associated with that output. There are four classical factors: land, materials, capital, and labor. Overriding these four factors is a fifth factor; namely, technology.

In the manufacturing industry productivity is usually described in quantitative terms, but in the construction industry it is usually understood in qualitative terms. Consequently, construction productivity cannot be analyzed scientifically at the working level. For productivity analysis it is essential to identify the significant productivity factors, quantify them, and establish the relationship of these factors with each other. The work situation in construction is project-oriented and much more dynamic than that in the manufacturing industry, introducing many complex factors affecting productivity in construction. These factors involve: managerial, technological, regulatory, labor related, engineering, and craft-related items.

Importance of productivity

In 2006, new-construction-put-in-place accounted for roughly 8.0% of the gross domestic product (GDP). In addition, over 7.6 million people were employed in the US construction industry in 2007. They were involved in design, new construction, renovation construction, equipment and materials manufacturing, and supply, thus making the design and construction industry the largest manufacturing industry in the United States. Consequently, it has a great impact on the state of the economy.

Productivity is an important component of the overall concept of performance. Although there are differences of opinion on what constitutes performance, several definitions refer to it as a combination of productivity, quality, timeliness, budget adherence, and safety. Productivity is primarily measured by cost. Satisfactory productivity is work accomplished at a fair price to the owner and with reasonable profit for the contractor. The other factors are:

- Quality: conformance to specifications, satisfying owner's needs;
- Timeliness: on schedule, everything on hand when needed;
- Safety: accident-free projects; and
- Budget adherence: completion within the contract price.

Traditional construction project management tools do not address productivity; they include schedule slippages and cost overruns. The construction industry as a whole measures performance in terms of completion on time, completion within budget, and meeting construction codes. Quality is defined very subjectively and by default is considered to be satisfactory if construction codes are met.

Many contractors assume that it is impossible to meet desirable benchmarks in all four factors simultaneously and that there is a zero-sum relationship between them. In other words, an accelerated schedule results in an increased cost and lower levels of safety and quality. By the same token, higher quality is seen as requiring a higher cost and a longer schedule, while safety levels may remain unchanged.

Thomas and colleagues (1990) define performance in seven dimensions: effectiveness, efficiency, productivity, profitability, innovation, quality of work life, and quality. They emphasize four of the dimensions: effectiveness, efficiency, productivity, and quality. Ballard and Howell (1994) defined performance from the perspective of lean construction as percentage of assignments completed (PAC) for construction work planned for execution in the field.

Construction organizations (designers and constructors) would benefit significantly by establishing formal performance improvement programs that use quantitative approaches to improve productivity and quality. These quantitative approaches are critical – they involve applying measurement techniques to construction processes; these measurements are an essential ingredient for process improvement. As the saying goes "you cannot improve what you cannot measure."

At the macro level, productivity measures provide a consistent reference (outputs divided by inputs) that facilitates comparison between all construction project types, as well as other industries. Contractors need a tool to drive performance improvement through internal or external benchmarking. Measurement may not directly lead to performance improvement; however, performance improvement over time can be achieved through recognizing the need for improvement.

Construction productivity is a major concern, especially when compared to other industries. Nationally, productivity in the construction sector has not kept pace with other industries. At the firm level, it has a direct impact on profitability. As reported by the US Department of Commerce, construction productivity has been rising at a much slower rate than other industries; between 1990 and 2000 it rose by approximately 0.8% compared to more than 2% for all US industries. However, productivity in the construction industry is largely unmeasured, and those measures that do exist are contradictory and conflicting. The BLS does not maintain an official productivity index for the construction industry; it is the only exception to the industries that are tracked.

Productivity trends in the United States

Construction productivity trends carry immense consequences for the economy. According to the BLS, during the past 40 years, the productivity of non-farm industries has increased by over 100%, while that of construction industry has been stagnating. Construction costs have been increasing at the same time. Raw materials such as steel, copper, and cement have been rising, especially in the face of escalating global demand. Labor costs are a major component of most construction projects-in the vicinity of 40%, yet on many construction sites a large percentage of the daily labor hours are unproductive. While it is difficult to have absolute measurements of construction productivity, it is still a concern considering how such measurements are used for other industries.

Two independent methodologies demonstrate that total construction productivity increased between 1966 and 2003, on the order of 33%, or 0.78% per year. Figure 1.3 shows one estimate of productivity growth between 1964 and 2003. This level of productivity improvement, however, is less than one-half that of US non-agricultural productivity gains during the same period, which averaged 1.75% annually between

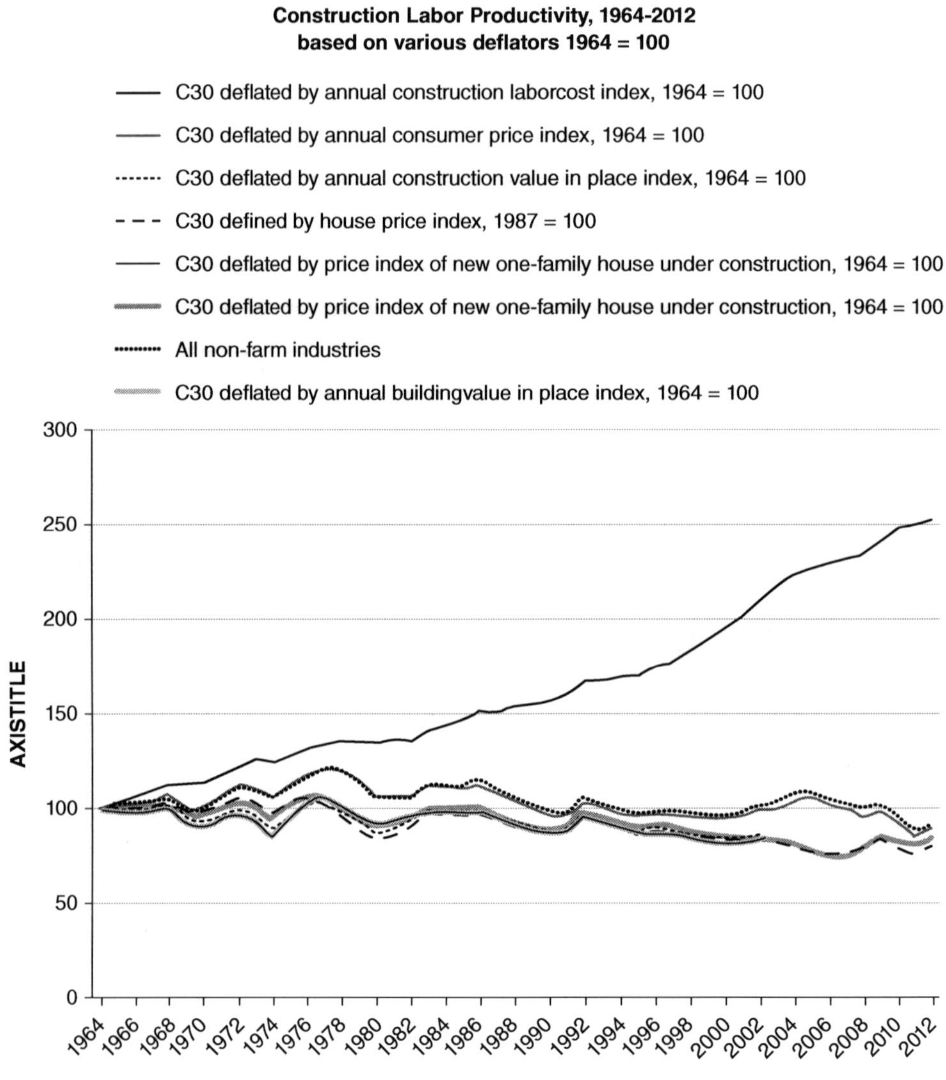

Figure 1.3 US productivity gap (1964–2012)

Source: Courtesy of Paul Teicholz, Ph.D., and AECbytes

1966 and 2003. Some of this differential in measurement is because construction output is affected by the conditions at the job site at that time, and the conditions are continually changing. Construction productivity measurement is complicated due to the heterogeneous conditions in the industry and the varying nature of the output or product, while other industrial products are factory-based, and assembly line produced. A follow-up study to 2012 was conducted.

Professor Emeritus Paul Teicholz of Stanford University calculated the construction industry's labor productivity for the publication AEC Bytes in 2013, covering the period 1964 to 2012. He found that productivity had been declining at 0.32% per year. Similarly, Prof Matt Stevens of the University of Melbourne observed in a study of productivity levels from 1993 to 2013 that productivity level increased in 2008 and 2009 but ended lower in 2013 than in 1993.

The McKinsey (2017) study observed that in the United States, construction labor productivity has declined by an average of 1.7% a year since 1968 while the productivity of the overall economy has grown by 1.6% over the same period. Construction lagged even further behind certain sectors that were improving their productivity sharply, including agriculture, which increased its productivity at a rate of 4.5% a year between 1947 and 2010, and retail, at a rate of 3.4% per year. A differential in productivity-growth rates among sectors of a few percentage points may seem insignificant, but the impact mounts up over many decades. The US Bureau of Economic Analysis estimated that an additional $0.86 of economic activity was generated by every $1 of construction sector GDP in 2012, making it one of the industries with the largest economic spillover effects.

Need for new approaches to construction

Global competition is increasing rapidly; in 2007 construction contracts won by Chinese companies increased in the United States and Europe by 160%. In the case of the United States, this trend would suggest that the competitiveness of domestic construction organizations is being challenged on its home turf. Construction spending is moving east to China, India, the Middle East, and Africa (CRIMEA). Construction organizations that are the most competitive are likely to benefit from this trend.

Clearly, new approaches are urgently required to address the construction industry's ills. Construction is said to need a greater sense of mission, focus, and industry support. There are several obstacles to its competitiveness in the United States:

- Declining productivity,
- A mindset for minimum first cost, and
- Prescriptive standards and codes that stifle productivity.

The industry is said to need more R&D, and to move beyond a reliance on historical surveys to identify new approaches for getting the job done. This problem is not confined to the United States. Sir John Egan's report in 1998 "Rethinking Construction," identified many shortcomings in construction practices in the United Kingdom and advocated radical change. The government in the United Kingdom appointed a construction task force that was comprised of a chairman and nine members representing senior representatives of construction organizations and their clients.

The task force studied several industries, including auto manufacturing and grocery retailing to accomplish the following objectives:

- Quantify the scope for improvement of construction efficiency;
- Examine current construction practice and assess the potential for its improvement by way of innovation;
- Identify specific actions and good practice that would help achieve more efficient construction; and
- Identify projects that might help to demonstrate the improvements that can be achieved.

Sir John Egan's (1998) report, also called "The Report," made positive observations about the UK construction industry, citing its ingenuity and capacity to handle difficult and innovative projects (Egan 1998; Table 1.2). At the same time, the report pointed to many industry weaknesses such as: low profitability, the urgent need to train the work force, inadequate R&D investment, and clients' misunderstanding of cost and value relationships. At the heart of the report was a desire to develop a change of style, culture, and process in the industry as opposed to recommending mechanistic activities. Proverbs *et al.* (2002) conducted a follow-up analysis of fifteen recommendations from the Egan report, using a sample drawn from the Managing Directors of 100 UK contractors. The study was intended to rank the recommendations with regard to the construction industry based on (a) Impact, i.e., potential effectiveness, (b) Effectiveness, i.e., the potential to achieve the intended purpose, and (c) Perceived potential for success, computed as the product of scores for impact and effectiveness.

Impact was correlated most highly with:

- Development of long-term relationships,
- Commitment of leadership from all parties, and
- Greater customer focus and more and better training.

Effectiveness was correlated most highly with:

- Development of long-term relationships,
- Commitment of leadership from all parties, and
- Improved quality and fewer defects.

Perceived potential for success was correlated most highly with:

- Development of long-term relationships,
- Commitment of leadership from all parties, and
- Greater customer focus and more and better training.

Table 1.2 Construction initiatives

Number	*Recommendation*
1	The commitment of leadership from all parties
2	Greater customer focus
3	Application of products and services from other industries
4	Promote best practice in health and safety
5	Provide decent working conditions and promote people as the construction sector's greatest asset
6	Measure and assess own performance, set performance targets, and share information
7	Development of an integrated process instead of reliance on convention
8	Improved quality and fewer defects
9	Use of demonstration projects to develop and illustrate innovations
10	Rethinking (i.e., reengineering) the construction process
11	Adopt lean thinking as a method of sustaining performance improvement
12	More and better training
13	Development of technology (e.g., standardization and prefabrication)
14	Development of long-term relationships
15	Greater use of performance improvement tools and techniques (e.g., value management, benchmarking, etc.)

Source: Adapted from Proverbs, Holt, and Cheok, Summary of J. Egan, Report, *Rethinking Construction*, 2002

These findings supported the recommendations of the Egan Task Force. In the United States, the need for innovative methods to improve construction performance is underscored by the ongoing and anticipated shortage of skilled labor in an industry that is highly labor-intensive. Studies have shown that the traditional sources of workers – vocational school programs and apprenticeships – have not been keeping up with the demand. In the area of technicians and equipment operators alone there is a projected shortage.

Private and public organizations in the United States have been concerned about construction cost and quality for many years. The Construction Users Anti-Inflation Roundtable was formed to investigate the causes of inflation in the cost of construction. In 1972, this organization joined with several others to form the Business Roundtable, represented by the presidents of the 200 largest US companies. This initiative led to the creation of the Construction Industry Cost Effectiveness (CICE) project to promote quality, efficiency, productivity, and cost-effectiveness in the construction industry. The CICE produced a January 1983 report titled "More Construction for the Money." The CICE report observed that construction was a $300 billion activity at that time, and that even a modest application of the teams' recommendations could save the industry at least $10 billion annually. The report revealed deficiencies in the planning, design, and construction of projects and that there was slow acceptance and use of modern management methods to plan/execute projects. Owners were said to be unwilling to bear the costs of modern systems. The CII was created in 1983 at the University of Texas–Austin. The CII has conducted many research studies on construction productivity during that time to the present.

The foregoing causes clearly need to be addressed to improve labor productivity. Activity sampling studies of construction operations in the field have shown that the working portion of activities generally consumes 40–60% of labor hours, and by the same token 40–60% of labor hours are unproductive. There are many reasons for lost time – poor communications, waiting on assignments, waiting on resources, double material handling, rework, accidents, late or inaccurate job status reports, lack of supervision, and so on. One-third of these losses reflect issues that are within management's control. Construction profitability is directly linked to labor productivity. Industry-wide studies suggest that most construction projects yield net profits of 2–3% of the total project cost. The calculations below are broad approximations, but they illustrate the potential impact of performance improvement initiatives.

A hypothetical example:

Contract price	= $10,000,000
Labor cost (40%)	= $4,000,000
Other costs, materials, overheads, etc.	= $5,700,000
Net profit	= $300,000

Assuming a 5% reduction in labor cost due to productivity improvement:

Savings in labor cost = $4,000,000 × 0.05 = $200,000
Revised net profit = $300,000 + $200,000 = $500,000.

Hence, a 5% improvement in labor productivity can improve profitability by 66.7%. Similarly, the value of lost labor hours due to management inefficiencies

= $4,000,000 × 1/3 = $1,333,333.

A 50% reduction in management-based losses would save $1,333,333/2 = $666,667:

Revised net profit = $300,000 + $666,667 = $966,667.

A 50% improvement in labor deployment would improve profits by

$$967 \times 100\%/300 = 322\%.$$

(This level of improvement is unlikely, but it shows the potential for improvements in profitability.)

Further gains can be derived by addressing other construction processes and supply chain management issues.

Recent surveys of the top 400 US contractors have indicated that cost control, scheduling, design practices, labor training, and quality control are perceived as having considerable room for productivity improvement. Identified as needing more improvement compared with previous surveys were prefabrication, new materials, value engineering, specifications, labor availability, labor training, and quality control, whereas those needing less improvement were field inspection and labor contract agreements. Also, respondents have indicated consistently over the years that they are willing to participate in activities related to improving construction productivity, but they are not interested in funding any such activities.

Unit labor costs in constant dollars and daily output factors were compared over decades for each task. Direct work rate data from 72 projects in Austin, Texas over a 25-year period were also examined. Increasing the direct work rate usually increases construction productivity. Depressed real wages and technological advances appear to be the two greatest reasons for an increase in productivity. The data also indicate that management practices were not a leading contributor to construction productivity changes over time. Subsequent studies are required to add weight to these observations and can be based on the approach presented here. Labor productivity is of central importance to the economic health of the United States economy as well as that of other countries. Due to the size of the construction industry, productivity changes within it have significant direct effects on the national productivity and economic well-being of the United States.

Productivity ratios

Productivity measurements may be applied at different levels, ranging from an individual work task to a firm, an industry, or a country. In the construction environment, productivity measurements evaluate the effectiveness with which management skills, workers, materials, equipment, tools, and working space are used to produce a completed building, plant, structure, or other facility at the lowest feasible cost.

Partial productivity is the ratio of output to one class of input. For example, material productivity is the ratio of output to the associated material input. Similarly, labor productivity is the ratio of output to labor input and is a partial productivity measure. These ratios can be configured in different ways; in physical terms, labor productivity for roofing may be calculated as the ratio of x square feet of roofing/y labor hours. Productivity may also be measured in financial terms, such as dollar value of roofing/dollar value of associated labor.

Value added productivity is based on output produced that adds value.

Total productivity

Total productivity is a ratio of output to all inputs. All input resources are factored in this principle. Tracking the productivity changes that occur at different time periods is the most useful application of total productivity. Sumanth (1984) points to the limitations of partial productivity measures, which are measured by the ratio of output to one class of input such as labor productivity. Such measures – if used alone can be misleading – do not have the ability to explain overall cost increases, and tend to shift blame to the wrong areas of management control.

Total productivity (TP) may be defined as:

$$TP = \frac{T(s)(\text{total sales or value or work})}{\text{labor cost}(M_1) + \text{ materials cost}(M_2) + \text{machinery cost}(M_3+) \\ \text{money cost}(M_4) + \text{management cost}(M_5) + \text{technology cost}(M_6)}$$

or

$$TP = T(s)/M1 + M2 + M3 + M4 + M5 + M6. \quad (1.1)$$

Since

$$P_i = T(s)/M_i$$

$$P_t = \frac{1}{1/P_1 + 1/P_2 + 1/P_3 + 1/P_4 + 1/P_5 + 1/P_6}. \quad (1.2)$$

The above-mentioned factors are expressed as constant dollars (or other currency) for a reference period. To increase total productivity, it is necessary to determine what partial productivity factor (P_i) has the greatest short-term and long-term potential effect on total productivity.

The terms productivity, effectiveness, and efficiency may be combined as follows:

$$\text{Productivity Index} = \frac{\text{output obtained}}{\text{input supplied}} = \frac{\text{performance achieved}}{\text{resources consumed}} = \frac{\text{effectiveness}}{\text{efficiency}} \quad (1.3)$$

Therefore, productivity is the combination of effectiveness and efficiency. To increase productivity, the ratio(s) mentioned in Equation 2.1 must increase. This can be achieved by increasing the output, reducing the input, or permitting changes in both such that the rate of increase in output is greater than that for input. An increase in productivity can be achieved in five ways as follows (Gerald 1997):

(1) Reduced costs: $\frac{\text{output at same level}}{\text{input decreasing}}$

(2) Managed growth: $\frac{\text{output increasing}}{\text{input increasing (slower)}}$

(3) Reengineering $\frac{\text{output increasing}}{\text{input constant}}$

(4) Paring-down $\frac{\text{output down}}{\text{input down (faster)}}$

(5) Effective working: $\frac{\text{output increasing}}{\text{input decreasing}}$

These strategies often have unintended consequences. Cost reduction usually involves decreasing expenses, services, training, and advertisement (Forbes, Ahmed and Azhar 2003). Management generally views people as a direct expense. Unless the respective costs are prioritized against the organization's objectives, this method may result in the wrong identification of cost-cutting items and can become more defensive and contraction-oriented.

Growth management may involve an investment that yields a greater return than the cost of investment. It may be in the form of capital, technological improvements, system redesigns, training and/or organizational restructuring. Increased productivity may also be achieved by reducing the cost of production inputs through a reengineering of the design and production processes. Paring down means reducing both output and input amounts, with the input diminishing at a higher rate. This "sloughing off" seeks to improve marginally unproductive facilities. If an organization has a number of production facilities that vary in productivity, it would progressively eliminate the use of low-producing facilities.

Working effectively involves a total output increase with a decrease in input. This can be accomplished through a combination of changes in work and management procedures and through proper training of workers. It also requires the motivation of all involved to "produce more with less" and perform to high-quality standards. Simply producing more may be detrimental to an organization if the increased outputs do not meet the quality expectations of its customers. Crosby (1979) points to the high cost of not doing jobs right the first time.

Although several publications exist on the subject of construction productivity, there is neither an agreed upon definition of work activities nor a standard productivity measurement system. Researchers have concluded that it is difficult to obtain a standard method to measure construction labor productivity because of project complexity and the unique characteristics of construction projects. The uniqueness and non-repetitive operations of construction projects make it difficult to develop a standard productivity definition and measure.

Reasons for low productivity

There is evidence that improved productivity is related to the levels of R&D in an industry, but the extent of R&D effort in the construction industry has remained very low in the United States. The overall productivity in construction has been greatly affected by regulatory controls, the environment, climatic effects, cost of energy, and other factors. Productivity improvement has never been the focus of the construction industry, probably due to the lack of a model that can fit together all the fragmented components of the overall construction process.

Productivity in the construction industry is largely unmeasured, and those measures that do exist are contradictory and conflicting. Research has pointed to a significantly high level of wasted resources in the construction industry – both human and material (Figures 1.1 and 1.2). An article titled "Construction and the Internet" in *The Economist*, dated January 15, 2000, noted that up to 30% of construction costs is due to inefficiencies, mistakes, delays, and poor communications.

- The Construction Industry Institute (CII) estimates that in the United States, 10% of project cost is spent in one rework.
- Between 25 and 50% of construction costs is lost to waste and inefficiencies in labor and materials control.
- Losses are incurred in errors in information when translating designs to actual construction. Inadequate interoperability accounts for losses of $17 to 36 billion per year. This problem is due to communication difficulties between software used by different operators in the design and construction supply chain.

The observations of the CICE project in 1983 are still relevant in 2019 and beyond. The study teams noted that "more than half the time wasted in construction is attributable to poor management practices." They concluded that "if only the owners who pay the bills are willing to take the extra pains and pay the often small extra cost of more sensible methods they would reap the benefit of more construction for their dollars." The waste in construction occurs mostly in the interaction between trades. The self-interest of parties makes them put themselves first.

The construction industry is not unified on the measurement of productivity or overall performance. Many researchers define construction project success in terms of (a) cost: within budget, (b) schedule: on-time completion within schedule, (c) safety: high safety levels – few or no accidents, and (d) quality: conformance with specifications and few defects. Some view quality as one point of a triangle with cost and schedule; quality is often the first to be sacrificed in favor of cost savings and schedule reductions. Many contractors believe that it is impossible to meet desirable benchmarks in all four factors simultaneously and that there is a zero-sum relationship between them. In other words, an accelerated schedule results in an increased cost and lower levels of safety and quality. Material waste is a significant factor in construction costs – 9% by weight in the Dutch construction industry and 20–30% of purchased materials in the Brazilian construction industry. Material wastes are caused by several sources such as design, procurement, material handling, operations, and so on.

At least four main elements are important to overall performance:

1 Productivity: primarily measured by cost, satisfactory productivity is work accomplished at a fair price to the owner and with reasonable profit for the contractor;
2 Safety: accident-free projects;
3 Timeliness: on schedule and everything on hand when needed; and
4 Quality: conformance to specifications and satisfying owner's needs.

In terms of the prevailing industry practices, the concept of construction performance does not emphasize productivity and quality initiatives. The work of many researchers has revealed an industry tendency to measure performance in terms of the following: completion on time, completion within budget, and meeting construction codes. Very little attention has been directed to owner satisfaction as a performance measure. Although the Malcolm Baldrige National Quality Award was established in 1987, very few, if any, construction-related organizations have won this coveted award during the more than 30 years elapsed since its inception.

There is a growing emergence of subcontracting. In effect, so-called general contractors perform the role of construction manager, with individual contracts, and with specialty subcontractors. The subcontractors are often priced in a manner that does not reflect the general contractor's agreement with the owner; even if the owner pays a high price, the subcontractor may still have to work with inadequate budgets, often compromising quality as a result.

Safety is a major source of waste. In the year 2006 alone, construction accounted for approximately 21% of all deaths and 11% of all disabling injuries/illnesses in the workplace (BLS 2006).

Poor communication inhibits productivity. Communication tends to be via the contract. Essentially, the designer is paid to produce a design expressed in the form of specifications and drawings. The contractor is expected to use these as a means of communication and produce the completed facility. This communication often does not work as well as it should.

Small contractors often lack the expertise or financial resources to adopt technological advances – adoption is inhibited further by fear and uncertainty. Roofing contractors, for example, tend to use the same time-honored methods to ensure that supplies and equipment are on-site each day. Expediters, at additional cost, deliver items that are frequently forgotten.

Needed training often does not get to the decision makers in the construction industry. Construction management (CM) programs around the country have been providing higher levels of training for managers; however, this training has not reached the ultimate decision makers in the industry. Efforts to enhance quality and productivity are likely to be frustrating under this scenario.

Owners have not specifically demanded productivity and quality. There is a general lack of productivity/quality awareness in the industry among all parties, including owners. Owners have come to accept industry pricing – they have not been able to influence the productivity of the industry – prices have simply trended upwards with fluctuations based on market conditions. By contrast, manufacturing activities have become cheaper over time on a per unit basis.

Architect/engineer contracts are said to be unclear with respect to professional standards of performance, often leading to unmet expectations. Construction owners feel that typical A/E contracts protect designers at the owner's expense. For example, prevailing contract language relieves designers of any role in the case of a lawsuit or arbitration between an owner and contractor. An outgrowth of this is the practice of "substantial completion" where a facility is usable but may have a small but significant percentage of remaining work in the form of a "punch list". An owner often has a very difficult time in persuading a contractor to finish that work.

Few large design or construction companies, and virtually no small companies, have implemented the concept of a quality or productivity manager – cost cutting trends have resulted in such a position being viewed as an unjustifiable luxury. On the positive side, there is a growing interest in lean construction techniques, and as of 2010, a small number of projects have been carried out based on a lean approach. However, the deployment of these methods is mostly driven by consultants, as "lean" is still in its emerging stages.

There is little, if any, benchmarking – many manufacturers and service organizations have become preeminent by adopting the best practices of benchmarked organizations. Construction has done very little of this, due to distrust, fear of losing a competitive advantage, but more likely, simply by being anachronistic.

Mistakes, rework, poor communication, and poor workmanship are part of an ongoing litany of deficiencies that seem to be accepted as being a natural part of construction activity. Safety is a major national concern. Construction has an abysmally poor safety record, worse than virtually all other industries.

Categories of construction

In order to understand how improved methods can be applied to the construction industry, it is helpful to understand that environment; it is truly diverse, so much so that its participants have found it easy to rely on such clichés as "the industry is like no other" and "no two projects are alike," to maintain the status quo in which long-established management traditions are seen as an arcane art that others cannot fully understand.

The BLS refers to three major headings: General Building Contractors SIC Code 15, Heavy Construction (except building) SIC Code 16, and Special Trade Contractors SIC 17. These are further subdivided into 11 SIC Code headings that include:

- Commercial building construction: offices, shopping malls;
- Institutional construction: hospitals, schools, universities, prisons;
- Residential: housing construction, including manufactured housing;
- Industrial: warehouses, factories, process plants; and
- Infrastructure: Road and highway construction, bridges, dams.

The parties involved in construction

These individuals and entities play very important roles in the planning and execution of design and construction projects:

1 Owners: They originate the need for projects and determine the locations and purpose of facilities. They arrange for design, financing, and construction.
2 Designers: They are usually architects or engineers who interpret the owner's wishes into drawings and specifications that may be used to guide facility construction. In the design/build concept, they may be a part of the construction team.
3 Constructors: These are contractors and subcontractors who provide the work force, materials, equipment and/or tools, and provide leadership and management to implement the drawings and specifications to furnish a completed facility.

4 The labor force: This is comprised of foremen, craftsmen, or journeymen, and skilled or semiskilled apprentices or helpers. Many different crafts are represented, such as masons, pipefitters, carpenters, electricians, and so on.
5 Major suppliers: Equipment and material manufacturers, transporters.
6 Financial institutions: banks, construction financial organizations.
7 Lawyers, insurers.
8 Federal and local regulators: code enforcement professionals.
9 Public services.
10 Utilities.
11 Safety professionals.
12 Quality assurance/quality control professionals.
13 Lean facilitators/coaches/mentors: These professionals provide support in the implementation of new lean approaches in design and construction.

Project delivery methods

The following information on delivery methods is intended to clarify some of the many options that available for executing construction projects, and to set the stage for later discussion on lean construction projects. It is worthy of note that, while lean project delivery is identified as a method, the lean methodology has been applied with varying degrees of success to traditional projects. Examples are identified in later chapters.

There are a number of models for the process of designing and constructing facilities. Several of these models have been in existence for many years and have been used with varying degrees of success, depending on the type of project and the skills required. They are considered to be "traditional" methods. The basic workings of these delivery methods are discussed below in items 1 through 5 simply to describe the current practices in the construction industry. As the emphasis of this book is on lean construction and its variants such as integrated project delivery, the conversation will be focused on them after this introduction.

The project delivery methods include:

1 Design-bid-build (DBB),
2 Design-build (DB),
3 Engineer-procure-construct (EPC),
4 Design-CM contracts,
5 Design-agency CM contracts,
6 Fast-track construction,
7 Partnering,
8 Design-build-operate, and
9 Relational contracting/lean project delivery.

Design-bid-build contracts

Design-bid-build contracts represent the most frequently used type of project delivery systems for most construction projects and are considered to be the "traditional" delivery method. DBB projects have the following characteristics (Figure 1.4):

1 The project is conceptualized by the owner.
2 Planning is carried out based on the objectives to be met and on economic and technical feasibility.

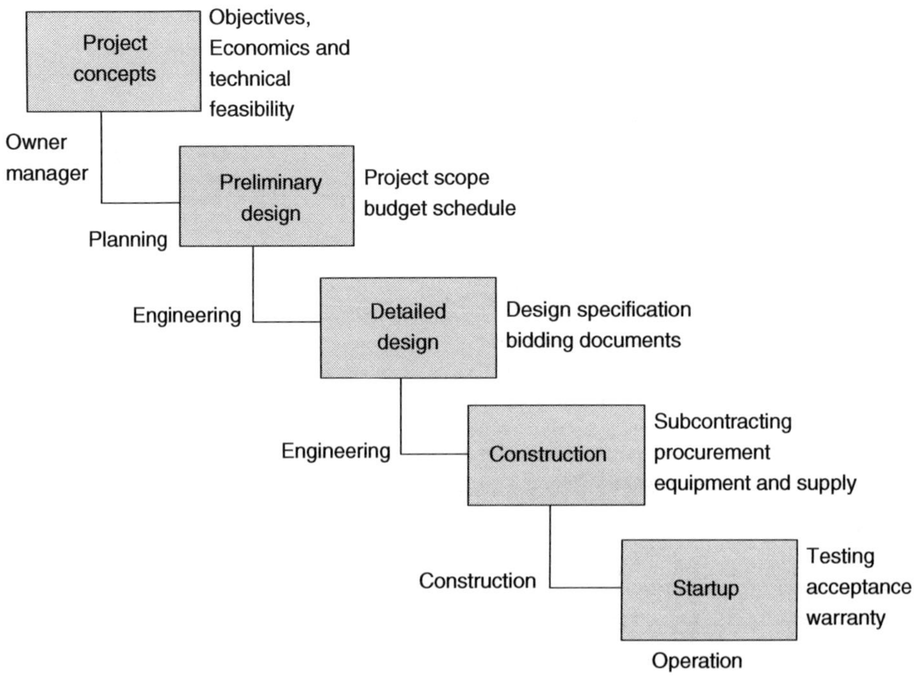

Figure 1.4 Traditional design-bid-build construction phases

3 Programming is carried out to identify the uses and desired sizes of various spaces, followed by a schematic design to identify relationships of these spaces relative to each other. The scope of the project, preliminary budget, and schedule are derived.
4 Detailed design is usually carried out in stages, with intermediate checkpoints for verification by the parties to the project.
5 The design work culminates in the preparation of completed drawings and specifications, representing bid documents as well as detailed cost estimates. The bid documents are used to solicit construction bids, or are otherwise used to negotiate a construction price.
6 Bid analysis is carried out and a legally binding contract is then awarded. The drawings, specifications, and signed documents then become construction documents.
7 The contractor is given access to the site and instructed to proceed, based on legally established time frames. A contract may contain incentives for timely completion, as well as penalties for avoidable delays or cost overruns.
8 The owner, or agents of the owner such as architects/engineers or construction managers, monitor(s) the progress of the construction, ensuring that interim payments to the builder/contractor match construction progress.
9 At completion, there are acceptance inspections, leading to the commissioning of the facility for the owner's use. The project is turned over to the owner.

Design-build projects

Design-build (DB) projects accelerate delivery through concurrent design and construction activities. As is typical of all types of projects, a DB project is conceptualized by the owner; planning is carried out based on the objectives to be met, and on the economic and technical feasibility of the project. Planning and schematic design are carried out by the owner's design professional and may include infrastructure and foundation details for the project. This information allows construction to start shortly after contract award, while the design builder continues the preliminary design to obtain a final design. Typically, their design professional develops a preliminary design and cost and schedule proposals for the overall project. In some DB projects the owner may review proposals from a number of design builders and enter into a legally binding contract with one that provides the most appropriate proposal. The design builder is given access to the site and instructions to proceed, based on legally established time frames. This type of contract may also contain incentives for timely completion, as well as penalties for avoidable delays or cost overruns. The DB organization initiates construction while finalizing the detailed design. The owner or agents, such as architects/engineers or construction managers, monitor the progress of the construction, ensuring that approvals for interim payments match the progress of the construction work.

Engineer-procure-construct (EPC) projects

These are configured in a manner very similar to DB projects. Most of the design and construction functions are performed or managed by one organization. This model, however, is used primarily for industrial projects that emphasize engineering design, as opposed to architectural design. The EPC projects typically have commissioning and maintenance phases included to allow for a plant to reach its designed operating capacity after acceptance.

Construction management (CM) contracts

The owner typically hires a CM organization, for a fee, to provide professional management services. Trade contractors contract directly with the owner on an individual basis and not through the construction manager, although the CM advises the owner on the formation and conduct of those contracts. The owner also contracts separately with a design concern (i.e., an architectural/engineering (A/E) firm) to obtain the design documents. In some instances, the A/E firm may play the role of the CM. This form of contracting places a heavy responsibility on the owner to coordinate the work, as the trade contractors do not have contracts with each other and have no contractual obligation to cooperate.

The CM agent typically has less clout to resolve design or construction issues than a general contractor and serves only as a mediator.

Design-agency CM contracts

In this type of contract, the owner hires a design team to prepare project construction documents, and hires a construction manager (CM) to oversee the construction phase of the project. This is often done based on a lump-sum or fixed-price contract. The CM may act as an agent of the owner, contracting directly with all the trade contractors. The CM prepares bid packages that are priced competitively by the trade contractors and reviews these bids to select the most appropriate ones.

CM-at-risk contracts

The construction manager in this type of contract assumes the risk of pricing and contracting directly with the respective trade contractors. In general, CM-type contracts are not as amenable to quality initiatives as DBB and DB contracts.

Construction Management at Risk (CM@Risk) is increasingly being used by the US public sector. Like design-build, it facilitates improved quality through a selection process based on factors other than the low bid. A scoring system is used to consider the previous performance of a contractor, based on various criteria. It is not the cheapest method – it is best used where there is uncertainty, such as renovation projects where the current state of a facility or its infrastructure may not be entirely known. This uncertainty is reduced by having the CM involved in managing the design phases of a project, in the selection of sub or specialty contractors, and then assuming the risk for successful completion. Bidders tend to build in safety cushions for the unforeseen, but a guaranteed maximum price (GMP) approach may be used to set a specific limit to the owner's project cost. Consequently, project budgets for CM@Risk are somewhat generous, resulting on less emphasis by contractors on cost reduction, and fewer compromises in the area of quality.

Fast-track construction

Fast-track construction is valuable in meeting accelerated schedules demanded by the owner. It allows a contractor to commence construction immediately after contract award, while a designer simultaneously completes the construction documents. It may be carried out with or without a design builder.

Partnering

Partnering is the formation of a project team to deliver a construction project; the team commits to open communications in a spirit of trust and works to accomplish mutual project goals. While the team members work supportively to meet mutual goals, they also focus on their individual goals. Yet, they recognize that their individual success is linked to the overall success of a project.

Partnering has resulted in projects that are far less reliant on litigation to ensure that project objectives are met. It significantly reduces the fragmentation that characterizes many construction projects, as a result of which the employees of the involved companies tend to work more on an individual level rather than an organizational level.

Design-build-operate (DBO)

Like design-build (DB) projects, DBO projects accelerate delivery through concurrent design and construction activities. DBO is an emerging collaborative concept, sometimes referred to as design-build-maintain. DBO are in the early phase of development and seem to have gained traction in public-private partnership projects.

Relational contracting/lean design and construction

Howell (2000) describes lean construction as a new way to design and build capital facilities. Lean theory, principles, and techniques jointly provide the foundation for a new form of project management. It uses production management techniques to make significant improvements particularly on complex, uncertain, and quick projects. Lean methods have reduced office construction costs by 25% within 18 months and in one case reduced the schematic design portion of the project design phase from 11 to 2 weeks.

Lean construction departs significantly from current project management practice. Processes are actively controlled, and metrics are used in planning system performance to assure reliable workflow and predict project outcomes. Performance is optimized at the project level. Whereas current project management approaches reduce total performance by attempting to optimize each activity, lean construction succeeds by optimizing at the project level, as opposed to the less effective current project management approaches, which reduce total performance by attempting to optimize each activity.

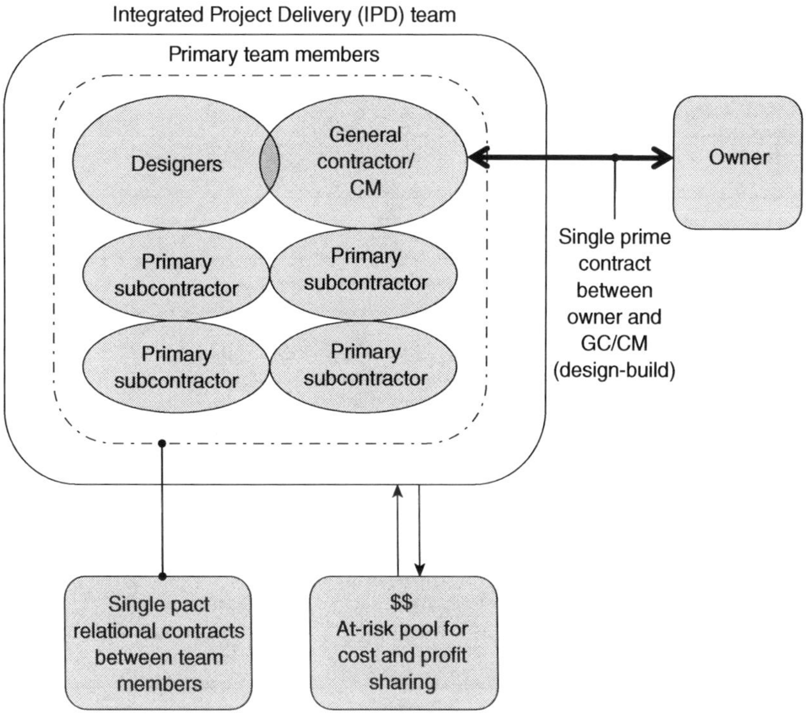

Figure 1.5 A representation of IPD

Lean applications in design and construction are continually evolving. The most successful applications have been observed with forms of contract that reward cooperation and collaboration between the parties that are actively involved in delivering design and construction. The Integrated Form of Agreement (IFOA) is one form of contract that has been successfully applied to lean construction.

Figure 1.5 represents one interpretation of lean project delivery, namely IPD. The team comprises the GC/CM working closely with designers and the primary subcontractors. The team members have relational contracts between them, as delineated by those entities encircled by the dotted line; these are described as "Single pact." The right-hand box (at-risk pool) "rescues" any individual firm that encounters financial difficulty; by the same token, team behavior is likely to leave a surplus in the account. This surplus is shared between the members of the integrated team.

The foregoing concepts have been embodied in variations in delivery methods termed IPD and lean project delivery. They are discussed in the chapters that follow.

Forms of contract

There are several categories of construction contracts that cover the spectrum of available approaches, all of which may be integrated with the foregoing construction delivery methods. Some categories of contract formats may be better suited to particular delivery methods, although the number of contract formats is almost infinite, just as there are no two projects exactly alike. The more frequently used formats are lump-sum contracting, GMP, GMP with cost-savings sharing, and cost-reimbursable contracts (cost-plus). The cost-plus compensation formats may include cost-plus with GMP, cost-plus with guaranteed maximum and incentive, and cost-plus with GMP and provision for escalation.

Fixed lump-sum price

The contractor receives a fixed sum for completion of the contract. This fixed sum may be adjusted for such things as owner changes in the work, differing site conditions, suspensions or delays, defective specifications, and so on.

Guaranteed maximum price (GMP)

The contractor provides cost estimates throughout the design process and, at some point prior to completion of the construction documents, submits a GMP proposal. The GMP proposal is based on a scope of work outlined in the preliminary design documents, and usually proposes that the contractor be reimbursed for the cost of the work, plus a predetermined fee with a maximum fee specified. Because of the incomplete nature of the design, contingencies and allowances are included in the GMP and the contractor is expected to build the job within the GMP limits, unless there is an owner-directed scope change.

Cost plus a fee

The contractor receives its actual "cost of the work plus a fee for overhead and profit." Definitions of cost (such as in AIA document A111) include the contractor's direct jobsite costs. Agreement on cost definitions is essential to avoid disputes over allocation of items to reimbursable cost or to fee. Since an open "cost-plus" arrangement is in effect a blank check, cost controls, and audits are essential. Fees may be negotiated either as a fixed lump-sum or as a percentage of cost. The latter encourages the contractor to spend more to increase its fee.

Fixed unit prices

The contractor receives a fixed sum for each unit of work completed. Because the units are not fixed, contracts may provide for renegotiation of the unit-price for substantial variations in unit quantities from estimated quantities.

Strategies for improving construction performance

Subsequent chapters address lean and productive practices to improve construction performance. These practices include a new paradigm for construction that includes a strategic management approach to focus the parties on optimizing performance at the project level, instead of seeking their own self-interest.

These practices include the application of performance improvement and industrial engineering techniques, including the lean methods that have been successfully applied in the manufacturing sector. The lean methods are based on a participative approach such as relational contracting and lean project delivery to improve the interaction between the members of the project team. Lean is the ultimate in collaboration and cooperation. In relational contracting, companies in the project team put the project first and gains and losses are shared.

Questions for discussion

1 What are the primary causes of low construction productivity?
2 Discuss the implications for construction quality when comparing the Design-bid-build method of project delivery with DB.

3 What are the factors that constitute construction performance?
4 What advantage does partnering offer in comparison with the traditional DBB method of project delivery?
5 How does optimization of work performance in lean construction differ from optimization in traditional construction?

References

Abdelhamid. 2013. *Lean construction.* https://www.msu.edu/user/tariq/Learn_Lean.html (March 21, 2013).

Alarcon, L., and H. Mesa. 2012. *A modeling approach to understand performance of lean project delivery system.* 20th Annual Conference of the International Group for Lean Construction, San Diego, CA, USA, July 18–20.

Ballard, G. 2000a. *The last planner system of production control.* Ph.D. dissertation. Birmingham, UK: University of Birmingham.

Ballard, G. 2000b. *LCI White Paper-8: Lean Project Delivery System.* Lean Construction Institute, September 23.

Ballard, G. 2008. The Lean Project Delivery System™: An Update. *Lean Construction Journal 2008*: 1–19.

Ballard, G., and Y.-W. Kim. 2007. Roadmap for lean implementation at the project level. *Research Report 234-11*, Construction Industry Institute, TX.

Ballard, G., and G. Howell. 1994. Implementing lean construction – Stabilizing work flow. *Conference on Lean Construction.* Santiago, Chile, September.

Bureau of Economic Analysis. 2019. *Construction gross output.* Online at www.bea.gov/. Accessed on May 1, 2019.

Bureau of Labor Statistics. 2019. *Construction employment 2009–2019.* Online at www.bls.gov. Accessed on May 1, 2019.

Crosby, P. B. 1979. *Quality is free: The art of hassle-free management.* New York: McGraw-Hill.

Forbes, L. H., M. Ahmed, and S. Azhar. 2003. Productivity management, improvement, and cost reduction. In *System-based vision for strategic and creative design*, ed. F. Bontempi Lisse. The Netherlands: A.A. Balkema Publishers, Swets and Zeitlinger, B.V.

Horman, M., and R. Kenley. 2005. Quantifying levels of wasted time in construction with meta-analysis. *Journal of Construction Engineering and Management* 131(1): 52–61.

Howell, G. A., A. Laufer, and G. Ballard. 1993. Interaction between subcycles: One key to improved methods. *Journal of Construction Engineering and Management, ASCE* 119(4): 714.

Howell, G. A. 1999. What is lean construction. *Proceedings of the 7th annual conference of the international group for lean construction.* UC Berkeley, CA: 2–10.

Koskela, L. 1992. Application of the new production philosophy to construction. *Technical Rep. No. 72*, Center for Integrated Facility Engineering (CIFE), Stanford University, Stanford, CA.

Koskela, L., and G. Howell. 2002. The underlying theory of project management is obsolete. In *Proceedings of PMI research conference 2002*, eds. D. P. Slevin, D. I. Cleland, and J. K. Pinto. Seattle WA: Project Management Institute.

Ohno, T. (1988). *Toyota production system: Beyond large scale production.* Cambridge, MA: Productivity Press.

Smart Market Report. 2013. Leveraging collaboration and advanced practices to increase project efficiency. *McGraw-Hill Construction.* www.construction.com.

Sumanth, D. J. 1984. *Productivity engineering and management.* New York: McGraw Hill, Inc.

Thomas, H. R., W. F. Maloney, M. W. Horner, G. R. Smith, V. K. Handa, and S. R. Sanders. 1990. Modeling construction labor productivity. *Journal of Construction Engineering and Management* 116(4): 705–726.

Tommelein, I. D. 2015. Journey toward Lean Construction: Pursuing a paradigm shift in the AEC industry. *Journal of Construction Engineering and Management*, 12.

Womack, J., and D. Jones. 1996. Lean thinking. New York: Simon & Schuster.

Bibliography

Adrian, J., and D. Adrian. 1995. *Total productivity and quality management in construction.* Champaign, IL: Stipes Publishing, LLC.

Alfeld, L. E. 1988. *Construction productivity.* New York: McGraw-Hill, Inc.

Alarcon, L. F., and A. Serpell. 2004. *Performance measuring, benchmarking, and modeling of construction projects.* Santiago, Chile: Pontificia Universidad Catolica de Chile.

Ashford, J. 1989. *The management of quality in construction.* London: E & F.N. Spon.

Bureau of Labor Statistics. 2006. *Fatal and non-fatal occupational statistic from 1992–2004.* Online at www.bls.gov.

Egan, Sir John. 1998. *Rethinking construction: The report of the Construction Task Force.* [The Egan Report]. London, UK: Department of Environment, Transport and Regions and HMSO.

Forbes, L. 1999. *An engineering management-based investigation of owner satisfaction, quality and performance variables in health care facilities construction.* Dissertation. Miami: University of Miami.

Forbes, L. H., and W. A. Golomski. 2001. A contemporary approach to construction quality improvement. In *The best on quality*, ed. M. N. Sinha, IAQ Book Series, 12: 185–200. Milwaukee, WI: ASQ Quality Press.

Gerald, F. 1997. *Building a strong economy: The economics of the construction industry.* Livonia, MI: Sharpe, Inc.

Howell, G. A., A. Laufer, and G. Ballard. 1993b. Uncertainty and project objectives. *Project Appraisal Journal* 8: 37–43. Guildford, England.

Howell, G. A. 2000. White paper for Berkeley/Stanford CE & M Workshop. *Proceedings of the construction engineering and management workshop.* Palo Alto, CA: Stanford University.

Industry Cost Effectiveness Task Force. 1983. More construction for the money. *Summary report of the Construction Industry Cost Effectiveness Project, Business Roundtable*, 11, USA.

Kubal, M. T. 1994. *Engineered quality in construction – Partnering and TQM.* New York: McGraw-Hill Inc.

Malcolm Baldrige National Quality Award. 2007. *Criteria for performance excellence.* Milwaukee: American Society for Quality.

Oberlender, G. D. 2000. Figure 9–15 in *Project management for engineering and construction*, 2nd ed. New York: McGraw-Hill, 228.

Proverbs, Holt, and Cheok. 2002. *Summary of J. Egan, Report, Rethinking Construction.* Construction.

Sanvido, V. E. 1988. Conceptual construction process model. *Journal of Construction Engineering and Management* 114(2): 294–311.

2
FOUNDATIONS OF LEAN CONSTRUCTION

This chapter provides an important foundation with an explanation of how lean concepts apply to the construction environment. The history of lean manufacturing is explained, and how the Toyota Corporation used lean techniques to minimize waste in its processes to accomplish its preeminence in automotive manufacturing. The chapter also describes the work of researchers who sought to adapt Toyota's approach to the construction environment. Lean principles are described in the context of construction projects as a foundation for the application of integrated delivery approaches and The Last Planner® System (LPS).

Defining lean construction

Lean construction has been defined in several ways as the concept continues to evolve. In fact, the term "Lean Construction" has come to signify both design and construction as these activities are closely interlinked. The activities that occur in the construction phase of a project are to a great extent predetermined by decisions that were made during the design phase, as can be seen in the lean project delivery system (LPDS) that is described later. For the sake of brevity, the term lean construction will be used in this book. As stated in Chapter 1, lean construction came into existence as a reaction to longstanding problems in the construction industry, evidenced by schedule delays, cost overruns, quality deficiencies, low productivity, high levels of waste and low levels of owner satisfaction with completed projects. The prevailing definitions of lean construction reflect the existence of these factors and refer to countermeasures to overcome these deficiencies.

The following descriptions are among the most established examples to date.

Koskela *et al.* (2002) described lean construction as a way to design production systems to minimize waste of materials, time, and effort, in order to generate the maximum possible amount of value.

Greg Howell and Glenn Ballard, co-founders of the Lean Construction Institute (LCI), described lean construction as a new way to manage construction. The objective, principles, and techniques of lean construction taken together form the basis for a new project delivery process. Unlike approaches such as design-build and programmatic improvement efforts (partnering and TQM), lean construction provides the foundation for an operations-based project delivery system. From its roots in the Toyota production system (TPS), this new way of designing and making capital facilities makes possible significant improvements in complex, uncertain, and quick projects.

Iris Tommelein (2015): "Lean stems from the recognition that the world is subject to all kinds of variation and that one can distinguish good variation from bad variation. Good variation is variation that customers intentionally want (e.g. people want buildings to not all look the same); all other variation is bad. To improve performance, lean starts by relentlessly driving out bad variation and then buffering where needed to protect the system from the impact of remaining variation."

The Lean Construction Institute (Abdelhamid 2013): "Lean construction is a production management-based project delivery system emphasizing the reliable and speedy delivery of value. It challenges the generally-accepted belief that there are always trade-offs between time, cost and quality."

The Associated General Contractors of America (AGC 2013) described Lean Construction as a set of ideas, practiced by individuals in the construction industry, based on the holistic pursuit of continuous improvements aimed at minimizing costs and maximizing value to clients in all dimensions of the built and natural environment: planning, design, construction, activation, operations, maintenance, salvaging, and recycling.

Abdelhamid (2013) defined lean construction as "a holistic facility design and delivery philosophy with an overarching aim of maximizing value to all stakeholders through systematic, synergistic, and continuous improvements in the contractual arrangements, the product design, the construction process design and methods selection, the supply chain, and the workflow reliability of site operations"

The Construction Industry Institute (CII) has defined it as " the continuous process of eliminating waste, meeting or exceeding all customer requirements, focusing on the entire value stream, and pursuing perfection in the execution of a constructed project" (CII Lean Principles in Construction Project Team – PT 191).

William Lichtig (2006) observed that lean construction aims to embody the benefits of the Master Builder concept. Essentially, Lean Construction recognizes that desired ends affect the means to achieve these ends, and that available means will affect -realized ends.

There are two slightly differing interpretations of lean construction

(i) The application of the methods of lean production to construction
(ii) A theoretical inspiration for the formulation of a new theory-based methodology for construction. This later model has prevailed with the International Group for Lean Construction (IGLC).

Lean construction practitioners seek to apply lean production theories to both design and construction practices. Its three tenets are: minimize waste and add value in all forms, continuously improve products and systems, and ***maintain respect for people***. The Lean Construction philosophy views a project as a "promise delivered by people working in a network of commitments" (Macomber and Howell 2005).

In clarifying the question "what is lean construction?" Glen Ballard (2016) points to the construction industry's special characteristics: All artifacts are first designed, then made in project production systems. It is differentiated from other forms of production:

Its products are rooted in the earth – they are designed for their conditions of usage – meteorological seismic cultural, etc.

As the product is assembled it cannot move through workstations because of its size, so it has to be addressed with fixed position manufacturing.

Construction's customers are individuals, not types. Products are usually prepared in response to a specific order. It is very important to understand the customer's needs and construction.

In Ballard's view, lean shares some characteristics with shipbuilding – ships involve fixed position manufacturing, but their products are not rooted in the earth. Their criteria based on different purposes, as they float.

As described by Ballard 2016, lean construction is effectively represented by a triangle as in Figure 2.1. The lean construction triangle symbolizes the components of the organizational relationships that characterize lean construction. The three sides include the organization and commercial terms which are both underlain by a unique operating system. For example, while traditional construction is *profit*-driven, LPD

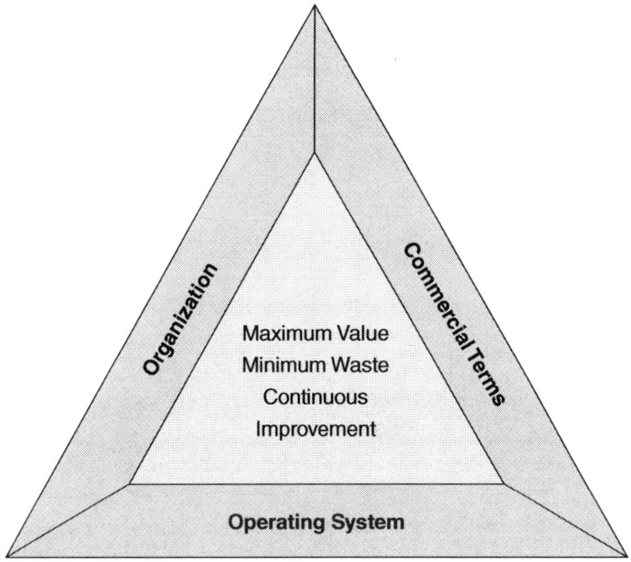

Figure 2.1 The lean construction triangle

Source: Adapted from Mossman (2014)

is *value*-driven. In traditional construction, the operating system is *push*-driven, whereas in LPD, it is *pull*-driven; the organization suboptimizes each *part*, operating by *command and control*, whereas LPD optimizes the *whole*, facilitating *collaboration;* and in traditional construction, the commercial terms are *transactional*, whereas in LPD they are *relational* (Mossman 2014).

Lean theory

In order to appreciate how lean techniques apply to the design and construction processes, we first need to understand how the lean approach improves performance in the manufacturing industry where it originated, and subsequently in the service industry.

Lean production is lean because it uses less of everything compared to mass production. Womack *et al.* (1990), in their book *The Machine That Changed the World*, chronicled the genesis of lean thinking in the automotive industry. They explain how lean production uses half of the resources that are typically consumed: the labor in the factory, the manufacturing space, the investment in tools, materials, and so on. It results in fewer defects and generates an increasing variety of products. Essentially, a lean system provides what is needed, in the amount needed, when it is needed.

Henry Ford pioneered the concept of continuous flow in his manufacturing plants but subsequently placed his company's emphasis on mass production to meet the huge consumer demand that came after World War II. Eiji Toyoda and Taiichi Ohno of the Toyota Motor Company perfected the just-in-time (JIT) concept as an alternative to mass production. The term "lean" is attributed to John Krafcik (1988), a researcher with the International Motor Vehicle Program (IMVP) at MIT. The IMVP was created in the 1980s to study the automotive industry internationally and to examine the differences between mass production and the innovative methods used by Japanese manufacturers.

The researchers used the term lean to designate what they saw as a new and superior type of production system for motor vehicles (Womack *et al.* 1990). The researchers were persuaded by the difference in comparative performance measures between Japanese, American, and European companies. Subsequently, it was found that the Japanese advantage was largely because of Toyota, and many academics and practitioners now believe that the Toyota Product Development System and the TPS are exemplars for a new and superior way of designing and making all kinds of goods and services.

Lean production can best be described by comparing it with craft production and mass production. Craft production was practiced in the preindustrial era – it was based on highly-skilled workers making products for customers, one at a time. Examples of craft production would be custom furniture, art works, and hand-built automobiles, products that are generally too expensive for most consumers. Mass production, by contrast, uses unskilled or semiskilled workers to produce standardized products in large quantities using highly specialized machines, based on processes designed by specialized professionals.

Mass producers are concerned with large production quantities with a narrow range of products and tend to generate a level of defects that they consider to be acceptable. By contrast, lean producers seek perfection, aiming for zero defects, zero inventories, declining costs, and a wide variety of products. Lean producers are focused on value-added process flow; that is, a continuous addition of value to a product or service (one-piece flow) by avoiding or eliminating interruptions or activities that do not add value. Examples of nonvalue-adding activities would be wasted material and wasted time waiting for value to be added. Lean producers address the root causes of wastes, such as defective processes, poor work layouts, lack of standardization, errors by operators, errors in communication, and poor decisions by management.

When lean first became popular and more widely known, there was a tendency to see it as a collection of tools and techniques, but it is now widely recognized as a fundamental business philosophy. "(Our) final conclusion is that lean cannot be reduced to a set of rules or tools. It must be approached as a system of thinking and behavior that is shared throughout the value stream" (Diekmann *et al.* 2004).

Henry Ford established many practices that are embodied in lean thinking even today. Ford's "Lean" Business Norms circa 1903–1947 included the following principles:

1. That work environments be spotlessly clean.
2. That business leaders think in terms of serving their communities and society at large.
3. That production techniques not be taken for granted but continuously improved.
4. That primary industries should help their suppliers and the service industries to produce cheaper and better products in less time.
5. That managers should not remain in their offices but should walk around, know their workers, and be capable of doing the work themselves.
6. That workers should be trained and have the opportunity to better themselves and make product improvements.

The Toyota experience: building on Ford's principles

The Toyota Corporation has been highly successful in automotive manufacturing and is recognized as an outstanding example of a lean organization. With regard to Toyota's recalls in 2009 and 2010, it must be noted that these problems were due to failures of management. The Toyota Production System is still a preeminent example of lean manufacturing. Toyota began as a loom company in 1926, founded by Sakichi Toyoda, a loom inventor. They began auto manufacturing in the 1930s, influenced by Henry Ford's continuous flow manufacturing. In the latter part of 1949, Toyota adopted statistical quality-control methods learned through courses provided by Japanese Union of Scientists and Engineers (JUSE). At that time Eiji Toyoda visited Ford Motor in Detroit to gain an understanding of their mass production system. Ford's operation was recognized as the world's most efficient, with a production rate of 7,000 cars per day, while Toyota had only produced 2,685 cars in 13 years up to that point. In effect, during 1948 a typical American autoworker produced 10 times the output of the Japanese autoworker.

Toyoda studied Ford's operation for three months and saw possibilities for improving the auto production plant in Japan. Taichii Ohno, Toyota's chief engineer, had also visited Detroit on a number of occasions. Toyoda and Ohno recognized that there was much waste or *muda* (Japanese term for much waste) everywhere in Ford's operation. For example, there was a high inventory cost to keep parts that were later found to be defective. There was *muda* of manpower, *muda* of waiting, *muda* of transport, and *muda* of facilities. They noted that only assembly line workers were adding any value to the production process. Specialists were responsible for designing the production process and directing workers to follow it. The foremen served only to ensure that assembly line workers followed orders, and they in turn simply performed repetitious tasks. Ohno realized that all these items of waste represented a financial loss and identified the seven wastes as follows: (Note that the sequence has been changed to provide the memorable mnemonic "TIM WOOD"):

1 T – Excess Transportation: Transporting product from one location to another increased energy costs.
2 I – Excess Inventory: Unnecessary stock resulted in excessive inventory carrying costs.
3 M – Excess Motion: Wasted motion led to nonproductive time (that incurred cost).
4 W – Waiting, idle time: Waiting time between stages in production resulted in lost labor.
5 O – Overproduction: Overproducing led to product quantities greater than the market needed.
6 O – Overprocessing: Processing using inefficient methods required more inputs than necessary to produce needed outputs.
7 D – Defects (in production): Defective production would have to be corrected before it could be paid for by the customer/client.
8 Other wastes.

These wastes are shown in Figure 2.2. An eighth waste is shown but will be discussed later in the chapter.

Toyota had the challenge of meeting diverse customer needs with a small capital base, as opposed to the Detroit strategy of having a large inventory of cars from which customers could pick and choose. The

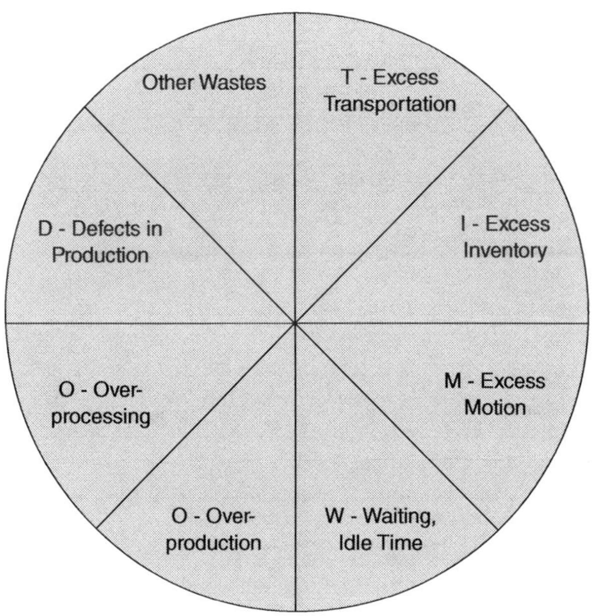

Figure 2.2 Production wastes: eight categories

Detroit strategy was also based on minimizing the cost of each car instead of minimizing wasted inventory. While mass production was effective in minimizing unit costs, any defects that developed in the fast moving process would be replicated in many cars before they could be detected. Toyota's efforts to improve productivity enabled them to produce larger production volumes at low cost, overly optimistic sales forecasts left them with large inventories of unsold vehicles in the 1960s that led to huge financial losses. Japanese consumers also preferred customized features that would prove to be uneconomic with mass production techniques and the long-setup times they required. Toyoda and Ohno decided that the human and material resources waste they saw in the United States was not appropriate for Toyota, or Japan, for that matter. They decided that a better strategy would be to enable the production system to change rapidly between different models and types of vehicles while producing minimal amounts of waste.

Ohno found a new way to coordinate the flow of parts in the supply system on a day-today basis so that parts would only be produced at each step to meet the immediate demand of the succeeding operation. This led to the development of the Toyota Production system (Ohno 1988). At each work station, teams of workers controlled their local operations-based on continuous observation and relationships instead of by specific metrics. The workflow of a Toyota factory regulated itself through "takt time," a beat like a metronome, based on a term used by German musicians. This beat matched the daily production demand.

Engineers were required to sense the needs of the operation and opportunities for improvement by standing in a chalk circle for several hours at a time. Later, a card called a kanban (Japanese for "card") in 1958 was used as a communication tool between different production lines, and became an important part of the JIT system. Ohno had been inspired with this idea by observing the US supermarket system in 1950. He wanted every member of the production system to pay attention to the prevention of potential problems and so reduce waste (muda). This system worked so well that in 1959 Toyota produced 100,000 cars per year for the first time – a major accomplishment.

Toyota was subjected to heavy pressure around that time with the failure of the new Toyota Crown in the United States, and Japan's decision to liberalize trade. This spurred a strategic decision – to compete on the basis of quality. Toyota instituted total quality control (TQC) with the "kanban" system in 1963. Their improved quality and productivity were recognized with the Deming Prize.

Toyota's production system

From the 1960s to the 1980s, this system evolved into the highly successful TPS that provided a competitive advantage through outstanding levels of production, high quality, and low costs. The TPS has been described as a combination of methods with consistent goals – cost reduction, quality assurance, and respect for humanity – to ensure sustainable growth. It has four main elements: JIT, autonomation, workforce flexibility, and creative thinking.

The Toyota Corporation has since evolved the TPS to become "The Toyota Way."

Toyota was able to produce a car to meet each owner's specifications in real time. Workers performed their own quality control without inspectors. With the JIT system, external suppliers delivered parts directly to work stations without being warehoused. In the mid-1970s, Toyota reduced the time needed to produce a car from 15 days to 1 day, using JIT. It is important to note that Ohno's improvement of Toyota's production process was not necessarily a new technology, but rather the result of involving all participants in a new philosophy of avoiding waste of any type. Ohno also developed a number of tools for operating his production format in a systematic framework. The best known of those tools is the kanban system, which provides for conveying information in and between processes on instruction cards. Ohno collaborated with Shigeo Shingo in training manufacturing engineers on implementing the efficiency processes, which Ohno had developed earlier for the Toyota Production System and which led to an integrated manufacturing strategy.

It is to Toyota's credit that they shared knowledge with the rest of the world through visitor tours of their facilities. They formed strategic alliances with competitors such as General Motors. They allowed their employees Ohno and Shingo to publish books about Toyota's Just In Time (JIT) success. Toyota's success was also due to strategic relationships with suppliers. Adversarial relationships were considered wasteful; instead, Toyota would pick two or three suppliers for each component and guarantee them a specific volume of business. This trusting relationship reduced waste, cut costs, promoted innovation, and increased profits for all stakeholders.

Just-in-time (JIT) concept

Whereas mass production is a so-called push system that dictates production volume based on market forecasts, JIT is a pull system that responds to actual customer demand and enables the important lean concept of "Pull in Production". In essence, products are pulled from the JIT system. While push systems are susceptible to over or under-producing based on the accuracy of forecasts, JIT only commits the resources needed to meet the actual demands. The customer is the driver of production. Mass production emphasizes the production of the largest possible quantities, as setups and adjustments of machines are time consuming. JIT, on the other hand, seeks to build small lots that are better capable of providing variety for the customer. Companies that have used the Toyota approach have reduced setup times by 90% or more. Omark Industries, one of the earliest US companies to adopt the TPS was able to reduce machine setup time from 8 hours to 64 seconds. Overall, JIT facilitates lower costs, shorten lead and cycle time, and leads to improved quality.

The kanban system minimizes inventories through backward requests. Kanban cards follow a product through various stages of production and provide visual control. They contain information such as the part name, description, process instructions, and quantity required. In keeping with the pull system, kanban cards signal "upstream" when production or the release of inventory is required. In effect, inventory is not created until it is needed. In effect, Toyota's underlying strategy was to produce a car to meet the requirements of a specific customer, instead of maintaining a large inventory of cars from which the customer could choose. The application of JIT results in a lower inventory of raw materials and of parts, less work in process, and shorter lead times. It also leads to a reduction in floor space, less overhead, and lower costs. However, there is a risk that inventories may fall to critically low levels. To be successful, JIT is dependent on very high quality and reduced setup times – the concept calls for raw materials and components to reach a production operation in desired quantities when needed and not before. The supplier and customer have to work closely for this to occur and there is little to no inventory.

Production leveling accommodates fluctuations in demand and the number of setup activities is reduced so that little time is required for changing over to different product mixes. (In modern manufacturing, the kanban card has been replaced by an electronic equivalent.)

The Toyota Way

The Toyota Corporation has continued to evolve beyond the foundation of the TPS. The TPS was represented by two pillars – JIT and autonomation. Liker (2004) describes 14 principles in the contemporary Toyota Way. The Toyota Way represents an important foundation for lean construction. Economics were a factor in the Japanese use of human resources; it proved too expensive to rely on few highly qualified people. Lean thinking as represented in *The Toyota Way*, sees the most cost effective way of improving product quality as empowering all people in organizations (Table 2.1). This varies significantly from the prevailing model in the construction industry, which emphasizes power and control.

Table 2.1 A representation of the Toyota Way

Toyota Foundations	*Principles*
Problem Solving (Continuous Improvement and Learning	Continual organizational learning
	View the situation firsthand to thoroughly understand it
	Make decisions slowly by consensus – consider all options: Implement rapidly
People and Partners (Respect, Challenge, and Grow Them)	Grow Leaders who live the philosophy respect, develop, and challenge people and teams
	Respect, challenge and help suppliers
Process (Eliminate Waste)	Create process "flow" to reveal problems
	Use pull system to avoid overproduction
	Level out workload
	Stop when there is a quality problem
	Standardize tasks for continuous improvement
	Use visual control – transparency
	Use only reliable, tested technology
Philosophy (Long-Term Thinking)	Base management decisions on a long-term philosophy, even at the expense of short-term financial goals

Origins of lean construction

Early work by Frank Gilbreth in the 1890s indicated the potential for manufacturing approaches to be used in construction to improve speed and labor efficiency. Gilbreth is considered to be the father of industrial engineering as he later built on Frederick Taylor's principles. Gilbreth saw an opportunity in bricklaying – he looked at the way bricklayers worked and felt that they could perform more efficiently. He developed the idea of what he called "speed work." Gilbreth noted that many of the motions made by bricklayers were unnecessary and did not really contribute to the job. They were accustomed to bending down and retrieving each brick from a pile – they would turn it over in their hands before plastering it in place, taking extra time. Frank made several improvements to the job. He had the working stock of bricks located on a platform that made it unnecessary for workers to bend excessively, thereby reducing retrieval time. He also arranged for lower-skilled (and lower-paid) workers to transport the bricks to the platform, allowing the craftsmen to focus on applying their expertise. He developed a "best method" that reduced the number of motions from 18 to 4-½, minimized fatigue and increased productivity.

Gilbreth experimented with system improvements such as finding the optimum load that a worker could manage with a shovel in order to safely carry out the maximum amount of work each day. He developed labor standards that could improve the predictability of work output. Gilbreth had his own contracting team and went on to compete very profitably in the contracting business in the early 1900s. With his wife, Lillian Gilbreth, he developed a body of knowledge that led to the field of Industrial Engineering.

In most of the twentieth century, construction productivity grew but more slowly than manufacturing productivity. As described in Chapter 1, construction productivity was observed to decline at a rate of roughly 0.6% annually over a 45-year period to 2012. Nonfarm productivity grew by 1.8% annually.

Construction practice transitioned from the Master Builder concept, in which one entity was responsible for coordinating both design and construction, to a highly fragmented industry. Designers wished to limit their liability that could result from construction failures, and developed contract language that insulated them from the construction responsibility. Productivity growth was primarily due to the application of technology; better tools and equipment increased labor efficiency. Innovations such as prefabricated building elements and tilt-up construction also improved the speed of the process. Computer applications for planning and scheduling also improved overall efficiency. Yet the industry was fraught with distrust and adversarial relations, resulting in behaviors that were both counterproductive and costly.

It has been known for many years that construction performance has much room for improvement. The US Business Roundtable was formed by a number of industry leaders to identify improved methods for industries including construction. Their findings pointed to major shortcomings in the construction industry, and the Roundtable appointed the Construction Competitiveness Committee to study it in more detail. The Construction Industry Institute (CII) grew out of the Committee, and was created in 1983 and based at the University of Texas–Austin. The CII has been studying industry shortcomings for many years in order to develop system improvements.

A relatively small percentage of total work hours in construction is spent productively. A 1990 CII-sponsored study by Michael Pappas (1990) noted that in steel erection, only 11.4% of recorded work hours represented value-added work. In process piping projects, the percentage of value-added time was even less, at 7.5%. Hammarlund and Ryden (1989) and Nielsen and Kristensen (2001) observed that transformation activities in construction occurred during only 30% of the working time available in projects. Of greater concern than these activities was the time spent in waiting and transporting – as these do not add value. In fact, material handling and transporting greatly increase material costs (Bertelsen 1993).

Lauri Koskela (1992) addressed the application of manufacturing techniques to construction. Koskela spent a year at Stanford University as a visiting scholar and conducted a study titled "Applications of The New Production Philosophy to Construction." He drew parallels between both fields by characterizing construction as a form of production. Koskela looked to the manufacturing industry for a new direction for construction. Specifically, he modeled this new production philosophy after the Toyota's highly successful production system (TPS). Researchers had recommended such solutions for the construction industry's underperformance as industrialization (prefabrication and modularization), automation and robotics, and information technology to reduce fragmentation. Koskela proposed a new approach that was not based on technology, but rather on the principles of a production philosophy. He noted its evolution through three stages:

1 Tools, such as kanban and quality circles,
2 A manufacturing method, and
3 A management philosophy (lean production, JIT/TQC, etc.).

Koskela inferred from a number of productivity-based studies on US and European manufacturing plants that the most successful methods were based on the JIT philosophy. Manufacturing studies by other researchers such as Schonberger (1986) and Harmon and Peterson (1990) reinforced this observation.

In a typical production process, material is processed (converted) in a number of discrete stages. It is also inspected, moved from one operation to another, or made to wait. Inspection and waiting are considered flow activities.

Conversion activities are considered to add value while flow activities do not. Koskela visualized lean construction as a flow process combined with conversion activities, and noted that only the conversion activities add value. This was the transformation-flow-value (TVF) theory of construction. Production improvements can be derived by eliminating or reducing flow activities while making conversion activities more efficient. Koskela proposed that production should be improved by reducing or eliminating flow activities, while conversion activities should be made to be more efficient. He pointed to earlier studies that showed that only 3–20% of the steps in a typical process add value and that their share of the total cycle time is only 0.5–5%.

Koskela attributed the predominance of nonvalue-adding activities to three root causes: design, ignorance, and the inherent nature of production. Design was due to the subdivision of tasks; each added subtask increased the incidence of inspecting, waiting, and moving. A natural tendency for processes to evolve over time without close analysis leads to ignorance of their inherent wastefulness. The inherent nature of production is that events such as defects and accidents add to nonvalue-added (NVA) steps and time between different conversion activities.

Koskela listed the following heuristic principles:

- Reduce the share of nonvalue-adding activities,
- Increase output value through systematic consideration of customer requirements,
- Reduce variability,
- Reduce the cycle time,
- Simplify by minimizing the number of steps, parts, and linkages,
- Increase output flexibility,
- Increase process transparency,
- Focus control on the complete process,
- Build continuous improvement into the process,
- Balance flow improvement with conversion improvement, and
- Benchmark.

Points to note in relation to the heuristic principles are:

- Nonvalue-adding activities can be reduced by identification, measurement, and redesign.
- Output value can be improved by identifying the supplier and customer for each activity, and clarifying customers' value needs.
- Variability of construction activity duration increases the relative volume of NVA activities. The adoption of standard procedures can streamline both conversion and flow processes.
- Process control requires measurement and designated controlling authority – that authority may be cross-functional process owners or self-directed teams. Team building and cooperation with suppliers are proposed for optimizing total flow when multiple organizations are involved.
- Flow improvement and conversion improvement should be balanced – with complex production processes, improvement of flow (NVA) activities has a higher payback than seeking technology-based enhancements to conversion activities.

Other researchers built on this foundation and determined that construction-specific tools would be needed in order to adapt construction to the lean production principle.

Glenn Ballard heard Koskela speak at University of California–Berkeley and they began a collaboration on lean production that led to the staging of a conference in Helsinki, Finland, in August 1993. That conference became the first meeting of the International Group for Lean Construction (IGLC). As Ballard reports, it was decided in that meeting to use the expression "Lean Construction." Ballard and Greg Howell subsequently co-founded The LCI in the United States in 1997. Other organizations dedicated to lean construction were formed in Chile, Denmark, the UK, and several other countries.

Ballard pioneered the development of The Last Planner® System (LPS) in 1992. The LPS was based on the concept of reducing the hierarchical layers of construction management and empowering field-based actors in the construction process to optimize the allocation of available resources in the weekly planning, scheduling, and execution of work. Ballard further refined the LPS in 1998, transitioning from weekly work planning to look ahead planning and to phase scheduling. This refinement focused on managing flows in the construction process, reducing flow variation from the plan, and using buffers to limit the impacts of any remaining variability in these flows.

Members of both IGLC and LCI performed studies to further develop the lean construction concept. Howell and Ballard sought to validate the results of their experimental field work with applying lean theory to construction processes and collected metrics to substantiate results obtained during the playing of live simulations. In collaborating with them Iris Tommelein, a UC Berkeley civil engineering academician played a pivotal role in validating the concept. Analyzing results obtained during the playing of live and computerized simulations confirmed the mathematical logic of lean principles (Tommelein *et al.* 1999). It gave assurance to practitioners that the successes enjoyed during case studies could be trusted as generalizable and repeatable.

Ballard set about developing what would later come to be called The Lean Project Delivery System™. Studies identified that traditional project management methods normally accomplished only approximately 50% of scheduled weekly work plans (Ballard and Howell 1994). Many white papers were generated by the LCI that furthered the development of the LPDS – Ballard's White Paper #8 introduced The Lean Project Delivery System™ in 2000 (2000b).

The LPS was applied to the design process in 1998, representing another milestone. The design process was further improved with the application of target value design (TVD). The first TVD paper was published in 2004.

Theory of production

Koskela *et al.* (2002) in their chapter titled "foundations of lean construction" discuss his earlier research on the theory of production and systems of production, classifying them as the transformation, flow and value (TFV) concepts. Koskela argued that all three aspects of construction must be managed simultaneously for successful project delivery (Tommelein 2015).

Traditional project management experience was seen as limited by trying to manage by using scheduling, cost and output measures. They point out that lean succeeds by attempting to manage the value created by the work processes mobilized between project conception and delivery. The paper advances lean construction as a novel theory-based approach to construction, while recognizing another interpretation i.e., the application of lean production. Koskela *et al.* explain the use of a theory as facilitating tests of validity, decomposing the total transformation into component tasks. It can also be applied in other settings and facilitates use by non-experts, as indicated below:

a) an explanation of observed behavior and enables prediction of future behavior
b) tools can be built for analyzing, designing and controlling systems
c) a common framework for collective undertakings by several people
d) direction in pinpointing sources of further progress
e) condensed knowledge that can empower novices

Koskela *et al.* would expect a theory of production to be prescriptive, explaining the outcomes expected from specific actions. In the production context, the possible actions would be the design of the production system; control of the production system to realize the production intended; and improvement of the production system.

Three kinds of goals are linked to production activity:

1 Having specific products generated, establishing internal goals that influence the characteristics of production, such as cost minimization and level of utilization; and external goals related to customer requirements such as quality, dependability and flexibility. Koskela *et al.* would expect the theory of production to explain all essential areas of production, especially the actual production and the design of products. Applying the theory should lead to improved performance, and lack of application should lead to inferior performance.
2 The transformation view of production has been dominant in the twentieth century and represents the conversion of inputs to outputs. It has greatly influenced the principles that underlie operations

management. The origins of the transformation view lie in economics; the value chain theory is linked to it (Porter 1985; Wortmann 1992). It is beneficial in identifying the tasks that are critical for a production activity.

3 The flow view aligns closely with lean production and emphasizes the removal of wastes from flow processes. Value generation view seeks the best outcome from the perspective of the customer. These three views are seen, not as competing theories, but are partial and complementary. The TFV theory of production is an integrated combination of the respective theories.

Koskela *et al.* point to the limitations of the "thermostat model" of project control, especially with today's projects which tend to be complex, uncertain and quick (CUQ) as described by Shenhar and Laufer (1995). Project controls determine whether a project activities are within their cost and schedule limits. Deviations are addressed with the thermostat approach, adding workers or re-sequencing activities. The traditional use of centralized scheduling is blamed for this shortcoming, as it has little capacity to manage work flow and the creation and delivery of value.

The thermostat approach is unreliable in this context – whereas a thermostat can address temperature variations by adjusting heating or cooling systems. On the other hand the reliable release of work from one crew to the next cannot be managed effectively with this approach.

In fact, the actions of one supervisor to increase production or reduce costs with one team may reduce the reliable release of work to the next team; in turn this may reduce the performance of the entire project team.

NOTE: These issues will be explored later with a discussion of the Lean Project Delivery System.

Adoption of relational contracting

Another milestone in the evolution of lean construction was the adoption of relational forms of contracting. At an LCI meeting in 2003 in Las Vegas, Nevada, the attendees discussed the need for a special type of contract for lean; while they determined that such a contract was not necessary, they agreed that prevailing forms of contract could be obstacles to lean. In November 2004, LCI held an international symposium on relational contracting in Atlanta, Georgia. Several types of contracts were discussed that were relational to varying degrees; from the cases presented, there was evidence of a synergy between relational contracts and lean construction. Will Lichtig, an attorney with McDonough, Holland and Allen, PC, and an LCI member, agreed to develop a form of relational contract that later became known as the Integrated Form of Agreement (IFOA). This form of contract was subsequently adopted in several major lean construction projects – notably those established by the Sutter Health System in California.

Bertelsen and Koskela (2004) addressed future directions for lean construction as a product strategy and a process strategy. The product strategy would involve simplifying construction to allow components to be fabricated off-site under relatively predictable conditions; this would facilitate the use of manufacturing techniques, with great increases in productivity. The process strategy would involve improving the on-site construction process.

With regard to the product strategy, Bertelsen noted a trend in construction toward prefabrication and modularization; the use of prefabricated structural steel and concrete slabs increasingly changing on-site work to assembly and observing tolerances. Despite the uniqueness of a given project or facility, the manufacture of modules and subassemblies in controlled manufacturing environments reduced the overall project complexity, placing less emphasis on the management of materials and craft workers on-site. The tradition of building in place had always been subject to high levels of inefficiency. On the other hand, the product strategy seemed to rely on supply-chain management, with various suppliers increasingly manufacturing and supplying components and subassemblies such as windows, beams, roof trusses, and so on.

Bertelsen described open systems and closed systems in the product strategy. Open systems are made to standard specifications and can be combined with systems from other manufacturers, resulting in high levels of architectural design flexibility. Conversely, closed systems are proprietary and offer less flexibility in terms of cross combination. Manufactured housing is a typical example of a closed system. He observed that closed systems seemed to benefit greatly from mass production approaches, but had limitations in the amount of flexibility that they offered. Overall, he saw the product strategy as being more adaptable than the process strategy in addressing special needs such as affordable housing – where uniformity was more tolerable.

The process strategy has become the primary focus of lean construction as it is most applicable to on-site construction. Koskela viewed construction as "unique products of art on a very large scale." Construction complexity originated from this unique nature and from associated undocumented production processes – a temporary production system with shared resources between subcontractors working on different projects simultaneously. It occurs in an ad hoc organization that is a complex social system.

Process improvements include enhancements to the flow of materials – this complements the last planner as described by Bertelsen and Nielsen (1997), and Arbulu and Ballard (2004). Information flow improvements may similarly offer opportunities for improvement. Macomber and Howell (2003) proposed management by conversations as a means of improving the flow of information to improve the construction process.

Bertelsen commented on the development of "construction physics" as analogous to factory physics described by Hopp and Spearman (2000), but noted that the latter was based on production as an ordered system, whereas construction runs counter to a direct application of these principles. Instead, construction physics should be based on maximizing value and minimizing waste. As the lean construction principles evolved, they have been tested and refined through both application and research.

Owen Matthews of Westbrook Air conditioning and Plumbing a mechanical contractor in Orlando, FL, coined the term "Integrated Project Delivery" (IPD) to represent the application of lean principles in a team-based framework. Westbrook Service adapted design build contracting to include lean principles in a chiller plant installation project in Orlando, FL. This was documented by Matthews and Howell (2005) and described in Chapter 9 of this book.

Sutter Health embarked on a major construction program, mandated by the State of California. They made a major commitment to lean construction and had their legal counsel, Will Lich develop an IFOA for their projects.

The IFOA was subsequently adopted by the construction industry in a standard form of contract termed "ConsensusDocs 300." Twenty two leading construction associations united to publish a consensus set of contract documents called ConsensusDocs in 2007. They succeeded in creating a standard contract for the industry that they felt was fair to all parties. When ConsensusDocs was released, two large stakeholder organizations, AGC and COAA converted their contract documents to comply with the consensus process. Several other variations on the theme of lean construction have been developed and successfully applied to construction projects. The American Institute of Architects has also developed IFOAs for different types of lean projects. The ConsensusDocs IFOA has been updated a number of times since its inception.

Currently, lean construction is still evolving and is gradually being adopted. The LCI's strategy is to mainstream the lean philosophy in the US construction industry through the formation of regional chapters. These chapters are associated with university-based research centers that serve as learning laboratories. The original model for this structure was LCI's Northern California chapter in association with University of California–Berkeley's Project Production Systems Laboratory (P2SL) is headed by Prof. Iris D. Tommelein who serves as Director. She is also a Professor of Engineering and Project Management Program, UC Berkeley. Glenn Ballard is Research Director of P2SL, and Greg Howell is a Research Associate. Greg Howell and Glenn Ballard were co-founders of the Lean Construction Institute in 1997. Members of the Construction industry are beginning to fund the needed research and is actively engaged in experimentation, the findings from which are shared through LCI and groups termed P2SLs. Chapters dedicated to lean construction have been formed in a number of other countries including: the Canada, United Kingdom, Denmark, Finland, Australia, Brazil, Chile, and Peru. Similar chapters are starting up in Singapore, Indonesia, Ecuador, and Colombia. (This list is by no means complete and continues to grow.)

Lean design and construction

Lean design and construction involve the application of lean methods/techniques to the design and construction processes, in order to derive the benefits that have been clearly established in manufacturing operations. Benefits include:

- Lower costs,
- Fewer delays,
- Less uncertainty,
- Less waste,
- More efficient buildings/facilities, and
- Higher user satisfaction.

The application of lean principles results in better utilization of resources – especially labor and material. It also results in better construction quality in completed facilities, greater owner/client satisfaction, higher levels of safety, and ultimately greater profitability for clients, builders, and design professionals. Lean construction uses production management techniques to make significant improvements particularly on complex, uncertain, and quick projects Howell (2000). There have been many examples of success with projects based on lean methods: a small office construction project has had building costs reduced by 25% and schematic design time cut from 11 to 2 weeks.

To not use lean construction means major project failures, resulting in both cost and schedule overruns. As shown in Table 2.2, many major projects in the United States have exceeded budgets by 57–460%.

Deficiencies in traditional construction methods

Conventional construction is based on craft production methods, hence it is slow and expensive in comparison to mass and lean production methods that have been used successfully by manufacturers of autos and consumer goods. These craft production methods are carried out by many different specialists – shell contractors, masons, roofers, mechanical and electrical subcontractors, and so on. While they may be highly skilled as specialists, they are interdependent, yet they typically have separate contracts with a central

Table 2.2 A representation of the TFV theory of production – Koskela *et al.* 2002

	Transformation view	*Flow view*	*Value generation view*
Conceptualization of production	Transforming inputs to outputs	Flow of material – transformation, inspection, moving and waiting	Clarify where customer value is created, meeting requirements
Main principle	Getting production to occur efficiently	Elimination of waste, (NVA activities)	Elimination of value loss, achieving best possible value
Methods and practices	Work breakdown structures, MRP Responsibility chart	Continuous flow, pull, production control, continuous improvement	Methods for requirement capture, quality function deployment
Practical contributions	Taking care of what has to be done	Avoiding unnecessary actions	Ensuring that customer requirements are best met
Suggested name	Task management	Flow management	Value management

Source: Adapted from Table 14.1

entity – a general contractor, construction manager, and so on. These contracts place the parties to construction projects in adversarial roles, with penalties for underperformance. In turn, construction profit margins are subject to many areas of risk such as material price increases or shortages, labor shortages, adverse weather conditions, and so on. Consequently, the parties tend to have little intercommunication and often proceed as quickly as possible, frequently resulting in work that interrupts the sequence of upstream tasks. There have been many examples of plumbers and HVAC subcontractors who competed for the same ceiling space, resulting in poor layouts for both disciplines; or the drywall subcontractor who completed an interior wall before the electrical subcontractor was able to place the required conduit and outlet boxes.

Typically, projects are subdivided into activities; in turn the resource requirements and time frames for each activity are placed on a CPM chart, guided by a master schedule. Constructors are in fact construction managers who subcontract activities, generally by discipline. Once the work starts, the CM attempts to control it by comparing progress with preplanned schedules, usually at weekly meetings.

- Costs are reduced by finding ways to increase productivity – such as use of special equipment or better methods.
- Activity durations are adjusted by work force or other adjustments.
- Attempts are made to improve quality and safety by inspection and enforcement.

These actions represent the "thermostat model" (Hofstede 1978); performance is measured at the output of the controlled process. The possible variance between a standard and the measured value is used for correcting the process until the standard can be reached. Unfortunately, these efforts to improve productivity have limited impact on the overall project. Traditional project management attempts to manage activities by centrally applied scheduling, which is limited in its ability to manage activities and workflow. As schedules could quickly go off track, CMs are often in a reactive mode, adjusting staffing or sequencing to maintain schedules. Errors often result from poor communication and unplanned events. Attempts to get back on schedule may lead to cost overruns, quality deficiencies, or lapses in safety practices.

Traditional project management is very limited in its ability to reduce project variability. It uses the critical path method (CPM) to establish overall project schedules and keeps track of whether projects are on time or not. Gantt charts show critical timelines. The master schedule defines construction milestones and tasks but does not give workers incentive to work together or define best methods. As defined by Shenhar and Laufer (1995), projects are increasingly complex, uncertain, and quick (CUQ); They challenge even the most detailed CPM schedules (Tommelein *et al.* 1999).

Earned value analysis helps project managers to evaluate performance. However, these tools do not control project variability. Even though these tools are available to project managers, projects are still delayed and have cost overruns. Traditional project management also involves a culture of "pushing" work assignments to subcontractors in order to meet the master schedule, whether or not these producers have all the needed resources to complete those assignments in a given week. This tendency is the very opposite of the downstream demand pull required by the Last Planner® System. It disrupts the reliable release of work to the next team.

Philosophical differences between lean construction and traditional construction

Lean construction departs significantly from traditional project management practice. Processes are actively controlled, and metrics are used in planning system performance to assure reliable workflow and predict project outcomes. With lean methods performance is optimized at the project level, whereas current project management approaches reduce total performance by attempting to optimize each activity. Traditional construction approaches reward individual crew performance – crews may focus on their tasks to the detriment of other crews. In the lean approach, all involved disciplines are rewarded for completing major sections of the project. Lean construction succeeds by optimizing at the project level, as opposed to the local optimization of an individual subcontractor.

Lean construction involves the basic infrastructure that is used in conventional construction. It requires an infrastructure of design professionals, construction organizations (builders/contactors), trades subcontractors, and project managers. These roles are described in the Project Management Body of Knowledge (PMBOK) established by the Project Management Institute (PMI). However, lean involves a philosophy that builds on the basic infrastructure and institutes a different approach to managing projects.

Lean construction involves better short-term planning and control that improves the timely completion of job tasks, and reduces the variability of work output that tends to happen with traditional project management methods. It emphasizes having work flow between crews without interruption. Consequently there is more cooperation between these crews and a joint focus on completing the overall project as opposed to self-interest in their own work tasks.

Lean construction does not replace CPM or other tools that define the overall work schedule, but it works within them to improve the delivery of short-term assignments. CPM is seen as playing a strategic role, identifying major project milestones and the sequencing of critical activities, while lean planning is seen as tactical in nature, planning the work flow for the next four or five weeks. *It is difficult for CPM to plan and track individual short duration tasks.* One important principle in lean construction is to have the smooth work flow handled by the trades that are in direct contact with it. The "Last Planner" concept relies on empowering the foreperson and crew leaders to decide on the specific tasks to be done in the following weeks, as they are closest to the work and (once attuned to the lean methodology) are more familiar with its relation to each person's capabilities.

The PMBOK views planning as the key to success, and management by control as most important to organize cost and time and achieve a specific scope of work. These requirements seem to be very firm and leave no room for deviation. They view projects as controllable and relatively predictable – based on the critical path analysis – if the long-established planning and control tools are utilized. They also view each project as unique and not very adaptable to the methods used in continuous operations.

Lean construction, on the other hand, recognizes the planning that is so important to the PMBOK is often limited in effectiveness by unplanned events that occur in almost every project. Lean construction promotes the application of cutting-edge approaches used in manufacturing. The Last Planner® technique for scheduling is effective because it focuses on a short-time horizon so it becomes a technique for continuous production.

Lean construction focuses on adding value as opposed to controlling cost and schedule. It also focuses on flexibility and learning in order to cope with uncertainty and unplanned events. This is especially important with projects that are not routine and are very complex. The CPM is really an approximation of how work should be done and is not as effective in handling the details of what can actually be done.

Overall, the PMBOK works well with projects that are relatively simple and predictable. When there is a lot of complexity and uncertainty, lean construction is likely to be more effective. Such projects require learning new lessons as the project progresses and using this knowledge in planning successive stages of the work.

Barriers to applying manufacturing methods to construction

Researchers have described the obstacles that limit the application of manufacturing techniques to the construction industry, and referred to the differences between the respective fields as a major factor. Consequently the lean tools and techniques that have been successfully applied in manufacturing have to be adapted to the construction environment. These differences include

- Construction "products" are made in response to individualized orders. The same design is rarely used more than once, whereas most manufacturing is repetitious.

- Manufacturing activity is usually carried out in a controlled environment – generally indoors, but the location of construction projects is not constant – it differs from one project to the next.
- Manufacturing processes can be designed in detail – this lends to optimization.
- Construction processes are determined by crews in the field where there is the potential for much variability.
- Manufacturing projects tend to be of short duration and are more easily quantified and more easily improved in real time. On the other hand, construction projects are generally of long duration and more complex.
- Owners often commission architectural design and construction separately – working relations between architects/engineers, contractors and subcontractors vary on a project-by-project basis.
- The construction industry has become highly specialized, partly because of liability concerns. Designers document the desired facilities with drawings and specifications that relegate construction details to the contractor.
- Organization of and facilities for production are dispensed with after project completion.
- At completion, each project is appraised differently.

Characteristics of lean construction

Lean methods are a departure from conventional methods and their adoption requires changing the behavior of people. Cultural change is the most compelling quest along with the physical transformation of an organization. One cannot force change on people. They have to be engaged so that the intrinsic satisfaction of outstanding performance will motivate them. Chapter 7 describes strategies for engaging construction stakeholders. In a lean culture, people have to be treated as the only appreciating asset. Lean enables organizations to have Responsiveness, Reliability, and Relevance.

Lean principles

Five lean principles described by Womack and Jones (1996) apply to any organization. The principles are:

1. Value: It is critical to identify the value actually desired by customers and provide it. Lean organizations resist the tendency to persuade customers that they desire what the organization finds easiest to provide.
2. Value stream: Mapping the value stream for each product or service exposes waste and facilitates its removal; establishing cooperation between the participants and stakeholders results in lean processes.
3. Flow: It is necessary to make value creating steps flow. Business, job site, and supply flows depend on an effective value stream with few or no impediments.
4. Pull: Under the lean transformation, the efforts of all participants are to stabilize pulls in keeping with the demands of the customer.
5. Perfection: Strive for perfection, although it may never be achieved. Develop work instructions and procedures and establish quality controls.

Value

Glenn Ballard views the lean ideal as delivering value (instrumental value)to the customer with no waste. Value-adding activities transform materials and information into products and services needed by the customer. Value is not necessarily economic; in the process environment it is obtained when desired products and services are delivered. Wandahl and Bejder (2003) differentiate between product and process values. Product value relates to tangible aspects of a product, such as material composition, price of construction, flexibility, and so on. Process value in the construction environment relates to stages in the building process and interaction between producers such as time, communication, and so forth. The interaction between producers can best be represented by process value. Nonvalue-adding activities use up resources but do not directly contribute to the product or service. One example of value-adding compared with nonvalue-adding activities can be illustrated by considering the installation of an electrical outlet in a preinstalled outlet box on a construction site.

- The box has a length of electrical cable protruding from it.
- The electrician takes hold of the cable with one hand and reaches with the other for a wire stripper/cutter.
- The electrician cuts the wire to the appropriate length, and strips the insulation to expose ¾ inch of bare conductor for each wire. The electrician uses a pair of narrow-pointed pliers to make a loop on the bare conductors, in anticipation of securing them to connecting screws on the outlet. He or she may have misplaced the pliers from a job done on the previous day. It could take five minutes to track down the tool.
- The electrician may then make a short search to find a new outlet device in the bottom of a tool box. On retrieving the outlet, the electrician may realize that it is not a ground fault interrupter (GFI) outlet specified in the plans, and then extends the search to find the right device. It does not take long – perhaps three minutes.

The steps required to attach the outlet and mount it to the outlet box could take two minutes, representing value-added activities. The time spent locating tools and the outlet took eight minutes – all of it nonvalue adding. In this case, a job duration of 10 minutes included 20% value-added and 80% NVA.

Value stream

Wherever a process exists, it is accompanied by a value stream that represents the actions needed to create a product or service from its inception as raw material until it reaches a customer in its completed form. Value streams include both value-added and NVA activities. Value stream maps illustrate processes, identify sources of waste, and distinguish them from the value-adding activities that are critical to customers' needs.

Value stream mapping

Value stream mapping (VSM) shows material and information flows required to produce outputs. It helps users to understand the process, identify sources of waste by distinguishing between NVA and value-added activities. The VSM requires identifying each step in a process, with an emphasis on where action takes place. In drawing the map it is necessary to collect data on such factors as cycle time, changeover time, working time, scrap rates, and production batch sizes. The completed value stream map of the current state reveals areas that can be improved. Typically, the map is studied by a team that understands the process, and can make changes to the map to reflect recommended improvements. The map is changed to create a future state map with improved processes that maximize value-added time and minimize waste. The VSM process is further explained in Chapter 14.

Flow

The goal of lean thinking is to have a continuous flow of a product from one activity to the next in a process without delays, stoppages, or storage as *work in process*. This concept is the so-called one-piece flow or single-piece flow. Examples of flow are:

- Business flow: Related with the information of a project (specifications, contracts, plans, etc.).
- Job site flow: Involves the activities and the way they have to be done.
- Supply flow: Refers to materials involved in a project. This is similar to any other supply chain.

Pull

In the JIT philosophy, production is "pulled" from the system by the customer demand. This is the antithesis of mass production that is based on large batch sizes – a "push" philosophy in which production forecasts project customers' needs and produce outputs based on those assumptions.

Perfection

Perfection may not be attainable in the construction environment but represents a desirable future state – defects would be minimized, as would other categories of waste that delay delivery, incur additional cost, and detract from meeting the customers' needs.

Systems perspective of lean

The time a product spends in a production system is an important measure of efficiency (Figure 2.3). Assumptions:

1 The cost of a product is related to the length of time in the system: the shorter the time, the lower the cost to the producer and customer.
2 The shorter the time in the system, the better the producer can meet the customer's delivery requirements.
3 The shorter the time in the system, the smaller the probability of operational problems.

Description of system components

Move time

It represents the time required to move a product or service from one work station to another or from a queue to a processing activity. Move time does not add value. Lean methods minimize move time by optimizing the layout of the production facilities.

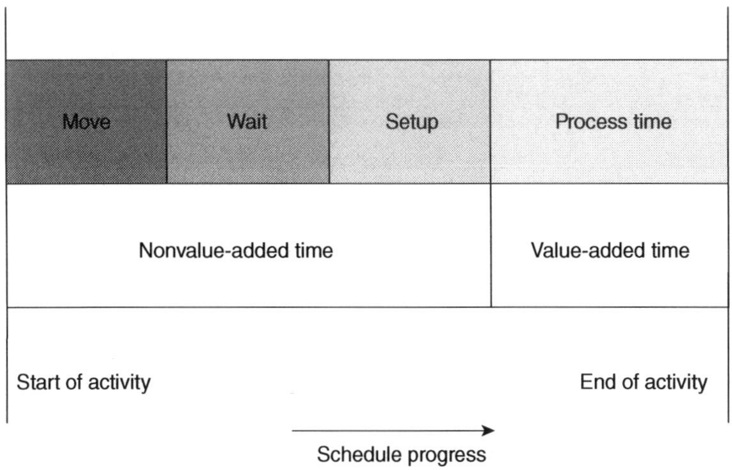

Figure 2.3 Impact of value-added vs. non-value-added time in a typical construction process

Wait time

It is the sum of all phases in a system in which a product or service is waiting to be transformed. Efficiency is negatively impacted the longer the wait time, as it does not add value. This wait time is directly influenced by such issues as equipment downtime, material shortages, and unbalanced lines. Wait time is generally the longest phase in most systems (i.e., the greatest NVA activity).

Setup time

Preparations are made for a process by adjusting equipment, material, procedures, and so forth in anticipation of processing activities. No value is added in this phase.

Process time

This represents the only value-added phase; that is, the time a product is actually undergoing transformation by equipment and/or operators. It is the total of all processing activities.

Reducing or eliminating waste

Lean construction seeks to reduce or eliminate waste in construction activities. Improving the efficiency of individual activities does not necessarily improve the efficiency of an overall process. Waste can occur in the handoffs from one activity to another, or one trade to another, in the form of delays, defects that need to be corrected, and so on. Leanness is increased by reducing or eliminating the waste that occurs between activities in order to increase overall process efficiency.

Adoption of the Toyota production system (TPS) applies the principles used in manufacturing to the construction environment. The TPS addresses:

1 Overproducing: When one discipline is so absorbed in self-interest, it may seek to carry out as much work as possible between payment cycles without considering the impact on other disciplines or trades. After all, the standard practice is for crews to request payment for work done within each payment cycle; the more work that has been completed, the greater the dollar value of payment requests. An overzealous drywall contractor may exceed a schedule and install drywall before an electrical subcontractor has completed conduit work for wall outlets or control devices. To get back on schedule, some partial demolition may be needed.
2 Idle time waste (waiting time/queue time): Poor coordination results in significant lost/idle time in projects. An impending concrete pour for a floor slab is planned by scheduling a team of journeymen and laborers who help to direct the pour from a concrete truck over the reinforcing bars and vibrate it to eliminate voids. A motorized float is used to level the surface. If the crew foreman did not have a firm commitment from the supplier, the crew could be left waiting for an hour or more because of a late delivery. Ten workers at an average of $2,000 per hour incur an idle time waste of $200 per hour. If the pumping equipment and the vibrators are rented, the idle time incurs additional cost as well.
3 Transporting/conveyance waste: Excessive transport of materials or equipment does not add value, but consumes resources. A poor site layout can result in long-travel distances to bring materials to the point of use. For example, if drain pipes for installation underground are stored at a location far from the planned excavation, then the retrieval and transportation could incur significant expense. A front-end loader and its operator could consume additional hours in the process, whereas a more thoughtful location would reduce the transportation costs.
4 Processing waste – waste in the work itself: Studies have pointed to high levels of material waste in construction, depending on the region involved. Bossink and Brouwers (1996) found that 9% by

weight of purchased materials are wasted. In the Dutch construction industry, from 1 to 10% by weight of each material item becomes solid waste. In Brazilian construction, 20–30% of purchased materials are not used and become waste. As material costs are 50–60% of project costs, material waste can cause cost overruns. Material waste could be due to:

Design	Poor anticipation of design impacts
Procurement	Wrong material or suboptimal order quantities
Material handling	Damage due to poor handling
Operation	Mistakes, poor use of material sizes
Inefficiencies	Poor waste management

5 Inventory waste (having unnecessary stock on hand): Holding excessive amounts of inventory (materials and supplies) has negative consequences:

a. The cost of material is significant and involves tying up capital; this could negatively impact cash flow.
b. Storing excessive inventory involves a carrying cost for storage.
c. Managing the storage of materials long before they are needed runs the risk of spoilage or "shrinkage" (i.e., unexplained loss of material).

6 Wasted operator motion (using unnecessary motion): Inexperienced trades may not be familiar with best practices. Poor job organization could cause the operator to make additional steps to carry out the job. For example, experienced trades use templates or jigs for positioning bathroom fixtures, and setup kits with the appropriate tools and fasteners so that each installation proceeds quickly without risk of error or delay. Inexperienced workers have difficulty in attaining similar levels of productivity.

7 Producing defective goods (waste of rejected production): Construction that does not meet codes may be rejected by an inspector. It then has to be corrected by extensive, costly rework. Construction errors require correction and inhibit downstream activities because of the time delays that usually result.

Other categories of waste

"Making do": This waste is caused when a task is initiated without ensuring that all needed resources are available, such as some materials, or workers that have a skill for the work involved. It also occurs when information or equipment needed by the task are unavailable. If the resources become unavailable after the task has started, then continuing the task is also considered to be making do. It is also seen as equivalent to negative buffering (Koskela 2004).

"Not speaking and not listening" are two great wastes described by Macomber and Howell (2004). These relate to the waste of human potential observed by Henry Ford in 1926 – he thought that nurturing human potential through education would tap workers' creativity with positive results for themselves and for his firm (Ford 1926). Ford thought that failing to nurture this potential would result in degrading performance.

Not listening is caused by the command and control style of management, so prevalent in traditional construction projects; it makes those in power disinclined to listen to those who are closest to the problem. Management structures are established with rigid procedures in the interest of efficiency; these procedures often dictate a chain of command that obstructs communication from those in the trenches to the decision makers. This results in institutionalized not listening.

"Not speaking" comes as a conditioned response to overt or implied signals from management that saying something that runs counter to prevailing beliefs may label one as a complainer or even troublemaker. Some children are culturally conditioned to not speak and that habit often continues into adulthood. Most individuals adopt the attitude that not speaking is a reasonable, prudent action. This inhibition denies management critical information about problems that can be averted, or about creative ideas that can benefit all stakeholders.

As Macomber and Howell imply, leaders who wish to be successful in lean construction should cultivate the art of listening as a master skill. They should also create the circumstances for others to speak.

Ed Anderson, President and CEO of Lean Implementation Services of Florida, has identified a waste termed Chase UP©. It represents the significant loss of time spent performing "Chase UP" with other project team members who have made promises but are tardy in meeting them. It also includes the time being chased by others He estimates that 60% of work hours are easily spent each week pursuing those people on the telephone or otherwise.

Lean construction fundamentals

The Construction Industry Institute (CII) identified five lean construction principles in their study PT 191:

- Customer focus
- Culture and people
- Workplace organization and standardization
- Elimination of waste
- Continuous improvement and built-in quality

An earlier CII-sponsored study headed by Professor Victor Sanvido, investigated project delivery methods (Sanvido and Konchar 1999). The study findings identified several critical keys to project success:

1 A knowledgeable, trustworthy, and decisive facility owner/developer;
2 A team with relevant experience and chemistry assembled as early as possible, but certainly before 25% of the project design is complete; and
3 A contract that encourages and rewards organizations for behaving as a team.

Relational contracting has emerged as a means of embodying the concepts and success factors listed above; it serves as an excellent vehicle for lean construction.

Three connected opportunities

Three connected opportunities in design and construction projects as a foundation for lean construction (Figure 2.4).

1. Impeccable coordination
2. Organizing projects as production systems
3. Projects are a collective enterprise–optimize the project, not the pieces

Figure 2.4 Three connected opportunities

Source: Courtesy of Lean Project Consulting

These opportunities are described as follows:

1 Impeccable coordination seeks to overcome the unpredictability that is typical of traditional construction projects; lack of coordination results in an average of only 55% or fewer of promised tasks being completed in a specific week as promised. Project success depends on the predictability of workflow that results when commitments are met between various disciplines and trades involved in a project.
2 Organizing projects as production systems align the roles of the parties in the project to maximize overall performance. It emphasizes productions system design to meet the owner's value proposition; conversations between contractors and designers inform the process of translating design to the built environment. Project execution strategies take advantage of technology or best practices such as prefabrication, modularization, and concurrent multitrade coordination.
3 Projects are a collective enterprise. Aligning financial incentives with project-wide optimization motivates project team members to adopt an investment mindset for improving performance. Sharing resources avoids expensive duplication and waste – the savings derived benefit both the team and the owner/client. Team orientation and trust are essential for mobilizing creativity and reducing waste.

Five big ideas™

The concept of these ideas proposed by Lean Project Consulting aligns the physics of work (how work gets done), organizations and contracts with lean methodology. It is especially valuable with large, complex projects. The concept is represented by five overlapping circles (Figure 2.5), and the development and application of these ideas are described in Chapter 9 (IPD) with regard to construction projects for Sutter Health. The following description is excerpted from an earlier paper on these projects (Lichtig 2006):

1 Collaborate, really collaborate,
2 Increase relatedness among all project participants,
3 Projects are networks of commitments,
4 Optimize the project, not the pieces, and
5 Tightly couple action with learning.

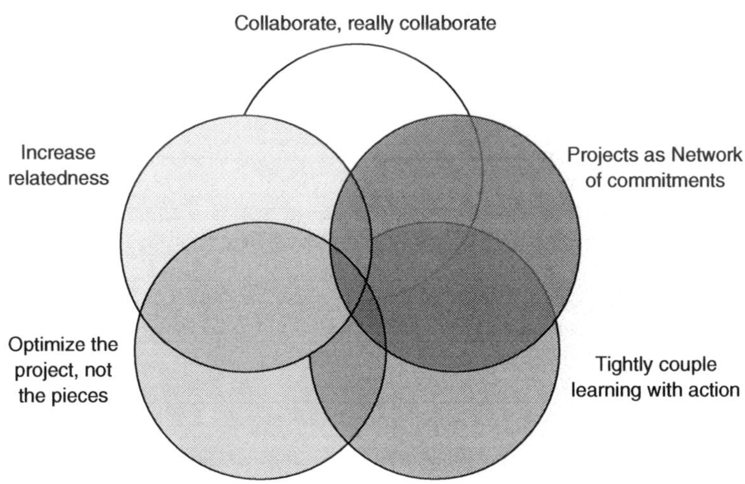

Figure 2.5 The Five Big Ideas

Source: The Five Big Ideas are the intellectual property of Lean Project Consulting

1 Collaboration overcomes the fragmentation that has come to characterize the design and building process. Expectation gaps between designers, builders, and owners lead to poor constructability and low quality. Close collaboration between team members secures the best decisions and outcomes in relation to design and construction choices. Design is an iterative conversation; the choice of ends affects means and available means affect ends. Collaborative design and planning maximizes positive iterations and reduces negative iterations.
2 Increase relatedness among all project participants. People come together on AEC projects as strangers. The chief impediment to transforming the design and delivery of capital projects is an insufficient relatedness of project participants. An active pursuit of relationships between team members promotes trust, openness, and learning that are essential for growth, as well as for creating a willingness to innovate.
3 Projects are networks of commitments. Projects are not processes. They are not value streams. Project management's role in the project environment is to activate and articulate unique networks of commitments between the many participants that form a project team. These commitments bind team members to manage and direct projects in real time. This contrasts with the common sense understanding that limits planning as predicting, managing as controlling, and leadership as setting direction.
4 Optimize the project, not the pieces. Conventional practice conditions project managers to push aggressively to maximize speed and minimize cost. Pushing for high productivity at the task level may maximize local performance but reduce the predictability of work released downstream. Lack of work reliability increases project durations, complicates coordination, and reduces trust. Project optimization involves having project teams collaborate to identify and implement solutions that are best for the overall project, as opposed to meeting only the self-interest of specific team members. This approach avoids conflicts between trades, and increases site safety by as much as 50%.
5 Tightly couple action with learning. This big idea proposes that work should be carried out in a manner that facilitates ready observation of the results of specific actions. In order for this to occur, assessments have to be made at intervals that provide actionable information; discovering deficiencies long after the fact makes corrective action difficult. The lessons so learned should be used for continuous improvement of costs, schedule, and overall project value. Doing work as single-piece flow avoids producing batches that do not meet customer expectations. The traditional separation of planning, execution, and control contributes to poor project performance and to declining expectations of what is possible.

The observation of the foregoing principles facilitates the deployment of lean design and construction. In particular, they focus construction stakeholders on common objectives and commit them to collaborating closely in order to secure project success. Two important systems represent a critical foundation for lean construction deployment: The Lean Project Delivery System™ and The Last Planner® System. These systems are described in detail in subsequent chapters.

Questions for discussion

1 How would you define lean construction?
2 How does lean construction differ from traditional construction delivery methods?
3 Why has the Toyota Production System been so successful in auto manufacturing?
4 How can the Toyota philosophy apply to construction, which lacks the uniformity and predictability of manufacturing cars in an enclosed building?
5 The "Command and Control" structure has worked well in the military and in many business situations. What are its shortcomings in construction?
6 Describe the types of waste found in construction projects. Provide examples from the Design Phase. Also provide examples from the Construction/Assembly Phase.

Appendix: ConsensusDocs

ConsensusDocs: (www.consensusdocs.org/) The Website explains that, before ConsensusDocs, there were a variety of construction associations that produced standard forms of construction contracts. Acceptance by other industry groups was low; standard contracts published by one association were perceived as ultimately favoring that association's membership. ConsensusDocs is widely accepted for their standard contract documents, especially those that relate to lean project delivery, because all the parties were invited to the drafting table and had a full vote in deciding final contract terms. These documents are available for fee on a project-specific basis. The American Institute of Architects also has standard contract documents that are based on lean project delivery.

References

Abdelhamid. 2013. *Lean Construction*. https://www.msu.edu/user/tariq/Learn_Lean.html (March 21, 2013).

Associated General Contractors of America. 2013. *Lean Construction Forum: What is Lean?* http://agcleanforum.org/about/(March 21, 2013). Accessed January 8, 2020.

Alarcon, L., and H. Mesa. 2012. A Modeling Approach to Understand Performance of Lean Project Delivery System. *20th Annual Conference of the International Group for Lean Construction*, San Diego, CA, USA, July 18–20.

Arbulu, R., and G. Ballard. 2004. Lean supply systems in construction. *Proceedings of the 12th International conference on lean construction*. Helsingor, Denmark, 547–559.

Ballard, G. 2000b. *LCI White Paper-8: Lean Project Delivery System*, Lean Construction Institute, Ketchum, ID, September 23.

Ballard, G. 2008. Ch.24. Lean construction. In *The Routledge companion to lean management*, 2016, eds. D. J. Powell, and T. Netland. London: Routledge.

Ballard, G., and G. Howell. 1994. *Implementing lean construction – Stabilizing work flow. Conference on lean construction*. Santiago, Chile, September.

Bertelsen, S. 1993. *Construction logistics I and II, materials management in the construction process*. (in Danish) Denmark: Boligministeriet, Bygge-og, Boligstyrelsen, Kobenhavn.

Bertelsen, S., and L. Koskela. 2004. Construction beyond lean: A new understanding of construction management. *Proceedings of the 12th annual conference of the international group for lean construction (IGLC-12)*. Elsinore, Denmark.

Bertelsen, S., and J. Nielsen. 1997. Just-in-time logistics in the supply of building materials. *1st International conference on construction industry development*, Singapore.

Bossink, B. A. G., and H. J. H. Brouwers. 1996. Construction waste: Quantification and source evaluation. *Journal of Construction Engineering and Management* (March): 55–60.

Diekmann *et al.* 2004. *Executive summary, research report RT191*, 4. Austin, Texas: Construction Industry Institute.

Ford, H. 1926. *Today and tomorrow*. New York: Productivity Press.

Hammarlund, Y., and R. Ryden. 1989. Effektivitetet i VVS-branschen, Arbetstidens utnyjande, (Effectivity in the plumbing industry – the use of the working hours, in Swedish), *Svenska Byggbranschens utvecklingsfond*, Sweden.

Harmon, R. D., and L. D. Peterson. 1990. *Reinventing the factory: Productivity breakthroughs in manufacturing today*. New York: The Free Press.

Hofstede, G. 1978 The poverty of management control philosophy. *Academy of Management Review*, July, pp. 450-61.

Hopp, W. J., and M. L. Spearman. 2000. *Factory physics: Foundations of manufacturing management*, 2nd ed. Boston: Irwin/McGraw-Hill.

Howell, G. A. 2000. White paper for Berkeley/Stanford CE & M workshop. *Proceedings of the construction engineering and management workshop*. Palo Alto, CA: Stanford University.

Koskela, L. 1992. *Application of the new production philosophy to construction*. CIFE Technical Report #72, pp. 15–17, 24. Stanford, CA: Stanford University.

Koskela, L. 2004. Making do-the eighth category of waste. *Proceedings of the twelfth annual conference of the international group for lean construction (IGLC-12)*. Copenhagen, Denmark.

Koskela, L., G. Howell, G. Ballard, and I. Tommelein. 2002. The foundations of lean construction. In *Design and construction: Building in value*, eds. R. Best, and G. de Valence. Oxford, UK: Butterworth-Heinemann, Elsevier.

Krafcik, J. 1988. Triumph of the lean production system. *Sloan Management Review* 31(1): 41–52.

Lichtig, W. A. 2005. Sutter health: Developing a contracting model to support lean project delivery. *Lean Construction Journal* 2(1): 105–112.

Liker, J. K. 2004. *The Toyota way*. New York: McGraw-Hill.

Macomber, H., and G. Howell. 2003. Linguistic action: Contributing to the theory of lean construction. *Proceedings of the 11th annual conference of the international group for lean construction (IGLC-11)*. Blacksburg, VA, USA.

Macomber, H., and G. Howell. 2004. Two great wastes in organizations. *Proceedings of the 12th annual conference of the international group for lean construction (IGLC-12)*. Elsinore, Denmark.

Matthews, O., and G. Howell. 2005. Integrated project delivery an example of relational contracting. *Lean Construction Journal* 2(1): 46–61.

Mossman, A. (2014). *Traditional construction and lean project delivery—A comparison*. The Change Business Ltd.

Nielsen, A., and E. Kristensen. 2001. *Tidsstudie af vaegelementmontagen pa NOVI Park 6* (Time study of the erection of concrete walls on the NOVI Park 6 Project). Part of a not-published master thesis. Aalborg University.

Ohno, T. 1988. *Toyota production system*. Cambridge, MA: Productivity Press.

Pappas, M. 1990. *Evaluating innovative construction management methods through the assessment of intermediate impacts*. Master of Science Thesis, University of Texas at Austin (CII supported).

Porter, M. 1985. *Competitive advantage*. New York: The Free Press.

Sanvido, V., and M. Konchar. 1999. *Selecting project delivery systems: Comparing design build, design bid-build and construction management at risk (CII)*. State College, PA: The Project Delivery Institute.

Schonberger, R. J. 1986. *World class manufacturing: The lessons of simplicity applied*. New York: Free Press.

Shenhar, A. J., and A. Laufer. 1995. Integrating product and project management – A new synergistic approach. *Engineering Management Journal* 7(3): 11–15.

Tommelein, I. D. 2015. Journey toward Lean Construction: Pursuing a paradigm shift in the AEC industry. *ASCE, Journal of Construction Engineering and Management*, June, 141(6): 1–12, https://doi.org/10.1061/(ASCE)CO.1943-7862.0000926

Tommelein, I. D., Riley, D., and Howell, G. 1999. Parade game: Impact of work flow variability on trade performance. *Journal of Construction Engineering Management*, 125(5): 304–310.

Wandahl, S., and E. Bejder. 2003. Value-based management in the supply chain of construction projects. *Proceedings of the 11th annual conference of the international group for lean construction (IGLC-11)*. Blacksburg, VA, USA.

Wortmann, J. C. 1992. Factory of the future: Towards an integrated theory for one-of-a-kind production. In *One-of-a-kind production: New approaches*, eds. B. E. Hirsch and K. D. Thoben. Amsterdam: Elsevier Science.

Womack, J., and D. Jones. 1996. *Lean thinking*. New York: Simon & Schuster.

Womack, J., D. Jones, and D. Roos. 1990. *The machine that changed the world*. New York: Harper Collins.

Bibliography

Ballard, G. 1994. *The last planner. Northern California construction institute spring conference*. Monterey, CA, April.

Ballard, G. 2000a. *The Last Planner® System of production control*. PhD Thesis. Faculty of Engineering. School of Civil Engineering. Birmingham, AL: The University of Birmingham.

Howell, G. A. 1999. What is lean construction. *Proceedings of the 7th annual conference of the international group for lean construction*. Berkeley: University of California, 2–10.

Howell, G. A., A. Laufer, and G. Ballard. 1993a. Interaction between subcycles: One key to improved methods. *Journal of Construction Engineering and Management, ASCE* 119(4): 714–728.

Howell, G. A., A. Laufer, and G. Ballard. 1993b. Uncertainty and project objectives. *Project Appraisal Journal* 8: 37–43.

Koskela, L. 2002. The theory of project management: Explanation to novel methods. *Proceedings of the international group for lean construction (IGLC-10), 10th Annual Conference*. Gramado, Brazil.

Macomber, H., and G. Howell. 2005. *Using study action teams to propel lean implementations*. Louisville, CO: Lean Project Consulting Inc.

Polat, G., and G. Ballard. 2004. Waste in Turkish construction: Need for lean construction techniques. *Proceedings of the 12th annual conference of the international group for lean construction (IGLC-12)*. Copenhagen, Denmark.

3
LEAN IN DESIGN

This chapter explains how lean principles are incorporated in the design phase of projects to provide a foundation for the deployment of lean best practices in the construction phase and the subsequent use of constructed facilities. It describes several important concepts and practices that are essential components of the lean philosophy under the following headings:

Part A: An overview and description of the Lean Project Delivery System and its components, Its interconnection with the Last Planner System is also described.
Part B: Lean in the design phase of projects. The transition from design to construction is also described. Special topics include the Last Planner System in design, target value design, Set-based design, Target Value design and validation studies
Part C: Examples of lean design applications, as performed by JE Dunn Construction Company, Inc. and Boulder Associates Architects. It also explains agile and and scrum approaches to design
Part D: Sustainability and Commissioning considerations

Part A: Overview and description of the Lean Project Delivery System and its components

The Owner, Architect, Engineer, and Contractor (OAEC) community has long been aware of deficiencies in the design and construction processes that are evidenced by cost overruns, project delays, and quality and performance shortfalls in the finished construction. Studies show that 70% of projects finish late, and 73% of projects are delivered over budget. Poor design and documentation quality have been identified as a major factor in reducing the overall performance and efficiency of construction projects. Consequently, they have directly caused many projects to run over budget, over time schedules, and to be plagued with rework, change orders (variations), and disputes. Various case studies have identified design and documentation deficiency as the major contributor for construction contract variations.

The Lean Project Delivery System™ (LPDS) provides a means of addressing these shortcomings and improving the entire design and construction process (Ballard 2000; Ballard and Howell 2003a). It was developed by Glenn Ballard in 2000 and subsequently refined. Whereas traditional industry practice has separated the roles of designers and constructors, the LPDS sees the activities of these professionals as a continuum for project management to achieve three fundamental goals (Koskela 2000):

- Deliver the product,
- Maximize value, and
- Minimize waste.

Fundamental to the LPDS is the deployment of the "Five Big Ideas" as described in Chapter 2:

1 Collaborate, really collaborate,
2 Increase relatedness among all project participants,
3 Projects as networks of commitments,
4 Optimize the project, not the pieces, and
5 Tightly couple learning with action.

The first big idea requires assembling and empowering of all the people and resources necessary to develop appropriate design solutions and explore their impacts on construction delivery. Inherent in the use of the LPDS is also the use of The Last Planner® System (LPS) from the perspective of design. It is explained later in this chapter and in Chapter 4.

In brief, it comprises

1 The master pull schedule,
2 The lookahead schedule,
3 The weekly work plan (WWP),
4 The daily work plan where applicable, and
5 Performance measurement, learning, and improvement.

Structure of the Lean Project Delivery System (LPDS)

The LPDS comprises a number of phases that capture the intent of the traditional project phases, but juxtaposes them in such a manner as to apply production system design principles to enhance the delivery of the entire project from predesign to completion and use. The phases are:

1 Project definition
2 Lean design
3 Lean supply
4 Lean assembly
5 Use/completion

The LPDS model represents these phases as a series of overlapping triads. As described by Ballard and Ballard and Howell (2003a), these phases influence each other "so a conversation is necessary among the various stakeholders." In the diagram (Figure 3.1), work structuring provides a foundation for the process and sets the stage for production control, which is represented by a horizontal bar. Work structuring is a term developed by the Lean Construction Institute to indicate process design. It is a process of subdividing work such that pieces are different from one production unit to the next to promote flow and throughput, and to have work organized and executed to benefit the project as a whole. Production is redefined from "monitoring results" to "making work flow according to plan, re-planning when it cannot." Plans to complete are continually updated in a proactive rather than reactive mode, looking ahead instead of backward.

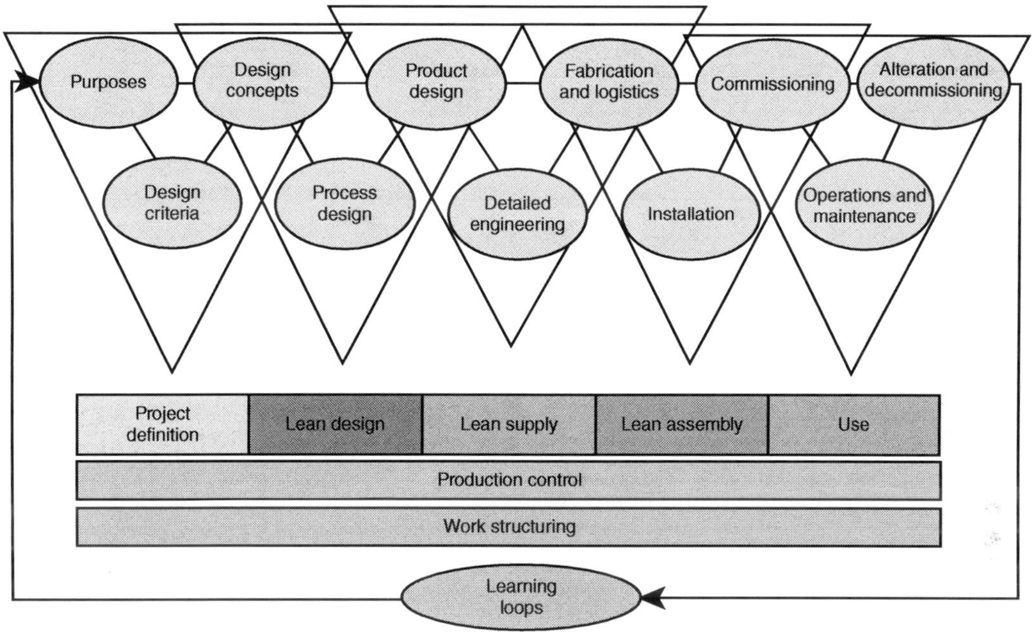

Figure 3.1 The Lean Project Delivery System™

Source: Adapted from Ballard, G., and G. Howell, *Building Research and Information*, 2003a

The LPDS was updated to include additional concepts and methods drawn from the Toyota product development system, especially target costing and set-based design. These have been adapted to the construction industry and integrated with computer modeling and relational forms of contract (Ballard 2008).

As described by Koskela *et al.* 2002 (page 214), the LPDS is structured to achieve the Transformation, Flow, and Value (TFV) goals The model improves project delivery through the following characteristics:

1 Downstream stakeholders are involved in upstream decisions – such as front-end planning and design through cross-functional teams.
2 Commitments are deferred to the last responsible moment.
3 Project control has the job of execution as opposed to reliance on after-the-fact variance detection.
4 Pull techniques are used to govern the flow of materials and information through networks of cooperating specialists.
5 Capacity and inventory buffers are used to absorb variability.
6 Feedback loops are incorporated at every level, dedicated to rapid system adjustment (i.e., learning).

Project definition

Project definition typically involves developing project alternatives at a conceptual level, analyzing project risks and the economic payoff, and developing a financial plan. A decision is made to proceed if the owner considers conditions to be favourable; a plan is developed for the organization and control of the project. Effective project scope definition enables all involved parties to understand the owner's needs, and to work toward meeting those needs. Traditionally, project definition has been established by the architect (or engineer for non-building projects) working alone with the owner. In lean project definition, representatives of every stage in the life cycle of a facility are involved, including members of the production team that will design and build it. (Koskela *et al.* 2002). The project definition phase comprises:

- Needs and values determination,
- Design criteria, and
- Conceptual design.

In the needs and values determination, design professionals assist the owner/client in clarifying a value proposition; that is, the purpose of the project and the needs to be served. A design criteria document describes specific needs to be met, such as size, space proximities/adjacencies, and energy efficiency requirements. The conceptual design uses the design criteria and value proposition to define an outline design that serves as a starting point for the design phase.

Establishing design criteria

The design criteria define the owner's basis of design (BOD). It reflects the owner's needs and wants that must be satisfied by the design of the project. This may include the use of spaces, their sizes, finishes, and activities to be performed within them. For example, a room that houses equipment with a high-sensible heat load will need power outlets/connections with adequate power capacity ratings. It will also need cooling designed to offset the respective heat loads. The aspect ratio – length versus width may influence the operations to be housed – a manufacturing assembly operation has different operational flow requirements than an office environment with a high-occupant load. Owners such as institutions – schools, universities, hospitals, and franchise retailers – may have formal design criteria developed through many years of experience. Other owners may require more intensive evaluation and analysis in order to develop project-specific design criteria.

Lean design phase

The lean design phase comprises: conceptual design, process design, and product design. It builds on the output from the project definition phase, but with a deviation from traditional design practice. Traditional design tends to follow a familiar pattern. In architectural projects (as opposed to EPC projects) an architectural firm serves as the principal designer, and is supported by design engineers such as civil/structural, mechanical, and electrical. The design team often starts by designing the design process; they may use Post-It notes on a wall and apply the activity definition model (ADM) to ensure that design assignments have necessary prerequisite work completed and that no constraints would delay the process.

In conventional projects, the architect provides preliminary drawings to the other disciplines at various stages in the design process and the engineers apply their respective design parameters to the architectural framework. Although this work may benefit from the cost efficiencies of 3-D CAD (or more advanced versions), the process is not generally interactive – it progresses forward and is additive as new information is included. It is often punctuated by design changes originating from scope changes or value engineering exercises necessitated when someone determines that the budget may not accommodate the design features that have been developed. There is very little involvement, if any, of contractors/builders in the design process in design-bid-build projects.

In design-build projects, a single entity contracts with the owner/client; there is an interplay between designers and builders, but the owner/client is not generally included in this process and does not have a fiduciary relationship with the designer of record. In CM@ Risk the builder is brought on board with the primary motive of becoming familiar with the project and its environment; the risk of unforeseen conditions is reduced.

The foregoing traditional methods separate design from construction – designers focus on product design and not process design in order not to assume the constructor's risk. This risk avoidance has other negative consequences. Research findings suggest that designers can in fact have a strong influence on construction safety. In 1985 the International Labor Office recommended that designers give consideration to the safety of workers who are involved in erecting buildings. In 1991, the European Foundation for the Improvement of Living and Working Conditions concluded that about 60% of fatal accidents in construction are the result of decisions made before the site work begins. A 1994 study of the United Kingdom's construction industry found a causal link between design decisions and construction safety.

With lean projects, as shown in the second triangle, lean design is a significant departure from the foregoing scenarios. Design is done with both the construction product and the process in mind. Constructability reviews and value engineering are not seen as tools to apply in a problem-solving mode, but rather, are continually integrated with decision-making in the design process. This is accomplished with cross-functional teams that include architects, engineers, contractors, subcontractors, and various specialists who collaborate with each other and interactively make decisions that are optimal for the product and the process. A cluster organization is one very effective way to bring knowledgeable team members together for joint activity.

Lean supply

The third triad, lean supply, comprises product design, detailed engineering, fabrication, (Ballard and Howell 2003a); it requires up-front product and process design to define what is needed and when it should be delivered. This is especially important with engineered to order components as utilized in EPC projects. Lean supply also includes reducing the lead time for project information requirements. It involves the use of supply chains to optimize the delivery of materials, components and major building sub-assemblies. It is described in greater detail in Chapter 4.

Lean assembly

Lean assembly is practiced in the actual construction of a project, putting materials, systems, and components in place to create a completed facility.

The lean assembly triad contains fabrication and logistics, installation, and commissioning. Prefabrication is a production technique that can enable a contractor to operate in a lean manner by reducing the many nonvalue-added steps that are required in field fabrication. Whereas field work contends with uncertain conditions such as weather, and limitations in the availability of skilled labor, material, and equipment, shop fabrication benefits from a predictable, controlled environment. Tweet/Garot, a specialty contractor in Wisconsin has improved efficiency by prefabricating many of its products for quick and efficient site installation. This includes plumbing work for bathrooms in a given building in congruence with the construction schedule. Over a period of time, Tweet/Garot has progressed to 70% installation in the field and 30% fabrication in the shop. With that new approach, workers spend a relatively short period of time doing installation work. Prefabricated piping is interconnected within minutes because of the precision of design.

The commissioning process overlaps in the last two triangles; it provides quality assurance before a facility is completed and accepted by the owner by ensuring that all systems have been installed as promised in keeping with the designer's plans and specifications. It improves the probability that the facility will meet the owner's project requirements (OPR) and performs as expected, to provide user satisfaction.

The fifth triad, "Use," refers to a completed facility. Following successful commissioning, the facility should undergo a protracted operations and maintenance phase as it is used. "Alteration and Decommissioning" refer to a future activity when the facility may be repaired, renovated, or taken out of service.

Learning loops

"Learning loops" span all the activities in the five triads, and refer to the application of root cause analysis to the Last Planner® System on a weekly basis to review PPC values and commitment reliability. Learning is also accomplished through ongoing retrospectives as well as the process of post occupancy evaluation (POE). In POE reviews a facility is surveyed after occupancy/acceptance to review the consequences of decisions made during the execution of the project. As mentioned by Koskela *et al.* (2002), it is important to know the extent to which design and construction were based on a correct determination of customer and stakeholder values. It examines both the product and the process used to create it. The POE enables project participants to learn from past decisions.

Production control

Production control is represented as a horizontal bar that extends from the very inception of the project to its conclusion (Figure 3.1). Koskela *et al.* (2002) refer to traditional project control as "trying to steer a car by looking in the rear view mirror". Instead, lean production control is achieved through a systematic process for making assignments ready to be performed, combined with explicit commitments by people at the production level to what work will be released to their customer processes in the next plan period. This is typically one week" It consists of work flow control and production unit control.

Work structuring

Work structuring and production control are used throughout a project to manage production. The term work structuring was developed by the Lean Construction Institute (Ballard 1999) to describe construction-related process design. It is a process of subdividing work so that pieces are different from one production unit to the next to promote flow and throughput, and to have work organized and executed to benefit the project as a whole. Work structuring is described in greater detail as a part of the LPS. Koskela *et al.* (2002).

Table 3.1 Comparison of the LPDS with traditional systems

	Lean project characteristics	*Traditional project characteristics*
1	The focus is on the production system	The focus is on transactions and contracts
2	The TFV goal	The T goal
3	Downstream players are involved in upstream decisions	Decisions are made sequentially by specialists,
4	Design coordinates both the product and the process	Process design is done separately after product design
5	Design considers product life-cycle stages	Design does not consider life-cycle stages
6	Actions are taken at the last responsible moment	Actions are taken as soon as possible
7	Supply chain lead times are reduced systematically	Organizations accept market conditions, availability
8	Learning is included in projects, firms and supply chain management	Learning occurs sporadically
9	Stakeholder interests are aligned	Stakeholder interests are not aligned
10	Buffers are sized and located to absorb system volatility	Participants protect their interests with large inventories.

Source: Adapted from Table 14.2 of Koskela *et al.* 2002

The LPDS and the Last Planner® System of production control

Several tools have been developed to manage the LPDS; some are conceptual, some are procedural, some are embedded in software. The Last Planner® System is a complex tool that includes multiple rules and techniques, namely the activity definition model (ADM), Constraints analysis, and Percent Plans Complete.

This varied tool set is very powerful and has yielded significant performance improvements in the design and construction processes for almost two decades. Bertelsen (2001) reported that Danish contractors had reduced project durations by 10%, increased productivity by 20%, and improved profitability 20–40% on projects where they applied lean principles.

The Last Planner® System of production control (LPS) is an important subset of the LPDS. It is a collaborative commitment-based planning system that integrates "should/can/will/did" planning with constraint analysis, and work planning based on reliable promising, and learning. It works in conjunction with it and facilitates continuous improvement of processes by keeping track of plan reliability as PPC values. It also tracks the reasons for non-completion (RNC), enabling project teams to adopt countermeasures and continuously improve performance during the course of a project.

One important simulation that is based on the LPS is the Parade of Trades. It demonstrates the importance of reducing variation to increase the speed of construction operations. Typically, seven players assemble around a table and roll a die in turn to move 35 chips around the table. In one iteration a standard die is used with face values from one to six. The average value per roll is 3.5. In another iteration other dies are used that have an average value of 3.5. One special die may have faces of 3 and 4. But its variability is much less than that of a standard die. It does not offer dramatic rolls of sixes. Nor the disappointment of aces. At the end of the game the "slow" die often clears all the chips faster than the other dies that have a higher variability. It provides a more reliable workflow, and faster work completion. In effect, it demonstrates the principle of "going slow to go fast".

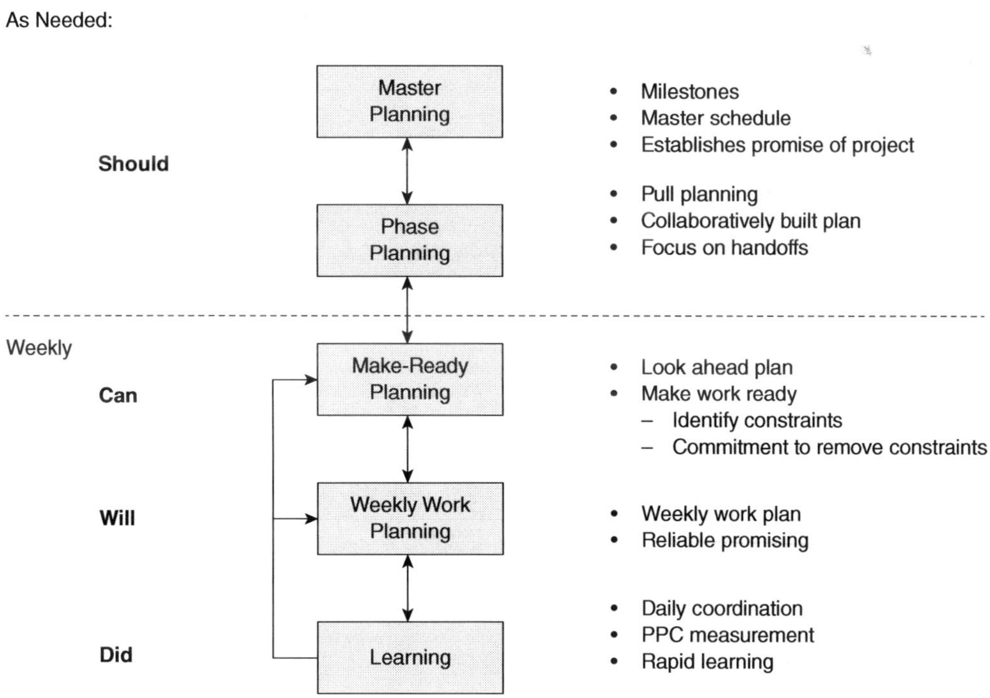

Figure 3.2 Schematic of the Last Planner® System

The Last Planner® System of Production Control will be discussed in upcoming sections from the perspective of both the design and construction phases. (The simulation is also described in Chapter 7 – Lean Culture and Training.)

A special note on the Last Planner® System

Following his early research that led to the development ofThe Last Planner System, Glenn Ballard has initiated further studies to update and improve the model. He has convened various teams of industry experts to study how the system has served in over two decades of application to both design and construction activity, and identify improvements that will update the system to LPS 2.0. Henry Nutt III, Preconstruction/Business Development Manager of Southland Industries and Klas Berghede, Director of Production for The Boldt Company have been working with other construction professionals to evaluate the impact of location-based planning on the LPS. As Henry indicates, the LPS has been in use for over 20 years, and has been gradually modified or customized by users. The research project will observe how it is being used, and determine what modifications are needed, especially with design teams.

This effort includes the participation of graduate students to review and collect data on the use of Location-Based Planning, in design and construction. A number of site visits will be conducted to observe projects and conduct interviews. Data will be collected on past projects and the findings will be assembled for review.

While the results of the study are not yet available, the location-based team has identified a critical need for a common language within the project team that everyone can understand. A key map for design and construction flow should identify work structures early in the planning stages and develop an ideal flow of work structures to optimize the execution of work at each stage.

Part B: Lean in the design phase of projects

The purpose of this section is to discuss project definition and lean design as the first two triads of the lean project delivery system (LPDS) and to explain how the related activities affect the delivery of lean projects. The design phase of projects has a major impact on project outcomes; the degree to which the design team interprets the owner's project requirements has a direct influence on the use of the facility. It directly correlates with the owner's satisfaction with a project. Decisions made during the programming and design phases determine the extent to which a facility is buildable or constructible. Traditional projects do not generally have a clear line of communication between designers and builders – except for design-build projects, and some applications of CM at Risk. Lean projects on the other hand promote this communication which ultimately benefits all stakeholders.

Impact on project outcomes

Building design is a complex undertaking. It involves developing a value proposition that represents the owner/client's needs, balancing those needs with available budgets that are frequently underfunded, and meeting codes that are increasingly restrictive. Design professionals have to work with complex technologies and pay close attention to environmental standards in order to configure high-performing buildings that meet twenty-first century standards.

Lean construction advocates simultaneous product design and process design. It contrasts with traditional construction that is based on specialization of the parties – architects/engineers design a facility, then convert the design to plans, specifications, and bid documents. These instruments are used as source information for builders/contractors to compete with each other for the award of a contract – and subsequently become contract documents.

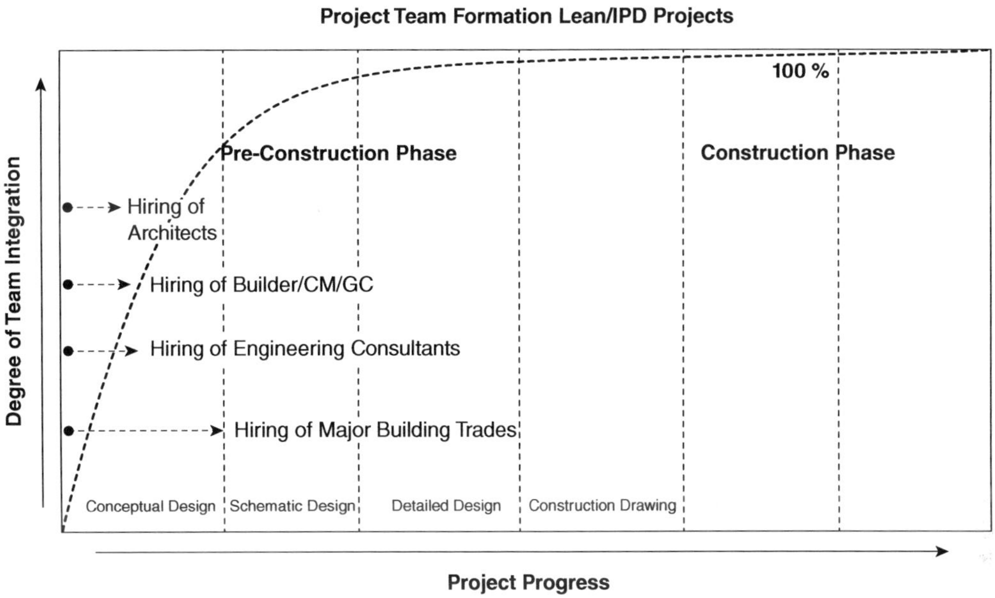

Figure 3.3 Representation of lean team formation

In traditional projects, contractors, subcontractors, and suppliers then determine how the design is to be converted into a working facility that serves the needs of the owner. This approach severely limits builders in exploring options for the production of the facility. It is prone to error and often results in costly rework. In lean projects a more collaborative approach is used to determine the owner's project requirements (OPR) and convert them to a functional design that can be built to meet the requirements.

In lean projects the team members are hired much earlier so that they can be available to contribute to decisions on important design features such as those mentioned below. As shown in Figure 3.3 the CM/GC, engineers and major trades may be hired as early as the schematic design phase; they can contribute actively to important design decisions very early when the impact of decisions is greatest, as stated by Crosby 1979.

In traditional projects the engineering consultants are often brought on board later than the conceptual design phase. Builders are usually not hired until the construction drawings are available for inviting bids from contractors. The early engagement of the project team in the lean construction approach makes it possible for constructors to provide input to the design process. Constructability is significantly improved with this approach.

The first triad of the LPDS contains three important attributes – purposes, design criteria, and design concepts. They provide valuable information on:

The owner's needs and values
The owner's business plan and its value proposition
Project alternatives and their associated risk
Target values and constraints for lean design

For customer/stakeholder satisfaction with the overall project to be achieved, value must be defined and agreed upon at the very beginning. While flow, pull, and striving for perfection may take different forms during the lean supply and lean assembly phases, early clarification in the project definition phase sets the stage for how well a facility serves its users during its entire life cycle. Understanding value enables designers to know which features to provide and which to avoid. While lean seeks to maximize value and minimize waste it must be understood that eliminating waste will not necessarily generate value.

This information enables designers and other project team members to establish a shared mind among themselves and the owner representatives such as end users, on such issues as allowable costs and time frames. The approach used draws on the first of the five big ideas: "Collaborate, Really collaborate."

The team members determine three variables – Ends, Means, and Constraints. Overall, the collaboration of many stakeholders in the project definition process provides a common understanding of the project that incorporates lean thinking and enables the project to meet the expectations of the primary and other stakeholders.

Figure 3.4 refers to a conversation that is held by the project team to define the project effectively. Starting from the left, the discussions in the project definition phase clarify the purpose of a project, reconcile it with the values to select the appropriate design criteria. The means are considered from the right side of the diagram, reconciled with facility design alternatives, and working within the constraints of available funding, time, space, and prevailing regulations. This information leads to the creation of the owner's Basis of Design (BOD), including size and space requirements. It also provides a design concept and an outline for the design of the product and the process. In considering owners' project requirements, it is important to recognize four main stakeholder groups as defined by Fischer *et al.* (2017): the design and build team, the professionals operating the building, the users of the building, and the managers of the building users. The design and build team needs to use the input from the other three groups to determine the owner's project requirements, and also inform it with their professional knowledge, and experience and observation gained from earlier projects that may be of a similar nature. Fischer *et al.* describe four components of building performance – buildability, operability, usability, and sustainability. They also state that a building should not only perform well on the four components, but that it must also be aesthetically pleasing.

From the standpoint of value all of the foregoing factors as well as the cost of a project and the duration determine the value to the primary stakeholder – the owner. There can be a major emphasis on the timing of a project to meet the owner's needs, such as in adding a wing to a hospital to provide additional beds by a target date. A buildable design is based on the best available construction systems and practices, i.e., that have high levels of constructability and safety. A buildable design also increases the likelihood of a building being completed on schedule and on budget. Examples of high constructability include prefabrication and modularization, and they should be incorporated in the design.

Operability refers to systems in a building and how easily they are kept in operation, and repaired when necessary. These systems include electrical, mechanical, structural, and the building envelope. They should maintain the conditions they are designed for, so that building users can enjoy high levels of comfort and productivity. Maintenance staff should be able to carry out standard functions of preventive maintenance with systems and materials that are readily obtainable without the risk of extensive downtime. There should be adequate access to systems that have to be maintained, or replaced. A design that produces cost savings by

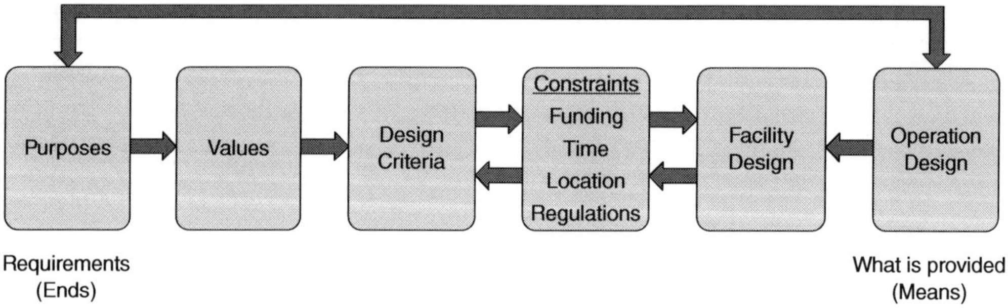

Figure 3.4 A representation of the lean project definition process

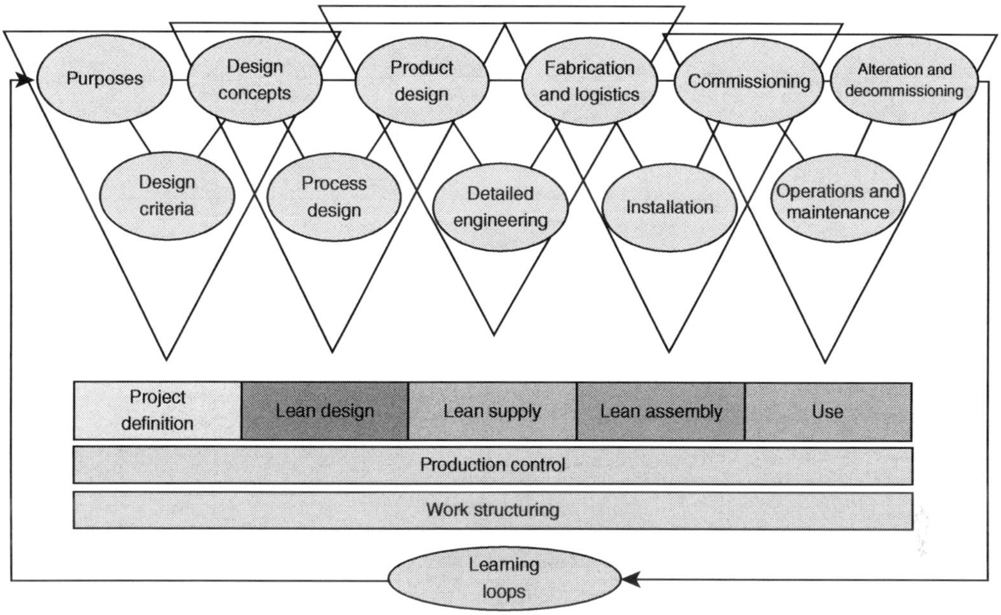

Figure 3.5 Project definition and lean design phases

using mechanical or electrical systems that are lower cost, but less reliable, does not meet the value proposition of an owner who seeks maximum uptime. A hospital ICU cannot afford downtime, even though backup systems are available. The skill level required of maintenance staff should be based on people hat can readily hired in the marketplace as much as possible.

Usability refers to the extent to which people can use a facility to perform their expected functions. In reality, a building exists to serve the needs of its users. A poor floor layout can easily result in congested corridors or a long waiting time for elevators that take users from one floor to the next. The locations of frequently accessed spaces should make it easy for building users to get to them with minimal delay or travel distance. Designers should focus on determining users' needs in order to optimize the assignment of room capacities and locations.

Sustainability: Fischer *et al.* 2017 describe a sustainable building as one that works in harmony with its natural, social, and economic contexts. Other definitions include concepts of green building design and construction, reuse and recycling of materials, reduced use of material and energy resources for building construction and operation, water conservation and responsible stewardship of the surrounding environment. A green building (or "sustainable building") is a structure that is designed, built, renovated, operated, or reused in an ecological and resource-efficient manner.

Green buildings are designed to meet such objectives as protecting the health of occupants; improving worker productivity; using energy, water, and other resources more efficiently; and reducing the overall impact on the environment. Studies confirm the benefits of green buildings, such as lower operating costs and measurable improvement in the productivity of occupants. The average length of stay (ALOS) has been shown to be reduced in hospital settings where environmental sustainability has been implemented.

Glenn Ballard references a UK study that indicates the cost of design as 0.1 unit, construction cost as as 1 unit, maintenance cost as 4.2 units, while operation over 20 years costs 42 units.

Crosby (1979) considered that the impact of rework at different stages of a project could be defined with the "1/10/100" rule. This rule in construction would mean that changes made during the predesign phase would have an impact of $1 to the project, $10 during the design phase, and up to $100 after the beginning

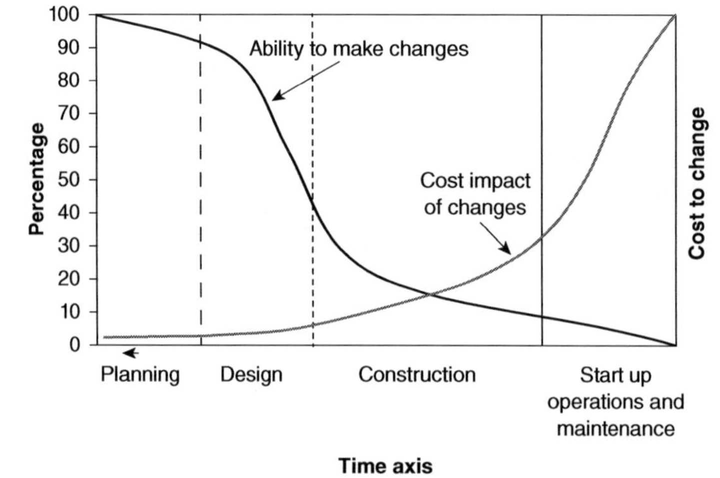

Figure 3.6 The relationship between time and the impact of design changes

of construction. If rework is deferred until after project completion, the cost to implement changes not picked up previously could be as high as $1000!

Opportunities to capture the expertise of the construction team and develop cost-effective solutions are more difficult after the design phase because design review becomes more costly (complex) and resistance to change becomes greater (Figure 3.2).

With the design process generally being managed by traditional project management methods, there is an obvious link between the way design and documentation are managed and the poor level of performance being achieved. For that reason, research is currently being carried out in relation to lean design management, a methodology that promotes the implementation of lean production philosophies within the design and construction process.

Building information modeling (BIM) is a highly productive tool for carrying out lean design. It generally uses 3-D, real time, dynamic building modeling software to increase productivity. A building information model is created for each facility, representing it in terms of building geometry, spatial relationships, geographic information, and quantities and properties of building components.

Lean design with BIM was successfully used on the El Camino Medical Group campus in Mountain View, California. A 250,000 square foot medical office building was included in the project, as well as a 420,000 square foot parking structure, at a cost of $94.5 million. The project schedule was highly accelerated for early occupancy, hence it was necessary to utilize concurrent engineering approaches to start construction activity while the design was still in progress. The general contractor and a number of mechanical subcontractors were engaged during the design phases at approximately the same time as the architect and the structural engineers. The contractors and designers collaborated on the design to derive maximum constructability, lower cost, and an aggressive construction schedule. The team collaborated with a virtual model of the project; design focused not only on the product but also on the construction schedule, including material supply and prefabrication activities. The results at the end of the project were:

- The project was completed six months earlier than would have been achieved using the traditional design-bid-build project delivery method without BIM and lean techniques.
- It was completed under budget.
- Labor productivity was 15–30% better than the industry standard.

- There were no change orders related to field conflict issues.
- There were no field conflicts between the systems coordinated using BIM.

As was experienced in this project, 3-D modeling is one of the tools for improving both the design and construction processes. It not only helps designers to visualize and avoid potential design conflicts between different trades; it simultaneously generates bills of materials. Three dimensional modeling can also be used to simulate the facility design and the construction and fabrication process. It is described in greater detail in Chapter 5.

The lean design phase

The lean design phase comprises: conceptual design, process design, and product design. It builds on the output from the project definition phase. It is represented by the second triad of the lean project delivery system.

The British Institute of Architectural Technologists (BIAT) defines the purpose of lean design: "to improve manufacturability through attention to information coordination and flows at the outset of the project." The BIAT suggests that lean design management should follow the lean thinking principles as follows:

1 Understanding how value is delivered
2 Identifying the customer's point of view of value
3 Achieving flow within the process as waste is removed
4 Achieving pull so information is delivered when needed
5 Constant pursuit of improvement (perfection)

It is absolutely important to have value defined at the beginning of the design process. It is achieved when the client's requirements are maximized through design and construction, and results in customer satisfaction. Some definitions of value are conformance to design and specifications, fitness for use, and the fundamental lean principle that drives the value stream, flow, pull, and striving for perfection. It should be the basis for all activities such as planning sessions and daily huddles. Without a clear definition there is a risk that some actions taken may actually reduce customer value.

Some tools and techniques suggested for lean design are:

1 Organize cross-functional teams, involve downstream players in upstream decisions.
2 Use Set-based design as concurrent engineering, to increase value via decisions at the last responsible moment.
3 Minimize negative iteration even though design is not a linear process. Use pull scheduling and a design structure matrix.
4 Apply production control: utilize a small design batch approach, evaluating against project requirements purpose.
5 Use technologies that facilitate lean design with a web-based interface such as BIM.

Very importantly, lean design addresses the desired systems that are scheduled for design, while evaluating the best methods that can be used to construct them. The primary factors considered are:
Designing the design process itself (design workstructuring)
Obtaining collaborative inputs from specialists
Assigning detailed designs to specialty contractors
Conducting cross-functional discussions to optimize product and process design
Utilizing BIM for unified modeling based on the inputs and needs of stakeholders
Designing for usability with owners' facilities maintenance staff
Using the Last Planner® System to reduce negative iterations and optimize delivery of design tasks meet schedules and commitments

The Last Planner® System in design

This system is an important subset of The Lean Project Delivery System™ (LPDS) and is critical to its effective deployment. It accommodates project variability and smooths work-flow so that design activities can be maximally productive. The LPS has been less frequently used in design than in construction so the Leadership of stakeholders needs to advocate strongly for the use of the LPS as it clearly offers improved timeliness and cost-effectiveness in delivering assignments. In particular, it meets the very important lean tenet of involving downstream project participants in downstream decisions. The design process is iterative, with backward loops to explore alternatives, as opposed to construction processes which are well documented and predetermined by various schedules in the LPS. The use of the LPS in construction has been shown to increase the completion of work assignments from historical average of 54% to over 80% as measured by PPC. Similar improvements in the design process can be highly beneficial to overall schedules.

Kerosuo *et al.* (2012) observed that projects that used the LPS had faster completion rates. It focused design teams on addressing actions needed to complete design-related tasks. The LPS was credited with securing reliable promises, addressing a major shortcoming of the design process. Earlier research had confirmed that pull planning increased the sense of ownership among design team members, encouraging them to be more reliable in fulfilling promises.

The Lookahead plan draws attention to making tasks ready in a proactive rather than reactive mode. PPC tracking and root cause analysis provide vital information on the effectiveness of the design production system. From the design perspective, a Lookahead is generally 2 to 4 weeks. In practice, a period of over 4 weeks' duration is too long; it may allow other distracting issues to emerge.

Flow variability greatly influences lean practices, as a delay in work completion by one discipline directly affects the downstream activities of the next discipline. The LPS uses lean methods to provide improved project control. In essence, the last planner is in the best position to match resources to accomplish assignments in response to downstream demand. Work planning may establish time frames but is not very effective in establishing that the tasks assigned are capable of completion. Stops and starts typically occur or making do (i.e., keeping occupied with noncritical tasks). The LPS technique as refined by Ballard (2000) decentralizes decisions and empowers the individuals that are in direct contact with the work to plan and schedule detailed tasks; in effect, they become the last planner.

The LPS is based on three or four levels of schedules and planning tools. Certain projects such as fast-paced EPC projects and/or projects involving engineered-to-order components may require a daily work plan (Figure 3.7):

1 The master pull schedule
2 The lookahead schedule
3 The WWP
4 A daily plan (when required)

Master schedule

The master schedule represents an overall view that identifies major project phases and milestones for these phases. In design, specific milestones would need to be developed for design deliverables. Packages of tasks are transferred from the milestone schedule by the act of "pulling" in response to the downstream demand of other team members. It is essential for a team to collaborate in preparing a milestone schedule of deliverables and their timing, with a high degree of accuracy. It clarifies the commitments needed to perform design activities in a timely manner, avoiding negative iterations, wasted time and resources, and schedule slippage.

In construction, the master schedule is typically included in the contractor's bid. Hence it is an optimistic forecast of the major stages in carrying out a project, based on the early information provided in the

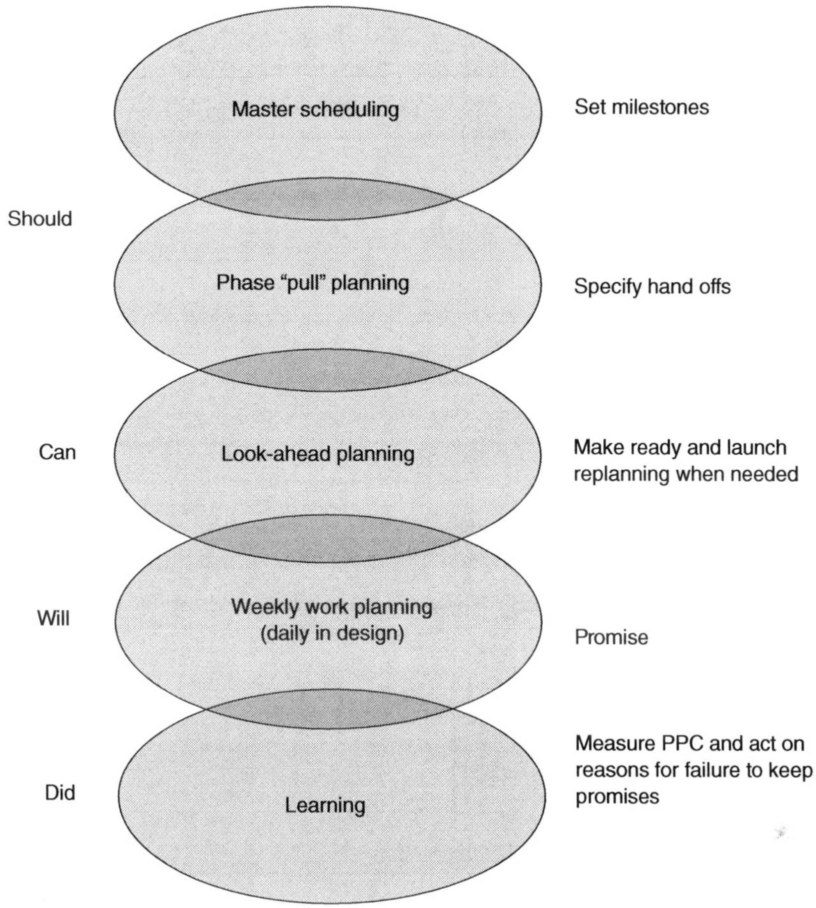

Figure 3.7 Interconnected conversations in The Last Planner® System

Source: With permission from the Lean Construction Institute. www.leanconstruction.org

contract documents. Contractors use it as a guide for the timing of procuring labor and materials. Design is less transparent as the work products are not initially known; they evolve though a process of discovery and experience many iterations. Time frames may be estimated through prior experience, but every project has a unique solution. Needed changes that may arise from owners' requirements or site conditions may require extensive study. Unexpected pulls may occur from random occurrences and lead to additional design iterations, whereas construction activities are more standardized and predictable. A countermeasure would be to assign a specialized design team for investigating such unexpected items expeditiously.

Later in this chapter, JE Dunn Construction served as CM on a project that used the LPS in Design. The team first developed a milestone schedule for design. In order to ensure that each milestone could be accomplished as planned, a Reverse Phase Schedule (pull plan) was developed. Three major milestones were developed –

(a) Having a drawing bid set for make-ready work,
(b) Installation of footings and foundations, and
(c) The remainder of the building.

Lookahead schedule in design

In the design phase, the lookahead schedule is derived from the phase plan, based on a time frame of 2 to 4 weeks, and is used for work flow control. It uses items pulled from the master schedule. In design, this time frame may be adjusted to fit the type of project. A re-use of a pre-existing design may be more reliable, i.e., it may have less variability and can serve adequately with a shorter time frame. By comparison, the construction phase of projects typically uses a 6 to 8 week time frame.

Weekly work plan

The WWP is derived from the lookahead plan. Weekly planning is especially important in the design environment. It provides a detailed work plan specifying hand offs at each phase between disciplines. It is updated weekly, based on discussion between the last planners. In the case of integrated project delivery some aspects of design may involve collaboration within clusters of both design and construction professionals. The team may include design team leaders as well as trade forepersons or supervisors, depending on the complexity and size of the project.

The weekly work plan is then highly beneficial to construction professionals, and the inputs from the designers are then an accommodation to them to provide needed structure.

Reverse phase scheduling in design

Reverse Phase Scheduling or "Pull Planning" involves working backwards from the deliverables of the completed project, determining what needs to be done at each stage to complete the preceding activity. In developing a pull plan for one week, for example, the discussion would occur at a planning board designed for the purpose with a separate column for each day of the week. The deliverables for the end of the following week would be defined as a starting point. In the case of design it is a detailed work plan that specifies the handoffs between specialists for each stage of design, indicating what SHOULD be done at each phase. In lean construction, the reverse phase schedule is developed through conversations with the primary project team representatives, including their last planners, typically with sticky notes on a display board. A senior designer representing a specific discipline would serve as a last planner, interacting with the last planners of other disciplines at a wall mounted planning/display board. Each discipline uses stickies (sticky notes) of a particular color. Structural/concrete may be gray, while electrical may be red, and so forth. Each note is marked to indicate the area being worked on, the requestor, what tasks one expects to be complete as a predecessor activity, and the required dates. In preparing a weekly work plan the starting point would be the end of that week. Working backwards, the last planners populate the planning board with stickies, matching them with others in order to get to the current date.

Daily work plans

Some projects such as engineer, procure, and construct (EPC) construction that comprise engineered-to-order components need more precise control than is offered by the WWP. Such engineered components are designated for installation by specific trades in specific locations. The emergence of constraints can quickly disrupt the construction schedule; in a fast moving project, a week's delay may result in significant schedule deviation. Establishing a daily commitment enhances coordination with other trades.

Last planners

The so-called last planner is the professional who prepares the weekly (or daily) planner schedule. This schedule also includes a buffer of work activities based on future work. The weekly accomplishment is measured as percent planned, complete (PPC). The PPC is based on 100% completion within a given week; tasks are considered to be either complete or not complete – there is no partial credit. The last planner also determines physical, specific work *assignments* for the next day. Work selection rules shield production from upstream uncertainty and variation. Production control is aimed at the avoidance of waste, such as waiting for someone to complete a task after the scheduled completion date or time.

Weekly meetings are used to evaluate the team's performance during the preceding week (i.e., what the team DID). Root cause analysis is carried out to determine the reasons for underperformance, and the knowledge gained is used to improve the process used in assigning work in subsequent weeks.

Communication between disciplines is an essential requirement for the LPS; many assignments require a close collaboration between various design specialists. For example, designing a structural wall that contains electrical wiring and plumbing requires having electrical and mechanical engineers collaborate with structural engineers.

As has been noted previously, improving the efficiency of individual activities does not necessarily improve the efficiency of an overall process. Waste can occur in the handoffs from one activity to another in the form of delays, defects that need to be corrected, and so on.

Team work planning

Team work planning is highly necessary to keep track of the work of many designers who have different specialties. They may also be working in clusters with different professionals such as quality assurance and commissioning specialists. Experience indicates that weekly pull planning and measurement of performance are most effective when done weekly. Mobile planning boards such as those used by Linbeck Group LLC (with wheels) are very effective as they can be moved to different physical locations where they provide visual management to work teams. Having lean boards in the work area facilitates ongoing updates and increased awareness of team performance.

The lean education program provided by the Associated General Contractors of America (AGC) recommends eight steps that design teams need to follow each week.

Teams should:

1. Lay out a minimum of three weeks' design work.
2. Check the commitments made in the previous week, and the PPC value based on those completed.
3. Do basic root cause analysis by using the "5 why" technique with the design tasks that were not completed.
4. Plan to complete the unfinished work, considering the reasons for non-completion.
5. Plan the upcoming week, using lessons learned from the past week to avoid over-committing.
6. Have a design team meeting that involves the entire team. Commitments need to be reciprocated; team members must make reliable promises to complete tasks that others will depend on to do their work. Team members should remember to "optimize the whole, not just the pieces".
7. Have daily confirmation of needed work. Keeping the work plan visible enables this.
8. Practice continuous learning; team members should determine what went well and what did not. They should use the lessons learned to plan for better performance in the future.

A number of Visual Management tools can be very effective in maintaining awareness of the team's performance: These may include:

Standard work lists that are visible to others informs them of the time a team member may have available for other activities.

Summary sheets that help to avoid excessive time use and cost overruns by documenting the hours spent to date.

A lookahead schedule that documents commitments and promises made that affect the work of downstream professionals.

A team summary that consolidates all planning sheets on one page, and displays accomplishments, activities, and commitments, etc.

Success in lean planning efforts begets further success. When teams see that the LPS provides a smooth and sustainable workflow it reinforces their adoption of the LPS principles, such as reliable promising and consistently meeting commitments.

Set-based design

Set-based design is an approach that defers design decisions to the "last responsible moment" to allow for the evaluation of alternatives that improve constructability. Toyota has used this technique to design new models in a much shorter time than the industry standard. They have learned that the best solutions are hybrids of original design options.

As described in Figure 3.8, set-based design begins in the project definition phase when designers begin to prepare conceptual designs to meet the owner's project requirements. Set-based design involves developing a

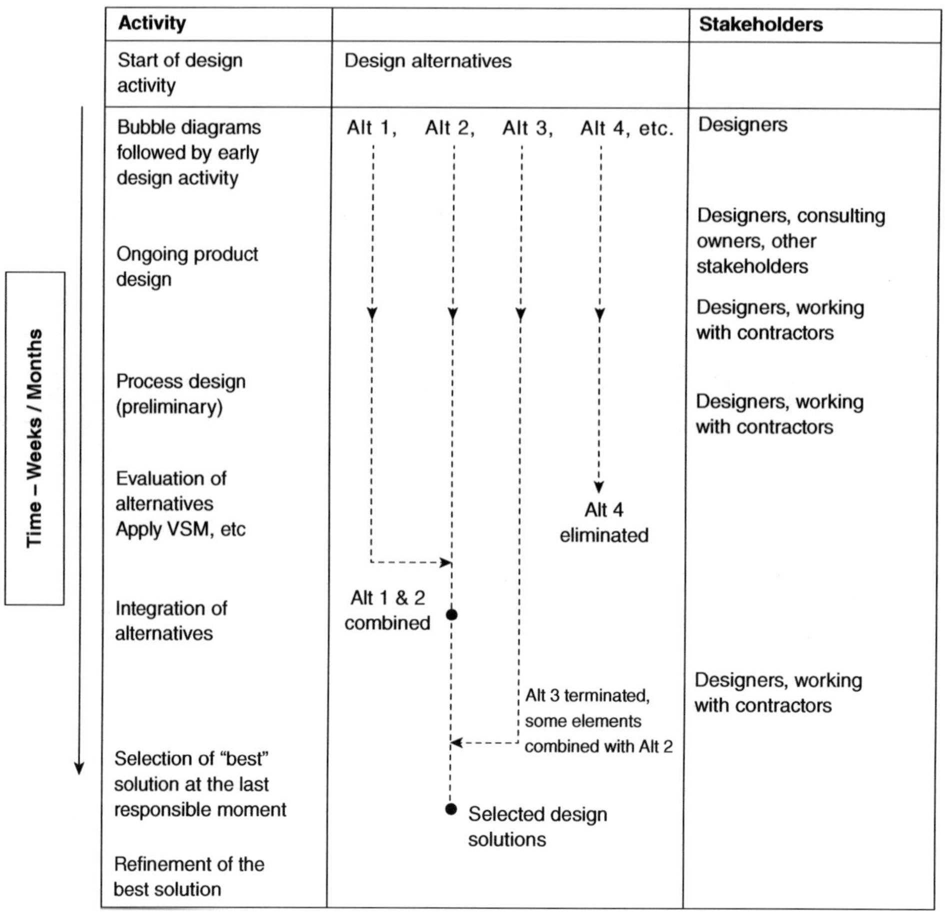

Figure 3.8 Set based design

set of design alternatives in parallel with each other until a point where a final choice has to be made. Designers engage in deep thinking at each level of design for systems, subsystems, assemblies, and components. They communicate and collaborate with other stakeholders such as contractors to determine the buildability and operability of each design alternative. The process needs to be carefully managed as additional professional hours and cost are spent in developing additional solution sets. This has to be done while avoiding sunk costs in alternatives that may not be viable for a project. In some instances two or more conceptual designs may be consolidated if the resulting design provides clear advantages over other alternatives. By having decisions made at the last responsible moment, set-based design avoids loop-backs that may need to be made later because of hasty decisions, as past issues have to be revisited. Opportunities may also be lost in the case of time-sensitive issues. Value stream mapping helps to identify the last responsible moment. While this approach consumes much time and effort from knowledgeable professionals, it derives the best collective decisions.

Designing for lean operations

As a design is developed, careful consideration should be given to facilitating lean operations in the completed facility. A poorly laid out facility could be relatively expensive to operate – excessive travel from one point to another is one of the seven wastes documented in the Toyota method. In the case of manufacturing facilities, it would be a good idea to include an industrial systems engineer with experience in facilities layout. In some cases the size of the expansion may be reduced – the expansion may not even be necessary.

By way of an example, the president of a window manufacturing operation felt a need for more square footage to house the equipment needed for increased production. The factory was located in a warehouse district with adjacent buildings. He acquired additional space from one of these buildings for expansion. Shortly after that he became interested in lean production, and contacted a consultant who specialized in Kaizen for help. Unexpectedly the consultant asked over the phone if he had recently bought additional space. The president was taken by surprise – why would this stranger care if he had bought more space? The consultant explained that by the time clients contacted him they were already having production bottlenecks and felt they needed to expand their facilities. After implementing Kaizen or other lean approaches they often discovered that the original square footage was more than enough! It could be reconfigured to meet the updated needs of a facility,

Once it is determined that increased building size is necessary, then all the stakeholders should be included to the greatest extent possible. They may include:

- Building users,
- Designers: architects and/or engineers,
- GC/CMs, subcontractors,
- Suppliers, or
- Specialty consultants: sustainability, and/or commissioning.

Lean operations and target value design (TVD): TVD involves designing to a specific estimate rather than reducing a design to meet a stated budget. It supports lean by avoiding the wastes incurred in scaling back a completed design. The TVD process was used in one of the projects for Sutter Health System in California. It was decided to apply TVD such that 90% of the program for the facility should be accommodated in 70% of the initially programmed space. Discussions with users (nurses, doctors, and other medical personnel) identified a more efficient layout that met operational needs than was originally thought. Through the TVD process, the project cost was reduced to 13% below the market cost average. In effect, the space requirement of the new facility was reduced by approximately 30% below the originally anticipated square footage. The savings derived from that exercise allowed the owners to include additional features that were excluded earlier due to budget limitations.

Target-value design

Target value design (TVD) involves designing to a specific estimate instead of estimating based on a detailed design. TVD applies Target Costing to building construction, Cooper and Slagmulder (1997), defined Target Costing as "a disciplined process for determining and realizing the total cost at which a proposed product with specified functionality *must* be produced to generate the desired profitability at its anticipated selling price in the future." It seeks to address the problem that affects many projects – various design disciplines work from a common schematic design to do design development in their areas of expertise. Working in their respective offices they are subject to "project creep" and "project growth"; with little cross-functional collaboration, project designs often become overpriced, unconstructable, and behind schedule. Corrective action may include a misuse of value engineering to radically cut the scope of a project, or to suppress certain features that are desirable but unaffordable. Furthermore, the lack of collaboration often results in early design decisions that are later found to be suboptimal, but may be difficult to change. Ultimately, much time and effort are wasted, and the design cycle is longer than it should be. These wastes run counter to the lean philosophy.

Glenn Ballard refers to four components of TVD: *Market Cost*, *Allowable Cost*, *Expected Cost*, and *Target Cost*. They are defined as follows:

> *Market Cost* is a benchmark cost; it consists of the cost per square foot that would be expected for comparable construction projects. *Allowable Cost* represents the maximum cost that must not be exceeded; if the project team cannot design to allowable cost, the project must be cancelled because it would, by definition, become financially unfeasible. *Expected cost* is the estimated cost of the project in its current state during the TVD process; the expected cost is continually recalculated with each new iteration of design. *Target Cost* is the stretch goal for the project, meaning it is usually set below Allowable Cost.
>
> *(Ballard 2009a; Rybkowski 2009)*

In traditional product costing, a manufacturer may add a profit markup to a product's production cost to establish its selling price. The problem with this method is there is no guarantee that buyers will be willing to pay the asking price. The process of Target Costing, by contrast, implements a reverse strategy; the market price is *first* established by determining how much buyers might be willing to pay by using focus group research or by looking to similar products on the market, for example. A desired profit is then subtracted to give product designers the cost to which they must design the final product:

- TVD proposes that designers should "do it right the first time" and build constructability into their designs, as opposed to designing first and then evaluating constructability later.
- TVD recommends concurrent design, with various disciplines in ongoing contact, as opposed to periodic reviews.
- Solution sets should be carried forward in the design process to ensure that good alternatives can be available later.

Ballard and Reiser (2004) documented one of the first projects in recent history to successfully apply TVD. The Saint Olaf College Fieldhouse project involved an athletic facility that was built between 2001 and 2002. The project was delivered on time, within budget, and with more value provided to the owner than comparable projects; the contractor (Boldt) also made a reasonable profit. A similar facility that was built in the region between 1998 and 2000, took 10 months longer to build, at a cost 18% higher.

Target value design was adopted as an essential part of the integrated form of agreement for lean project delivery used in projects for the Sutter Health System. The TVD process is intended to have a project designed within a desired budget in accordance with a detailed estimate, and establishes value, cost, schedule, and constructability as basic components of the design criteria. With the Sutter projects, there was a provision in the contract documents that a project's target cost should never be exceeded without the express approval of the owner. Design activity included a core group of contractors and subcontractors that blended considerations of constructability and the construction process. According to Do (2004) a statistical analysis of capital projects delivered by TVD revealed cost 15 to 20% lower than traditional projects.

There are a number of foundational practices for target value design (Macomber, Howell, and Barberio 2005). These practices promote design conversation, as they see design as a social activity that involves several professionals focusing on meeting the needs of the client. This approach is especially effective in light of the fact that the client's needs can change over time and value assessments need to be repeatedly made to ensure that design decisions meet these needs.

1. Conduct design activities in a Big Room (*Obeya* in Japanese). Toyota has used it successfully, especially in product development, to enhance effective and timely communication. The Obeya is similar in concept to traditional war rooms, and will contain charts, pictures, and graphs on display boards that visually represent program timing, milestones, and progress to date. The boards also display actions or recommendations to resolve delays or technical problems. The Obeya houses project leaders and key staff in close but comfortable proximity to shorten the communication cycle and promote an effective plan, do, check, and act (PDCA) cycle. Spontaneity comes easily as specialists can collaborate readily in key design or construction decisions.
2. Create a multi-disciplined team that consists of designers – architectural, structural, MEP, etc, an estimator, contractors, and a facilitator.
3. Determine the market cost for the facility using available benchmarks such as cost per square foot, etc.
4. Work closely with the client to establish the *Allowable Cost; it* represents the maximum cost that must not be exceeded. The target value is a stretch goal below the allowable cost. Designers should guide clients to establish what represents value and how that value is produced. They should ensure that clients are active participants in the process, not passive customers.
5. Once the target value is established, use it to work with a detailed estimate. Have the design team develop a method for estimating the cost of design alternatives as they are developed. Deviations should not continue unchecked; if a particular design feature exceeds the budget allocated for it, then that design should be adjusted promptly in order not to abort further design work that cannot be accepted. The expected cost is continually recalculated with each new iteration of design.
6. Apply concurrent design principles to design both the product and the process that will produce it. This work should be done as a collaboration between architects/engineers, specialty designers, contractors/subcontractors, and the owner. Be flexible to include innovation in this process. Practice reviewing and approving design work as it progresses.
7. Working in small, diverse groups is best. Groups of eight or fewer people establish better group dynamics; it is easier to create a spirit of collegiality and trust that lead in turn to more innovation and learning. Design with the customer in mind. Focus on designing in the sequence of the discipline that will use it. Use the "pull" approach with each design assignment to serve the next discipline. Lean is obtained by meeting downstream needs as opposed to producing what is convenient. Overproduction increases the possibility that the work so produced may not be what is needed for the next discipline to maintain its schedule, and it may lack the collaboration necessary for constructability.
8. Collaboratively plan and re-plan the project. Planning should involve all stakeholders to continually maintain an actionable schedule. Joint planning will refine practices of coordinating action. This will avoid delay, rework, and out-of-sequence design.
9. Lead the design effort for learning and innovation. Expect the team to learn and produce something surprising. Also expect surprise events to upset the current plan and require more replanning.

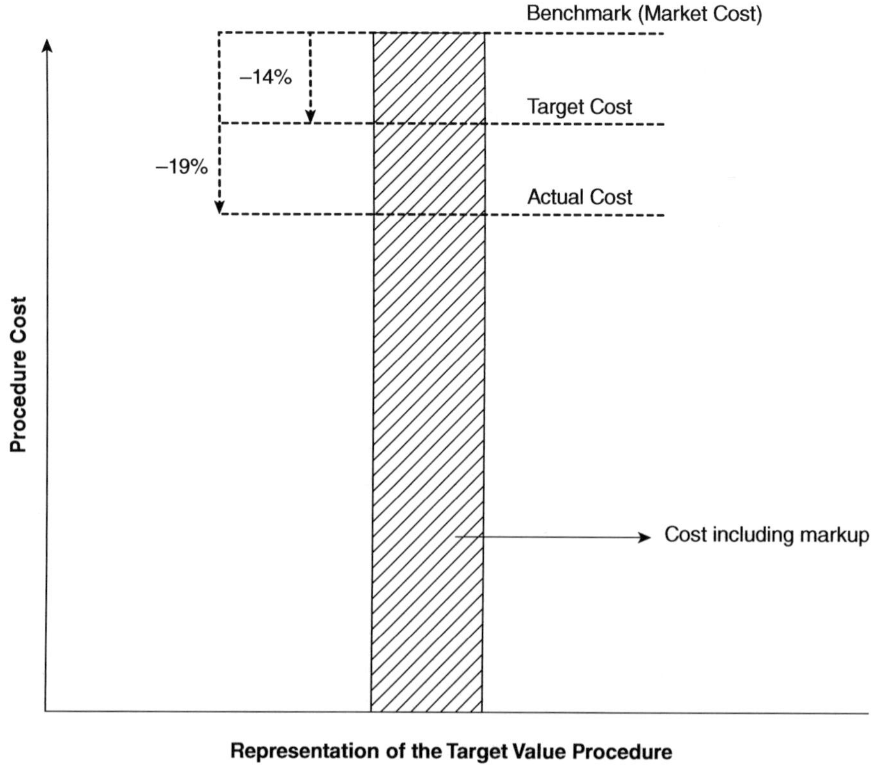

Figure 3.9 In this example the target cost was set at 14% below the market or benchmark cost (not to scale). The actual cost was 19% below market

10 Learn by carrying out conversations on the results of each design cycle. Include all project participants in order to capture knowledge on success factors. Use this information as a part of the PDCA cycle, and use formal measurement systems, if possible.

A major effort has to be directed at the front end of the process by working with the client and determining what represents value and what it might cost. Subsequently the budget is allocated to various facility systems and cross-functional teams collaborate to work within those limits. A critical requirement is for cost estimating to keep up with the design work as it progresses – with very frequent real-time updates. This approach varies substantially from traditional practice, which is highly linear. The budget becomes an influence on the design, rather than an outcome of design.

Target Value Design/Delivery (TVD) relies heavily on the expertise of people organized as clusters; a work cluster is a select small group of people drawn from different companies to work together on a project. An MEP cluster, for example, includes representatives for mechanical, electrical, and plumbing disciplines. They usually include MEP engineers, representatives from the owner and architect, each of the MEP trades, and estimators.

In general, for this process to work the construction staff must include members who can: estimate from preliminary drawings instead of contract drawings and specifications. Designers need to be receptive to critical reviews of their designs with regard to both pricing and constructability. The owner (or representatives) must be accessible and be equipped to provide feedback on balancing project costs with specific building features. This may involve value engineering compromises of life-cycle costs versus form, function, and time. Target value design brings a standardized process to the team (Figure 3.9).

Validation studies

The purpose of validation studies is for the owner and the project team to establish the feasibility of a project. It also enables the team to visualize the features of a successful project. It is imperative for the team to understand the business case for a project and the outcomes expected. Value definition must be established in order to reveal the specific value that will be expected.

Having clearly articulated values simplifies the process of achieving the value, and provides an ongoing reference point. As indicated by Hill *et al.* (2016) value may be defined by four types of factors:

- Objective factors such as a specific capacity or square footage.
- Subjective factors that are difficult to measure such as space for discussion groups.
- Project constraints such as time, cost and complexity.
- Operational constraints including staff availability and the capacity to attract a particular clientele.
- Project preferences: The owner may have a preference for certain design solutions or systems that reflect the owner's values, such as Feng Shui or net zero concepts.

The application of validation studies at Sutter Health

Before Sutter Health funds a project that it knows will cost more than $10 Million, it funds a validation study to arrive at the goals, scope, budget and risks of a project, and then puts that before the board for funding approval. To initiate the validation effort, Sutter assembles an integrated team of designers and builders. Sutter retains the right to not use members of this team for the approved project, but usually the team is retained and awarded the project, once it is approved. Across the 23 Sutter Health IFOA projects (at the time of this writing), only in rare cases have team members involved in the validation study not been retained for the full project. (Reference Chapter 9 on IPD at Sutter Health)

Any project in excess of $10 Million must have a Validation Study. This study turns Sutter's goals for a proposed project into a functional narrative that describes how different spaces will be used, and a space program that shows how much of each clinical function is required to meet the operational goals. It includes a proposed budget and schedule. It also lists all the assumptions that the validation effort has made and the risks to the scope, budget, and schedule.

Sutter selects an architect and builder as an integrated team to carry out the validation study. Typically the cost of such studies has been in the region of 0.5 to 2% of the total project budget. The team also includes specialty design, consultant and trade partners where justified by project complexity and risk warrant such.

Typical consulting and preconstruction services contracts are used for the Validation Study, and teams earn their expected overhead and profits. However, a mandatory condition of being selected to be on the validation team is the company's agreement to the terms and conditions of the Sutter IFOA without exception, so if the project is approved, and the EMP is then finalized, they would have already agreed to sign the IFOA. Having the team involved in the process that creates the Validation Study for the project, creates transparency and prepares the project for future success.

Figure 3.10 An overview of the ISU facility under construction

Source: Courtesy of JE Dunn Construction Company Inc.

Part C: Examples of lean design applications - JE Dunn Construction Company, Inc., Boulder Associates Architects, Agile and Scrum approaches

The JE Dunn Construction Company, Inc. is a family-owned General Contractor headquartered in Kansas City, Missouri. It engages in the construction of multiunit residential, commercial office, healthcare, and education projects. JE Dunn is #25 in Engineering New Record's (ENR) 2018 list of Top 400 Contractors. Since 2018, Rebecca Snelling, a former lean consultant, has been serving as s VP and National Lean Director. Under her leadership, JE Dunn has embraced the lean methodology and lean thinking as much as possible. This involves projects that are traditional in nature and do not have a lean-oriented form of contract. Although the majority of the firm's projects are design-build or CM@Risk, project teams including the trade partners, are given a full day of training that focuses on lean thinking and on the specific tools that a project will use.

Project example 1 – The Iowa State University Student Innovation Center
This facility was designed as a CM – Agency project
Owner/end user: Iowa State University (ISU), Ames, Iowa
Construction manager: JE Dunn Construction
Construction budget $59 million
It was started in 2016 and the expected completion is in 2019
Design team: Substance Architecture, KieranTimberlake
IMEG, Saul Engineering; Snyder & Associates, C-Wise
Bakergroup, Roofmeadow

Project Description:

The ISU Student Innovation Center is intended to a collaborative learning space to serve each of Iowa State University's seven colleges in a single facility, promoting exploration, experimentation, and higher learning in a hands-on environment. The building includes lab spaces for student organizations such as glass blowing, solar cars, and lunar excavation. It also has a demonstration kitchen, an auditorium, a step-atorium, various green roofs, and several group maker spaces. With a construction budget of $59 million, completion is expected in the fall of 2019.

ISU selected JE Dunn as the construction manager for the project. This enabled the JE Dunn team to begin creating relationships with the design team and the owner earlier in the project than usual. The team was able to provide critical feedback to the design team in terms of the schedule and budget. JE Dunn was also required to use the lean methodology in the construction phase of the project to deliver value. The information in this case is a good example of the use of the Last Planner® System in design.

The Last Planner® System in design

According to VP Rebecca Snelling the Last Planner® System (LPS) is used in the Design phase somewhat differently to the construction phase.

Whereas construction sequences are more defined, design is unclear, and there is hidden waste in many design activities. Deliverables are established for the design team, but lead designers work with others to sort out activities and the sequence needed. Phone calls and checks are made to ensure that follow up is done. The approach is one of facilitation rather than applying pressure. Co-location helps because of the personal contact and communication that it provides.

The LPS for Design was used in the ISU project starting In October 2016. JE Dunn's representatives met with the owner's project manager and other representatives as well as key players from the design team to begin the Last Planner process. As the company points out, that presented a great challenge because of the large number of stakeholders within ISU – namely seven colleges and many student organizations. Representatives from these groups formed a steering committee, and a joint decision would be needed to define an acceptable scope of work.

A kick-off meeting was held to "get everyone on the same page", and establish the importance of accountability. It also clarified the overall project goals and the logistics of the design process.

The team first developed a milestone schedule for design. In order to ensure that each milestone could be accomplished as planned, a Reverse Phase Schedule (pull plan) was developed. Three major milestones were developed:

(a) Having a drawing bid set for make-ready work,
(b) Installation of footings and foundations, and
(c) The remainder of the building.

An Excel Work Register was set up to document the commitments, and was also used to track constraints that would hinder the meeting of the commitments that had been made. An excerpt from the register is pictured below.

Another challenge was the fact that most of the stakeholders had not been exposed to using pull planning during the design phase. Yet another limitation was the fact that the project team members were not able to co-locate, yet they needed quick, responsive communication in order to make decisions and keep the project on track. The project architect was located out of state, and could not easily attend weekly meetings in person. The project team was innovative, using Skype to share the work register, and weekly check-ins to provide a cohesive team throughout both the design and the construction phases. JE Dunn's representatives provided input to design decisions with regard to cost implications, and conversations on constructability issues.

Check-ins were held on Monday mornings and allowed the team to communicate on the status, and understand how its commitments affected counterparts as well as the overall ability of the team to meet the common goal. A team that is co-located could hold the weekly check-in in person, but the JE Dunn team used Skype to share the work register and talk through its commitments.

Table 3.2 The Excel Work Register

WORK PLAN/COMMITMENT LOG

Projec Name: low State University – Student Innovation Center: Issuance 2 and 3 *Last Update: 2/17/2017*

Reason For Variance: 1) Over-committed 2) Miscommunication 3) Previous work not completed 4) Change in work plan 5) Outside Constraint 6) Resource not available 7) Materials/Equipment not available 8) Safety concern 9) Work not authorized by contract/submittal/etc. 10)Other

Item #	*Mile stone*	*Promise from Stickie*	*Performer*	*Requester*	*Due Date*	*Constraints/Comments*	*Done?*	*Reason For Variance*
13	Issuance 2	50% CD Issuance 2 MEPT	Ryan LePere	Design Team	2/28/2017	Need subslab drainage scope deliniation between Civil/ Mechanical		
14	Issuance 2	Loading Dock Structure	Paul Taylor	(8)	2/28/2017			
15	Issuance 2	Design for Special Equipment	Paul Taylor	(9)	2/28/2017			
16	Issuance 2	Footing Elevations for MEP lines	Paul Tayfor		2/28/2017	Need MEP lines & Elevations from KJWW		
17	Issuance 2	ISU Dining Decision	Liz Clarke		3/1/2017	Per Steering Committee Mtg		
18	Issuance 2	Issuance 2 50% CD Project Manual	Ann Sobiach Munson		3/3/2017	Need Consultant Specs and Internal Design Decisions		
19	Issuance 2	Issuance 2 50% COs	Cody Knop	Lit Clarke	3/3/2017	Consultant Dwgs due 3 days prior		

Figure 3.11 Design features of an upper floor area

Through the use of pull plans, the work register, weekly check-ins, and monthly updates, the Last Planner® System promoted a collaborative environment early in the project which helped maintain a team-oriented attitude throughout design and into construction. It is worthy of note that most of the team members had not been involved in pull planning during the design phase before, but they viewed it as a very positive experience.

Various team members helped to monitor the process from start to finish, assembling the work register and leading weekly check-ins. They helped to hold team members accountable for their commitments and supported the necessary collaboration between them.

At the end of the design phase, several important lessons were learned:

a) The right people need to be included in the kickoff meeting, to ensure that the right questions can be asked, and that there is accountability.
b) Face to face meetings contribute to the accountability that is absolutely important for having commitments met.
c) It is best to have a lean specialist lead design pull planning sessions; they are significantly different from construction pull planning.
d) A "Parking lot" is most important for keeping important issues in view and in focus to avoid distractions during meetings.

The project was continued to the construction phase with the oversight of the JE Dunn team.

Lessons learned from team performance enabled countermeasures: The LPS was seen to be highly applicable. A suggested change versus the construction phase is to indicate the time of handoff rather than task duration. Design time is not readily measurable with work standards the way that construction tasks are.

The Boulder Associates experience

Several architectural firms have successfully applied lean construction methods to design thinking, and. Boulder Associates is one such organization. They specialize in architecture, interior design and graphic design, maintaining offices in California, Colorado, and Texas. Their many years of lean experience include early lean projects such as the Sutter Davis Medical Office Building in 2004 and an IPD project in 2008 with a medical office building in San Francisco. Todd Henderson, a principal at Boulder, notes that their mission statement is "Do good work, and enjoy the journey". In their view, good work is not just technically important, but also socially beneficial. It is driven by a lot of heart, which is typically the case in healthcare projects. Boulder has a collaborative culture – they enjoy working with their clients, and they are very concerned about customer service. According to Henderson, Romano Nickerson, also a principal at Boulder, began experimenting with applying the Last Planner® to design in the early 2000's. Nickerson was an early advocate of lean who sought to remove waste from his own work. For Henderson, a turning point came when Nickerson shared a graph of staff working hours with his colleagues; three-quarters of the staff were accustomed to working on Saturdays and Sundays. Following the application of the Last Planner®, workflows steadied and staff found they were then able to spend weekends away from the office. The method spread laterally at a grassroots level via staff members assigned to multiple projects who then "mentored up" to other managers. Company directors recognized the positives in lean thinking and adopted the "Work Plan" – Boulder's lean initiative. Despite the 2009 recession and a downsizing of nearly one-fourth of their staff, Boulder doubled their profits. Boulder's Team Weekly Work Plan program was developed and implemented in 2012.

Henderson explained that the Last Planner® System has improved productivity, but that the use of PPC calculations has not seemed to add much value; the LPS seems more suitable for the field while design is more organic. Rather, discovery and false starts are viewed as a part of the design process. As a design firm, Henderson admits they are discovering that agile and scrum, which are heavily practiced by the software design industry, may be a better fit for the "loopy" iterative process of architectural design than the LPS. The LPS seems better suited for the more linear processes used by general contractors, although there is some disagreement about this (Todd Henderson, *Personal Communication*, November 11, 2017).

Agile and scrum approaches

Demir and Theis (2016) refer to the waterfall model of project planning that is currently used for scheduling. A scheduler or planner uses it sequentially to forecast activities before the start of a project. The waterfall model does not work effectively with the design phase as the design is very unclear in the early part of the design phase, such as preliminary design. Design and software development share the characteristic of being iterative-as ideas are developed and incorporated incrementally in a design they are shared with a client to obtain feedback. Ideally, this iterative approach progresses to completion, but may not provide breakthroughs in thinking that can greatly increase customer satisfaction. On the other hand, agile methods are appropriate for the construction environment (Koskela and Howell 2002; Owen and Koskela 2006).

Schwaber and Sutherland (2017) developed scrum in the 1990s and define it as a framework within which people can address complex adaptive problems while productively and creatively delivering products of the highest possible value. The word scrum is derived from the game of rugby; a scrum is used to restart activity after a ball has gone out of play. Used since the 1990s, scrum is not a process, technique, or definitive method. Scrum is based on empirical process control theory, Empiricism views knowledge as based on experience and making decisions on what is known. It uses an iterative, incremental approach to assure predictability and minimize risk. Empirical process control is based on three pillars – transparency, inspection, and adaptation. For transparency, there must be a common understanding of significant aspects of the process by all participants, so there can be clarity on the completion or "Done" phase of an activity.

As a framework scrum consists of scrum teams, their associated roles, events, artifacts and rules. Each component within the framework serves a specific purpose and is needed for the successful use of scrum.

Team size should be 5 to 9 people; a scrum master helps the team work through the process. By way of detail, the typical length of a sprint is 1 week; this is considered the best duration. Note that people can be members of different sprint teams. Daily scrums keep people clear on what is happening. Fifteen minute meetings help to keep people aligned, so their actions will move in the right direction. This time frame is short, but very important. When people meet there should be definitive outcomes and decisions. The scrum master may ask: What did you do yesterday to finish the sprint? What do you need to finish the sprint? Are there any obstacles blocking the sprint goal?

Sprint goals are critical; If by Friday of a given week the goals are not reached, then the goal is considered to have been missed. It is best to address one issue at a time, as it is in keeping with the principle of 1-piece flow. With scrum a percentage is not goal – a binary choice is best. Value assignments must be delivered on time, and must meet specifications.

Agile design management

Agile design management (ADM) has evolved from scrum. The goal of ADM is to increase coordination, interface management, collaboration and transparency through all design phases. It offers significant advantages over the Scrum methodology. Scrum typically includes a daily sprint. In a design phase there are typically many work packages and deliverables that have to be managed, and this can be challenging for the methodology. Design teams are often separated geographically and are not accessible for the daily sprint. Design teams involve many disciplines and may exceed 20 members. It is often not feasible to break construction projects into modules that can be worked on separately.

At that point the scrum team can be very difficult to manage. Overall, the scrum approach used in software development may be not be practical for certain categories of design projects. Agile design Management may fill that need.

The recommended agile approach

a) Project setup
b) Planning phase (overall process analysis)
c) Process planning
d) Task management

Project setup:

Four levels are created in a typical workshop setting:

Decision making body
Project management
Planning team
Overall process analysis.

Workgroups and technical planners are scheduled in a takt-based sequence to meet and discuss design issues with subproject leaders. The purpose of this is to identify work packages; they are then prioritized.

Process planning:

It is scheduled every 3 or 4 weeks with a subproject manager. It defines the completion timelines for each work package.

Task management: Each work package is assigned to a work group, That group defines the tasks needed for each package, and adjusts as needed. New tasks may become necessary as the design evolves; some existing tasks may become unnecessary. This status information is posted on a planning board that is visible to the team.

Whereas scrum works best with daily meetings, ADM participants may be situated in distant locations where such meetings would not be feasible. Agile Design Management requires meetings between designers for a minimum of two days. As these meetings are process-focused their time requirements are reduced. Demir and Theis (2016) observed a time reduction during the preliminary design phase from 5 hours to one. As discussion is held at the planning board, the key issues are identified with a red dot; resulting in focused decisions. These dots are placed on the planning board to reflect concerns, comments or proposed actions.

ADM is credited with improved transparency of the status of design tasks, improved integration of design team members. Problems have been identified on a timely basis, allowing early corrective action. Team morale improves – especially with less experienced designers. Takt-based scheduling of the ADM meetings is said to provide better use of resources and a less stressful environment for the design team.

Table 3.3 LEED categories n

Location and transportation	LT
Sustainable Sites	SS
Water efficiency	WE
Energy and atmosphere	EA
Materials and resources	MR
Indoor environmental quality	EQ
Innovation	ID
Regional priority	RP
Integrative process	IP

Part D: Sustainability and Commissioning considerations

Sustainability issues[1]

Contemporary design criteria should include responsible sustainability practices. This design must be based on reducing the carbon footprint, including lower energy usage and less utilization of potable water. Buildings account for one-sixth of the world's fresh water usage, one-quarter of its wood harvest, and two-fifths of its material and energy flows. In the United States, buildings and their landscapes are a major source of pollution, using one-third of all energy consumed, producing more than one-third of the carbon dioxide emissions and almost one-half the sulphur dioxide emissions. A typical construction project generates 2-½ pounds of waste per square foot of completed floor space; construction waste accounts for 40% of the total US waste stream.

So-called green buildings use energy, potable water, and other resources more efficiently. They also yield improved indoor air quality and higher worker productivity. The US Green Building Council's leadership in energy and environmental design (LEED) certification system rewards sustainable practices in design and construction with a ranking system that reflects each facility's compliance with "best" environmental practices. Research findings suggest that LEED-certified buildings have a 3.8% higher occupancy than non-LEED-certified buildings. Buildings that meet the EPA's Energy Star standards are said to have a 3.6% higher occupancy than unrated buildings. Buildings that meet either LEED or Energy Star standards have been shown to have lower energy usage and command higher rental rates.

The adoption of Green Building standards may significantly influence a facility's design criteria and the OPR (see also Chapter 9). It may also increase the number of decision makers involved in the design and construction process, for beneficial reasons. In order to fully apply sustainability principles to project design, the design team should include a professional with recognized credentials in that field. The designation of a LEED accredited professional (LEED® AP) is one of the best established.

LEED projects generally require an accredited professional to serve as a resource in a project from the definition phase through to the occupancy phase. They also generally require a commissioning agent (CxA) for either fundamental commissioning or enhanced commissioning, to serve as an owner's representative and participate in policy-level project decisions.

Sustainability standards are an important component of a facility's design criteria and the owner's project requirements (OPR). These standards require decisions and support structures as early as the project definition and preliminary design phases. The LEED standards, for example, are subdivided into nine categories:

[Special note to readers – The US Green Building Council manages these standards and may change them at any time. If the need arises, they should be contacted for the most current information.]

In order for a project to comply with LEED standards, many early decisions have to be made to set the stage for the conduct of the project. For example, the EA category has the potential for a high-point score because it includes HVAC equipment. Enhanced commissioning contributes to this score; it involves hiring a CxA during the design phase to oversee all commissioning activities. The CxA reviews the OPR, BOD, and design documents prior to the mid-construction documents phase.

Under the MR category, designers have to lay the groundwork for additional points through the choices of materials. LEED encourages the use of existing sites and buildings, including the reuse of building elements such as walls, floors and roofs. The use of rapidly renewable materials is also rewarded – these include bamboo, cotton insulation, agrifiber, linoleum, wheatboard, strawboard, and cork.

Design approaches for LEED compliance

The USGBC advocates design integration as a means of creating buildings that serve the needs of users, owners, the community, and the environment. LEED should be integrated in the design at the very beginning in order to not only obtain the desired certification but to derive superior building performance as well. The design process needs to include nontraditional team members, a LEED-accredited professional (LEED® AP) and a commissioning agent/authority. In order to incorporate lean principles in a project, a professional who is knowledgeable in lean methodology such as a consultant or lean coach should be included in the design team as well.

The USGBC visualizes the design process as a collaboration of many disciplines with owners and other stakeholders to accomplish holistic design solutions that maximize building performance and a return on investment. It proposes design charrettes to bring multifaceted teams together in the predesign phase of a project to explore economic, environmental, and social issues. Very importantly, contractors and end users should be included in the collaboration, unlike the approach used in many traditional projects. This approach closely parallels lean design, in which contractors and designers resolve concerns that affect constructability and cost effectiveness. LEED charrettes include a wide range of professionals to develop a shared vision of a project, so that the extensive requirements of LEED credits can be incorporated in a design from the very outset. The attainment of LEED credits is by no means accidental; the process is highly structured and involves well-researched and deliberate actions with regard to design parameters.

The LEED standards embody such factors as urban planning, community connectivity, ecology, energy efficiency, materials, air quality, recycling, daylighting and artificial lighting. Consequently, buildings that attain high levels of LEED certification have experienced a process that integrates a very wide range of issues that not only secure the best environment for building users, but that ultimately reduce each facility's impact on the carbon footprint. Energy modeling and life-cycle cost analysis also facilitate important long-term decisions on the way each facility is designed and expected to operate in the future.

Commissioning plays a vital role in this collaboration. LEED standards require commissioning in order to ensure that the credits desired are not only programed into the design of a facility, but are also validated in actual building performance through exhaustive tests. Early use of the Commissioning Agent or Authority (CxA) clarifies the owners' project requirements (OPR) and basis of design (BOD). The CxA also supports lean design by bringing to the team the experience gained on earlier projects that results in better constructability and in the avoidance of errors that are likely to require increased rework in the project. The CxA serves as a resource to facilitate the analysis of design choices to determine their short-term and long-term impacts.

In the earlier years following the promulgation of LEED standards, there was a misperception in the design and construction community that LEED standards resulted in higher initial costs and complicated procedures. With each passing year an increasing number of design and construction professionals have become LEED accredited, and have developed a better understanding of the procedures and processes involved in applying the standards. This has resulted in a reduction of the so-called Green Premium for LEED-based projects. Increasingly, LEED has become a part of the business model.

Lean construction benefits greatly from the foregoing collaborative approach; the design and construction team should also promote lean by avoiding the following "examples of waste related to sustainable construction"

- Ignoring the contractor's knowledge during the design
- Late recognition of green goals – when costs have escalated
- Demolition of a building that could otherwise be renovated
- Failure to reuse/recycle building elements and materials
- Lack of a "Green" design – leads to overdesign

Checklist for environmentally sustainable design and construction

- Smaller is better
- Design an energy-efficient building
- Design buildings to use renewable energy
- Design water-efficient, low-maintenance landscaping
- Look into the feasibility of gray water and roof-top water catchment systems
- Design for future reuse
- Avoid potential health hazards: radon, EMF, pesticides
- Renovate older buildings
- Evaluate site resources
- Locate buildings to minimize environmental impact
- Pay attention to solar orientation
- Situate buildings to benefit from existing vegetation
- Minimize transportation requirements
- Avoid ozone-depleting chemicals in mechanical equipment and insulation
- Use building products made from recycled materials
- Use salvaged building materials when possible
- Minimize use of old-growth timber
- Avoid materials that will off gas pollutants
- Install high-efficiency heating and cooling equipment
- Install high-efficiency lights and appliances
- Install water-efficient equipment

Commissioning and design

The National Conference on Building Commissioning has officially defined "Total Building Commissioning" as: a "Systematic process of assuring by verification and documentation, from the design phase to a minimum of one year after construction, that all facility systems perform interactively in accordance with the design documentation and intent, and in accordance with the owner's operational needs, including preparation of operation personnel."

As described in the second part of Chapter 12, commissioning is a critical component of the lean construction processes. The CxA is a third-party professional that reports to the owner directly, and acts as an extra pair of eyes and ears on the owner's behalf. Commissioning is a planned, systematic, quality control process that involves the owner, users, occupants, O&M staff, design professionals, and contractors. It is most effective when begun at the project's inception. It is essential for a project to be LEED-certified. An important benefit of the commissioning process is the quality assurance provided by ensuring that installation work is faithful to the plans and specifications. Problems and conflicts are detected early when the CxA asks installers to verify installation details at a time when corrective work can be done cost-effectively if it is needed. In the case of HVAC systems, for example, this can be very beneficial; changes to ductwork configuration may avoid conflicts, but airflow quantities may be very difficult to compensate for once ceiling spaces are filled with various plumbing, sprinkler systems, and electrical raceways. The Cx review process ensures that user/client satisfaction is secured with the final product, in keeping with lean principles.

Building performance can be validated by using the commissioning process for quality assurance beyond the requirements of LEED standards. For example, CxA review of the building envelope is not required by LEED, but obligates the contractor to demonstrate that there are no leaks or early points of failure.

Questions for discussion

1 Describe the components of and the process flow for The Lean Project Delivery System™.
2 Describe a staff support structure that can be used for managing lean construction projects.
3 What are the steps in Target Value Design? What are its advantages?
4 Who is the so-called Last Planner? What value does that individual bring beyond the traditional method of project planning and scheduling?
5 What is work structuring? What steps does it involve?
6 What is the purpose of Reverse Phase Scheduling?
7 How can one ensure that the weekly work plan represents the most appropriate work to be performed in a given week?
8 What role does constraint analysis play in The Last Planner® System?

Note

1 This subject is covered in detail in Chapter 12.

References

Ballard, G. 1999. *Work structuring white – Paper #4*. Las Vegas, Nevada: Lean Construction Institute.

Ballard, G. 2000. *The Last Planner® System of production control*. PhD Thesis. Faculty of Engineering. School of Civil Engineering. Birmingham, AL: The University of Birmingham.

Ballard, G. 2008. The Lean Project Delivery System™: An update. *Lean Construction Journal*, 1–19.

Ballard, G., and G. Howell. 2003a. Lean project management. *Building Research and Information*, 119–133.

Ballard, G., and G. Howell. 2003b. An update on last planner. *Proceedings of the 11th annual conference of the international group for lean construction (IGLC-11)*, July. Blacksburg, VA.

Ballard, G., and Z. K. Rybkowski. 2009. *Overcoming the hurdle of first cost: A case study in Target Value Design*. The 2009 Construction Research Congress (Construction Institute of ASCE), Seattle, WA, April 5–7, 10 pp.

Bertelsen, S. 2001. Lean Construction as an Integrated Production. *Proceedings of the 9th annual conference of the international group for lean construction (IGLC-9)*. Singapore.

Ballard, G., and P. Reiser. 2004. The St Olaf fieldhouse project: A case study in designing to target cost. *Proceedings of the 12th annual conference of the international group for lean construction (IGLC-12)*. Elsinore, Denmark.

Cooper, R., and R. Slagmulder. 1997. *Target costing and value engineering*. Portland, UK Productivity Press.

Crosby, P. 1979. *Quality is free: The art of making quality certain*. New York: Mentor.

Demir, S. T., and P. Theis. 2016. Agile Design Management – The application of scrum in the design phase of construction project. *Proceedings of the 24th annual conference of the International Group for Lean Construction (IGLC-24)*, Boston, MA, USA.

Fischer, M., H. Ashcraft, D. Reed, A. Khanzode. *et al.* 2017. *Integrating project delivery*. Hoboken, NJ: John Wiley and Sons, Inc.

Hill, C., K. Copeland, and C. Pikel (eds.). 2016. *Target Value Delivery-Practitioner Guidebook to Implementation – Current State 2016*. Arlington, VA: Lean Construction Institute.

Kerosuo, H., T. Maki, H. Macomber, R. Codinhoto, L. Koskela, and R. Miettinen. 2012. "In time at last – Adaptation of Last Planner tools for the design phase of a building project". *Proceedings of the 20th annual conference of the International Group for Lean Construction (IGLC-20)*, San Diego, CA, USA.

Koskela, L. 2000. *An exploration towards a production theory and its application to construction.* VVT Publications; 408. Online at www.inf.vtt.fi/pdf/publications/2000/P408.pdf.

Koskela. L., and G.A. Howell. 2002: *The underlying of project management is obsolete*, Project Management Institute.

Koskela, L., G. Howell, G. Ballard, and I. Tommelein. 2002. The foundations of lean construction. In *Design and construction: Building in value*, eds. R. Best, and G. de Valence. Oxford, UK: Butterworth-Heinemann, Elsevier.

Macomber, H., and G. Howell. 2003. Linguistic action: Contributing to the theory of Lean Construction, *Proceedings of the 11th annual conference of the international group for lean construction (IGLC-11),* July 22–24. Virginia Tech., Blacksburg, Virginia, USA.

Owen, R., and L. Koskela. 2006. *Agile Construction Project Management.* 6th International Postgraduate Research Conference in the Built and Human Environment (pp. 22–33). Delft: University of Salford, UK.

Rybkowski, Z. K. 2009. *The application of root cause analysis and target value design to evidence-based design in the capital planning of healthcare facilities.* Doctoral Dissertation, University of California, Berkeley.

Schwaber, K., and J. Sutherland. 2017. The Scrum Guide™ – The Definitive Guide to Scrum. https://trello.com/b/518rby88/most-basic-trello-scrum-board [presented at the 2019 LCI Design Forum Scrum Workshop].

Bibliography

Alarcon, L. F., A. Grillo, J. Freire, and S. Diethelm. 2001. Learning from collaborative benchmarking in the construction industry. *Proceedings of the 9th annual conference of the international group for lean construction (IGLC-9),* August 6–8. Singapore, University of Singapore.

Al-Sudairi, A. 2004. Simulation as an aid tool to the best utilization of lean principles. *12th annual conference on lean construction,* August 3–4. Helsingor, Denmark.

Alves, T. C. L., and I. D. Tommelein. 2003. Buffering and batching practices in the HVAC industry. *Proceedings of the 11th annual conference of the international group for lean construction (IGLC-11),* July 22–24. Virginia Tech, Blacksburg, Virginia, USA.

Arbulu, R., and A. Koerckel. 2005. Linking production-level workflow with materials supply. *Proceedings of the 13th annual conference of the international group for lean construction (IGLC-13),* July. Sydney, Australia.

Ballard, G., and G. Howell. 1994. Implementing lean construction – Stabilizing work flow. *Conference on lean construction,* September. Santiago, Chile.

Ballard, G., and G. Howell. 1995. Toward construction JIT. *Proceedings of the 1995 ARMCOM conference, association of researchers in construction management.* Sheffield, England.

Bertelsen, S. 1993. *Construction logistics I and II, materials management in the construction process (in Danish).* Kobenhavn, Denmark: Boligministeriet, Bygge-og, Boligstyrelsen.

Bureau of Labor Statistics. 2006. *Fatal and non-fatal occupational statistic from 1992–2004.* Online at www.bls.gov.

Fiallo, M., and V. Revelo. 2002. Applying the last planner control system to a construction project: A case study in Quito, Ecuador. *Proceedings (IGLC-10),* August.

Gwynn, P. 2008. *ULSD lean production management (LPM).* Lean Implementation Services, Schereville, IN: Author.

Howell, G. A. 1999. What is lean construction? *Proceedings of the 7th annual conference of the international group for lean construction.* Berkeley: University of California.

Howell, G. A. 2000. White paper for Berkeley/Stanford CE & M workshop. *Proceedings of the construction engineering and management workshop.* Palo Alto, CA: Stanford University.

Howell, G. A., A. Laufer, and G. Ballard. 1993. Interaction between subcycles: One key to improved methods. *Journal of Construction Engineering and Management, ASCE* 119(4).

Koskela, L. 1992. *Application of the new production philosophy to construction.* 87. Technical Report #72. Stanford University: CIFE.

Koskela, L. 1999. Management of production in construction: A theoretical view. In *Proceedings of the 7th annual conference international group for lean construction (IGLC-7),* ed. Iris Tommelein, July 26–27. Berkeley: University of California.

Lichtig, W. A. 2005. Sutter health: Developing a contracting model to support lean project delivery. *Lean Construction Journal* 2(1): 105–112.

Macomber, H., and G. Howell. 2005. *Using study action teams to propel lean implementations.* Louisville, CO: Lean Project Consulting Inc.

Macomber, H., and G. Howell. 2005. 5 necessary actions for change, 1–2. *Lean project consulting,* Louisville, CO 80027.

Macomber, H., G. Howell, and J. Barberio. 2005. Target value design: Seven foundational practices for delivering surprising client value, 1–2. *Lean project consulting,* Louisville, CO 80027.

Matthews, O., and G. Howell. 2005. Integrated project delivery an example of relational contracting. *Lean Construction Journal* 2(1): 46–61.

Moser, L., and A. dos Santos. 2003. Exploring the role of visual controls on mobile cell manufacturing: A case study on drywall technology. *Proceedings of the annual conference of the international group for lean construction (IGLC-11)*. Blacksburg, VA. Online at http://strobos.cce.vt.edu/IGLC11

Salem, O., J. Solomon, A. Genaidy, and I. Minkarah. 2006. Lean construction: From theory to implementation. *Journal of Management in Engineering, ASCE,* 168–175.

Spearman, M. L., D. L. Woodruff, and W. J. Hopp. 1990. CONWIP: A pull alternative to Kanban. *International Journal of Production Research* 28: 879–894.

Tsao, C., I. Tommelein, E. Swanlund, and G. Howell. 2004. Work structuring to archive integrated product, process design. *Journal of Construction Engineering and Management* 130(6): 780–789.

Tsao, C. C. Y., I. D. Tommelein, E. Swanlund, and G. A. Howell. 2000. Case study for work structuring: Installation of metal door frames. *Proceedings of the 8th annual conference of the international group for lean construction (IGLC 8)*, July 17–19. Brighton, UK.

Vrijhoef, R., and L. Koskela. 1999. *Role of supply chain management in construction (IGLC 7)*. July 26–28. Berkeley: University of California.

Womack, J., and D. Jones. 1996. *Lean thinking*. New York: Simon & Schuster.

Womack, J., D. Jones, and D. Roos. 1990. *The machine that changed the world*. New York: Harper Collins.

4A

LEAN IN THE CONSTRUCTION PHASE OF THE LPDS

The previous chapter provided an overview of the lean project delivery system (LPDS) with an emphasis on the programming and design phase of projects. It also presented the Last Planner® System from the perspective of design, in preparation for the construction phase of projects. This chapter presents lean in the construction phase of the LPDS, including lean supply, lean assembly, and use. The Last Planner® System is described extensively with regard to the joint activities of different specialists in the project team as they plan project execution typically from week to week; using reverse phase scheduling or pull planning to improve work flow predictability and reliability.

Whereas traditional industry practice has separated the roles of designers and constructors, the LPDS sees the activities of these professionals as a continuum for project management to achieve three fundamental goals (Koskela 2000):

- Deliver the product
- Maximize value
- Minimize waste

Fundamental to the LPDS is the deployment of the "Five Big Ideas" as described in Chapter 3:

1 Collaborate, really collaborate
2 Increase relatedness among all project participants
3 Projects as networks of commitments
4 Optimize the project, not the pieces
5 Tightly couple learning with action

The first big idea requires assembling and empowering of all the people and resources necessary to develop appropriate design solutions and explore their impacts on construction delivery. Inherent in the use of the LPDS is also the use of The Last Planner® System (LPS). It comprises

1 The master pull schedule
2 The lookahead schedule
3 The weekly work plan (WWP)

As described in Chapter 3, part A The LPDS comprises a number of phases that capture the intent of the traditional project phases, but juxtaposes them in such a manner as to apply production system design principles to enhance the delivery of the entire project from predesign to completion and use. The phases are:

1 Project definition
2 Lean design
3 Lean supply
4 Lean assembly
5 Use/completion

The LPDS model represents these phases as a series of overlapping triads. As described by Ballard and Howell (2003a), these phases influence each other "so a conversation is necessary among the various stakeholders." In the diagram (Figure 4A.1), work structuring provides a foundation for the process and sets the stage for production control, which is represented by a horizontal bar. Work structuring is a term developed by the Lean Construction Institute to indicate process design. It is a process of subdividing work such that pieces are different from one production unit to the next to promote flow and throughput, and to have work organized and executed to benefit the project as a whole. Production is redefined from "monitoring results" to "making work flow according to plan, re-planning when it cannot." Plans to complete are continually updated in a proactive rather than reactive mode, looking ahead instead of backward.

The LPDS was updated to include additional concepts and methods drawn from the Toyota Product Development System, especially target costing and set-based design. These have been adapted to the construction industry and integrated with computer modeling and relational forms of contract (Ballard 2008).

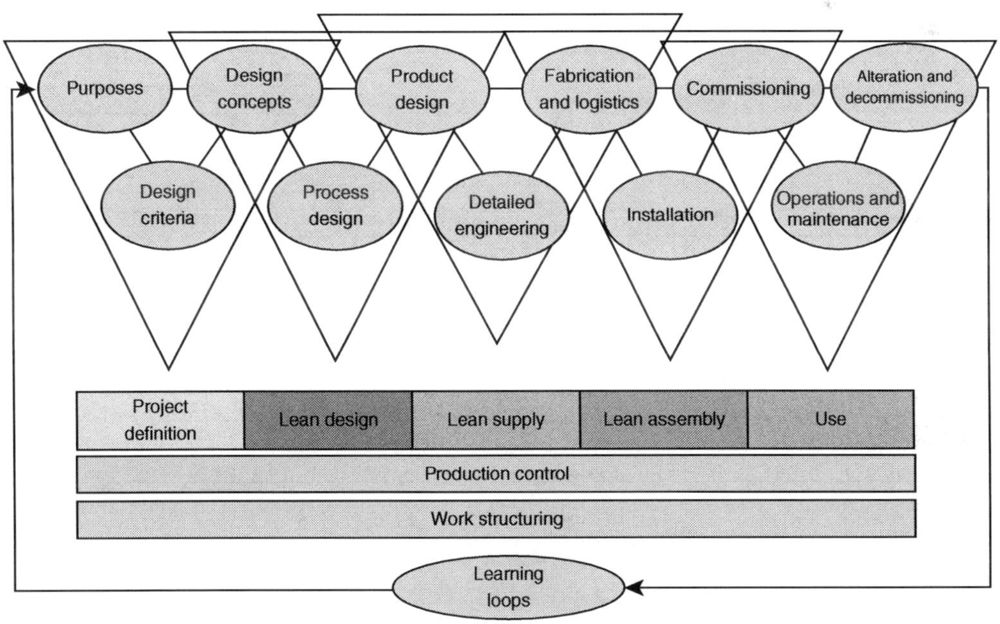

Figure 4A.1 The Lean Project Delivery System™

Source: Adapted from Ballard, G., and G. Howell, *Building Research and Information*, 2003a

The LPDS model improves project delivery through the following characteristics:

1 Downstream stakeholders are involved in front-end planning and design through cross-functional teams.
2 Project control has the job of execution as opposed to reliance on after-the-fact variance detection.
3 Pull techniques are used to govern the flow of materials and information through networks of cooperating specialists.
4 Capacity and inventory buffers are used to absorb variability.
5 Feedback loops are incorporated at every level, dedicated to rapid system adjustment (i.e., learning).

Lean supply

The first two triads of the LPDS, i.e., project definition and lean design have already been discussed in the preceding Chapter 3. The third triad, lean supply, comprises product design, detailed engineering, and fabrication and logistics, Product design overlaps with the lean design triad; it represents the conversations inherent in the LPDS between different actors to ensure a smooth transition from one phase to the next. The model requires up-front product and process design to define what is needed and when it should be delivered. This is especially important with engineered to order components as utilized in EPC projects. Lean supply also includes reducing the lead time for project information requirements.

Traditional projects depend on procurement specialists or buyers to ensure that materials are available for installation; they operate as a functional silo that is decoupled from project workflow. Flaws in this process result in material shortages or incorrect materials on site when they are needed that often cause serious project delays and losses in construction quality. Alternatively, material inventories accumulate, creating an unsafe condition on-site, while tying up scarce capital. These situations represent examples of construction waste. Lean supply addresses these problems through three main approaches (Arbulu and Koerckel 2005): (1) improving workflow reliability – maintain constraint identification and removal, (2) using web-based project management software to increase transparency across value streams, and (3) linking production workflow with material supply.

Supply chain management and lean construction

Supply chain management (SCM) has saved hundreds of millions of dollars in manufacturing while improving customer service through taking a systems view of production activities of autonomous units. It is proposed that similar savings could be derived in construction, especially since subcontractor and supplier production account for the largest percentage of project costs. Studies have indicated project cost increases of up to 10% because of poor supply chain practices.

The term supply chain encompasses all the activities that lead to having an end user provided with a product or service – the chain is comparable to a network that provides a conduit for flows in both directions, such as materials, information, funds, paper, and people. The main elements of SCM are information flow, order fulfillment, and product development with faster response times, less waste, more effective information flow, and smaller amounts of inventory. A lean supply chain eliminates waste by using efficient methods to procure materials and equipment. The physical nature of material used in construction makes it very susceptible to several categories of waste, such as excessive transportation, excessive waiting on deliveries, defective material that cannot be used, excessive inventory to have material available, overproduction, and safety problems in excessive handling and site congestion. In many cases 30 to 50% of a project's schedules is dedicated to having materials and equipment delivered to a site. The design process can support the supply chain by using standardization to eliminate waste as well. Studies by Bertelsen (1993) indicated project cost increases of up to 10% because of poor supply chain design. Comparisons

Table 4A.1 A comparison between lean supply chains and traditional procurement

	Lean Supply Chains	*Traditional Procurement*
1	Based on systems thinking	Purchasing function not systems based
2	Factored in product and process design	Standard procedure –
3	Tries to match resources with site needs	Seeks lowest price through competitive bidding
4	Seeks overall cost savings via value stream strategy.	Narrow focus on balancing price quantity and quality
5	Early supplier involvement	Late supplier involvement
6	Pulls resources where possible	Pushes resources at the site
7	Material consolidation through centralized logistics centers	Independent warehouses/inventories
8	Multi-project view	Individual contract view
9	Supply channelized to project needs	Project constrained by supply
10	Long term strategy for procurement	Short sighted thinking and control
11	Material supply is synchronized through the LPS.	Material supply decoupled from workflow

Source: Adapted from AGC course Unit 5

with previous research (Jarnbring 1994) on SCM of construction projects in Sweden indicated a potential for cost savings of 10–17% due to inefficiencies caused by lack of coordination between contractors and suppliers.

Supply Chain Management analyzes the impact of facility design on the construction process and enables superior project planning and management, avoiding the fragmented approach of other methods. Through SCM, all parties are kept aware of commitments, schedules, and expedites all work as a virtual corporation that can source, produce, and deliver products with minimal lead time and expense. The SCM application needs to be tailored to the conditions in the geographic area and environment in which projects are executed. Construction supply chains are subject to inefficiencies caused by interdependency with causes in various stages of the chain, primarily due to self-serving actions by individual companies and organizations, as well as divisions of the same organization.

The implementation of SCM in construction is challenging. The industry trend toward subcontracting has resulted in specialization and fragmentation; each subcontractor tends to act in its own interest and the relationships between the various parties often becomes adversarial. The sharing of information is critical for both SCM and lean construction. Central to their business model is the importance of making and keeping commitments. A general contractor (GC), for example, should be able to extract a reliable commitment from an air conditioning subcontractor that a chiller has been ordered from a specified supplier. Using an online information system, it should be possible for the GC and the owner's project manager to verify the supplier's delivery schedule, although pricing information could be kept confidential. The dimensions and capacities of the chiller could be verified long before delivery to ensure that it will be delivered and installed in seamless agreement with the construction schedule. This system allows a JIT approach to be used, so that the chiller is not delivered too far in advance, when on-site storage requirements may become costly and inconvenient. By the same token late deliveries would be avoided. Unavoidable deviations from the construction schedule would be shared with all stakeholders and needed adjustments made.

Supply chain systems facilitate the sharing of resources, staff and expertise, problem solving, improved economic performance, and increased innovative capacity. Supply chain systems also facilitate the tracking of work performance, resource utilization and provide feedback on workforce productivity. This feedback is indispensable for enabling continuous improvement efforts and corrective actions to be taken on a timely basis.

Three case studies were examined in the Netherlands and Finland, to apply different methods of analysis (Vrijhoef and Koskela 1999). The breakdown of the cases was as follows:

Case 1: The installation of concrete wall elements in residential construction. Time buffers in the supply chain were quantitatively assessed.
Case 2: The façade elements in residential construction. The control systems in the supply chain were qualitatively evaluated.
Case 3: A study of the cost impact of the contractor's method of materials acquisition.

The results were informative. Large time buffers developed that were 70–80% of the total lead time. Controllability problems were caused by the parties near the beginning of the chain separate from the activities being studied. This excessive value was caused by adversary relations between the parties and lack of goal congruence. These results pointed to self-interested actions by the parties.

Vrijhoef and Koskela (1999) propose a four-stage approach similar to the Deming cycle to improve construction supply chain performance: Supply chain assessment, supply chain redesign, supply chain control, and continuous supply chain improvement. This involves first evaluating a supply chain to identify waste and problems and determine root causes. The supply chain is then redesigned to address shortcomings, using analytical and feedback systems to monitor system performance, measure waste, and apply corrective measures. This is followed by continuous improvement.

One important function of the supply chain is to ensure the appropriate timing of material acquisitions to avoid excess material inventories; ideally, forecasts of materials requirements should be accurate enough to facilitate JIT in construction (Arbulu and Koerckel 2005). Material costs may exceed 50% of project costs, hence supply chain improvements that benefit material management greatly improve project performance.

Materials fall within three categories that must each be handled differently:

1 Made to stock items (MTS). These are commercial items that are made to meet market forecasts. Examples are consumables such as nails and small tools.
2 Made to order items (MTO). These are standard items such as rebar and concrete that are easily produced to meet project needs.
3 Engineered to order (ETO). Items are customized to the project; they include steel trusses, HVAC ductwork, systems piping for chillers, and plumbing.

A study by Arbulu and Koerckel (2005) on the Heathrow Terminal 5 (T5) project in the UK identified several systems improvements in material supply practices.

A concrete supply operation experienced untimely deliveries throughout the day because of a lack of transparency between site operations and the concrete batch plant. A coordinator was responsible for placing orders but based his requests on guesstimates; reliability worsened during his absences. The ordering process was unreliable (50–80%) Overlapping orders from different project members at the same time led to overtime operations for both the plant and the project. A computerized system was implemented to integrate concrete orders across the project, significantly reducing the need for ad-hoc orders. Productivity improvement during the first weeks of implementation saved 5,000 man hours or 100,000 British pounds.

The T5 project used expanded polystyrene panels to minimize movement in floors installed over soil with high clay content. These panels varied in thickness and were made-to-order materials. Lack of transparency between site operations and the fabrication operation found production levels greatly exceeded requirements and resulted in excessive inventory of a fragile product that could not be double handled. The excessive storage was maintained in trailers or returned to the factory that also held excessive quantities.

Value stream mapping identified the deficiencies in the supply chain; system improvements included: storage trailers used as a kanban system – empty units generated orders, full trailers indicated no need to produce.

- Order sizes were reduced to trailer capacity.
- A rolling 2-week lookahead of orders was maintained.
- Shorter lead production times were reduced to 5 days.
- Trailers were reallocated weekly between project teams based on forecasts of demand.

The kanban system yielded significant benefits. Inventory capacity was reduced to one day's usage, down from several weeks. This resulted in overall cost reductions of hundreds of thousands of British pounds for the duration of the project.

Kanban and lean construction

As described above, Pull/Kanban is a system of cascading production and delivery instructions from downstream to upstream activities in which the upstream supplier does not produce until the downstream customer indicates a need. It is an essential part of the just-in-time (JIT) process. In JIT, production is driven by customer demand; that is, the needs of the next trade downstream as opposed to "pushing " to meet schedules. Essentially, nothing is built until there is an order for it. JIT avoids wastes such as inventory of raw materials, work in process, or finished assemblies that are not yet needed. The JIT system produces only what is needed, when it is needed, and in the quantity that is needed. Therefore, no goods are produced that cannot be sold at a price that is cost effective for the producer.

In its original form, Kanban was a card-based system that improved process management through visual control. Kanban cards follow a product through various production stages, maintaining information on the product name, number, due dates, and so on. Kanban is a visual system, providing status information at all times. In contemporary applications, the card has been replaced by a computer-based system, but the purpose is the same. The kanban tells the supplying process exactly what it wants and how many. The supplying process is not authorized to make more product until the kanban advises it to do so.

The Toyota system utilizes two different types of kanban – the withdrawal kanban and the production kanban. The withdrawal or "move" kanban authorizes the movement of materials or work in process from one process to another. The production kanban authorizes a process to produce another lot of the required size. It also specifies the description of the items, the materials required, the production work center, and so on. The following rules enable the kanban system to work successfully:

1 Do not send defective product to the subsequent process. Stop the process and determine the reason for the defects.
2 The subsequent process withdraws only the specific items needed, when they are needed.
3 Produce only the exact quantity required by the following process. Do not produce more than is authorized.
4 Smooth the production load.
5 Adhere to kanban instructions while fine tuning.
6 Stabilize and rationalize the process.

The supermarket concept in lean supply

This concept is based on having a logistics center close to a construction site that is comparable to a supermarket that understands its customers' needs and maintains necessary supplies and subassemblies to be used by installers as needed. This is very cost effective in eliminating the need to consume valuable workers' time in locating material – several times daily in some cases. By understanding inventory needs based on the type of project and the stage in construction, logistics centers provide optimal project support. Logistics centers can also be used for pre-assembly of components like faucets to minimize installation time when they are needed. They can also serve as a "decoupling point" to receive items that are not yet needed on a site. This also reduces clutter on a site and the risk of damage or loss. Coordinating these activities with BIM can enable these strategies to approach a Just-in-time capability.

Lean assembly

Lean assembly is practiced in the actual construction of a project, putting materials, systems, and components in place to create a completed facility. The objective of lean assembly is to assemble building elements in the most efficient manner possible, both onsite and offsite. It can be accomplished in a number of ways, such as through prefabrication offsite where the work can be done in a controlled environment like a factory, and then transported to the site. Piping systems, for example, can be assembled in sections in a prefabrication facility, tested under high quality standards, then shipped to a site on a just-in-time basis. As described later in this chapter, in The Last Planner® System of production control (LPS), work structuring culminates in the form of schedules that represent specific project goals. Schedules are created for each phase of the project, beginning at the design phase and ending at project completion. The production control provided by the LPS deploys the activities necessary to accomplish those schedules. As shown in the diagram (Figure 4A.1) production control and work structuring refer to the management of production throughout the project.

The lean assembly triad contains fabrication and logistics, installation, and commissioning. Prefabrication is a production technique that can enable a contractor to operate in a lean manner by reducing the many nonvalue-added steps that are required in field fabrication. Whereas field work contends with uncertain conditions such as weather, and limitations in the availability of skilled labor, material, and equipment, shop fabrication benefits from a predictable, controlled environment.

Preassembly is a very productive method; the greater the number of components that are pre-assembled prior to their final installation, the more straightforward the final assembly process becomes, provided, of course, that assemblies can be managed logistically.

Standardized and interchangeable parts provide significant process improvements. The repeated use of standardized parts greatly eases assembly; not only will crews be familiar with the parts, they will also be able to learn from their repeated use.

Limbach LLC is an integrated building systems provider – managing all components of mechanical, electrical, plumbing, and control systems from systems design and construction through service and maintenance. Limbach engineers, constructs and services the above mentioned systems in both new and existing buildings and infrastructure. They work for building owners in the private, not for profit, and public/government sectors. A Limbach prefabrication project is described in Section C of the chapter.

Tweet/Garot, a specialty contractor in Wisconsin has improved efficiency by prefabricating many of its products for quick and efficient site installation. This includes plumbing work for bathrooms in a given building in congruence with the construction schedule. Over a period of time, Tweet/Garot has progressed to 70% installation in the field and 30% fabrication in the shop. With that new approach, workers spend a relatively short period of time doing installation work. Prefabricated piping is interconnected within minutes because of the precision of design.

The commissioning process provides quality assurance before a facility is completed and accepted by the owner by ensuring that all systems have been installed as promised in keeping with the designer's plans and specifications. It improves the probability that the facility will meet the owner's project requirements (OPR) and performs as expected, to provide user satisfaction.

Use

The "Use" triad of the LPDS refers to a completed facility, and contains the activities Commissioning, Operations and Maintenance, and Alteration and Decommissioning. Following successful commissioning, the facility should undergo a protracted operations and maintenance phase as it is used. "Alteration and Decommissioning" refers to a future activity when the facility may be repaired, renovated, or taken out of service.

"Learning loops" refer to the application of root cause analysis to construction activities to review PPC values and commitment reliability. In this review project teams determine the reasons why the tasks planned for a particular week may not have been accomplished. If errors are detected in this review, the team endeavors "not to make the same mistake in succeeding weeks". The causes are identified so that they can be avoided in the future. This behavior differs significantly from that in traditional projects, where errors tend to be punished instead.

Learning is also accomplished through retrospectives which are scheduled events (Seed 2018). They involve reflecting on past events and considering how they affect project outcomes. Convenient times for retrospectives may be after the first two months, at significant points in a schedule, when major team members leave or join a project. The reaching of major project milestones is a good opportunity for a retrospective to evaluate how well the project or program objectives are being met. The time frame involved may vary from five minutes for a "plus Delta" review to several hours for a more detailed review. Longer events should be carefully planned and facilitated in order to capture the needed information in an effective manner. Post occupancy evaluation is a structured and systematic method for learning lessons from past construction projects. As described in Chapter 15, it can be designed to cover a very wide of topics, including how well design decisions have met the needs of building users and other stakeholders. It has the disadvantage of being administered approximately one year after the completion of a facility.

Production control and work structuring

Production control is represented in the LPDS as a horizontal bar that extends from the very inception of the project to its conclusion (Figure 4A.1). It consists of work flow control and production unit control. Work structuring and production control are used throughout a project to manage production (Ballard 1999). The term work structuring is a process of subdividing work so that pieces are different from one production unit to the next to promote flow and throughput, and to have work organized and executed to benefit the project as a whole. Work structuring is described in greater detail as a part of the LPS.

Sustainability issues in the construction phase[1]

The adoption of Green Building standards may require a number of procedures to be integrated with the construction activities and schedules for a project in order to ensure that a facility attains the energy initiatives it has been designed to meet. For example, on-site practices may in some cases require incomplete ductwork for HVAC systems to be sealed off to guard against contamination. The Commissioning Agent (CxA) must be equipped to verify that such precautions have been taken.

It is highly recommended to have a commissioning plan that provides for a review of documentation during the design phase. It should also provide for a commissioning authority or agent to attend specific site meetings, review contractor submittals, and verify that the LEED standards are being met.

Depending on the size and complexity of a project, a commissioning budget may need to be established. The CxA essentially works for the project owner, essentially serving as extra eyes and ears and providing quality assurance in a support role in order to avoid any possible conflicts in the administration of the project. LEED projects generally require an accredited professional to serve as a resource in a project from the definition phase through to the occupancy phase. They also generally require a commissioning agent (CxA) for either fundamental commissioning or enhanced commissioning, to serve as an owner's representative and participate in policy-level project decisions.

LEED requires fundamental commissioning in order to verify that related inspections have been carried out during the construction phase. Enhanced commissioning may also be required, depending on the level of LEED compliance involved. The contractor may find it advantageous to have a staff person such as a LEED-AP in a support role to ensure that required sustainability standards are met.

The Last Planner® System in construction

This system is an important subset of The Lean Project Delivery System™ (LPDS) and is critical to its effective deployment. It accommodates project variability and smooths work-flow so that labor and material resources can be maximally productive. In conventional construction management, there is a tendency for project managers to assign work schedules beyond crews' ability in the hope they can complete it. Research has shown that in traditional construction projects only approximately one-half of the tasks assigned for a given week are likely to be completed in that week. Ballard and Howell (1998) published measurement data on work flow variability; their studies of a wide range of projects identified an average plan failure rate of 54%.

Flow variability greatly influences lean practices, as a delay in work completion by one trade directly affects the downstream activities of the next trade. The LPS uses lean methods to provide improved project control. In essence, the last planner is in the best position to match labor and material resources to accomplish assignments in response to downstream demand. Work planning may establish time frames but is not very effective in establishing that the tasks assigned are capable of completion. Stops and starts typically occur or making do (i.e., keeping occupied with noncritical tasks). The LPS technique as refined by Ballard (2000) decentralizes decisions and empowers the crews that are in direct contact with the work to plan and schedule detailed tasks; in effect, they become the last planner. Foremen are empowered; they have a duty to say no to an assignment that fails to meet agreed criteria (Howell 1999).

Principles:

- Assumption: Construction project work plans are really forecasts, on the other hand forecasts are often wrong.
- Planning is more detailed the closer one gets to doing the work.
- Plans are made in collaboration with those who will do the work.
- Constraints should be identified and removed by a team effort.
- Construction team members must make reliable promises.
- Plan failures/breakdowns should be treated as an opportunity for learning, not for negative actions.

The LPS is based on three or four levels of schedules and planning tools. Certain projects such as fast-paced EPC projects and/or projects involving engineered-to-order components may require a daily work plan (Figure 4A.4):

1 The master pull schedule
2 The lookahead schedule
3 The WWP
4 A daily plan (when required)

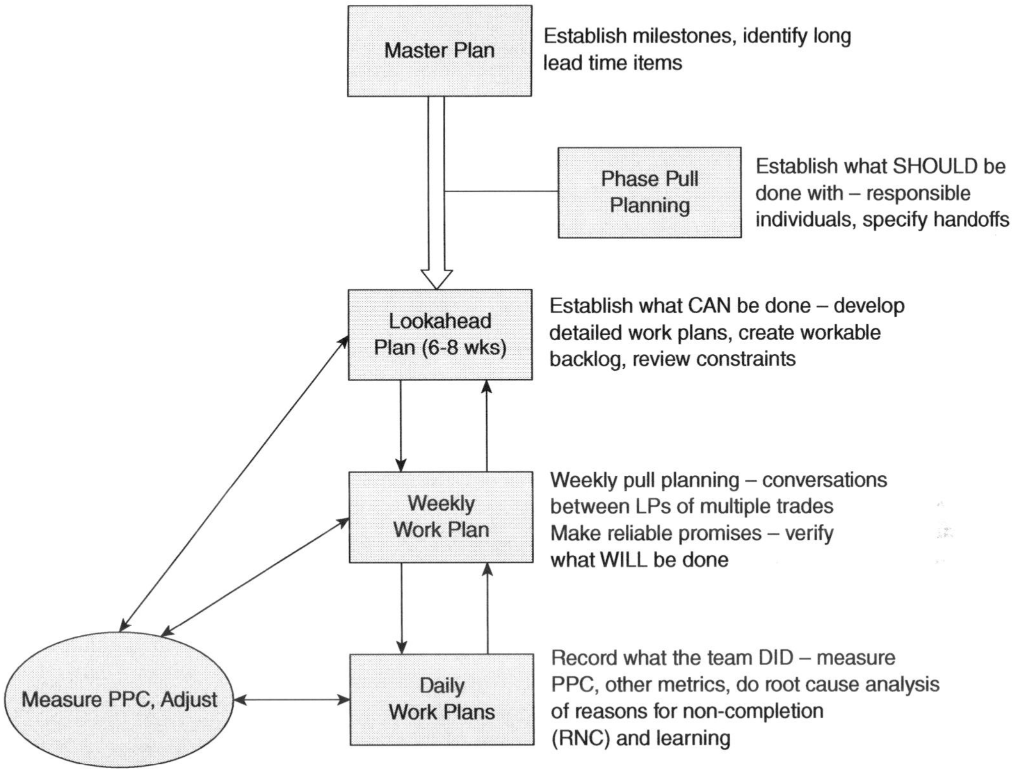

Figure 4A.2 Activities and information flows in The Last Planner® System

The master schedule

The master schedule represents an overall view that identifies major project phases and documents milestones for these phases. In traditional construction, the master schedule is typically included in the contractor's bid. Hence it is an optimistic forecast of the major stages in carrying out a project, based on the early information provided in the contract documents.

Lookahead schedule

The lookahead schedule is derived from the phase plan, based on a time frame of 6–8 weeks for construction tasks, and is used for work flow control. It uses items pulled from the master pull schedule.

Weekly work plan

The WWP is derived from the lookahead plan. It provides a detailed work plan specifying hand offs at each phase between trades. It is updated weekly, based on discussion between the last planners for various trades/disciplines; they may be design team leaders, trade foremen, or supervisors, depending on the complexity of the project.

Daily work plans

Some projects such as engineer, procure, and construct (EPC) construction that comprise engineered-to-order components need more precise control than is offered by the WWP. Such engineered components are designated for installation by specific trades in specific locations. The emergence of constraints can quickly disrupt the construction schedule; in a fast moving project, a week's delay may result in significant schedule deviation. Establishing a daily commitment enhances coordination with other trades.

The so-called last planner is the foreperson or other professional who prepares the weekly (or daily) planner schedule. This schedule also includes a buffer of work activities based on future work. The weekly accomplishment is measured as % planned, complete (PPC). The PPC is based on 100% completion within a given week; tasks are considered to be either complete or not complete – there is no partial credit. The last planner also determines physical, specific work *assignments* for the next day. Work selection rules shield production from upstream uncertainty and variation. Production control is aimed at the avoidance of waste, such as waiting for someone to complete a task after the scheduled completion date or time.

Weekly meetings are used to evaluate the team's performance during the preceding week (i.e., what the team DID). Root cause analysis is carried out to determine the reasons for underperformance, and the knowledge gained is used to improve the process used in assigning work in subsequent weeks.

In the construction phase communication between trade specialists is an essential requirement for the LPS; many assignments require a close collaboration between various trades. For example, erecting a structural wall that contains electrical wiring and plumbing requires having electricians and plumbers work closely with form work carpenters and masons to ensure that conduit and piping are positioned and secured in time for a concrete pour. If either the electrical or mechanical trade fails to carry out their prescribed assignment on schedule, the concrete pour may be delayed – at great cost. Completing the pour without one of those trades may require time consuming corrective work, such as cutting a chase in the wall to retrofit the necessary pipes and/or conduits.

As has been noted previously, improving the efficiency of individual activities does not necessarily improve the efficiency of an overall process. Waste can occur in the hand offs from one activity to another in the form of delays, defects that need to be corrected, and so on. It must also be remembered that removing waste does not necessarily add value.

Requirements for successful application of the LPS in the construction phase

The lean construction philosophy views a project as a promise delivered by people working in a network of commitments supported by linguistic action (Macomber and Howell 2003). Smooth work flow is dependent on having the parties make and keep promises to carry out assignments. Reliable promises downstream enable others (especially upstream) to make reliable promises as well, in anticipation of predictable workflow. When all the parties keep their promises, waste is reduced, productivity is increased, and projects can be completed more rapidly. Promise-keeping by the parties is based on their ability to meet their promises. At the level of the WWP, the trades in question must be aware of the scope of upcoming assignments. They must use this knowledge to determine the resource requirements – labor, equipment, materials, information, and so on. Above all, they should ensure they can deliver on their promises.

The reliability of the LPS and lean construction in general, hinges on informed commitments to maintain the trust that is essential for avoiding waste. If subcontractor (Sub) B depends on Sub A to have a work area available by a given day, then Sub B should be able to mobilize crews for that day without the reservation that Sub A might not keep its promises. Should that occur, Sub B could incur significant waste by having workers report for duty without being able to carry out the scheduled tasks. This waste would not only penalize Sub B, but would contribute to an ongoing spiral of distrust that characterizes traditional construction. Each trade schedules hesitantly, cautiously optimistic that other supporting trades may keep to the schedule, but anticipating more often than not that they will have to fall back on other tasks in order to keep the work force busy. This is the waste of "making do". Being busy does not necessarily advance the progress of work on the critical path.

The LPS of production control sets up and manages schedules in order that:

- The flow of work is structured for the project.
- Workflow is planned within the structure of the work.
- Work is controlled to accomplish the plan.
- Weekly measurement evaluates accomplishment and permits fine tuning for improvement.

Creating a support system for managing the lean process

A support system with assigned responsibilities is necessary for managing and sustaining the lean process; its size would depend on the size and complexity of the construction project. On a medium to large-sized project, the PM, field engineers, schedulers, and so on serve in a staff role and become facilitators and guardians of the lean process. They remove constraints and focus the team on continuous performance improvement. They maintain the master schedule and lookahead schedules. They track the attainment of the master schedule by removing constraints and supporting the forepersons/last planners, and they enable the daily work plans to have a higher level of reliability. They track major measures such as earned value. Forepersons commit to performing work based on confirmation that resources or materials are available. They also need to confirm that prerequisite work has been done.

In the case of projects that involve a high percentage of engineered-to-order materials or systems, assembly is a very carefully ordered process that may require a fourth level of scheduling (i.e., a daily schedule; Materials and/or systems are designated for specific locations, with very little interchangeability. Examples of this type of project include processing plants and petroleum refineries, which involve large quantities of pipe sections that are engineered for a specific purpose and location. Items such as process piping and ductwork are configured to fit with other building elements. It is critical to have the appropriate systems available for installation in their respective locations. For this reason, daily planning is important for that type of project. Daily planning can also benefit AEC projects.

Information technology support would be needed for contemporary projects requiring the management of large quantities of data. Close collaboration is needed between multiple trades to ensure that hand offs from one work activity to another occurs without extensive delays. Premature starts on a successor activity also represent waste; hence the information on all is voluminous and requires computer support. Several software systems can be customized to the lean environment. Strategic Project Solutions is one of the systems that have been used successfully.

Lean construction applies the principle of autonomation that is an important ingredient of the Toyota Production System; the people closest to the work are empowered to stop production if they determine that the upstream production is defective. The last planners and their crew fulfill this role.

Work structuring

Work structuring and production control are used throughout a project to manage production. It is a process of subdividing work so that the pieces are different from one production unit to the next to promote flow and throughput, and to have work organized and executed to benefit the project as a whole. Work structuring may be applied to both design processes and the construction processes.

In The Last Planner® System of production control, work structuring culminates in the form of schedules that represent specific project goals. Schedules are created for each phase of the project, beginning at the design phase, and ending at project completion. The production control provided by the LPS deploys the activities necessary to accomplish those schedules. In work structuring, a project is viewed as being comprised of production units (i.e., an individual or team performing construction tasks), and units of work that are capable of being handed from one production unit to another. Production units add value to each unit of work until it is completed (Tsao *et al.* 2000; Al Sudairi 2004). Work structuring requires determining:

- How work will be subdivided into units and assigned to specialists
- The sequence of these units
- The rules for releasing work from one production unit to the next
- Whether work will be performed in continuous flow or decoupled
- The locations and sizes of decoupling buffers The timing of various work units

Work structuring influences the size of work packages; that is, the quantity of work tasks planned for execution and this in turn affects the amount of work in progress (WIP). Koskela (1999) used the formula: Cycle time = WIP / Throughput.

If the throughput is maintained constant, then reductions in WIP reduce cycle time.

This relationship was studied with a construction project at King Faisal University, Saudi Arabia. It was observed that reduction of WIP resulted in reduced cycle time; however, diminishing returns could result by having too many tasks. As the number of tasks increases, crews may need to spend a disproportionate amount of waiting time between tasks or in moving from one task to another. In a project involving cast-in-place concrete processes, for example, the curing time caused excessive delays between steps (Al Sudairi 2004).

Workable backlog

Workable backlog is used to describe assignments that have met all quality criteria, but may need other prerequisite work to be done before they can be started. For example, the paving of the tennis court for a new university sports center is scheduled for 3 months' time. It can be done now, as it is on a remote part of the site that has already been compacted and graded. It is not on the critical path. This assignment will be maintained as workable backlog.

Workable backlog enables crews to continue working productively if there is a constraint that prevents the completion of an item on the WWP. As a rule of thumb it is recommended in committing crews to no more than 75% of their capacity when developing the WWP. The crew can maintain its reliability in passing work on to the downstream crew, when called to do so as they "pull" activities, even if the weekly (or daily) schedule is interrupted.

Process steps in the Last Planner® System of production control

The LPS as depicted in Figure 4A.5 consists of a number of activities that are noted in the respective symbols. These symbols represent three components that work together to provide production control: (1) lookahead planning, (2) commitment planning, and (3) learning.

Comparison with traditional planning and scheduling

Planning and scheduling are a standard requirement for construction projects in general. As construction activity is comprised of many activities and tasks, planning is a formal process that identifies what needs to be done, what resources are needed (such as labor, materials, and equipment), and when they are required. A schedule represents the plan in graphical form; typically a bar chart or critical path method (CPM) chart documents each activity on a horizontal time scale. This scale is graduated to represent days, weeks, months, or even years depending on the size and complexity of the project.

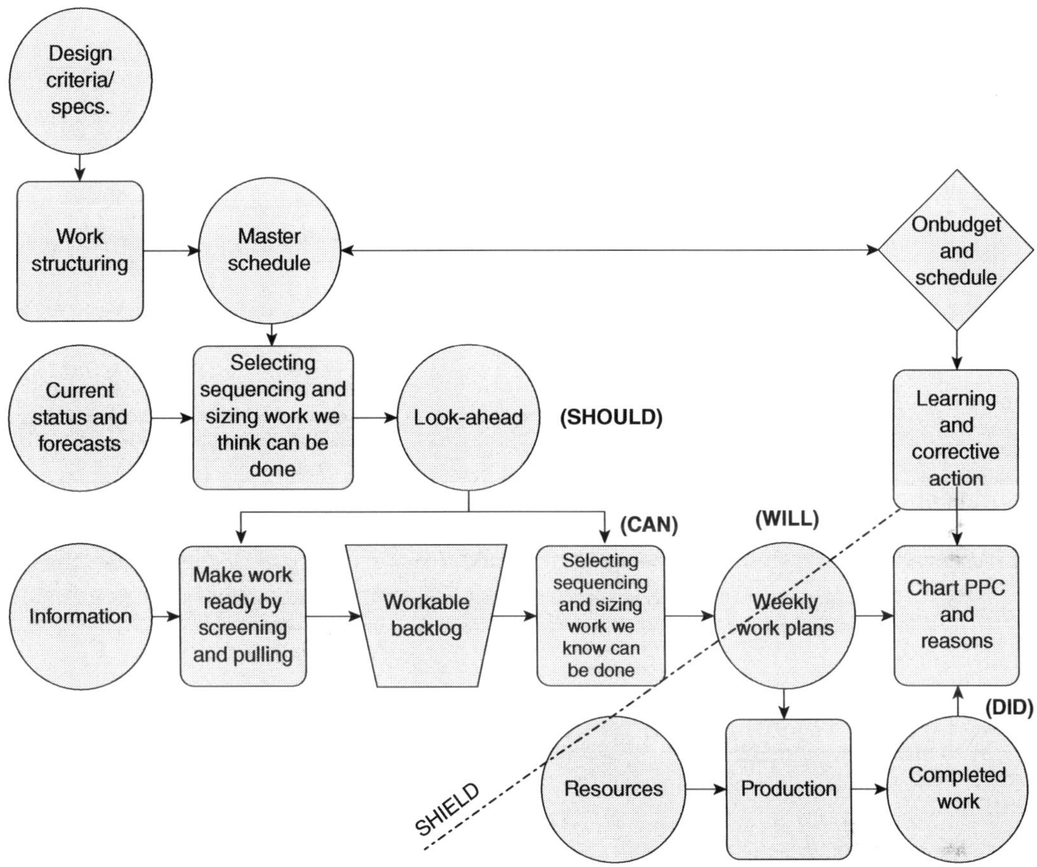

Figure 4A.3 The Last Planner® System: process flow

Procedurally, planning and scheduling to prepare a master schedule may involve the following steps:

1 The project is subdivided into major work activities.
2 The duration and work force requirements are determined for each activity.
3 The activities are placed in a logical sequence to form the master schedule. For example, an excavation has to be made and reinforcement placed before a concrete foundation can be poured.
4 Project duration is determined from the overall sequence of activities and their respective durations.
5 Overall duration is reconciled with the contractual requirements of each project. Activity durations are adjusted through resource utilization. For example, the time required for placing reinforcement can be accelerated by increasing the number of crews carrying out that activity.
6 The master schedule is further refined, using the CPM to (a) identify the critical path (i.e., the sequence of activities that directly affect overall project duration) and (b) identify "float" or "slack time" that determines how long an activity can be delayed without jeopardizing the schedule. This float enables construction managers to accommodate delays or unforeseen conditions such as bad weather without affecting the completion date.

CPM scheduling involves conducting calculations through the CPM diagram as forward or backward passes. A forward pass determines the earliest start time for each activity (EST). Adding the expected duration of the activity yields the earliest finish time (EFT). A backward pass through the CPM diagram yields the latest start time (LST). Adding the expected activity duration yields the latest finish time (LFT). Each activity is examined to determine the amount of float available; that is, the difference between the earliest start and latest start, or earliest finish and latest finish. In practice, some activities are interdependent and influence the overall schedule. For example, placing concrete blocks for the walls of a one-story structure has to be preceded by pouring the foundation, columns, and the floor slab. Placing the blocks cannot begin until those respective activities are completed, hence the EST for wall construction would be the EFT for the preceding activities. The overall schedule may leave some activities with a float of zero days (i.e., EST – LST = 0). Such activities cannot be delayed without delaying the overall project; they represent the critical path. The critical path is that sequence of activities throughout the entire project that have the minimum amount of total float available.

In traditional construction management production, control relies on the experience of construction managers. Master schedules are often generated preliminarily in the process of bid preparation, using estimators' takeoffs in conjunction with past experience. These schedules have significant limitations, yet they provide the basis of many projects. The available project management tools address what SHOULD be done to meet a master schedule or CPM schedule, as opposed to verifying what CAN be done. Decision makers at a level that is detached from the day-to-day activities in the field often lack the ability to ensure that scheduled work is within a crew's capabilities. They often present crews with optimistic "stretch" goals, which are clearly beyond their capabilities, in the hope that this will motivate them to above average performance. Workflow reliability on these projects often range between 30 and 60% and is linked to low industry performance.

In contrast, the LPS uses process-driven approaches for project control that improve workflow reliability. Improving workflow reliability is synonymous with increasing the accuracy of site demand, and enables planners to better match the supply of resources to that demand. This results in the accomplishment of a higher percentage of planned tasks. Figure 4A.5 shows the process starting with work structuring leading to the master schedule. As indicated above, work structuring subdivides the project into manageable production units (Ballard and Howell 1994).

Lookahead planning

Lookahead planning involves taking activities from the master schedule and placing them in the Lookahead schedule. This schedule represents a time frame that is generally 6–8 weeks in length for construction tasks. Lookahead planning:

- Shapes work flow sequence and rate
- Matches work flow to system capacity
- Creates workable backlog: a backlog of work that is ready to be scheduled
- Prepares detailed work plans

Commitment planning

This identifies work that CAN and WILL be done. It enables quality assignments. The criteria for these assignments are:

- Definition: The work should be specific such that its requirements and completeness are clear.
- Soundness: Prerequisites are identified and constraints are removed.
- Sequence: The sequence of activities should be best for constructability.
- Size: The assignments should be matched to the capacity of the crew.

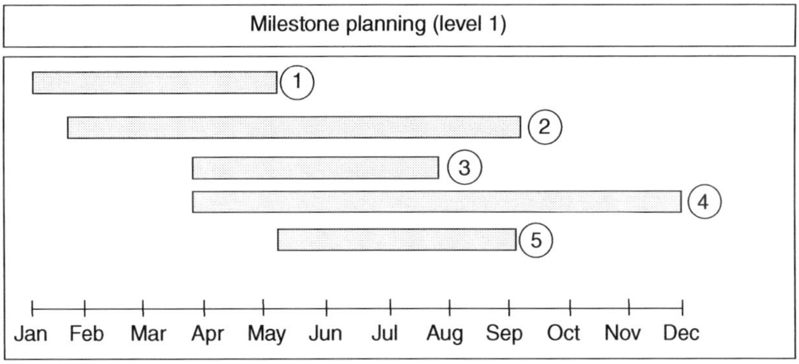

Figure 4A.4 The master pull schedule

Constraint-free assignments are kept in the block marked "workable backlog" and sequenced for the WWP. "Shield" indicates how this process shields downstream activities from upstream uncertainty and variation. Learning is based on using performance feedback to improve the system.

From the diagram in Figure 4A.5, after WWP have been implemented and completed through the "production" block, the completed work is analyzed in the "Chart PPC and Reasons." Where assignments have not been completed, they are examined and studied through root cause analysis in order to improve future assignments.

The LPS is based on three levels of schedules and planning tools (Figure 4A.6):

1 The master pull schedule: Level 1
2 The lookahead schedule: Level 2
3 The WWP: Level 3
4 Certain projects may require a daily work plan: Level 4

Master schedule

The master schedule may be used to establish project feasibility and likely duration. It identifies long lead-time items. If it is prepared before design work is completed, it establishes packages for action by designers and plans their sequencing and timing. The master schedule also includes infrastructure items based on:

- Estimated quantities
- Estimated craft density
- Standard rates for labor
- Labor distribution curves (industry based, or company based) and the appropriate craft density to complete the work safely and cost effectively.

The example illustrated above refers to an oil-refinery project. Much of the work involved installing pre-fabricated process piping that was stored in spools, each designated for a specific location.

There are significant limitations in the master schedule. It is generally prepared by people who are not going to do the work. It is tantamount to preplanning. While the master schedule is followed closely in traditional projects, it is further modified in lean construction projects. Phase scheduling, on the other hand, gets people involved who are actually going to do the work. For that reason it is generally more realistic to break down a master schedule into a lookahead schedule and subsequently into a weekly or daily schedule.

Master and phase planning

This planning occurs in a meeting of project participants and stakeholders. The steps in the process are:

1 Have an agenda ready for the meeting;
2 Introduce the entire team so that stakeholders will know each other;
3 Ensure that all required project data are available;
4 Use "Post It" notes on a wall-mounted board;
5 Use a backward (Pull) approach for assignments;
6 Promote creativity (with controlled chaos) to generate a broad cross section of ideas;
7 Identify float and verify with stakeholders;
8 Document the plan (SPS or MS project are examples of software that can be used for this purpose); and
9 Review and fine tune the plan.

Reverse phase scheduling

Reverse phase scheduling (RPS) (often termed "pull planning" is used as a starting point for the LPS. It is a detailed work plan that specifies the hand offs between trades for each project phase (Ballard and Howell 2003a, 2003b). In practice, subcontractors participate extensively in RPS as projects are generally comprised of activities performed by a number of subcontractors. They typically use Post-It notes on a wall-mounted board to plan their schedules in concert with the schedule for each project phase.

Traditional construction emphasizes pushing tasks to meet an optimistic master schedule, with little thought to what CAN and WILL be done in a given time frame. In LPS, the master schedule is refined as a reverse phase schedule by using a team to work backward from the expected completion date. This team comprises "last planners" from different trades/subcontractors. The master schedule and reverse phase schedule indicate what work SHOULD be done in order to meet the schedule.

The critical path is identified and an acceptable amount of float can be introduced to accommodate risk and uncertainty. This approach is based on the pull technique. The resulting phase plan provides a framework for dovetailing the work of various specialists that are needed to carry out their respective work plans. It is more realistic than the master schedule, but does not make allowances for conditions in the field that may fluctuate from week to week.

Comparison between push systems and pull systems

Production control systems are typically divided into two categories: push and pull systems. Push systems may be described as having production activities that are scheduled. Pull systems are those that have the start of one activity triggered by the completion of another (Spearman *et al.* 1990). In the manufacturing environment, push systems work best with predictable rates of production (or cycle time), and with material supply systems that can keep pace. In practice, cycle times are often random while supply systems are designed with fixed lead times. With push systems the variability in cycle time often leaves the supply system incapable of keeping pace. Alternatively, it may result in large quantities of WIP in order to compensate.

Pull systems do not schedule the start of jobs, but rather authorize production. The kanban system is a pull system; as parts or components are used in downstream work stations, the kanban card is passed upstream to authorize the production of more parts or components. Studies have shown that there is a less variance in cycle time for a pull system than in an equivalent push system.

The disadvantage of the push philosophy is that most construction tasks are interdependent – a completed project has components that are interwoven to create a functioning system; many disciplines have to have an ongoing interaction with each other, and cannot work uninterrupted at full capacity throughout the project. While gains can be derived by improving the performance of individual trade work assignments, this improvement may not be reflected in the overall completion of a project; hence, lean construction seeks to optimize the entire project.

The flow of work between producers is a hallmark of lean manufacturing and is a major factor in its success. In the just-in-time (JIT) concept, tasks are released from each work station when they are "pulled" by a downstream station and not before; this maintains a one-directional flow of work assignments that takes into account the variability that inevitably occurs in matching workloads and capacities. The progress of cars on an expressway is an example of the principles of dependence and variation. There are gaps between the cars and if their speeds are similar these spaces become small. A slight hesitation or delay results in an immediate effect on the flow of cars. An individual car cannot progress until the car in front yields roadway. Lean construction seeks to adapt these principles to the construction environment. In effect it pushes for reliability, responding to downstream demand, rather than pushing for speed.

Lookahead schedule

The lookahead schedule is derived from the phase plan, based on a time frame of six to eight weeks for construction tasks and is used for work flow control (Figure 4A.7). That duration has been identified by experience as the most flexible. Longer durations have been found to be unwieldy and subject to unexpected events. It uses items "pulled" from the master pull schedule. It reflects major work items that need to be completed for the milestone dates in the master pull schedule. It identifies the work activities that CAN be done, within the constraints that have been indicated. In effect, it reduces uncertainty by developing assignments that have a high likelihood of successful completion. The trade foremen are specifically asked what CAN be done in light of issues such as weather conditions, availability of crews, availability of materials, and completion of prerequisite work. In the case of design work, the lookahead schedule should be from 2 to 4 weeks' duration, based on teams' experience with lean projects. Constraints are identified so they can be planned for ahead of time. The purpose of lookahead planning is

- To shape the sequence and rate of work flow,
- To match work flow and capacity,
- To maintain a workable backlog, and
- To develop detailed plans for how work is to be performed.

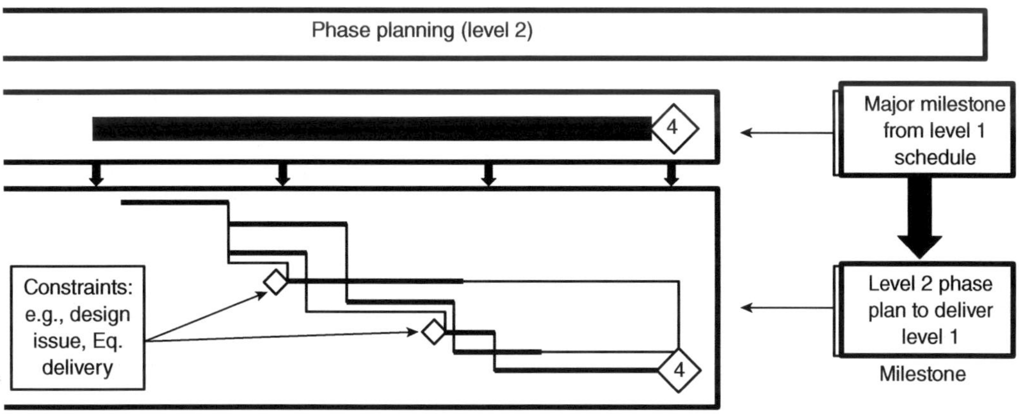

Figure 4A.5 Level 2 phase schedule and six to eight week lookahead

Lookahead planning includes:

- Constraints analysis,
- The ADM, and
- First run studies.

Constraints analysis involves observing the rule that no activity can be kept to its schedule date unless the planners are sure there are no outstanding constraints. This approach ensures that only "sound" assignments are made that can be depended upon, be free of limitations, and will not cause problems downstream. Constraints analysis applies to the production aspects of projects, in design, fabrication, or construction.

In the case of construction, the supervisors who are tasked with executing the schedule are assigned to developing and maintaining the lookahead schedule. They obtain input from the owner, designers, and material and equipment suppliers. They reconcile the work requirements with the contractor's labor estimates for the project. If there is a discrepancy between the lookahead schedule and the Level 1 schedule, then re-planning is done to bring them in line with each other.

Each week, the project team reviews the upcoming items to ensure that they are not obstructed by constraints and can be carried out predictably. If there is evidence of incomplete preparatory work, insufficient labor and/or materials, or unsatisfied requests for information (RFIs), such work assignments are considered defective and should be rejected. This review also includes: current progress, remaining duration estimates, and forecasted deliveries of equipment and materials. It results in a continual update on a workflow plan that is realistic and credible to all project stakeholders.

Lookahead schedule items should meet the following criteria:

- Manageable size: Schedule items are small enough, so they can be detailed to show downstream tasks that prompt work release.
- Readily measurable: Progress and remaining durations should be measurable.
- Free of constraints: It should be made clear where those constraints that have not been resolved are so as not to obstruct the flow of work.

Weekly work plan

This project included the installation of metallic piping in a refinery to convey liquids and gases from one process location to another. The piping is supplied to the site wound on large-diameter spools; it is unrolled and secured in place with hardware that is cut and welded in place.

The WWP is derived from the lookahead plan. The weekly planner schedule delineates the work activities or assignments pulled from the lookahead schedule that must be initiated to meet the completion dates in that schedule. Eligible activities or assignments are those that have no current constraints, and that have resources available and assigned. For example, responsible supervisors enhance the Level 2 activity into a detailed work-flow or plan. The plan is laid out by geographic area and the materials, sequence, and tasks are listed for each area. The pipe spools are identified within each work package and sequenced for optimal installation. Required tasks for each pipe spool are based on experience and documented best practices for installation. This may involve using a library of standard processes, developed and improved over time.

Level 3 activity contains work packages for physical/geographic boundaries; packages are small enough to be

- Managed by one foreman,
- Rapidly re-planned when change occurs, and
- Updated quickly and accurately by the foreman for completion.

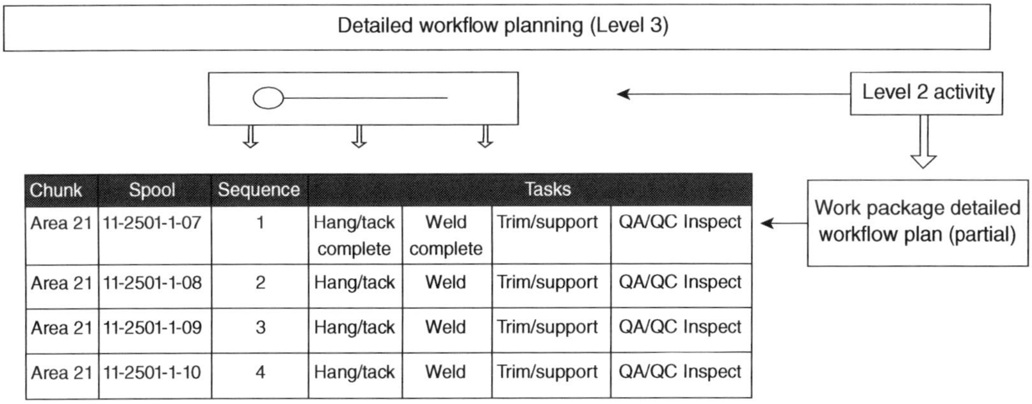

Chunk	Spool	Sequence	Tasks			
Area 21	11-2501-1-07	1	Hang/tack complete	Weld complete	Trim/support	QA/QC Inspect
Area 21	11-2501-1-08	2	Hang/tack	Weld	Trim/support	QA/QC Inspect
Area 21	11-2501-1-09	3	Hang/tack	Weld	Trim/support	QA/QC Inspect
Area 21	11-2501-1-10	4	Hang/tack	Weld	Trim/support	QA/QC Inspect

Figure 4A.6 Level 3 weekly work plan

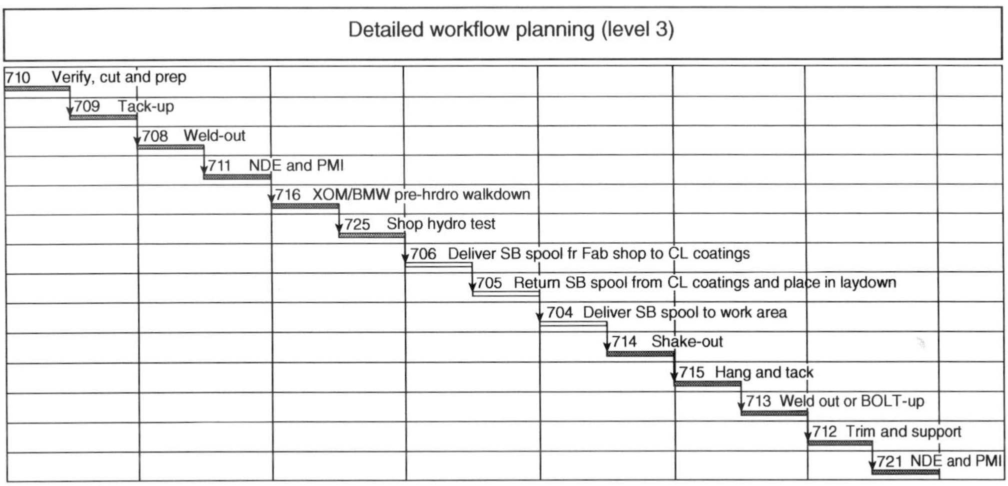

Figure 4A.7 Detailed workflow planning. Detailed workflow planning using standard processes on a complex project involving ~5,000 engineered-to-order pipe spools. Standard workflow plan (Level 3) incorporating each of the ~5,000 pipe spools within appropriately defined work packages. In this standard process example, a pipe spool's workflow requires: 14 tasks, five teams, six hand-offs between teams, and transport between four site locations

In Figure 4A.8, four chunks (areas) and spools are shown. They are listed on the sheet in a sequence (1–4) and the tasks involve welding, trim and support, and finally QA/QC inspection. In the WWP, constraints identified in the lookahead schedule are an important consideration. In some projects, the format for updating may include the project manager and the superintendent, or combinations thereof. Each trade foreman develops a weekly work schedule; based on the input from its last planner (the trade foreperson may be the last planner).

The detailed workflow plan represents work performed to install pipe spools for a complex project involving approximately 5,000 spools (Figure 4A.7). The work on these spools is organized into work packages.

Commitment planning is an essential component of weekly or daily planning. It is a method for defining criteria that lead to the selection of quality assignments (Ballard and Howell 1994). It results in commitments to deliver that other actors in the production system can rely on, by following a prerequisite that only sound assignments should be made or accepted. The foremen are empowered to reject any assignments that do not appear to be sound, as they are likely not to be completed and will reduce the PPC accomplishments. Despite this, foremen should be asked to confirm that each assignment that is scheduled WILL be completed in the coming week, based on preparations that are needed for this to occur.

Weekly work plan (WWP) meetings are held to review the work accomplished for the past week and plan the work for the following week. It is important to note that planning is really a *conversation*, not just a written document. Each WWP meeting is an essential part of the process – it depends on open and frank communication between all trades and their last planners. In some projects, it is appropriate to include suppliers such as rebar and truss and concrete suppliers. The issues discussed include the actual schedule, site conditions, safety issues, construction methods, manpower needs, and capabilities.

The WWP emphasizes what WILL be done in the upcoming week – from those tasks that CAN be done, based on the informed commitment by the involved trades. It considers "workable backlog" if it is needed as a buffer. The LPS is based on promises – it is imperative that the last planners promise what they WILL do only when they have verified beyond a doubt that they can carry them out.

Criteria for weekly planner schedule items:

- Detailed information: Level 2 tasks are subdivided into tasks that can be managed by one foreman.
- Soundness: Are the assignments workable? Do you understand what is required? Examples of constraints are:
- Best sequence: Are the assignments selected from those that are sound in priority order and in order of workability? Will doing these assignments release work needed by someone else? Are additional lower priority assignments identified as workable backlog whenever possible?
- Size: Are the assignments sized to the productive capability of each individual or group while still being achievable?
- Learning: Are incomplete assignments tracked, reasons for noncompletion identified, and action taken?

Commitment reliability

Completed work (DID) is compared with planned work (WILL). The PPC values are calculated. The reasons for noncompletion are recorded and analyzed in order to improve the process. Learning should be a positive experience. It should not be an occasion for casting blame!

Daily work plan

The daily workflow plan (Level 4) is appropriate for projects where there is extensive prefabrication and little or no interchangeability of materials/equipment. The Level 4 diagram is similar to the Level 3 weekly diagram, but the foremen commit daily to completing tasks based on the crew's capability and capacity. (See Chapter 7, Case C for a large EPC project.)

Constraint analysis

A constraint is an obstacle that inhibits the execution of a task that is required in the lookahead plan. It should also be beyond the control of the last planner in learn the forepersons project managers, schedulers, and other staff guide the scheduling process and remove constraints so work tasks can be assigned. They have to maintain communication with the foremen, who as last planners, coordinate the work with their counterparts in other trades. For construction processes to be lean:

- Weekly work plans must be accurate.
- Completed work must be correctly reported.
- Constraints must be indicated early.

Examples of constraints are:

- Adverse weather conditions,
- Unexpected events,
- Open RFIs, and
- Design modifications.

Constraints should be formally recorded in a log to indicate

- Date of record,
- Requestor,
- Responsible party – who should take action on the constraint,
- Description of the constraint,
- Tasks/activities affected by the constraint,
- The date required for constraint removal (based on the lookahead plan),
- Commitment date by the promissory,
- The cost impact of the constraint,
- Prerequisite work completed, and
- Resources that are unavailable, if any.

Benefits of the constraint analysis and management process

- If constraints are identified early, the chances of removal or mitigation are significantly improved (Table 4A.1).
- It clarifies the corrective actions that are needed and is linked to the lookahead plan.
- It serves as a positive course of action for problem resolution.
- It facilitates open, honest discussions by all parties on obstacles for schedule compliance.
- It documents events/actions for future reference and learning.
- It maintains a record that can be used for performance analysis and tracking.
- A formal process of constraint analysis and management improves the levels of PPC accomplished by work crews.
- It is also an important tool for continuous improvement by maintaining information on actions taken.

Removing constraints

By the time a task is committed during a Weekly Work Plan meeting, the expectation is that it will be completed as scheduled to maintain a predictable flow of work through the network of specialists. Therefore, anything that might hinder completion of the task needs to be cleared before it is assigned. During lookahead planning (two to six weeks before weekly work plan assignments are made), tasks are *made ready*. In a landmark paper on shielding, Ballard and Howell proposed five *quality criteria* against which a task must be checked before it is allowed into the weekly work plan (Ballard and Howell 1998). These are:

(1) *Definition:* Is the task specific? Will it be clear when it has been finished?
(2) *Soundness:* Are all materials available, including completed prerequisite work, for the task to be performed?
(3) *Sequence:* Is the task being performed in the correct order?
(4) *Size:* Is the task sized to the capacity of the crew?
(5) *Learning:* When assignments are not completed, are they tracked and reasons identified?

The facilitator of the meeting checks for these conditions in order to ensure that the *customer* of any task (the trade that immediately follows) is furnished with all that is necessary to complete it successfully. Because all downstream work suffers when a task cannot be completed, it is crucial that the facilitator rigorously honors this checklist. Once a task has been made ready, it can safely be assigned.

The quality criterion "soundness" is satisfied through constraint analysis and removal; the quality criterion "soundness" refers to having all necessary preparations being made ahead of time in order for an activity to be carried out effectively. A weekly work plan signals a request for a task to be completed for the customer that follows. It includes critical information such as: description of the task, a final check that all prerequisite tasks have been completed and all quality criteria have been met, and an indication as to when the task will be performed that week.

Activity definition model

The ADM is an important component ofThe Last Planner® System; it governs assignments that are moved from the phase schedule to the lookahead plan. It represents design tasks or construction processes and enables planners to enhance scheduled tasks to a level of detail where their readiness for execution can be determined. This evaluation requires a determination of constraints by examining directives to carry out tasks and comparing them with available resources as well as the status of prerequisite work. If the prerequisite work for a task has not been carried out, then the task cannot be carried through to completion even if the resources are available.

As shown in the diagram of Figure 4A.8, directives, prerequisite work, and resources are the three main sources of constraints. Directives may be assignments, design criteria, or specifications. Resources may be labor, tools, equipment, or space. Materials and information are not considered to be resources. These directives advise workers what tasks to carry out and how to execute them. The process ensures that prerequisite work has been carried out before passing transactions through to the output where it is examined in the decision node. Ready assignments are released for scheduling. Assignments that do not meet the criteria are sent to be redone.

Use of buffers

Ballard and Howell (1995) propose the use of buffers to adapt the JIT concept to construction. Two types of buffers may be used to shield downstream construction processes from flow variation. Plan buffers are inventories of workable assignments. Schedule buffers are materials, tools, equipment, manpower, time, and so on.

Alves *et al.* (2003) describe two types of buffers: Passive buffers are linked to process flow, such as materials, work in process, documents, time, and space. Active buffers relate to operations where work is

Table 4A.2 Constraint Analysis Worksheet, Benefits of the Constraint Analysis and Management Process

Item No.	*Date*	*Requestor*	*Responsible Party*	*Description of Constraint*	*Date Removal Required*	*Commit Date*	*Cost Impact?*	*Prerequisite Work Complete?*	*Resources Available*
1	3/11	JB	DB	Begin process of determining when control room floor plans will be available for planning electrical work	4/7	4/3	None	Yes	Yes

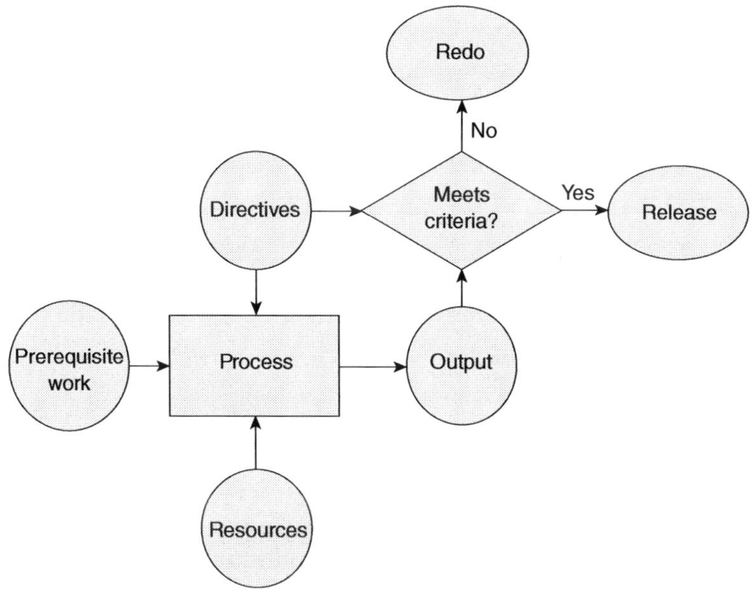

Figure 4A.8 The activity definition model

Source: From Ballard, G., and Howell, G., *Building Research and Information*, 2003a. With permission

performed on a product, such as excess capacity in equipment or labor. There should be a buffer of tasks that are appropriate (sound) for each crew as part of the last planner (Koskela 1999).

In the application of the Last Planner® technique, buffers are built into work plans to maintain the flow of work and to avoid interruptions. These buffers include a "workable backlog" of tasks that are available to be done but do not need to be done; that is, they are not on the critical path. This backlog allows crews to do productive work in case there are interruptions to the WWP. Plan buffers are backlogs of workable assignments. They are based on what CAN be done as opposed to what SHOULD be done. The work that CAN be done is used as a source of the work that crews commit to doing – WILL be done in the next week. This varies from traditional project management that insists on what SHOULD be done – without fully considering whether it CAN be done.

It is recommended to base WWPs on only 75% of the capabilities of the crew, in order to maintain the flow of work to the next crew. Workable backlog provides a source of work assignments that can maintain a smooth flow. As work is pulled from each crew by the downstream crew that leaves spare capacity and increases the probability of completing required tasks. On the other hand, traditional scheduling tends to assign as much work as possible to each crew in the hope that it can be completed resulting in a lower level of performance. An early start on a task that is not critical to the schedule does little to accelerate the project completion date. It simply builds inventory, which is a form of waste.

Prefabrication and modularization

Limbach Holdings, Inc.

A modular construction strategy provides several benefits: as production occurs off-site, there is little on-site construction waste. As production occurs in a controlled environment, there is no disruption from weather conditions or other distractions that may occur on a typical site. This can be especially advantageous during winter months or times of uncertain weather conditions. Quality is improved because closer tolerances can be obtained in a manufacturing type of environment. Sites are less cluttered, and less storage space is required. Site traffic is reduced, resulting in safer site activities. Very importantly, costs can be significantly improved and schedules can be more closely adhered to. As cited in McGraw-Hill's Smart Market Report (2013) Victor Sanvido, VP of Southland Industries Inc. comments on the safety benefits of prefabrication. Components or assemblies that are too heavy for manual lifting are routinely scheduled for mechanical lifting; this minimizes the possibility of scaffolding accidents and back injuries. Also cited are Rebecca Snelling and Mike Dennis, Vice presidents of JE Dunn Construction who indicate that early decisions need to be made on the elements that will be prefabricated. The design process should conform to the required measurements and tolerances as early as possible; this may not be feasible with delayed design decisions. With regard to logistics, there are no clear guidelines or limitations. While having a plant close to a project may simplify Just-in Time deliveries, opinions vary on the location of the prefabrication facility. One manufacturer points out that transportation cost is less than 2% of the cost of prefabricated systems if the distance is less than 1,000 miles, and is generally not an issue.

Limbach Holdings, Inc. is an integrated building systems provider who offers many of these benefits – managing all components of mechanical, electrical, plumbing, and control systems from systems design and construction through service and maintenance. Limbach engineers, constructs and services the above mentioned systems in both new and existing buildings and infrastructure. They work for building owners in the private, not for profit, and public/government sectors.

Headquartered in Pittsburgh, they serve several cities in New England, Michigan, Ohio, and Southern California. Harper, a Limbach Holdings company operates in the State of Florida. The various systems prefabricated by Limbach include modular mechanical rooms, equipment skids, multi-trade above ceiling modules, piping spools, plumbing carriers, head walls and wet walls, pre-assembled plumbing fixtures, ductwork, and other HVAC & plumbing components.

The pictures relate to the construction of a health care facility, the McLaren Port Huron South Tower. The project was a 209,000 square foot hospital addition with patient rooms, operating rooms and other support-related spaces. Limbach self-performed the installation of Heating, Ventilating, Air Conditioning (HVAC), plumbing and temperature controls. They prefabricated ceiling racks for the project that had heating, hot water, domestic water and medical gases for the patient room floors.

Other items Limbach prefabricates include modular mechanical rooms, equipment skids, multi-trade above ceiling modules, piping spools, plumbing carriers, head walls and wet walls, preassembled plumbing fixtures, ductwork, and other HVAC & plumbing components.

General practices:

> Limbach becomes involved in projects in different ways. Some owners are aware of prefabrication and its benefits, and they ask for it in many cases. If Limbach becomes aware of the project and is able to meet with the owner's representatives or the contractor, they will encourage the use of prefabrication as much as possible. Many of Limbach's projects are contracted with the owner and not necessarily with contractors. Limbach prices the job as a subcontractor, for supply and installation for specific systems in a project.

A great deal of precision is required in laying out prefabricated modules, and in determining the respective material quantities. In new projects a building information model (BIM) model is usually available and is used as a baseline. The coordinates of the various systems are available in 3D. Limbach prepares its own BIM in order to have a reliable model. With existing buildings site measurements are necessary and are done with laser scanning.

With regard to the lead time required for projects, Limbach works easily with the schedules they are usually given. What would need a week to build on site can often be done in two days in the factory, so the constraint is really the storage space in the factory. Quality assurance checks are conducted before and after delivery. FabPro 1 software is used to set up the work flow. The shop manager does a Quality Assurance check on material leaving the factory. Piping is often pre-charged with medical gases in the case of a hospital. This simplifies the start-up processes as less purging has to be done.

Systems such as plumbing and gases are typically staged on carts which are mounted on trucks and are then taken to the respective project site on a JIT approach. At the site they are hoisted in place on delivery so that there is minimal downtime, if any. Limbach coordinates with the contractor on site to schedule the delivery and installation work. The custom racks are delivered and installed on the same day. Typically there may be two trucks – one with the racks and another with the workers. Travel distance could be 30 to 70 miles and is typically covered in a couple of hours. In the case of Limbach's jobs in Michigan, they involve union labor; some workers may come from the shop, others from other places within reach of the jobsite. A building.

With regard to the benefits of prefabrication, Tom Polygois, general service manager, mentions the ability to level manpower requirements. He points out that one of the keys to successful prefabrication is early involvement of the trade partners.

The following pictures illustrate Limbach's activities on one of their projects.

Figure 4A.9 shows an order of pipe sections loaded in a shipping cart at the prefabrication facility.

Figure 4A.9 Prefabricated pipe sections mounted on a custom shipping cart

Figure 4A.10 Custom racks being loaded into a building

The shipping carts are then trucked to the construction site.

In Figure 4A.10 the shipping carts are loaded into a building under construction. The contractors remove the pipe sections from the carts and quickly install them in designated locations.

Questions for discussion

1 Describe the components of and the process flow for The Lean Project Delivery System™.
2 Describe a staff support structure that can be used for managing lean construction projects.
3 Who is the so-called Last Planner? What value does that individual bring beyond the traditional method of project planning and scheduling?
4 What is work structuring? What steps does it involve?
5 What is the purpose of reverse phase scheduling?
6 How can one ensure that the weekly work plan represents the most appropriate work to be performed in a given week?
7 What role does constraint analysis play in The Last Planner® System?
8 How can prefabrication benefit lean projects?
9 What are the main attributes to consider when considering prefabrication for a lean construction project? Why are they important?

Note

1 This subject is covered in detail in Chapter 12.

References

Al-Sudairi, A. 2004. Simulation as an aid tool to the best utilization of lean principles. *12th annual conference on lean construction*, August 3–4. Helsingor, Denmark.

Alves, T. C. L., and I. D. Tommelein. 2003. Buffering and batching practices in the HVAC industry. *Proceedings of the 11th annual conference of the international group for lean construction (IGLC-11)*, July 22–24. Virginia tech., Blacksburg, Virginia, USA.

Arbulu, R., and A. Koerckel. 2005. Linking production-level workflow with materials supply. *Proceedings of the 13th annual conference of the international group for lean construction (IGLC-13)*, July. Sydney, Australia.

Ballard, G. 1999. *Work structuring white – Paper #4*. Las Vegas, Nevada, Lean Construction Institute.

Ballard, G. 2000. *The Last Planner® System of production control*. PhD Thesis. Faculty of Engineering. School of Civil Engineering. Birmingham, AL: The University of Birmingham.

Ballard, G. 2008. The Lean Project Delivery System™: An update. *Lean Construction Journal*, 1–19.

Ballard, G., and G. Howell. 1994. Implementing lean construction – Stabilizing work flow. *Conference on lean construction*, September. Santiago, Chile.

Ballard, G., and G. Howell. 1995. Toward construction JIT. *Proceedings of the 1995 ARMCOM conference, association of researchers in construction management.*, Sheffield, England.

Ballard, G., and G. Howell. 1998. Shielding production: essential step in production control. *Journal of Construction Engineering and Management, ASCE* 124(1): 11–17.

Ballard, G., and G. Howell. 2003a. Lean Project Management. *Building Research and Information*, 119–133.

Ballard, G., and G. Howell. 2003b. An update on last planner. *Proceedings of the 11th annual conference of the international group for lean construction (IGLC-11)*, July. Blacksburg, VA.

Gwynn, P. 2008. *ULSD lean production management (LPM)*. Lean Implementation Services, Schereville, IN: Author.

Howell, G. A. 1999. What is lean construction. *Proceedings of the 7th annual conference of the international group for lean construction*. Berkeley: University of California.

Jarnbring, J. 1994. Translated – *Material flow costs on the building site*. Rappot 94:01. Lund, Sweden: Lunds Teknisa Hogskola, Institutionen for Teknisk Logistik.

Koskela, L. 1999. Management of production in construction: A theoretical view. In *Proceedings of the 7th annual conference international group for lean construction (IGLC-7)*, ed. Iris Tommelein, July 26–27. Berkeley: University of California.

Koskela, L. 2000. *An exploration towards a production theory and its application to construction*. VVT Publications; 408. Online at www.inf.vtt.fi/pdf/publications/2000/P408.pdf.

Macomber, H., and G. Howell. 2003. Linguistic action: *Contributing* to the theory of Lean Construction, *Proceedings of the 11th annual conference of the international group for lean construction (IGLC-11)*, July 22–24. Virginia tech., Blacksburg, Virginia, USA.

Spearman, M. L., D. L. Woodruff, and W. J. Hopp. 1990. CONWIP: A pull alternative to Kanban. *International Journal of Production Research* 28: 879–894.

Tsao, C. C. Y., I. D. Tommelein, E. Swanlund, and G. A. Howell. 2000. Case study for work structuring: Installation of metal door frames. *Proceedings of the 8th annual conference of the international group for lean construction (IGLC 8)*, July 17–19. Brighton, UK.

Bibliography

Alarcon, L. F., A. Grillo, J. Freire, and S. Diethelm. 2001. Learning from collaborative benchmarking in the construction industry. *Proceedings of the 9th annual conference of the international group for lean construction (IGLC-9)*, August 6–8. Singapore, University of Singapore.

Ballard, G. 1994. The last planner. *Northern California construction institute spring conference*, April. Monterey, CA, 1–8.

Ballard, G., and P. Reiser. 2004. The St Olaf fieldhouse project: A case study in designing to target cost. *Proceedings of the 12th annual conference of the international group for lean construction (IGLC-12)*. Elsinore, Denmark.

Bertelsen, S. 1993. *Construction logistics I and II, materials management in the construction process (in Danish)*. Kobenhavn, Denmark: Boligministeriet, Bygge-og, Boligstyrelsen.

Fiallo, M., and V. Revelo. 2002. Applying the last planner control system to a construction project: A case study in Quito, Ecuador. *Proceedings (IGLC-10)*, August 2002.

Howell, G. A. 2000. White paper for Berkeley/Stanford CE & M workshop. *Proceedings of the construction engineering and management workshop*. Palo Alto, CA: Stanford University.

Howell, G. A., A. Laufer, and G. Ballard. 1993a. Interaction between subcycles: One key to improved methods. *Journal of Construction Engineering and Management, ASCE* 119(4): 714–728.

Koskela, L. 1992. *Application of the new production philosophy to construction*. 87. Technical Report #72. Stanford University: CIFE.

Lichtig, W. A. 2005. Sutter health: Developing a contracting model to support lean project delivery. *Lean Construction Journal* 2(1): 105–112.

Macomber, H., and G. Howell. 2005a. 5 Necessary actions for change, 1–2. *Lean Project Consulting*, Louisville, CO 80027.

Macomber, H., and G. Howell. 2005b. *Using study action teams to propel lean implementations*. Louisville, CO: Lean Project Consulting Inc.

Macomber, H., G. Howell, and J. Barberio. 2005. Target value design: Seven foundational practices for delivering surprising client value, 1–2. *Lean Project Consulting*, Louisville, CO 80027.

Matthews, O., and G. Howell. 2005. Integrated project delivery an example of relational contracting. *Lean Construction Journal* 2(1): 46–61.

Moser, L., and A. dos Santos. 2003. Exploring the role of visual controls on mobile cell manufacturing: A case study on drywall technology. *Proceedings of the annual conference of the international group for lean construction (IGLC-11)*. Blacksburg, VA. Online at http://strobos.cce.vt.edu/IGLC11

Salem, O., J. Solomon, A. Genaidy, and I. Minkarah. 2006. Lean construction: From theory to implementation. *Journal of Management in Engineering, ASCE*, 168–175.

Tsao, C., I. Tommelein, E. Swanlund, and G. Howell. 2004. Work structuring to archive integrated product, process design. *Journal of Construction Engineering and Management* 130(6): 780–789.

Vrijhoef, R., and L. Koskela. 1999. *Roles of supply chain management in construction (IGLC 7)*. Berkeley: University of California Press.

Womack, J., and D. Jones. 1996. *Lean thinking*. New York: Simon & Schuster.

Womack, J., D. Jones, and D. Roos. 1990. *The machine that changed the world*. New York: Harper Collins.

4B
LOCATION-BASED PLANNING AND CONTROLLING METHODS

*Olli Seppänen**

Introduction

Location-based management approaches break a project down to physical locations and use the Location Breakdown Structure (LBS) to manage design and production in construction projects. This makes sense in construction because locations are stationary and thus easy to monitor, whereas crews flow from one location to the next. Because crews are on the move, they are harder to track, although new technology such as indoor positioning has the potential to change this in the future.

Location is the basic unit of planning and control in location-based planning. Activity-based systems are based on activities, which can have location coding. In contrast, location-based tasks are formed of similar work in multiple locations. This enables the planning of continuous work and thus emphasizes productivity. Optimally, the same crew can work on the same task continuously without any breaks after finishing work in a location. This is very valuable for productivity because of learning, clarity of schedule to crew members and smaller risks of subcontractor return delays or lost productivity claims. Subcontractors suffer greatly from the standard practice of discontinuous "starts and stops" and prefer to prioritize projects that have more continuity.

The location-based management system (LBMS) is a combination of the traditional critical path method and location-based concepts enabling easy planning of continuous work and also easier planning of the logic network by utilizing locations to automate some of the logic relationships. The LBMS default planning decision is continuous work, and a conscious decision must be made to break continuity. In contrast, CPM schedules by activities as soon as possible by default and much effort has to be used to plan continuous work. In addition to including an improved CPM algorithm, the LBMS also refers to guidelines for planning and controlling which emphasize the productivity of crews and optimize schedules by aligning production rates. The LBMS process starts with the definition of the Location Breakdown Structure of the project, defining tasks and their relationships with other tasks. Then the production rates of dependent tasks are aligned and the schedule is optimized. Finally, time buffers can be added between tasks to decrease the impact of variability on the schedule.

As an alternative to the LBMS planning process, takt planning and control has been proposed. Takt is otherwise very similar to LBMS but has a slightly different emphasis. Rather than adding time buffers between tasks, capacity buffers are preferred. As an example of a capacity buffer, if the takt time is five days each operation is given five days to complete but in normal circumstances the work should be able to be completed in a shorter duration, for example four days, leaving Friday as a buffer. Additionally workable backlogs outside of takt areas are used to reduce the risk of crews leaving.

In contrast with activity-based methods, both location-based management methods emphasize the importance of controlling production proactively during construction. Location-based controlling methods emphasize using real-time information and reacting immediately to any problems. The LBMS has more of a technical approach and attempts to forecast production to warn of upcoming production problems well before they happen. Takt operates more socially and issues are elevated immediately when the teams notice that they are unable to meet the takt time. Both approaches put a lot more emphasis on production control and are thus clearly lean techniques based on pull controlling.

The components of location-based planning are presented first, focusing on similarities and differences between the LBMS and takt approaches. Then location-based controlling systems are presented. Finally the methods are compared and contrasted to evaluate their strengths and drawbacks based on the latest evidence at the time of writing.

Location-based planning

The Location-based management methods (LBMS and takt) both work on the foundation of a location-based plan. This section starts by describing the various components of the location-based plan and then proceeds to visualization of the plan. Logic and calculations related to the LBMS algorithm and takt are briefly described. Risk management and buffers are important parts of location-based plans to protect against variability and they are described next.

Location breakdown structure

The Location Breakdown Structure (LBS) is one of the most important up-front planning decisions in location-based planning. LBMS allows a hierarchical LBS with unlimited hierarchy levels. For example, the project can first be divided into separate buildings which can then be subdivided into areas, which can be subdivided to floors and then to interior areas. Although most examples of takt in construction focus on single hierarchy levels of locations (takt areas), there is no reason why takt would not work also with a hierarchical LBS. However, LBMS has technical calculations which use the different hierarchy levels to create CPM logic automatically but such logic automation has not been described for takt.

There are some differences in how locations are defined between the LBMS and takt methods. Although LBMS guidelines (Kenley and Seppänen 2010) state that at the lowest hierarchy level, the locations should be so small that only one contractor can productively work at an area, takt has in practice taken a more aggressive approach towards batch size reduction. Several documented takt examples include very small takt areas. The smaller the areas, the larger the duration reduction that is achievable by takt. This is one of the key areas of difference: LBMS primarily targets improved productivity and/or decreased risk and takt primarily targets shorter project durations and/or production stability achieved by a scheduling visualization more understandable for the workers. The takt approach also seems to pay more attention to getting areas that have similar work densities. The same principle is also included in the LBMS literature but practical use cases have preferred more logical locations even if the quantities change between locations.

The LBS is best determined as a team effort with trade contractors. Especially in interior construction, the definition of LBS benefits from the knowledge of functional testing areas, distribution of cabling from switchboards, the location of shafts etc. The constraints of locations are best known by the MEP contractors, whereas most of the architectural trades are more flexible with their locations. Optimally, the location breakdown structure has already been taken into account during design, so that functional testing areas match the structural areas. Figure 4B.1 shows an example where LBS has been sketched on a floorplan in collaboration with the MEP trades.

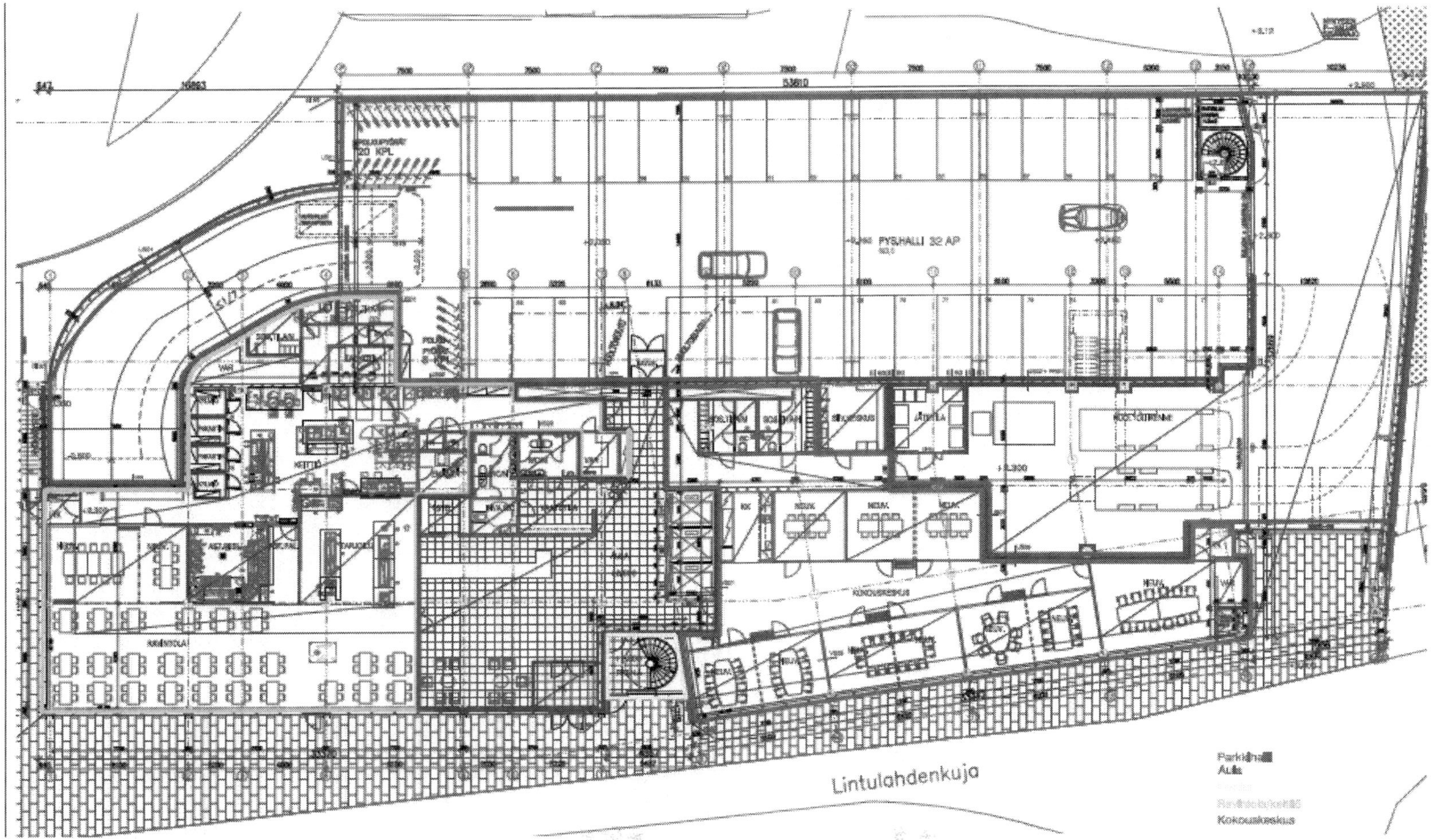

Figure 4B.1 The LBS sketched on a floor plan

Note: The lower left area is a restaurant, the area to the right of it is a lobby. The lower right area is a meeting room area (with tables); a stairwell to its left. The entire upper area is a parking garage

Tasks, location-based quantities and duration calculations

In LBMS, tasks are packages of work, which can be completed in a location by the same crew with no breaks and share the same external dependencies to other tasks. Wagons in takt are very similar in concept although takt allows also the work of multiple crews or contractors to be included in the same wagon if the tasks do not interfere with each other. Both tasks and wagons include work over several locations of the LBS. This is a key difference to activity-based systems, where each activity belongs to a single location.

The work content of a task or wagon is often based on quantities or work densities. The quantities can be defined physically by performing a quantity take-off for each location. This can be automated by using BIM-based quantity take-off systems. Alternatively work densities can be defined in manhours for each location. Regardless of whether quantities or manhours are used as a starting point, the schedule calculations are based on manhours in the end by multiplying the quantities by a labor consumption (manhours/unit).

There are slight differences in how to get from manhours to duration between the systems. Both systems recognize that the duration is a function of manhours, crew size, shift length and potentially some unique difficulty factors allocated to a location. However, the LBMS approach prefers to keep the crew size constant and allows the duration to change if there are quantity variations between locations. The takt approach emphasizes uniform takt times in each location. Therefore the duration is kept constant and the crew size or capacity buffer is allowed to adjust if there are quantity variations. The LBMS approach theoretically maximizes productivity (time workers are waiting for work) while the takt approach minimizes the time work is waiting on workers. However, takt durations are typically scheduled with less than 100% resource loading (capacity buffers), meaning that if there are no surprises, there is a potential for loss of resource efficiency. This potential resource efficiency loss is mitigated by including in the schedule off-takt areas, which typically include non-repetitive components. Excess resources can work on these off-takt areas which function as workable backlog. In order to minimize the loss of resource efficiency in takt projects, some preplanning effort should be directed towards identifying suitable workable backlog areas with enough work for each trade. However, takt itself is also somewhat flexible, with several different control mechanisms, so in practice it has been found that excessive workable backlogs are not required.

Flowline and takt visualizations

Location-based schedules can be visualized in flowline or takt diagrams. Both have the same principle. Locations are shown on the vertical axis and time flows horizontally. Time can be either calendar time (default in LBMS) or indicate the takt number (takt 1, takt 2) which is the default approach in takt. Flowline figures show tasks as diagonal lines and the slope of the line signifies the production rate of the task. If the locations do not have any special difficulty factors, the slope of flowline reflects quantity variation between locations. Normally flowlines are drawn from bottom left to top right (Figure 4B.1) although some prefer to draw flowlines from the top left to bottom right. The flowlines should normally not intersect because this indicates several trades working in the same location at the same time. Thus, any missing logic can be easily seen in the diagram. A takt diagram shows the same information but rather than using lines, the takts are shown as boxes. Because the takt times are typically constant, the visual look of a takt schedule (Figure 4B.2) is often simpler than in a typical flowline graph where the slopes of lines are changing). In both visualizations, a work sequence can be read horizontally and any empty areas indicate optimizational opportunities. Compared to Gantt Charts, both location-based visualizations offer a very efficient way of showing information. The quality of schedules can be immediately seen from the visual schedule. Even large schedules can be represented in a single flowline or takt diagram.

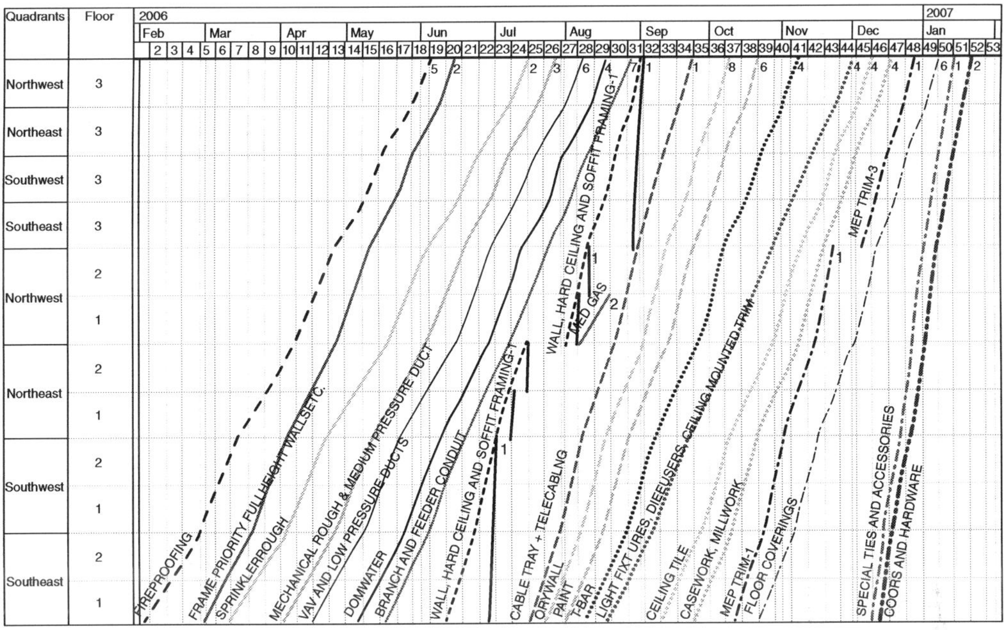

Figure 4B.2 A Flowline figure of a simple medical office building schedule

Source: Adapted from Seppänen 2014

Note: It is possible to see optimizational opportunities by looking at empty spaces between flowlines and work sequences by reading horizontally

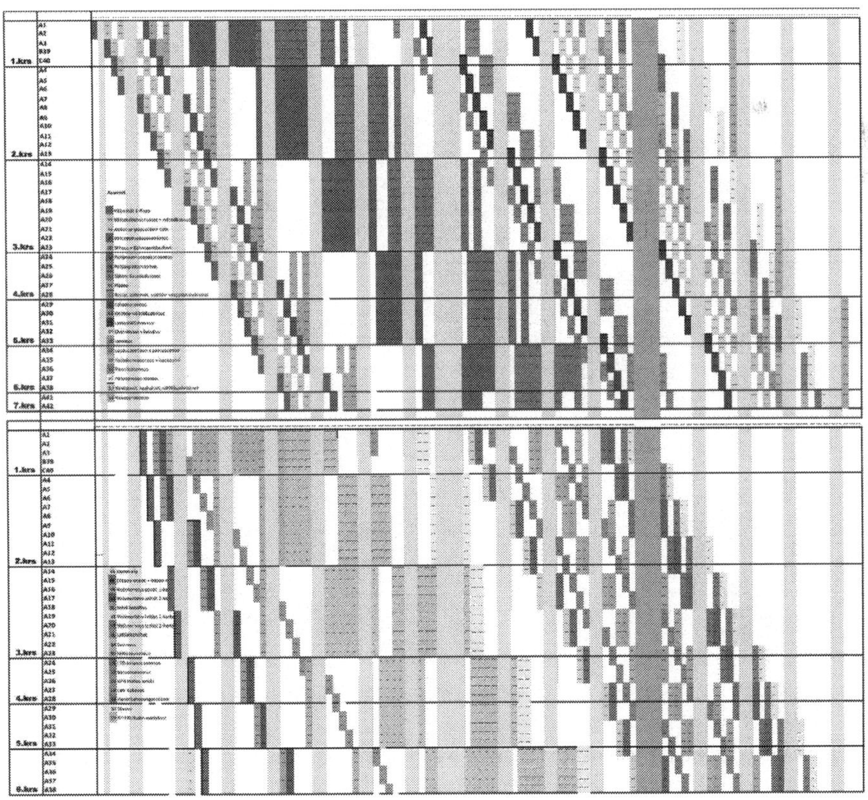

Figure 4B.3 A residential takt schedule

Note: The wagons are colorcoded and shown as boxes. Each apartment is its own location

Location-based controlling

Both LBMS and takt approaches give more weight to controlling production than just planning it. Plans are based on assumptions before starting production. The only really accurate data can be observed during the production phase. This is in contrast to the "after-the-fact" approach of controlling which is at the core of a CPM controlling model. The traditional CPM-based model has been called the *thermostat* model of controlling which is overly simplistic.

Controlling in LBMS tends to be a more technical process than in takt. It involves collecting progress data of locations and actual labor on site to calculate actual labor consumption and calculating projections of progress based on actual production rates. LBMS control attempts to prevent clashes between different tasks. Takt control is based on a more bottom-up approach. There is social pressure to complete work within the takt time. The crews are working very close to each other because time buffers are not typically used in takt. They know that missing the takt will cause problems for the next wagon and tend to self-correct their production, raising the issue right away if they know they cannot complete a task within the takt time. However, takt lacks the situational awareness given by LBMS projections. LBMS updates automatically, based on calculations and logic, whereas takt is normally updated by using a manual process. Because both approaches have strengths, the combination of LBMS technical calculations and the power of takt's social process should be explored in the future.

Controlling in LBMS

LBMS is more of a technical system which is based on collecting progress data in the form of actual start and finish dates, resources and actual shift lengths to continuously update actual labor consumption of each task. Actual labor consumption is used to calculate forecasts assuming that work will continue at the currently achieved production rate. When upcoming clashes between tasks are noticed, alarms are shown as visual cues. LBMS controlling is based on preventing the alarms from turning into production problems and potential cascading delays. Alarms are prevented by planning control actions which typically include removing waste to improve productivity or changing of crew size to speed up or slow down tasks or working more or fewer hours. Figure 4B.4 shows an updated flowline schedule with forecasts and an alarm.

Key metrics during controlling include production rate index for each task (actual production rate/planned production rate), labor consumption index (actual labor consumption/planned labor consumption), resource index (actual resources on site/planned resources) and degree of completion index (actual percentage completed/planned percentage completed). Degree of completion is often analyzed separately for each trade contractor, construction phase, as well as the whole project.

Controlling in takt

Controlling in takt relies less on technical calculations and more on social process between actors. Regular, often daily, takt meetings result in immediate actions when takts are not met. Control actions have wider variety than those of LBMS and often include moving scope from one wagon to the next – for example if a trade contractor is constantly missing its takt time, it is possible to leave less important parts of the scope to later wagons. By moving scope from one wagon to the next, it is often not necessary to update the schedule at all, just the wagon descriptions. Other possible actions include jumping over takt areas with design issues or stopping the train altogether for major issues. Adjustment actions related to takt have been well described by Binninger *et al.* (2017). As with LBMS, the plan only offers a framework for performance and is adjusted when there are deviations. However, a technical automatic adjustment process is missing and could be powerful in combination with the more technical LBMS approach. A typical takt controlling KPI is the share of takts completed on time compared to total planned takts.

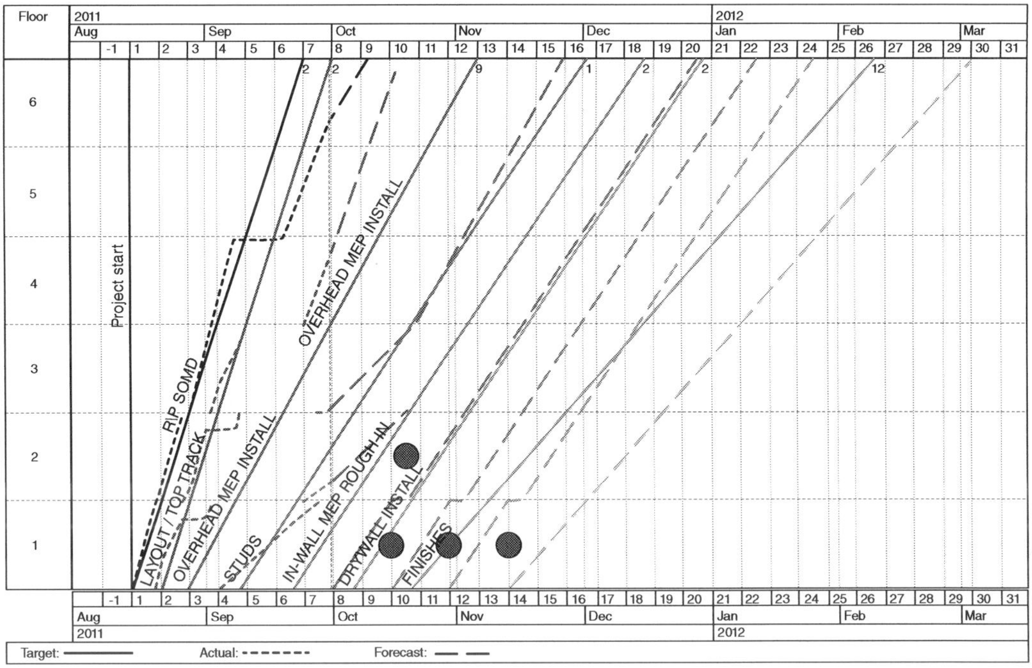

Figure 4B.4 An updated flowline schedule with progress (dotted lines) and forecasts (dashed lines) and alarms

Figure 4B.5 The takt schedule controlling meeting is more social than the LBMS updating process

Note: Pekka Kujansuu leads the takt schedule controlling meeting. Photo by Antti Talo, courtesy of Skanska Construction

Location-based methods with the Last Planner® System

Both the LBMS and takt methods are often combined with the social process of Last Planner® System. As was mentioned before, the Location Breakdown Structure is best planned together with the trade contractors. After the locations are known, the pull scheduling process of the Last Planner® System can be used to identify tasks. The trade contractors can then determine their work density (manhours/area) for each task in each location. Schedule optimization can also be performed collaboratively by starting with each contractor's preferred crew and making adjustments until all contractors achieve the same production rate or takt.

Similarly, during controlling, using lookahead planning and constraint screening of the Last Planner® System together with the forecasting of the LBMS or social peer pressure and transparency of takt, is a powerful combination. Lookahead planning can be used to ensure that the location-based schedule can be met, and updates can be triggered early if it is impossible to achieve the schedule. Weekly commitments can be compared to production rate forecasts and problems can be identified already during weekly planning if the commitments fail to meet the forecast.

Training requirements

In order to implement LBMS and takt, some training is required for project teams. Because LBMS is a more technical method, with more centralized control, typically the training focuses on the user of the LBMS software and users of various reports coming from the system. A few days of software training is typically required. The users of the reports (PMs and field superintendents) typically get familiar with the system in a few hour initial training session and the software user (typically a project engineer) helps in interpreting the output in meetings with PMs, field staff, and trade contractors. Everyone will gradually get familiar with the system in schedule optimization meetings and when control actions are discussed.

Takt is more decentralized and requires the participation and understanding of everyone to be successful. Typically takt is taught by using various simulation exercises which provide the real "feel" of a takt project, comparing it to traditional projects. A lot of effort is spent to ensure that everyone understands the requirements, including daily huddles where the takt schedules and daily objectives are reviewed with workers. Software training is not that important because there is currently no commonly used takt software. It is more important that everyone understands the principles and how they are expected to behave in a takt environment.

The LBMS and takt approaches compared

The performance of the LBMS has been mostly compared with CPM approaches. For example, Evinger *et al.* (2013) analyzed a project where similar work was carried out on different floors of the same project and other floors were managed using CPM and others using LBMS. LBMS floors had 18% higher productivity and achieved 10% higher production rates. The efficiency of LBMS controlling approach was evaluated by Seppänen *et al.* (2014). The research evaluated the impact of production alarms generated by LBMS on production team's decisions in three hospital projects in California. Some 39% of LBMS alarms resulted in control actions and 65% of actions trying to achieve a higher production rate were successful and achieved an increase of 37% on average. Production problems were also possible to be prevented. The research concluded that by implementing active LBMS based on control, it is possible for the General Contractor to impact trade contractors' production rates and thus decrease a project's duration. Olivieri *et al.* (2018) evaluated the planning capabilities of LBMS compared to CPM. They found that CPM schedules are not optimal because by just a few modifications in an LBMS environment, it was possible to achieve the same duration as in the comparison CPM schedule but using fewer resources to achieve more continuous and productive work.

Increasingly, takt case studies are also being documented. It seems that takt projects can achieve significantly shorter durations than LBMS projects due to the different type of buffering used and the social control mechanisms that emphasize peer pressure. The documented projects have not shown increased costs or the waiting hours that could be expected with theoretical comparison. Takt's benefits include easier visualization for crews to understand. For the first time, the entire team can understand the schedule. This makes takt planning and control a very powerful system which is easy to implement and does not require special software for calculations.

The key difference between takt and LBMS is the preferred buffering mechanism. In the LBMS, time buffers are planned between tasks so that there is enough time to react if there are problems. There is enough evidence now to say that this was based on a flawed theory. Time buffers seem to be causing problems because they are separating the workers which decreases social interaction and trust between the teams. Additionally, time buffers are used even if they are not needed because it is difficult to start new tasks earlier than scheduled due to design and material delivery constraints. This needlessly increases project durations. Because the buffers are there, delays are not considered critical and no one needs to take action. By contrast, in takt projects, there is a constant sense of urgency. Everyone knows that they are expected to hit the takt every time and they need to raise a flag if they cannot do so.

Although the takt social process seems to be superior, the LBMS technical process could still have a role to play if it were updated to include takt principles. Takt currently lacks a technical system that can automatically calculate the impact of different control actions when there are deviations from plans. The social process would benefit from some calculations and software that could be used as needed. In the future, the strengths of the systems should be combined. There is software for the LBMS (Vico Software Schedule Planner) and there are some early attempts to implement takt in software but the currently available software titles still function on top of a CPM algorithm. Most companies prefer to use Excel and manual methods.

Conclusion

The LBMS and takt planning and control are examples of location-based planning and control systems. The LBMS has been shown to outperform CPM in both planning and controlling. In fact, the comparison is not fair because LBMS includes the CPM algorithm as a key component. Takt has many similarities to LBMS but emphasizes social components while LBMS emphasizes technical components. Both systems have their strengths and weaknesses. The takt way of buffering against variability has been shown to be stronger but LBMS technical functions are missing from takt. By augmenting the LBMS technical system to include takt calculations, it would be possible to come up with a combined system where the planners could make choices on which strategy is best for each project, or even use two approaches in parallel in the same project.

Questions for discussion

1. What are the key differences between traditional activity-based planning and control systems and location-based systems?
2. What are the key differences and similarities between the location-based management system (LBMS) and takt planning/takt controlling systems?
3. What are the benefits of using the Last Planner® System together with location-based planning?
4. How is location-based controlling different from controlling in activity based systems?

Note

★ Olli Seppänen is an Associate Professor at the Aalto University School of Engineering, Department of Civil Engineering. His field of expertise is operations management in construction. His research interests include lean construction, real-time production control, location based management systems, lean design management, construction logistics and digitalized construction operations. Olli has authored the textbook *Location-Based Management System for Construction* with Professor Russell Kenley.

References

Binninger, M., J. Dlouhy, D. Steuer, and S. Haghsheno. 2017. Adjustment mechanisms for demandoriented optimisation in takt planning and takt control. *Proceedings (IGLC 25)*. Heraklion, Greece.

Evinger, J., C. Mouflard, and O. Seppänen. 2013. Productivity effects of starting as early as possible in hospital construction. *Proceedings of the 21st annual conference of the international group for lean construction*. Fortaleza, Brazil.

Kenley, R., and O. Seppänen. 2010. *Location-based management in construction: Planning, scheduling and control*. London and New York: Spon Press.

Olivieri, H., O. Seppänen, and A. D. Granja. 2018. Improving workflow and resource usage in construction schedules through location-based management system (LBMS). *Construction Management and Economics* 36(2): 109–124.

Seppänen, O. 2014. A comparison of takt time and LBMS planning methods. *Proceedings of the 22nd annual conference of the international group for lean construction*. Oslo, Norway.

Bibliography

Binninger, M., J. Dlouhy, M. Müller, M. Schattmann, and S. Haghsheno. 2018. Short Takt time in construction – A practical study. *Proceedings (IGLC 26)*. Chennai, India.

Dlouhy, J., M. Binninger, S. Oprach, and S. Haghsheno. 2016. Three-level method of Takt planning and Takt control – A new approach for designing production systems in construction. *Proceedings (IGLC 24)*. Boston, USA.

Frandson, A. G., K. Berghede, K., and I. D. Tommelein. 2013. Takt time planning for construction of exterior cladding. *Proceedings (IGLC 21)*. Fortaleza, Brazil.

Frandson, A. G., and I. D. Tommelein. 2016. Takt time planning of interiors on a Pre-Cast hospital project. *Proceedings (IGLC 24)*. Boston, USA.

Frandson, A., O. Seppänen, and I. Tommelein. 2015. Comparison between location based management and takt time planning. *Proceedings of the 23rd annual conference of the international group for lean construction*. Perth, Australia.

Kala, T., C. Mouflard, and O. Seppänen. 2012. Production control using location-based management system on a hospital construction project. *Proceedings of the 20th annual conference of the international group for lean construction*. San Diego, CA.

Seppänen, O. 2009. *Empirical research on the success of production control in building construction projects*. Doctoral thesis. Department of Structural Engineering and Building Technology, Faculty of Engineering and Architecture, Helsinki University of Technology, Espoo, Finland.

Seppänen, O., G. Ballard, and R-U. Modrich. 2015. Integration of Last Planner® System and location-based management system. *Proceedings of the 23rd annual conference of the international group for lean construction*. Perth, Australia.

Seppänen, O., G. Ballard, and S. Pesonen. 2010. The combination of Last Planner® System and location-based management system. *Lean Construction Journal*, 44–54.

Seppänen, O., J. Evinger, and C. Mouflard. 2014. Effects of the location-based management system on production rates and productivity. *Construction Management and Economics* 32(6): 608–624.

Seppänen, O., and J. Kankainen. 2004. An empirical research on deviations in production and current state of project control. *Proceedings of the 12th annual conference of the international group for lean construction*. Helsingor, Denmark.

Tommelein, I. D. 2017. Collaborative Takt time planning of non-repetitive work. *Proceedings (IGLC 25)*. Heraklion, Greece.

5

INFORMATION AND COMMUNICATION TECHNOLOGY/ BUILDING INFORMATION MODELING (BIM)

Introduction

New approaches to construction management such as relational contracting and lean design and construction are built on a foundation of team integration and open sharing of project-related information. Information and communication technology (ICT) and building information modeling (BIM) are an effective mechanism for providing construction stakeholders with the informational and analytical tools for better management of the construction delivery processes. This chapter includes three cases that illustrate the benefits of BIM in construction projects.

SPECIAL NOTE: It must be noted that the purpose of this chapter is to provide a very basic understanding of ICT and BIM in relation to the deployment of lean design and construction. For more detail, readers should refer to publications such as the BIM Handbook by Sacks *et al.* 2018.

ICT – a description

Information and communication technology encompasses computer hardware, software, and communications devices that allow the sharing and access of information conveniently, locally, and worldwide. The rapid growth of the Internet confirms the benefits of global communications and the need for information sharing. This technology and the Internet have affected many aspects of people's lives, including how they work and live. Almost all industries are affected and can benefit from the development of ICT. The A/E/C industry in particular has shown vast interest in adopting the new technologies in the area of 3-D visualization, data analysis, communications and collaboration, information sharing, and BIM. Building information modeling is the process of generating and managing building data during its life cycle. It is a tool as well as a process, and increases productivity and accuracy in the design and construction of buildings.

Animated 3-D/4-D visualization concepts

The use of animation to depict three-dimensional views of a building product/facility provides a useful view of what a finished product may look like, as shown in Figure 5.1. The core concept of animated 3-D visualizations is to provide a "near-real" replica of the finished product or facility from the basic design to the operation of the facility. Not only does this provide a better platform for the architect/designer and more imaginative flexibility, but also a better feel to the customer of their desired product/facility. They

Figure 5.1 Use of 3D visualization in design and construction

also provide a useful visual description to the designer/contractor. As the technology has improved, the 3-D models have been extended into 4-D (3-D + time representation), where construction schedules are integrated with 3-D displays to simulate the progress of a construction project. The 4-D system provides improved work sequencing. Spatial conflicts between different trades are avoided and resource utilization is maximized through visual monitoring. 5-D improves the accuracy of virtual estimates.

The basic technology required for animated 3-D consists of high performance personal computers or workstations connected over a network. Additionally a video studio is needed for digital editing and animation recording with graphic cards. The applications extend from project development to the facility operation and management stage. All project participants can increase the quality of both the design and construction process. Moreover, instead of having to browse through pages of documents, users can navigate through a model and then through simple mouse clicks, access the relevant information.

Virtual design studios and concept

Virtual design studios allow designers and experts from different locations to interact using audio and video conferencing. Virtual studios allow designers to discuss matters related to design or construction without leaving their offices. They can share the same screen and the same program thereby entering the same virtual reality space. The technology includes high-speed computers or workstations with fast Internet access. The main benefit lies in the fact that experts and designers from any location and at any time can discuss a design issue and propose an immediate solution.

Building information modeling

Building information modeling (BIM) is the process of generating and managing building data during its life cycle. It is also a tool as well as a process, and increases productivity and accuracy in the design and construction of buildings.

BIM uses three-dimensional, dynamic building modeling software and operates in real-time. It supports the continuous and immediate availability of project design scope, schedule, and cost information that is high quality, reliable, integrated, and fully coordinated.

The process produces the building information model, also known as the BIM, which encompasses building geometry, spatial relationships, geographic information, and quantities and properties of building components. Though it is not itself a technology, it is supported to varying degrees by different technologies. The BIM is a data-rich, object-oriented, intelligent and parametric digital representation of a facility, from which views and data can be extracted and analyzed to generate information that can be used to make decisions and to improve the process of delivering a constructed facility (AGC 2005).

The American Institute of Architects (AIA) has defined BIM as "a model-based technology linked with a database of project information." A BIM carries all information related to a facility, including its physical and functional characteristics and project life cycle data, in a series of "smart objects." For example, an air conditioning unit within a BIM would also contain data about its supplier, operation, and maintenance procedures; flow rates; and clearance requirements.

History of BIM

The term BIM appears to have been adopted by Autodesk to describe "3-D, object-oriented, AEC-specific CAD." This medium was rich in architectural information and distinguished from the traditional 2-D drawing. Professor Charles M. Eastman at Georgia Institute of Technology has done extensive work on the process starting in the 1970s under the heading of a building product model. He has described BIM and documented its application in several publications (Eastman *et al.* 2008). Jerry Laiserin is said to have popularized the term BIM as a digital representation of the building process to facilitate exchange and interoperability of information in digital format. In 1987, Graphisoft's ArchiCAD first implemented BIM under the virtual building concept.

Parametric building modeling technology

This technology currently provides the highest level of BIM with the least amount of effort – it is far more sophisticated than CAD and Object CAD technology, and requires a completely different mindset in its users. It is an integrated system that embodies geometry and data representation and the relationships between different elements of the model are user-defined. In a manner analogous to an electronic spreadsheet, as one element is changed, this change is reflected appropriately to all elements of the model. This facility makes the design process (and subsequent construction) faster, more cost effective, less prone to errors, and higher quality.

There are a number of well-established software titles that have been continually upgraded, such as ArchiCAD by Graphisoft, Bentley Systems, an integrated project model based on a number of related application modules, such as Bentley Architecture, Bentley Structures, and Bentley HVAC.

Nemetschek has been available primarily in Germany and Europe.

Autodesk Revit software is based on parametric building modeling technology and is designed specifically for BIM. It has a central project database that contains representations of all building elements. Design revisions are immediately reflected throughout the entire project, and errors are readily detected. A number of software companies have competing products. It has been subjected to many upgrades.

As described by AEC Bytes 2017 (edited by Lachmi Khemlani, Ph.D.), Autodesk's BIM 360 is a unified platform connecting project teams and data in real-time, from design through construction, supporting informed decision-making and leading to more predictable and profitable outcomes. It provides design collaboration solutions to accelerate project delivery, reduce rework, and improve productivity by enabling teams to securely co-author BIM designs in real-time and streamline deliverable coordination. Autodesk has developed a cloud platform termed Forge, introduced in 2015, to provide program interfaces for cloud-based applications. One of those applications is Lens, a model-based estimating tool developed by JE Dunn Construction Company Inc.

Forge has enabled the addition of several features of Autodesk's BIM 360 platform, such as BIM 360 Glue for model management, BIM 360 Field for managing construction logistics on site, BIM 360 Layout for site positioning, BIM 360 Plan for construction planning, and BIM 360 Docs for document management.

BIM 360 can be enhanced with the power of immersive visualization, that allows users to realistically evaluate and understand a design model by "walking through it" to get a much more accurate understanding of the final product before moving into construction.

Practical applications of BIM

A building information model can be used for the following purposes:

- *Visualization:* 3-D renderings can be readily generated.
- *Fabrication/shop drawings:* It is easy to generate shop drawings for various building systems, for example, the sheet metal ductwork shop drawing can be quickly produced once the model is complete.
- *Automated Fabrication:* In projects that involve technologically advanced suppliers, data from a BIM file can be used as input to program numerically controlled fabrication equipment.
- *Code reviews:* Fire departments and other officials may use these models for their review of building projects.
- *Forensic analysis:* A building information model can easily be adapted to graphically illustrate potential failures, leaks, evacuation plans, and so on.
- *Facilities management:* Facilities management departments can use BIM for renovations, space planning, and maintenance operations.
- *Cost estimating:* BIM software has built-in cost estimating features. Material quantities are automatically extracted and changed when any changes are made in the model.
- *Construction sequencing:* A building information model can be effectively used to create material ordering, fabrication, and delivery schedules for all building components.
- *Conflict, interference and collision detection:* Because BIM models are created to scale in 3-D space, all major systems can be visually checked for interferences. This process can verify that piping does not intersect with steel beams, ducts, or walls.

Building information modeling makes a reliable digital representation of the building available for design decision making, high-quality construction document production, construction planning, performance predictions, and cost estimates. Having the ability to keep information up-to-date and accessible in an integrated digital environment gives architects, engineers, builders, and owners a clear overall vision of their projects, as well as the ability to make faster informed decisions. BIM makes available more design options and alternatives than traditional methods with similar time schedules and budgets. BIM provides two major benefits:

- As important design information is kept in digital format, it can be easily updated and shared with all the involved parties.
- The maintenance of real-time design data in digital form, using parametric building modeling technology, provides major savings of time and money. This leads in turn to an increase in project productivity and quality.

BIM represents a major departure from the traditional computer-aided drafting method of drawing with vector file-based lines that combine to represent objects. It models the actual parts and pieces being used to build a building. While line drawings can be interpreted by people, they cannot be readily interpreted by computers. BIM represents object parametrically in 3-D; these objects reconfigure themselves automatically when software instructions are invoked to show different views in accordance with defined rules. As 3-D objects are machine-readable, errors can be avoided at all phases of design and construction in real-time, as opposed to the time honored method of post facto drawing review. In fact, design or fabrication work can be reviewed remotely through web conferencing tools such as Webex or GoToMeeting.

BIM can be used to demonstrate the entire building life cycle including the processes of construction and facility operation. Quantities and shared properties of materials can easily be extracted. Scopes of work can be isolated and defined. Systems, assemblies, and sequences are shown in a relative scale with the entire facility or group of facilities.

BIM enables a virtual information model to be handed from a design team to the contractor and subcontractors and then to the owner, each adding their own additional discipline-specific knowledge and tracking of changes to the single model. This approach greatly reduces the loss of information that occurs in the transfer of a project from one entity to the other.

These attributes enable BIM to reduce waste and inefficiency in building design and construction. It has proven to be attractive to a number of both public and private users that are concerned about managing completed facilities and wish to reduce or avoid the operational shortcomings that plague many such facilities. In particular, the reduction in design errors and construction errors has a major impact on the users' experience. Whereas the traditional approach involved producing contract drawings that were subsequently converted to more detailed drawings for shop fabrication purposes, a BIM model can readily generate such shop drawings. Furthermore, it provides instructions in machine-readable form that can be used by fabrication equipment for more cost-effective production and greater accuracy than craft labor.

BIM addresses a major shortcoming of the design and construction processes – that of the interoperability of construction documents and information between the various parties to construction projects. This lack of interoperability leads to errors, misinterpretations of building system conflicts, and delays as the design and construction process passes through different phases. Effectively, the owners' project requirements and the designer's intent are often not fully preserved during the transition. BIM improves interoperability by integrating computer models into project coordination, simulation and optimization – the information provided by such models is used to generate feedback.

BIM can serve to prevent the information loss associated with handing a project from the design team to the construction team and to the building owner/operator, by allowing each group to add or reference back to all the information they acquire during their period of contribution to the BIM model. By way of an example, a building owner could use BIM to track down the source of a building problem without leaving a computer terminal. A water valve can be identified in a location of interest, and checked to see whether it is a likely cause of the observed problem. A system user can readily verify the specific valve size, manufacturer, part number, and any other information that resides in the BIM database.

BIM for architects

The project architects can take advantage of BIM in schematic and detailed design; and construction detailing phases. Following are some of the main benefits of BIM for project designers: (1) Better design by rigorously analyzing digital models and visual simulations and receiving more valuable input from project owners; (2) Early incorporation of sustainability features in building design to predicts its environmental performance; (3) Better code compliance via visual and analytical checks; (4) Early forensic analysis to graphically assess potential failures, leaks, evacuation plans and so forth; and (5) Quick production of shop or fabrication drawings (Kymmell 2008).

The use of BIM in the project planning phase allows project designers to analyze space and understand the complexity of space standards and land regulations, which saves time and provide the team with opportunity of doing more value-added activities (CICRP 2009). Some researchers have investigated the integration of BIM with GIS (Geographical Information Systems) which can aid project planners in selecting appropriate site and conducting project feasibility and marketing studies (Berlo and Laat 2011; Isikdag *et al.* 2008). The architects and engineers can take advantage of BIM applications at different stages of project design namely schematic design (SD), detailed design (DD) and construction detailing (CD). Table 5.1 illustrates specific BIM applications in each stage of project design.

Economic benefits

Studies by Stanford University's Center for Integrated Facilities Engineering (CIFE) based on several major projects attribute several savings to the use of BIM: (CIFE 2007)

- Up to 40% elimination of unbudgeted change,
- Cost estimation accuracy within 3%,
- Up to 80% reduction in time taken to generate a cost estimate,
- A savings of up to 10% of the contract value through clash detections,
- Up to 7% reduction in project time,
- An increase in field productivity in the range of 20–30%, and
- A tenfold or greater reduction in requests for information (RFIs) and change orders.

Benefits of BIM at each construction project phase

Building information modeling supports the continuous and immediate availability of project design scope, schedule, and cost information that is high quality, reliable, integrated, and fully coordinated. Among the many competitive advantages it confers are:

- Faster delivery speed resulting in time savings.
- Better design – design proposals can be analyzed in detail, using simulation and other evaluation tools to identify the best solutions.
- Lower costs, which represents financial savings.
- Better coordination between different disciplines resulting in fewer errors.
- Higher work productivity and quality.
- Provides benefits in the design, construction, and management phases.

BIM Benefits in the design phase

During the design phase the scope, schedule, and cost of a project have to be carefully balanced (Figure 5.2). Changes in any of these factors affect the other two. Scope changes directly impact project costs, and are

Table 5.1 An Illustration of Cost and Time Savings via BIM in Hilton Aquarium Project

Schematic Design	*Detailed Design*	*Construction Detailing*
• Options Analysis (to compare multiple design options) • Photo Montage (to integrate photo realistic images of project with its existing conditions)	• 3D exterior and interior models • Walk-through and fly-through animations • Building performance analyses (e.g. energy modeling) • Structural analysis and design	• 4D phasing and scheduling • Building systems analysis (e.g. clash detections) • Shop or fabrication drawings

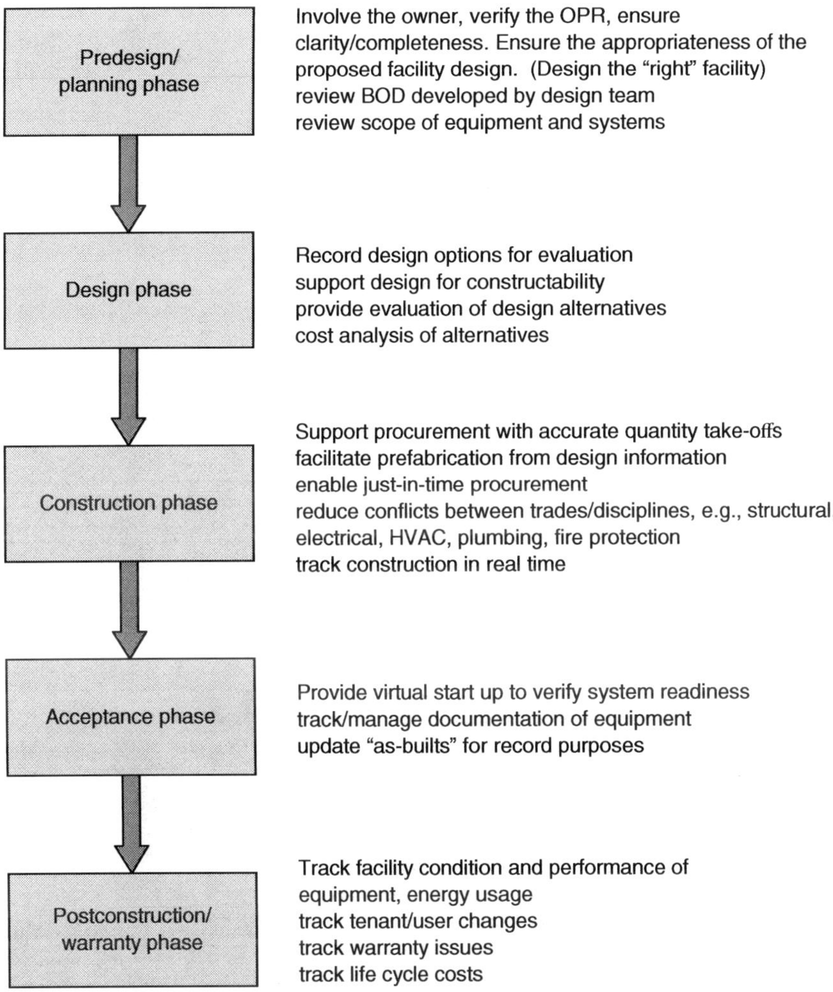

Figure 5.2 Process flow for BIM design and construction support

likely to affect the schedule of the project. In the traditional approach, designers readily manipulate information relating to spatial layouts and building geometry. On the other hand, this work is detached from cost and schedule considerations. Cost and schedule information is time-consuming and often difficult to assemble. It is also based on the design configuration available, and subsequent design changes may easily invalidate cost and schedule estimates.

BIM makes design information readily available to all stakeholders that are involved in the design process. The design team is empowered to make desired changes in real time without the need to do expensive and time-consuming coordination and checking by hand. The design concept is incorporated in the BIM model – as design work progresses its evolution is captured in it. Changes are automatically coordinated instead of requiring manual verification by the design team. The near elimination of errors enables the quality of design work and its documentation to be greatly improved.

The speed of delivering the design and accompanying permit or other documentation is greatly enhanced, with less individual effort. BIM provides the design team with a versatile tool for representing the design concept to the client – renderings can be generated automatically and manipulated electronically for improved client communication.

Three-dimensional modeling and target cost

This modeling supports the target value method of design. As design work progresses, cost estimating needs to keep pace with the development of the design. The traditional estimating method may prove to be too slow for projects that are fast-paced. Three-dimensional modeling can automatically generate bills of materials, and the associated pricing keeps designers informed of the overall project cost.

Three-dimensional enabled project delivery features

- Clash detection
- Construction/installation drawings
- Procurement (bill of materials)
- Structural steel detailing for fabrication

There is no need for data transfer for installation. There is a direct exchange of A/E 3-D model information to steel fabrication. In one project, this eliminated the need for 12,000 drawings. Four-dimensional CAD involves the simulation of construction activities – it provides improved work sequencing, resolution of spatial conflicts between different trades. It also maximizes resource utilization and visual monitoring of building sites.

BIM application in the construction phase

Constructors can use BIM models to provide much more information than the hard copy construction documents of yesteryear. Take-offs for cost estimating can be readily obtained and work planning can be integrated with the constructor's schedule to see how well it meets the project master schedule. Architectural design details can be readily converted to construction shop drawings and prefabrication instructions for subassemblies such as HVAC ductwork, piping, and so on. Overall, this makes constructors more responsive to the needs of a design in a shorter time frame with less impact on administration costs. Fewer errors are committed in the interpretation of contract documents, and the interaction of various building systems is checked for clash detection long before installation or erection begins. Given the amount of waste that typically occurs in translating the design to the built product, BIM is highly cost-effective for constructors and owners/clients alike.

BIM kiosks

According to Hans Thomas Holm of Norway, Head of Project Life Science in Oslo, Statsbygg/Owner Dept. BIM kiosks offer great benefits in terms of digital interaction on construction sites. They have been used on the construction site of the Bergen Academy of Art and Design, the KHiB project. A key benefit of BIM kiosks is mobility; it ensures optimum accessibility for all ongoing work in the building. During the KHiB project's shell phase, the BIM kiosks were used the most by the people working on the reinforcement. In a project with a lot of prefabrication, ordinary drawings will also provide the necessary information. For straight installation work on pillars, beams, hollow-core floor slabs, and steel, therefore, supplementary information from a BIM model would not necessarily have the same benefit to the project team. BIM kiosks were made in place, using a powerful laptop computer with a large screen. During the shell phase of the project, a movable container positioned so that an operator could stand before it and access needed information. During the interior construction phase the kiosks were set up in a plywood enclosure equipped with wheels so it could be moved around easily from one area to another. In the KHiB project, the kiosks were programmed to automatically download updated models from a secure server to the construction site's network. A BIM coordinator was able to verify the operational status of the kiosks and whether updates had been performed. The coordinator role was performed by either a member of the design team or a dedicated BIM site manager.

As reported by Hans, important lessons were learned from this application:

> Good web access on the construction site is an important prerequisite. Setting up a building site network can be very demanding, and can require significant technical expertise before it works satisfactorily.

Mobile data solutions might be a good alternative. BIM kiosks can be set up on all building floors with wireless connection, which works well as long as care is taken to ensure an almost 100% signal coverage. Automated updates for the BIM kiosks would simplify their operation.

Data export from the design software has to be organized such that the needed construction information is always available in the BIM kiosk's model.

The utility of the system is enhanced by the use of an artifact termed a "room diamond". It is a geometric object placed in each room layout and contains information such as room diagrams or other technical details.

Clicking on the room diamond brings up a hyperlink to the relevant information in the database. The KHiB project also created direct links from the model's various control area objects to specially organized drawing packages prepared by the architect for the respective area. The success of BIM kiosks led to several other applications in Statsbygg.

BIM during occupancy

Details of building materials, equipment, systems and "as built" information are readily available through BIM to facilitate the management of a facility for optimum performance. As opposed to the hard copy "as built" drawings that are usually maintained for completed facilities, BIM maintains building information in digital form and is more cost-effective for maintaining the platform for keeping current records. In rented space, for example, tenant-related changes can be accurately recorded. The reconfiguration of interior spaces can readily be tracked – not just as repositioned wall panels, but as HVAC, electrical, communications, fire safety, and other layouts. Energy consumption and air quality can be closely monitored and compared with the specified design conditions. Information about tenant-related issues such as square footage, repainting, rental cost per square foot, lighting levels, furniture, and so on are easily tracked and used for financial decision making. Subsequent projects in a facility can be managed for optimum performance by using the accurate records on the facility as well as the lessons learned over a period of time.

Case study 1: BIM for the detailed design phase

Professor Salman Azhar and his research teams of Auburn University, Alabama describe the following to illustrate the cost and time savings available with BIM on an actual construction project.

This case study outlines the use of BIM at the detailed design phase. The project is a $35 million LEED® Silver certified, 110,000 square academic building on the campus of Emory University in Atlanta, GA. It is a multi-purpose structure designed to provide instructional and research space and constructed by Holder Construction Company, of Atlanta, GA.

The project details are as follows:

> The project architect developed the building information model of the facility at the early design phase to determine the best building orientation and evaluate various skin options such as masonry, curtain wall and specific window styles. The building information model was also used to perform daylight studies, which, in effect, helped to decide the final positioning of the building on the site. In order to achieve this, views of the facility were established within a BIM software model using the software's sun positioning feature. Subsequently, the shading and lighting studies, as well as right-to-light studies were conducted in order to determine the effects of the sun throughout the year as well as the effects the proposed facility would have on surrounding buildings. Right-to-light studies were also conducted to evaluate lighting conditions at the proposed facility's courtyard space, and those spaces adjacent to the courtyard.

As a direct result of these studies, the building's design was adjusted as follows: (I) window openings on the west façade were reduced; (ii) the penthouse, which is located on the roof of the building, was reduced in overall square footage; and (iii) the overall height of the building was reduced. As all of these design adjustments were able to be incorporated during the design phase, the analyses prevented costly and time-consuming redesign at later stages in the project life cycle (Azhar 2011).

Case study 2: Using BIM with the Hilton Aquarium

The review teams collaborated with Holder Construction Company, Atlanta, Georgia. (Azhar *et al.* 2008a, 2008b).

The project details are as follows:

Project name: Hilton Aquarium, Atlanta, Georgia
Project scope: \$46M, 484,000 SF hotel and parking structure
Delivery method: Construction manager at risk
Contract type: Guaranteed maximum price
Design assist: GC and subcontractors on board at design definition phase
BIM scope: Design coordination, clash detection, and work sequencing
File sharing: Navisworks used as common platform
BIM cost to project: \$90,000, 0.2% of project budget (\$40,000 paid by owner)
Cost benefit: \$600,000 attributed to elimination of clashes
Schedule benefit: 1143 hours saved

The contractor created 3-D models of the architectural, structural, and MEP systems of the proposed building as shown in Figure 5.3. These models were created during the design development phase using detail level information from subcontractors based on drawings from the designers. This method allowed project team members to perform their work in the comfort of their traditional 2-D, drawing-based delivery process and eliminated the potential risk that is often associated with open sharing of digital models across stakeholders. Using the developed models, the project team achieved the following benefits:

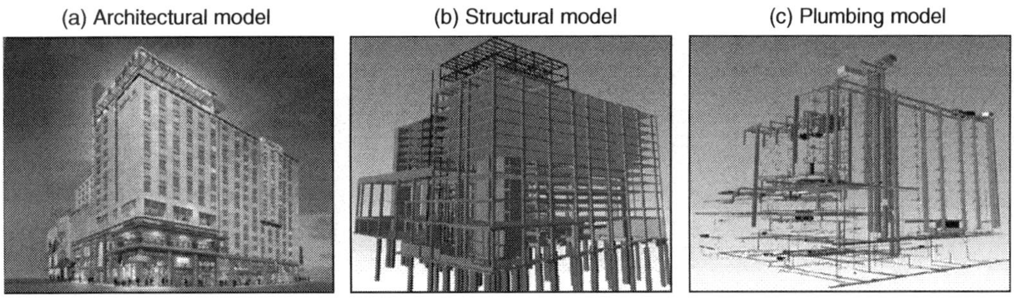

Figure 5.3 Building information modeling for Hilton Aquarium, Atlanta, GA

Source: Courtesy of Holder Construction, Atlanta, GA

- Proactively identified 590 conflicts between structural and MEP components and resolved them prior to field installations.
- Design coordination enhanced.
- Additional cost increases avoided.
- Enabled owner to view revisions without issuing change orders.
- Accommodated design changes.

As shown in Table 5.2, through frequent coordination sessions, the project team was able to quickly identify and resolve system conflicts, saving an estimated $800,000 in extras and avoiding months of potential delays. A net saving of $200,000 was calculated using the assumption that 75% of the identified collisions could be pointed out through conventional practices.

During the construction process, non-BIM-savvy stakeholders made use of Holder's visualization models through a free viewer (i.e., Navisworks). The collaborative 3-D viewing sessions also improved communications and trust between stakeholders and enabled rapid decision making early in the process. Finally, Holder's commitment to updating the model to reflect as-built conditions provided the owner, Legacy Pavilion, LLC, a digital 3-D model of the building and its various systems to help aid O&M procedures after occupancy.

Case study 3: Savannah State academic building

This case study illustrates the use of BIM in the project planning phase to perform the options analysis (value analysis). The project details are as follows:

Project: Higher Education Facility, Savannah State University, Georgia
Cost: $12 million
Project delivery: CM-At-Risk/GMP
BIM scope: Planning, preconstruction services, value analysis
BIM cost to project: $5000
Cost benefit: $1,995,000

The contractor coordinated with the architect and the owner in the predesign phase to prepare building information models of three different design options. For each option, the cost estimates were also prepared using three different cost scenarios (normal, midrange, and high range). The owner was able to walk through all the virtual models to select the option that best fit his requirements. Several collaborative 3-D viewing

Table 5.2 An Illustration of Cost and Time Savings via BIM in Hilton Aquarium Project

Collision Phase	*Collisions*	*Estimated Cost Avoided*	*Estimated Crew Hours*		*Coordination Date*
100% Design Development Conflicts Construction (MEP Collisions)	55	$124,500	NIC		June 30, 2006
Basement	41	$21,211	50	hrs	March 28, 2007
Level 1	51	$34,714	79	hrs	April 3, 2007
Level 2	49	$23,250	57	hrs	April 3, 2007
Level 3	72	$40,187	86	hrs	April 12, 2007
Level 4	28	$35,276	68	hrs	May 14, 2007
Level 5	42	$43,351	88	hrs	May 29, 2007
Level 6	70	$57,735	112	hrs	June 19, 2007
Level 7	83	$78,898	162	hrs	April 12, 2007
Level 8	29	$37,397	74	hrs	May 14, 2007
Level 9	30	$37,397	74	hrs	July 3, 2007
Level 10	31	$33,546	67	hrs	July 5, 2007
Level 11	30	$45,144	75	hrs	July 5, 2007
Level 12	28	$36,589	72	hrs	July 5, 2007
Level 13	34	$38,557	77	hrs	July 13, 2007
Level 14	1	$484	1	hrs	July 13, 2007
Level 15	1	$484	1	hrs	July 13, 2007
Subtotal construction labor	590	$564,220	1143	hrs	
20% MEP material value		$112,844			
Subtotal cost avoidance		$801,565			
Deduct 75% assumed resolved via conventional methods		($601,173)			
Net adjusted direct cost avoidance		$200,392			

Source: Courtesy of Holder Construction, Atlanta, GA

sessions were arranged for this purpose. These viewing sessions also improved communications and trust between stakeholders and enabled rapid decision making early in the process. The entire process took two weeks and the owner achieved a $1,995,000 cost savings at the predesign stage by selecting the most economical option.

BIM and energy/exergy analysis in design

Contemporary design trends place a major emphasis on the energy performance of buildings – facilities that meet the energy and environmental criteria of reference systems such as LEED and Energy Star in the United States. In Europe and Switzerland, building regulations such as The European Buildings Initiative and Minergie, respectively, also specify the requirements for high environmental performance. Buildings configured through these systems have been shown to have a reduced carbon footprint and provide a better working environment. The traditional approach of carrying out simulations is based on a completed design, and provides limited opportunities to make design changes later that improve energy performance. One of the reasons for this practice is that architectural design and energy-related design are performed separately. Architects are usually expert in building design but not in building performance simulation, hence this work is done by other experts.

Building information modeling can facilitate the integration of energy level calculations with the design process – the impact of design changes with these energy levels can be examined in real-time. BIM models store multidisciplinary information within one virtual building representation. This information includes geometric, semantic, and topological information – building shapes are 3-dimensional and are represented as geometric information. Semantic information defines material properties such as thermal resistance. Topological information relates to the dependencies of the components of the building.

Based on the thermoeconomic concept of energy, exergy analysis relates to the quality of energy and provides a holistic representation of a building. It requires an interaction between the factors that represent geometric, semantic, and topological information. The parameters that define this information during the design process can be used for performance calculations. Designers can gauge the impact of design changes by invoking the performance calculations almost in real-time. Adoption of BIM early in the design process facilitates interactivity in the design and performance measurement activities. In the United States, the government's procurement agency – the General Services Administration requires designers to utilize BIM for construction projects at the outset of the design process (Schlueter and Thesseling 2008).

BIM for green building design

The most effective decisions related to the sustainable design of a facility can be made in the early design and preconstruction stages. Traditional CAD planning environments, however, do not support the possibility of such early decisions (Azhar and Brown 2009). Building performance analyses are typically performed after the architectural design and construction documents have been produced. This lack of continuous analysis of sustainability in the design process leads to an inefficient process of retroactively modifying the design to achieve a set of performance criteria (Schlueter and Thesseling 2008). In order to assess building performance in the early design and preconstruction phases realistically, access to a comprehensive set of knowledge regarding a building's form, materials, context, and technical systems are required. Because building information modeling (BIM) allows for multi-disciplinary information to be superimposed within one model, it creates an opportunity for sustainability measures and performance analyses to be performed throughout the design process (Autodesk 2008; Schlueter and Thesseling 2008).

For example, architects can use a building information model to analyze a building's mass and form to optimize envelope and balance glazing ratios. Engineers can use this model to reduce energy demands through energy modeling, which uses the 3D model to calculate light reflectance and penetration. Contractors can analyze site conditions, including wetlands and protected habitats, and use the site model to coordinate logistics better to eliminate potential issues. Subcontractors can use BIM technology to reduce waste and combine shipments to further reduce carbon footprints (Hardin 2009). Figure 5.4 describes the typical flow of information taken into account when conducting performance analyses using a building information model. The building information model contains basic information such as building geometry, building materiality, building systems, and internal loads. Based on this information, as well as additional information, which is entered into the performance analyses software, environmental evaluation of a building can be performed. Such an evaluation could greatly help designers to optimize the environmental performance of the building.

BIM can reduce the costs associated with sustainability analyses by making the information required for sustainable design, analysis and certification routinely available simply as a by-product of the standard design process. BIM provides the opportunity to realize numerous benefits throughout

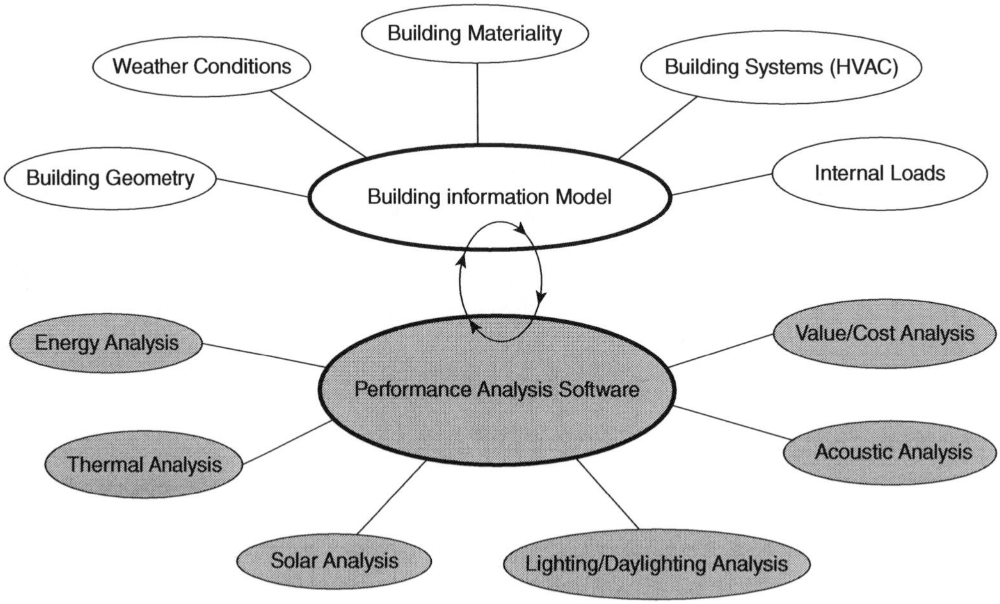

Figure 5.4 Typical information flow in BIM-based building performance analyses

Source: Azhar and Brown 2009

the project conception, design, construction and post occupancy phases of a building (Azhar *et al.* 2008). Linking the building information model to energy analysis tools allows for evaluation of energy use during the early design phase. This is not possible using traditional 2D tools, which require that a separate energy analysis be performed at the end of the design process, thus reducing the opportunities for the early modifications that could improve the building's energy performance (Azhar *et al.* 2009).

Hardin (2009) established three main areas of sustainable design with a direct relationship to BIM. These areas are "material selection and use", "site selection and management" and "systems analysis". In addition, Kriegel and Nies (2008) indicated that BIM can aid in the following aspects of sustainable design.

1 Building orientation (to select the best orientation that results in minimum energy costs),
2 Building massing (to analyze building form and optimize the building envelope),
3 Daylighting analysis,
4 Water harvesting (to reduce water needs in a building),
5 Energy modeling (to reduce energy needs and analyze renewable energy options such as solar energy), and
6 Sustainable materials (to reduce material needs and to use recycled materials).

The combination of sustainable design strategies and BIM technology has the potential to change the traditional design practices and produce a high-performance facility design. One such effort in the Columbia campus of the University of South Carolina resulted in approximately $900,000 savings over the next ten years at current energy costs (Gleeson 2008).

BIM and sustainable design

BIM tools facilitate sustainable design. They provide for:

- Structural design analysis,
- Energy analysis,
- Day lighting and lighting analysis,
- Model viewing tools for design review,
- Material database,
- Database tools for construction management,
- Database tools for facility management, and
- Monitoring and controls.

BIM should therefore be viewed as one aspect of purpose-driven software. Construction stakeholders should understand this current limitation. As BIM is primarily a geometric model with parametric variables, it is capable of operating in tandem with many other purpose-driven software systems. Every effort should be made to ensure that interoperability is established in order to have the best possible results from the BIM technology. The IFC-based model exchange provides an ability to transfer objects, their relationships, and their associated property sets.

BIM and lean construction

BIM supports many initiatives that are critical to lean design and construction. As the fundamental principle of lean is to reduce or eliminate waste, BIM addresses many aspects of waste that occur first in the design phase, and later in the construction phase. Lean design promotes the active participation of construction stakeholders as early as the project definition phase. As the design concept is developed, designers, owners, and constructors can work interactively to make decisions that influence the overall project – in real time, concurrently. Traditional design reviews are treated as sequential events, long after significant design decisions have been made. At that point, changes can quickly become time-consuming and expensive. Furthermore, as many disciplines are involved in the design of a project, changes in some elements of the design may not be fully represented by all the disciplines. In traditional projects that oversight often manifests itself as errors and omissions – classic examples of waste.

The "Big Room" or *obeya* concept is adopted in lean construction from the Toyota Production System, and brings together cross-functional teams under one roof to explore problems. Team members generate synergy by collaborating on not just the design, but on the construction process required to bring it to fruition (Macomber and Howell 2005). BIM provides a critical platform for Big Room meetings. "What-if" games can be played with various design approaches and the results can be evaluated immediately. Target value design is enhanced with BIM. Cost impacts of design are quickly determined in a concurrent manner, instead of relying on the traditional estimating approach.

Clash detection is easily accomplished with BIM – design errors often include having different building systems compete for the same limited space in ceilings and building penetrations from floor to floor. Air conditioning ducts and plumbing/fire protection piping typically compete for that space. When these clashes are detected during field installation, corrective action can have significant consequences. Bends in ducts and pipes that were not part of the original design increase their equivalent length, and may restrict the flow of air or water below the design levels defined by the mechanical engineer(s). That in turn leads to suboptimal building performance.

BIM has features that promote prefabrication – an important component of lean construction. As the BIM data are machine-readable, CNC machines can be readily provided with instructions for making sections of ductwork, piping, or other building elements under controlled conditions in fabrication shops. Such off-site work is more cost-effective and more accurate than on-site work, yet the accuracy provided with BIM literally allows such prefabricated components to be unloaded from a truck and mounted in place with fewer work hours than otherwise possible with traditional methods.

Waste due to lack of interoperability between different design and construction management software systems is improving with BIM. As reported by NIST and the Construction User's Roundtable, this lack of interoperability is a cause of project delays, cost overruns, and construction errors.

Factors to consider with BIM

Besides numerous benefits of BIM for project stakeholders there are many risks and challenges to implement BIM which can be divided into two broad categories: (1) Technology-related (2) Process-related (Azhar *et al.* 2012); and (3) Legal issues. Following sections present brief discussion on each category.

Technology-related risks and challenges

While the principle of BIM points to significant improvements in many facets of construction, there are still many opportunities for improvement. This may very likely occur as hardware capabilities improve and BIM software development progresses. Each project should have a written BIM plan to say who draws what and when. As BIM is both a tool and a process the owner needs to be educated on the impact of BIM. BIM systems create very large files, especially for large projects. The technology has not yet caught up with the need. The sharing of a single BIM between many users in real time will be a challenge until hardware capabilities increase. Existing BIM models represent architectural information readily. They do not yet have adequately detailed modeling capabilities for such systems as energy performance, fire safety, and structural integrity. The specialized software used by consultants such as mechanical and structural design may not yet be compatible with available BIM software. Detailed analysis and "what if" games may not be feasible with a common BIM. Construction stakeholders have at their disposal a number of specialized or "purpose-driven" software products that enable them to optimize-specific aspects of building systems.

- NavisWorks is used for design coordination.
- Graphisoft's Virtual Construction solutions facilitates construction planning and scheduling.
- Primavera may be used for construction resource planning.
- Integrated Environmental Solutions, Inc. produces software that provides building performance modeling (BPM). It requires users to have an understanding of BPM in order to use it effectively – many users may not have that expertise.

Many contractors develop their own software models that incorporate their historical experience with various project categories.

The first technology-related risk is lack of BIM standards for model integration and management by multidisciplinary teams. Integrating multidisciplinary information in a single BIM model requires multiuser access to the BIM model. This requires establishment of protocols in the project programming phase to ensure consistency in information context and formatting styles. At the moment, since there are no standard protocols available, each firm adopts its own standards. This could create inconsistencies in the model, which if not properly detected, could lead to inaccurate and inconsistent BIM model. The project team should perform frequent "model audits" to ensure avoidance of any such issues (Weygant 2011).

The interoperability issues, though significantly reduced during the last 5 years, still pose considerable risk. Interoperability is the ability to exchange data between applications to facilitate automation and avoidance of data re-entry. The introduction of Industry Foundation Classes (IFC) and XML Schemas has significantly helped to solve interoperability issues (Smith and Tardif 2009). However, both of these approaches have their inherent limitations. The users must research interoperability while selecting BIM software applications.

When project team members other than the owner and architect/engineer contribute data that are integrated into the building information model, licensing issues can arise. For example, equipment and material vendors offer designs associated with their products for the convenience of the lead designer in hopes of inducing the designer to specify the vendor's equipment. While this practice might be good for business, licensing issues can arise if the designs were not produced by a designer licensed in the location of the project (Thompson and Miner 2007).

Process-related risks and challenges

The process-related risks include legal, contractual and organizational risks. The first risk is the lack of determination of ownership of the BIM data and the need to protect it through copyright laws and other legal channels. For example, if the owner is paying for the design, then the owner may feel entitled to own it, but if team members are providing proprietary information for use on the project, their proprietary information needs to be protected as well. Thus, there is no simple answer to the question of data ownership; it requires a unique response for every project depending on the participants' needs. The goal is to avoid inhibitions or disincentives that discourage participants from fully realizing the model's potential (Thompson 2001). To prevent disagreement over copyright issues, the best solution is to set forth in the contract documents ownership rights and responsibilities (Rosenburg 2007).

Another contractual issue to address is who will control the entry of data into the model and be responsible for any inaccuracies. Taking responsibility for updating building information model data and ensuring its accuracy entails a great deal of risk. Requests for complicated indemnities by BIM users and the offer of limited warranties and disclaimers of liability by designers are essential negotiation points that need to be resolved before BIM technology is used. It also requires more time spent inputting and reviewing BIM data, which is a new cost in the design and project administration process (Ghaffarianhoseini *et al.* 2017). Although these new costs may be dramatically offset by efficiency and schedule gains, they are still a cost that someone on the project team will incur. Thus, before BIM technology can be fully used, not only must the risks of its use be identified and allocated, but the cost of its implementation must be paid for as well (Thompson and Miner 2007).

One of the most effective ways to deal with such risks is to have collaborative, integrated project delivery contracts in which the risks of using BIM are shared among the project participants along with the rewards. The American Institute of Architects has released an exhibit on BIM to help project participants define their BIM development plan for integrated project delivery (Building Design and Construction 2008). This exhibit may assist project participants in defining model management arrangements, as well as authorship, ownership, and level-of-development requirements, at various project phases.

Legal issues relating to BIM

RIGHTS OF OWNERSHIP

The first legal risk to determine is ownership of the BIM data and how to protect it through copyright and other laws. For example, if the owner is paying for the design, then the owner may feel entitled to own the data, but if team members are providing proprietary information for use on the project, their propriety information needs to be protected as well. Thus, there is no simple answer to the question of data ownership; it requires a unique response to every project depending on the participants' needs. The goal is to avoid inhibitions or disincentives that discourage participants from fully realizing the model's potential.

CONTROL OF DATA ENTRY/USAGE

Responsibility for updating BIM data and ensuring its accuracy entails a great deal of risk. As designers are subject to liability claims, such issues need to be clarified before BIM technology is utilized. It also requires more time spent imputing and reviewing BIM data, which is a new cost in the design and project administration process. Although these new costs may be more than offset by efficiency and schedule gains, they are still a cost that someone on the project team will have to bear. The cost of BIM implementation must be paid for as well.

RESPONSIBILITY FOR ERRORS

The integrated concept of BIM blurs the level of responsibility so much that risk and liability will likely be enhanced. Consider the scenario where the owner of the building files suit over a perceived design error. The architect, engineers, and other contributors of the BIM process look to each other in an effort to try to determine who is responsible for the matter raised. If disagreement ensues, the lead professional will not only be responsible as a matter of law to the claimant but may have difficulty proving fault with others such as the engineers (Rosenburg 2007).

As the dimensions of cost and schedule are layered onto the 3-D model, responsibility for the proper technological interface among various programs becomes an issue. Many sophisticated contracting teams require subcontractors to submit detailed CPM schedules and cost breakdowns itemized by line items of work prior to the start of the project. The general contractor then compiles that data, creating a master schedule and cost breakdown for the entire project. When the subcontractors and prime contractor use the same software, the integration can be fluid. In cases where the data are incomplete or is submitted in a variety of scheduling and costing programs, a team member – usually a general contractor or construction manager – must re-enter and update a master scheduling and costing program. That program may be a BIM module or another program that will be integrated with the 3-D model. At present, most of these project management tools and the 3-D models have been developed in isolation. Responsibility for the accuracy and coordination of cost and scheduling data must be contractually addressed.

Applications of mixed-reality technology

The recent rise of immersive technologies provided users with Virtual, Augmented, and Mixed Reality platforms. These platforms provide requested information to the users through a fully immersive environment, a partial immersive environment, or a mixture of the two (Azhar *et al.* 2018). While Virtual Reality (VR) is a fully immersive environment, the user is closed off from the outside environment. With AR technology (e.g. adding computer vision and object recognition) the information about the surrounding real world can be overlaid on the real world objects (Agarwal 2016). However AR technology does not allow users to manipulate virtual objects in the real environment. Mixed reality (MR), sometimes referred to as hybrid reality, is the integration of real and virtual worlds to produce new environments and visualizations where physical and digital objects co-exist and interact in real time. Mixed reality takes place not only in the physical world or the virtual world but is a mix of *actual-reality* and *virtual-reality*, encompassing both augmented reality and augmented virtuality features (Milgram *et al.* 1995).

Similar to basic computing in the 1960s, MR has traditionally been reserved for entities holding both specialized equipment and personnel such as the oil and gas industry or medicine. However, recent advancements in this field, such as the Microsoft HoloLens®, have simplified MR hardware to the point where almost anyone can take advantage; no specialized knowledge is necessary to perform most tasks (DeValle and Azhar 2018). The HoloLens® is a self-contained head-mounted display with specialized components such as a HoloLens Processing Unit (HPU) and advanced sensors to enable holographic computing. These components enable HoloLens users to engage with digital content and interact with holograms in the world around them. The first generation HoloLens® was released by Microsoft in 2016 while the second generation HoloLens® 2 was scheduled for release in July 2019. More information about the HoloLens® can be found at: www.microsoft.com/en-us/hololens.

The MR technology has long been explored as a potential asset to the Architecture, Engineering and Construction (AEC) and Facility Management (FM) industries. In 2011, Dunston and Wang published a hierarchical taxonomy of AEC operations for MR applications. They found that MR-based technology is suitable for information-intensive tasks where MR can augment a human's ability to access information and documentation in the course of performing the work and enhance the individual's decision-making ability (Dunston and Wang 2011). In this opinion paper, we will discuss four applications within the AEC and FM industries where the MR technology can be effectively used. Scenario 1 explores design visualization in a real space, scenario 2 depicts the use of MR for construction assembly, scenario 3 is based on construction site hazards identification, and scenario 4 illustrates MR applications in building inspection and maintenance.

Enhanced visualizations

MR is very effective in enhanced design visualizations. Imagine seeing exactly where an element will be placed on the site, ahead of time, or seeing how furniture fits in every room before it is even ordered. Designers are already taking clients on 3D virtual tours of their facilities using the HoloLens® to explore various design options (Blackmon and Azhar 2018). Retailers such as Lowe's have even partnered with Microsoft to work on an in-store headset where you can pick out kitchen equipment layout, paint colors, cabinet types and more. Lowe's utilizes two blank walls that mimic an unfinished kitchen, they then overlay a room hologram that, connected with a tablet app, can be manipulated to whatever finish the owner desires (Erickson 2016). Visualization tools are now used by design and marketing professionals on a daily basis. The question that lingers is how to adopt these tools as an asset to the construction phase. A good example might be seeing machinery, in a space designed to house it, before dropping it in to make sure it fits. This can be done by simply viewing a hologram of the machinery on a 1:1 scale via the HoloLens®. Specialized equipment rooms, such as MRI rooms in hospitals, need to be built to exact standards. In most cases, this specialized equipment is not even ordered until construction is sufficient and the machinery is ready for immediate installation. Once the specifications are determined and the room size is calculated, the machine can be exported as a hologram to the HoloLens® and placed inside the room before delivery. If any size restrictions arise, the room can be changed before it is too late. Another example might be viewing the airflow of an air diffuser in a room. Being able to visualize how a diffuser will distribute air in a room would make it easy to determine if the equipment is sized correctly (Blackmon and Azhar 2018).

Construction assembly

A second scenario where the AEC industry may begin to see the adoption of the MR technology is in a controlled modular construction environment (DeValle and Azhar 2018). An ENR article mentioned a contractor out of California, Martin Bros., who utilized the HoloLens® to frame a bathroom pod (Rubenstone 2016). The contractor relied solely on digital holograms to build the bathroom from the floor up. Studs were put in place without a level or a tape measure since the dimensions were preprogrammed into the hologram. As long as the hologram is spatially mapped to the area, errors would be minimized and corrected in comparison to traditional stud layout. The layout of interior partitions and exterior studs, on top of a slab on grade, would be a great area to start exploring. A modeled hologram could have all interior and exterior wall lines preprogrammed into the model so the studs could be put in without measurements or calculations in the field. The same idea would go for framing and placement of door and window openings. All the information would be preprogrammed by designers, or virtual design and construction coordinators, without ever stepping foot on site. Technology such as the HoloLens® has a potential to minimize human errors by allowing workers to have the exact measurements and layout points at their fingertips. DeValle and Azhar (2018) performed a controlled experiment to compare the productivity rates of construction crew installing plumbing and electrical lines in a bathroom pod. They found 9% improvement in productivity for crew using the HoloLens® as compared to those that were using the 2D drawings. Hologram technology even has the potential to highlight areas that do not line up after installation, with a simple scan of the area based on programmed spatial cues and light reflection off objects – much like a total station does with the laser light (Rubenstone 2016).

Construction site hazards identification

Safety is a huge area of concern in the construction industry and the MR technology could play a potential role in hazards identification and safety compliance at the jobsite (Blackmon and Azhar 2018). For example, if the HoloLens® could be preloaded with safety information that can be worn by a safety manager on site, he/she may be able to view potential areas of safety hazards and have a warning symbol that indicates where this potential hazards are and help him to "check off" the safety list. For example, any building over 2 stories would need a fall hazard remedy implemented by way of railings, any low hanging structure would need to be marked for equipment managers to determine access issues, and any workers on a roof would need to be tied off (Sacks *et al.* 2013). The safety manager could wear the HoloLens®, which would show a number of predetermined hazards with colored markers that could be located, via GPS, and overlaid on the model. One could go down a virtual checklist to make sure those major hazards have been mitigated. In addition, a safety manager could use the HoloLens® to review all the marked safety hazards overlaid directly onto the site. This would be one of the most effective ways to utilize the MR technology without added safety risk to workers (Blackmon and Azhar 2018).

Building inspection and maintenance

Globally, for facilities operations and maintenance, a facility manager has to repeatedly analyze and then solve decision situations in which it is necessary to determine a type of procedure and a way of solving a problem. These decisions are taken at different levels from strategic to operational levels. Any delay in these decisions can cause significant losses. It has been revealed that most risks associated with maintenance undertakings are attributed to a lack of proactive and preventive measures such as workforce training, risk source identification and control, safety awareness and education, waste recognition and mitigation measures and so forth (Li *et al.* 2018). In particular, the existing inspection practices primarily conducted and employed in the facility maintenance are outdated and results in inefficiencies coupled with time and cost losses. Through a MR system, the building inspectors can get quick on-demand information about routine and preventive building inspection and maintenance. For example, the information overlay technology within the HoloLens® can provide systematic inspection workflow to the building inspector inspecting a complex HVAC unit. He can visually see all necessary check points and identify areas of concern. Similarly in case of a utility breakdown such as a hot water pipe, the inspection crew can quickly pull up the 3D building model of the system showing all plumbing lines, and identify the their routes to determine the repair area without looking into 2D drawings or opening up the walls.

The abovementioned scenarios clearly indicate that the MR technology will have a significant impact on the AEC and FM industries in the coming years. This technology addresses some of the major inefficiencies during the design, construction and operation stages. We can observe that over the years, the AEC and FM industries have become more complex with the advent of various new technologies, systems, and materials. The designs are not confined to conventional strategies anymore. This is the reason MR is poised to become an important part of this sector in the times to come.

There are numerous opportunities for blending these technologies into today's AEC and FM industries. As the cost of the visualization technologies is becoming cheaper day by day, it can be envisaged that even the small firms may find it easier to procure and implement them in their projects. One of the greatest advantages MR provides is its ability to display information at the exact spot it is required. This equates to a reduction in time spent thinking about components layout, as the workers would automatically know the exact spot where the components are to be assembled. Specific improvements are needed to increase the use of the MR technology in the AEC and FM firms. The most immediate need for improvement is spatial anchoring technology so that the holographic model hold its position in a 3D space without jumping out of place as pointed out by several researchers and practitioners (DeValle and Azhar 2018; Rubenstone 2016). This is especially true in the realm of construction, where installation of components frequently require accuracy to be within 1/16 of an inch from perfect. Other shortcomings include occluded visuals in sun-lit conditions.

The MR and many similar technologies are still in the development stage and not readily available for consumer use. Substantial time is required in learning these technologies and related hardware/software and creating demonstrations. Many firms may find this process too much time consuming and lacking in added value. In addition, these technologies may not be attractive to firms that are not well versed with the technology. So, it is very important for the users to invest in those technologies which seem sustainable and be available in the long run. The MR technology still has a long way to go before being fully integrated, at a higher functional level, in any industry. The manufacturing and industrial sectors, where environments can be more controlled, may be the first to fully integrate this technology. The AEC and FM industries may then follow as has happened in the past with other technologies.

Drones in construction

Drones have many uses in the construction industry including inspecting roofs, surveying land, and observing projects. Using a drone to gather data about construction projects also protects the drone operator from having to enter a hazardous situation on site.

The limitations of ground-based surveying have led to the growth of aerial surveying using drones because it allows for easily accessible, reliable spatial information for large scale measurements. Drones are able to provide reliable data with high resolution, low cost, and great flexibility. The technology still has room for improvement; the battery life of a drone and wind resistance can become an issue depending on the size of the area that needs to be surveyed. A study by Kim and Lee (2017) identified discrepancies in the information provided by drones surveying coastal and reclaimed land, so it might be necessary to combine drone and ground survey techniques for more accurate information.

Drones have recently been introduced to help with the management of properties and real estate because they can be used to obtain 3D views and photographs for data collection. Drones allow managers to show customers views of properties that may be otherwise unavailable without the use of drone technology. Drones with high-resolution cameras and wall-climbing abilities have been able to change the perspective of viewing properties. They have the ability to provide wider, comprehensive images for consumers which allows them to make more informed decision about properties. (Ullah *et al.* 2018).

There are 5 goals for drone monitoring: 1) to collect images and videos on a project site, 2) to analyze progress and quality during construction, 3) to monitor productivity and safety on a construction site, 4) to document the current conditions of civil infrastructure systems, and 5) to quickly and frequently communicate the most up-to-date state of progress to offsite project participants. Current research on Simultaneous Localization and Mapping (SLAM) may make it easier to automatically collect data on and around project sites. Drones can monitor projects in many ways including: progress monitoring, site monitoring, building inspection, building measurement, surveying, safety inspection, assessment of structural damage, infrastructure inspection, urban monitoring, road assessment, solar power plant investigations, and geohazard investigations (Ham *et al.* 2016).

There are three types of drones (fixed wing, rotary wing, and hybrid wing) giving companies the flexibility to use the drone that best suits their needs. Wingcopter, a hybrid wing drone is one of many available brands. It can carry a load of up to 6 kilograms and can travel up to 100 kilometers without the need for a large landing space. It can map large areas using survey grade sensors and produce high resolution live video as well as thermal imaging. The Wingcopter can be customized to fit many specific needs. More information can be found at the following website: https://wingcopter.com/technology/

Summary

Information and communication technology (ICT) holds tremendous promise and potential to bring greater integration in the construction industry. The scenarios mentioned above clearly indicate that the MR technology will have a significant impact on the AEC and FM industries in the coming years. This technology addresses some of the major inefficiencies during the design, construction and operation stages. Significant benefits can be realized by developing appropriate ICT solutions. It is expected that with greater availability of cost-effective ICT tools, the traditional concept of project management will be replaced by revolutionary ideas. This should result in significant savings in time, cost, and resources. Fewer disputes and better coordination are expected as a consequence of increasing use of ICT solutions. Building information modeling (BIM) is the process of generating and managing building data during its life cycle. It is also a tool as well as a process, and increases productivity and accuracy in the design and construction of buildings. BIM uses 3-D, dynamic building modeling software and operates in real time. It supports the continuous and immediate availability of project design scope, schedule, and cost information that is high quality, reliable, integrated, and fully coordinated.

The building information model encompasses building geometry, spatial relationships, geographic information, and quantities and properties of building components. Though it is not itself a technology, it is supported to varying degrees by different technologies. BIM supports many initiatives that are critical to lean design and construction. As the fundamental principle of lean is to reduce or eliminate waste, BIM addresses many aspects of waste that occur first in the design phases and later in the construction phase.

However, there are several technological and managerial challenges ahead. The technological challenges can be broadly classified into three categories (Azhar 2011): (1) The need for well-defined transactional construction process models to eliminate data interoperability issues; (2) The requirement that digital design data be computable; and (3) The need for well-developed practical strategies for the purposeful exchange and integration of meaningful information among the building information model components. The management challenges cluster around the implementation and use of BIM. Currently, there is no clear consensus on how to implement or use BIM. Unlike many other construction practices, there is no single BIM document providing instruction on its application and use. Furthermore, little progress has been made in establishing model BIM contract documents (Post 2009). Several software firms are cashing in on the "buzz" of BIM and have programs to address certain quantitative aspects of it, but they do not treat the process as a whole. There is a need to standardize the BIM process and to define the guidelines for its implementation. Another contentious issue among the AEC industry stakeholders (i.e., owners, designers, and constructors) is who should develop and operate the building information models and how the developmental and operational costs should be distributed. To optimize BIM performance, either companies or vendors, or both, will have to find a way to lessen the learning curve of BIM trainees. Software vendors have a larger hurdle of producing a quality product that customers will find reliable and manageable and that will meet the expectations set by their marketing. Additionally, the industry will have to develop acceptable processes and policies that promote BIM use and govern today's issues of ownership and risk management. As a number of researchers, practitioners, software vendors, and professional organizations are working hard to resolve these challenges, the future of BIM is exciting and promising.

Questions for discussion

1 What is building information modeling (BIM) and how does it differ from 3-D modeling?
2 How does BIM differ from 2-D CAD?
3 What are the benefits of BIM?
4 Name three or four practical applications of BIM.
5 What precautions have to be taken when implementing BIM?
6 How can drones be best used to improve a) design and b) construction practices?

References

Agarwal, S. 2016. Review on application of augmented reality in civil engineering. *International conference on inter disciplinary research in engineering and technology*, 68–71, New Delhi, India.

Associated General Contractors of America. 2005. *The contractors guide to BIM*, 1st ed. Las Vegas, NV: AGC Research Foundation.

Autodesk, Inc. 2008. *Improving building industry results through integrated project delivery and building information modeling*. Online at http://images.autodesk.com/adsk/files/bim_and_ipd_whitepaper.pdf.

Azhar, S. 2011. Building information modeling (BIM): Trends, benefits, risks and challenges for the AEC industry. *ASCE Journal of Leadership and Management in Engineering* 11(3): 241–252.

Azhar, S., and J. Brown. 2009. BIM for sustainability analyses. *International Journal of Construction Education and Research* 5(4): 276–292.

Azhar, S., J. Brown, and R. Farooqui. 2009. BIM-based sustainability analysis: An evaluation of building performance analysis software. *Proceedings of the 45th annual ASC conference*, April 1–4. Gainesville, FL.

Azhar, S., M. Hein, and B. Sketo. 2008. Building information modeling: Benefits, risks and challenges. *Proceedings of the 44th ASC national conference*. Auburn, AL.

Azhar, S., M. Khalfan, and T. Maqsood. 2012. Building information modeling: Now and beyond. *Australasian Journal of Construction Economics and Building* 12(4): 15–28.

Azhar, S., J. Kim, and A. Salman. 2018. Implementing virtual reality and mixed reality technologies in construction education: Students' perceptions and lessons learned. *Proceedings of the 11th international conference of education, research and innovation*, November 12–14. Seville, Spain.

Azhar, S., A. Nadeem, J.Y. N. Mok, and B. H.Y. Leung. 2008. Building information modeling (BIM): A new paradigm for visual interactive modeling and simulation for construction projects. *Proceedings of the 1st international conference on construction in developing countries (ICCIDC-I)*, August 4–5. Karachi, Pakistan.

Berlo, L.V., and R. D. Laat. 2011. Integration of BIM and GIS: The development of the city GML GeoBIM extension. In *Advances in 3D Geo-information sciences*, eds. T. H. Kolbe, G. König, and C. Nagel. Springer Publishing Co. New York, NY.

Blackmon, T., and S. Azhar. 2018. Applications of augmented/mixed reality head mounted displays in the construction industry. *Proceedings of the 54th ASC annual international conference*, April 18–21. Minneapolis, MN.

Building Design and Construction. 2008. *New AIA documents include BIM agreement and two new IPD contracts*. Online at www.bdcnetwork.com/article/CA6600255.html.

CICRP. 2009. *BIM project execution planning guide, Ver 1.0.* University Park, PA: The Computer Integrated Construction Research Group, The Pennsylvania State University.

CIFE. November 22, 2007. *CIFE technical reports.* Online at http://cife.stanford.edu/Publications/index.html.

DeValle, A., and S. Azhar. 2018. An investigation of mixed reality technology for onsite construction assembly. *Proceedings of the 9th engineering, project, and production management conference (EPPM 2018)*, September 24–26. Cape Town, South Africa.

Dunston, P., and X. Wang. 2011. A hierarchical taxonomy of AEC operations for mixed reality applications. *Journal of Information Technology in Construction* 16: 433–444. Online at www.itcon.org/paper/2011/25. Accessed on December 18, 2017.

Eastman, C., Teicholz, P., Sacks, R., and Liston, K. (2008). *BIM handbook: A guide to building information modeling for owners, managers, designers, engineers, and contractors.* New York: John Wiley and Sons, Inc.

Erickson, S. 2016. *Microsoft HoloLens and Lowe's, working to redefine your next home renovation*, July 28, 2017. Online at https://blogs.windows.com/devices/2016/03/18/microsoft-hololens-and-lowes-working-to-redefine-your-next-home-renovation/

Ghaffarianhoseini, A., J. Tookey, A. Ghaffarianhoseini, N. Naismith, S. Azhar, O. Efimova, and K. Raahemifar. 2017. Building information modelling (BIM) uptake: Clear benefits, understanding its implementation, risks and challenges. *Renewable and Sustainable Energy Reviews* 75: 1046–1053.

Gleeson, J. 2008. *Computer-aided green design* [WWW Document]. Online at www.architectureweek.com/0330/tools_1-2.html.

Ham, Y., K. K. Han, J. J. Lin, and M. Golparvar-fard. 2016. Visual monitoring of civil infrastructure systems via camera-equipped Unmanned Aerial Vehicles (UAVs): A review of related works. *Visualization in Engineering* 4(1): 1–8.

Hardin, B. 2009. *BIM and construction management.* Indianapolis, IN: Wiley Publishing.

Isikdag, U., J. Underwood, and G. Aouad. 2008. An investigation into the applicability of building information models in geospatial environment in support of site selection and fire response management processes. *Advanced engineering informatics* 22: 504–519.

Kim, J. O., and J. K. Lee. 2017. UAV application for process control of the reclamation project. *Journal of Coastal Research* SI(79): 309–313.

Kriegel, E., and B. Nies. 2008. *Green BIM.* Indianapolis, IN: Wiley Publishing.

Kymmell, W. 2008. *Building information modeling: Planning and managing projects with 4D CAD and simulations.* New York, NY: McGraw Hill Construction.

Li, X., W. Yi, H-L. Chi, X. Wang, and A. P. C. Chan. 2018. A critical review of virtual and augmented reality (VR/AR) applications in construction safety. *Automation in Construction* 86: 150–162.

Macomber, H., and G. Howell. 2005. Using study action teams to propel lean implementations. *Lean Project Consulting Inc.*, Louisville, CO 80027.

Milgram, P., H. Takemura, A. Utsumi, and F. Kishino. 1995. Augmented reality: A class of displays on the reality-virtuality continuum. In *Photonics for industrial applications.* International Society for Optics and Photonics, 282–292. Bellingham, WA.

Post, N. 2009. Building team members see progress and problems. *Engineering News-Record* 22(12): 28.

Rosenburg, T. L. 2007. *Building information modeling.* Online at www.ralaw.com/resources/documents/Building%20Information%20Modeling%20-%20Rosenberg.pdf.

Rubenstone, J. 2016. Mixing reality on site. *Engineering News-Record* 227(16): 20–24.

Sacks, R., C. Eastman, G. Lee, and P. Teicholz. 2018. *BIM handbook: A guide to building information modeling for owners, designers, engineers, and contractors*, 3rd ed. New York: Wiley Publishing Co.

Sacks, R., A. Perlman, and R. Barak. 2013. Construction safety training using immersive virtual reality. *Construction Management & Economics* 31(9): 1005–1017. Online at https://doi.org/10.1080/01446193.2013.828844

Schueter, A., and F. Thessling. 2008. Building information model based energy/exergy performance assessment in early design stages. *Automation in Construction* 18(2): 153–163.

Smith, D. K., and M. Tardif. 2009. *Building information modeling: A strategic implementation guide for architects, engineers, constructors, and real estate asset managers.* Hoboken, NJ: John Wiley and Sons.

Thompson, D. B. 2001. e-Construction: Don't get soaked by the next wave. *The construction law briefing paper.* Online at www.minnlaw.com/Articles/68553.pdf.

Thompson, D. B., and R. G. Miner. 2007. *Building information modeling – BIM: Contractual risks are changing with technology.* Online at www.aepronet.org/ge/no35.html.

Ullah, F., S. M. E. Sepasgozar, and C. Wang. 2018. A systematic review of smart real estate technology: Drivers of, and barriers to, the use of digital disruptive technologies and online platforms. *Sustainability* 10(9).

Weygant, R. S. 2011. *BIM content development: Standards, strategies and best practices.* Hoboken, NJ: John Wiley and Sons.

Bibliography

Ahmad, I., and S. M. Ahmed. 2001. Integration in the construction industry: Information technology as the driving force. In *Proceedings of the 3rd international conference on construction project management*, ed. R. L. K. Tiong. Singapore: Nanyang Technical University Press, 429–434.

Akinsola, A., N. Dawood, and B. Hobbs. 2000. Development of an automated communication system using internet technology for managing construction projects. In *Proceedings of the 1st conference on implementing IT to obtain a competitive advantage in the 21st century*, eds. H. Li, Q. Shen, D. Scott, and P. E. D. Love. Hong Kong: Hong Kong Polytechnic University Press, 835–854.

Anumba, C. J., and K. Ruikar. 2002. Electronic commerce in construction: Trends and prospects. *Automation in Construction* 11: 265–275.

CRC Construction Innovation. 2007. *Adopting BIM for facilities management: Solutions for managing the Sydney opera house.* Brisbane, Australia: Cooperative Research Center for Construction Innovation.

Kim, K., H. Kim, and H. Kim. 2017. Image-based construction hazard avoidance system using augmented reality in wearable device. *Automation in Construction* (83): 390–403.

Thompson, D. B., and R. G. Miner. 2007. Building information modeling – BIM: Contractual risks are changing with technology. Online at www.aepronet.org/ge/no35.html.

6
PERFORMANCE MEASUREMENT

This chapter has two sections:

Part A: Construction Progress Measurement
Part B: Lean Process Measurement

Construction Progress Measurement is typically applied with traditional construction projects, but can also be useful with lean projects. The topic of lean process measurement is developed in greater detail in Chapter 16 of the book.

Part A: Construction progress measurement

Several methods are proposed for measuring construction performance that include a productivity component: units completed, incremental milestone, start/finish, supervisor opinion, cost ratio, weighted unit, and earned value.

Units completed

This method assumes that construction progresses linearly. If an activity requires 1000 cubic yards of concrete and 300 have been poured, then the percentage of completion is calculated as 300/1000 = 0.3 or 30%. This method falls short as the assumption of linearity does not hold true in many project types.

Incremental milestone

Percentage of completion is calculated based on a construction organization's experience. A database that catalogs previous projects can provide a reasonable estimate of job progress. For example:

If work on a building foundation has progressed as far as the placing of rebars, then the percentage in completion of the foundation milestone would be (5% + 50% + 20%) = 75%.

Other milestones can be evaluated by the same method (e.g., for piping, structural steel, etc.). Productivity calculations can be developed by comparing the labor hours utilized with the projected/budgeted labor hours.

Table 6.1 Incremental Milestone

Description	*Completion by Activity*
Foundation work:	
Excavation	5%
Form work	50%
Place rebars	20%
Pour and finish	10%
Strip forms/finish	10%
Backfill	5%
Total	100%

Start/finish

The status of activities can be estimated based on their initiation and completion. This method is suitable for activities that have a shorter duration than the reporting period. The incremental milestone method is convenient when activities have clear sequential subtasks. However, when an activity does not have clearly defined subtasks or they are not entirely sequential, the start/finish method is an effective way to measure project performance. For example, the work involved in painting a building exterior can be measured in stages using this method:

Activities	*Cumulative values*
• Work started:	10% completed
• Building pressure cleaned and prepped:	25% completed
• Main color applied:	65%
• Trim and details applied:	95%
• Clean-up and project completion:	100%

Although similar to the incremental milestone method, the primary difference is that some tasks can be started before other tasks are completed.

Supervisor opinion

Experienced supervisors have a wealth of knowledge that can provide reasonable estimates for certain categories of jobs that can be readily visualized. The most subjective method of measuring performance is opinion. This method is highly biased by the personality and benefit of the estimator. Some supervisors are overly optimistic and may provide inflated performance measurements while a company engineer may be overly critical of the current progress. Ideally, this method should be used for minor tasks. Examples of these jobs would be: painting, landscaping, dewatering, and installing architectural trim.

Cost ratio

The ratio of costs or work hours expended to the forecasted values can be used for estimating progress on construction management or administration activities. Examples of appropriate activities are quality

Table 6.2 Earned Value Analysis (Simplified Structure)

Weight	*Subtask*	*Unit*	*Quantity Total*	*Equivalent Steel Tons*	*Quantity To Date*	*Earned Tons*
0.02	Foundation bolts	each	200	1	200	1
0.28	Trusses	each	100	14	50	7
0.30	Columns	each	30	15	20	10
0.20	Beams	each	50	10	30	6
0.20	Roofing-steel deck	square foot	5000	10	1000	2
1.0		Steel	ton	50		26

Note: % complete = Earned tons to date/equivalent tons = 26/50 = 52%.

assurance, construction administration, and project management. This method is not reliable for work-in-place activities.

$$\%\text{Complete} = \frac{\%\text{actual cost of work to date}}{\text{forecasted cost at completion}}.$$

The cost incurred to date can also be used to estimate the work progress. For example, if an activity was budgeted to cost $20,000 and the cost incurred at a particular date was $10,000, then the estimated percentage complete under the cost ratio method would be 10,000/20,000 = 0.5 or 50%. This method provides no independent information on the actual percentage completed or any possible errors in the activity budget: The cost forecast will always be the budgeted amount. Good judgment is required especially when dealing with assemblies that have a large portion of the materials in the warehouse but no assembly has been started.

Weighted or equivalent units

This method calculates percentage of completion based on the concept of "earned value" (Table 6.1). In the case of erecting a steel building, subtasks are defined that are quantified by the unit of measure.

Earned value

Earned value management (EVM) is a project management technique for measuring progress in an objective manner. EVM has the unique ability to combine measurements of scope, schedule, and cost in a single-integrated system. When properly applied, EVM provides an early warning of performance problems. Additionally, EVM promises to improve the definition of project scope, prevent scope creep, communicate objective progress to management, and keep the project team focused on achieving progress. The earned value concept uses worker hours or dollars to compare a given activity with others (Griffis, Farr, and Morris 2000). The foregoing methods can be used to determine the percentage of completion for each activity. Multiplying estimated completion for each activity by its budgeted values provides earned value.

Percent complete = (Earned work hours or dollars on all accounts)/budgeted work hour or dollars.

Earned value management (EVM) application

Essential features of an EVM application include:

1 A project plan that identifies work to be accomplished.
2 A valuation of planned work, called planned value (**PV**) or budgeted cost of work scheduled (**BCWS**).

3 Predefined "earning rules" (also called metrics) to quantify the accomplishment of work, called earned value (**EV**) or budgeted cost of work performed (**BCWP**).

The EVM application in large or complex projects includes additional features, such as indicators and forecasts of cost performance (over or under budget) and schedule performance (behind or ahead of schedule). The most basic requirement of an EVM system, however, is that it quantifies progress using PV and EV. Definitions of these are:

- BCWS (PV): Budgeted cost of work scheduled (planned value): Sum of the budgeted costs for all planned work scheduled to be completed to-date or on any given date.
- BCWP (EV): Budgeted cost of work performed (earned value): Total of the budgeted costs of the activities (that make up a job or project) completed to-date or on any given date; also called achieved cost or earned value.
- ACWP (AC): Actual cost of work performed: Actual cost incurred for the work performed in a given period, including all labor, materials, other direct costs, overheads, and so on.
- BAC: Budget at completion: The original projected cost of the project.
- CV: Cost variance: Difference between an actual cost and the associated budgeted or estimated cost.

  $$CV = BCWP - ACWP$$

- SV: Schedule variance: Any deviation from the baseline plan of a project, measured by comparing budgeted cost of work scheduled with budgeted cost of work performed.

  $$SV = BCWP - BCWS$$

- CPI: Cost performance index: A measure of cost efficiency on a project. It is the ratio of earned value (EV) to actual cost of work performed (ACWP). A value of 1.0 means the project is on budget. A value greater than 1.0 means the project is under budget, while a value less than 1.0 means the project is over budget.

  $$CPI = BCWP/ACWP$$

- SPI: Schedule performance index: A measure of schedule efficiency on a project. It is the ratio of earned value (EV) to planned value (PV). A value of 1.0 means the project is on schedule. A value greater than 1.0 means the project is ahead of schedule while a value less than 1.0 means the project is behind schedule.

  $$SPI = BCWP/BCWS$$

Productivity estimation based on worth

Alfeld (1988) equates performance (worth) to the ratio of accomplishment (value) to methods (cost). Finished work is regarded as an accomplishment and a construction technique is a method. Performance is a combination of methods and accomplishment. Worthy performance occurs when the value of accomplishment is greater than the cost of methods used.

Alfeld developed another measure of performance, i.e., the "performance ability ratio" (PAR) as the ratio of exemplar performance to current performance. Exemplar performance is the historically best instance in which the value of accomplishment exceeds the cost of the methods. Performance ability ratio

(PAR) is calculated as exemplar performance/current performance. As the PAR value exceeds the ideal of 1.0, the need for improvement increases. Selecting an exemplar rate the question arises: Whose exemplar should be used?

1 This job's?
2 The company's best?
3 The industry's best?

Any of these measures can be appropriate. In most cases, construction organizations prefer to set the exemplar as their historical best performance. In pricing projects, this approach can provide a competitive advantage; bids that are based on industry norms may fail to distinguish a company from other bidders. This approach is valid only if the company's historical performance is better than the industry's performance.

Computing PAR values

Tables 6.2 and 6.3 provide examples of calculations to establish PAR values. In Table 6.2 productivity is calculated for installing gypsum board (drywall) of different thicknesses and in different configurations. For reference purposes, this calculation is built on data from the *RS Means Estimating Manual.* It is based on a crew of two carpenters working for an 8-hour day. The productivity values are calculated as the ratio of square feet of gypsum board installed to the number of worker hours required. The productivity values derived are used as a starting point in productivity improvement by using them as exemplar rates; that is, the best rate for performing each work task.

Table 6.3 shows calculations of work productivity for each of the four reference codes in Table 6.2. Productivity values are determined in column C by dividing production by work hours. Those values are compared with the exemplar values in Table 6.3 to calculate the PAR for each work activity.

Table 6.3 Productivity calculations for gypsum board installation

Reference Code	*Gypsum Board Installation (RS Means)*	*Crew Size (Carpenters)*	*Work Hours Per Day*	*Production (Square Feet Per Day)*	*Exemplar Rate Productivity (Square Feet Per Work Hour)*
0150	3/8" wall installation (standard) unfinished	2	16	2000	125
0200	3/8" ceiling installation unfinished	2	16	1800	112.5
0350	V" walls (standard) – taped and finished	2	16	965	60.31
1050	V" ceilings – taped and finished	2	16	765	47.81

Table 6.4 PAR value calculations for gypsum board installation

	Work Activity	*A*	*B*	*C = A/B*	*D*	*E = D/C*	*F*
Reference Code	*Gypsum Board Field Installation*	*Production (Square Feet) Observed*	*Work Hours*	*Productivity Square feet Per Work Hour*	*Exemplar Rate*	*PAR Values*	*Priority for Action*
0150	3/8" wall installation (standard) unfinished	2105	32	65.78	125	1.9	1
0200	3/8" ceiling installation unfinished	4153	48	86.52	112.5	1.3	3
0350	V" walls (standard) – taped and finished	689	20	34.45	60.31	1.75	2
1050	V" ceilings – taped and finished	637	16	39.81	47.81	1.2	4

Using PAR values to prioritize corrective action

The PAR values in excess of 1.0 indicate that a task has room for improvement relative to the exemplar values on which it is based. Those tasks that have the largest PAR value are the ones that have the greatest improvement potential and should be addressed first in planning for productivity improvement. In column E the exemplar rate (in column D) is divided by the job productivity (C). In this case, the PAR values range between 1.2 and 1.9; they are prioritized in column F. The PAR value of 1.2 for ½" ceilings – taped and finished – indicates performance that is slightly worse than the exemplar rate. By contrast, the PAR value for 3/8" wall installation is 1.9 and indicates great improvement potential.

The ranking in column F shows how these work activities should be prioritized for performance improvement. This exercise should be carried out for all work activities, especially those on the critical path of a project.

Using PAR values in project management

A project management team can use the methodology above to decide how best to increase crew performance to be in line with the initial estimates. To do this, changes in the methods may be needed. Simply adding crews to a job may defeat the purpose by increasing labor costs disproportionately. Whereas the example shown uses industry-wide values from the Means manual, foremen can work with crews to identify best methods that have higher levels of productivity. These best methods can then become the new exemplar rates for a contractor and for the job. These values are critical. Low-productivity levels are likely to result in schedule delays as the rate of work production will fall short of the estimator's projections. On the other hand, the attainment of higher exemplar rates will lead to earlier job completion, at lower labor costs, and very likely higher profits.

In the deployment of The Last Planner® System, weekly work plan (WWP) meetings may be used to continually improve exemplar rates and work performance. Quite often, performance below exemplar rates is due to poor planning, poor coordination, and promises not kept. The LPS process seeks to improve coordination through a detailed, ongoing evaluation of reasons for non-completion (RNC).

Labor productivity measurement

The rate of accomplishment of construction work in the field is highly dependent on the efficiency of workers and the level to which they are involved in assigned tasks.

Observations in the field in the form of activity sampling studies can be done readily by trained observers. At the most basic level crafts' activities can be classed as either working or not working; alternatively they can be classed as doing effective work, essential contributory work, or as idle, not doing useful work. The theory involved in sampling studies is beyond the scope of this book, but Oglesby, Parker and Howell (1989) offer valuable guidelines on the subject. They suggest that such studies should have a confidence level of 95% and a limit of error of 5 percent, i.e, plus or minus 5 percent. The working portion of activities should be between 40 and 60 percent, which is typical of construction activities. On that basis, 384 observations would be needed to meet the required confidence level and limit of error.

Trained observers are reportedly able to make 85 to 90% of their evaluations before being noticed by the people being studied.

Field ratings

These are computed as the ratio of people classified as working to the total number of people observed. An adjustment of 10% is added for foreperson and personal time.

Example:

Number of people on site = 120
Number observed = 100
Number noted as working = 50
Percentage working = 50%
Adjusted field rating index = 50% plus 10% = 60%

Labor Utilization Factor (LUF)

The calculation recognizes that work may be effective or essential contributory in nature.

$$\text{LUF} = \left[\text{effective work} + \tfrac{1}{4}\ \text{essential contributory work}\right] / \text{total observed.}$$

Setting up a performance measurement program

Performance measurement is essential for a construction organization's leaders to gauge performance on a number of selected indicators. These indicators provide measurements that serve as a foundation for improvement – as the old adage goes "what gets measured, gets done." Performance measurement systems should track several factors representing components of organizational performance such as: productivity indices, quality measurements, safety measurements, schedule compliance, and budget compliance. Improvement can be achieved by a variety of initiatives that include lean construction, total quality management (TQM), six sigma, and ISO 9001:2000. While most organizations have systems that keep track of financial performance, construction productivity calls for a number of specific measures. They include:

- Labor productivity
- Crew productivity
- Exemplar rates
- PAR
- Nonproductive time
- Safety performance
- Lost time accidents
- Rework labor hours
- Schedule productivity
- Subcontractor productivity
- Where applicable: Lean indicators
  - Percentage of projects complete (PPC)
  - Project waste index (PWI)

Labor productivity is a critical component of construction productivity. Labor costs account for approximately 40% of total costs for a broad range of projects. These costs may account for an even greater percentage in the case of small projects. Work sampling studies of construction operations indicate unproductive labor activity in the range of 40–60%. Figure 6.1 shows typical reasons for non-productive labor hours. In light of the probability of lost labor hours, the performance measurement of labor productivity offers perhaps the highest return on investment (ROI) of all productivity measurements.

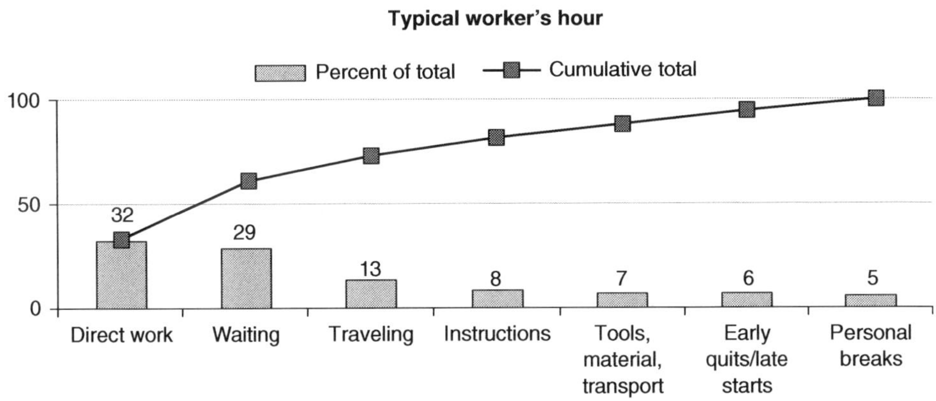

Figure 6.1 Workers' time utilization. Typical worker's hours

A successful program starts with the top management's vision. They must have a commitment to excellence, with an emphasis on a culture of continuous improvement and on management by fact. It should be included in their strategic management program, integral with their vision, mission, and values statement. In that context, productivity measurement provides the organization with the facts that it needs to make the tactical changes that are required to achieve and sustain higher levels of performance. Leaders must communicate to the entire organization its vision, mission, and values so that the continuous improvement philosophy can be incorporated in all activities, especially with construction projects.

At its simplest level, a productivity measurement program may be headed by an analyst that reports to a senior management level, such as a VP. In turn, the VP should maintain an active productivity agenda in executive meetings to keep all decision makers aware of the trends in the organization's productivity through its respective measures. Productivity measures should be selected based on key performance indicators (KPI). Critical success factors (CSF) should be identified and these should be closely linked with the organization's KPI.

Lean construction measurement

If the organization adopts lean construction as its protocol, performance measurement is an absolute requirement in order to determine the extent of accomplishment for a WWP. It is also a necessity to determine the reasons for not completing a task as planned. The analyst role may be assigned to a scheduler or other staff member who has been trained in the lean construction methodology. Other traditional measurements such as labor and crew productivity are still relevant as a means of tracking overall project performance.

Transitioning from productivity measurement to performance improvement and lean construction

The foregoing discussions emphasize the importance of productivity measurement as a component of the greater issue of performance improvement. In this context, the underlying purpose of measurement is to support initiatives to improve construction performance. These initiatives can include lean construction and integrated project delivery, TQM, ISO 9000, or six sigma. Construction organizations have a particular challenge to adopt this paradigm. Traditional construction project management tools do not address productivity; they include schedule slippages and cost overruns. Forbes and Golomski (2001) observe that the construction industry as a whole measures performance in terms of completion on time, within budget, and meeting construction codes.

Code enforcement officials ensure that minimum standards are met for building reinforcement and mechanical and electrical systems that have life-safety implications, but these standards may be met while the quality of finish and workmanship can be below expectations. Owner/client satisfaction is rarely considered under this scenario. Contracts are often awarded on the basis of low bids; most construction activity is subcontracted, and as a whole there has been a tendency to offer minimally acceptable quality in order to be price-competitive.

Construction organizations (designers and constructors) would benefit significantly by establishing formal performance improvement programs that build on the knowledge gained from the measurement approaches that have been discussed above. Lean construction can be deployed through this organizational structure. To be effective in improving the organization, management needs to commit to having a support staff for one or more individuals trained in the use of productivity/quality improvement techniques. Having them trained in lean construction methodologies would be a plus. Industrial engineers could be appropriate candidates for this position, provided they understand the construction environment.

This position should report to top management, and should be versed in the construction process as well as in the use of such tools as Pareto charts, cause and effect diagrams, activity sampling, time studies, histograms, and stratification. This position should also have experience with facilitating and leading team efforts such as assisting workers with identifying process improvement approaches. Top management should empower this position to develop and conduct programs and activities that infuse productivity and quality thinking in the workforce from the lowest to the highest levels. Productivity and quality reporting should become part of the organization's operating procedures, to the same extent as project progress and financial status reporting. These efforts can only succeed if management sets a clear example for the importance of productivity and quality and, very importantly, the use of measurement information to continuously improve the efficiency and effectiveness of procedures and activities. It is good to remember the following words of Albert Einstein:

> "Not everything that counts can be counted, and not everything that can be counted counts."
> (Sign hanging in Einstein's office at Princeton – documented by K. Harris, 1995)

Productivity measurement is both quantitatively and qualitatively vital to the profitability of a construction company. The central task is the systematic collection of data based on measured levels of output of different workers performing similar tasks, across sectors, and within particular sectors. This exercise will establish norms or reference output rates. On each specific construction project, measurements should be taken and measured rates should be compared with reference rates. Management must keep up with technological innovations worldwide and seek to adapt as well as motivate employees with improved management systems.

To facilitate performance measurement and improvement, constructors need to develop a culture of "building in" quality, and convey that philosophy to management and workers alike. They should begin to adopt Deming's fourth point "End the practice of awarding business on the basis of price tag alone."

Overall, as has been experienced in the manufacturing and other service industries, top management of all involved organizations – owners, designers, constructors, and construction management companies – must be committed to the concept of performance and quality improvement.

Questions for discussion

1 How do you define productivity in construction?
2 How does lean construction impact construction productivity?
3 What are the methods available for measuring construction progress?

4 What are their advantages and disadvantages?
5 How are exemplar rates determined?
6 Briefly explain how the terms "exemplar performance" and "PAR" values can be applied to crew production rates to identify productivity improvement opportunities. To help your explanation, consider a tile-setting task in which crews produce 700 square yards per week. Exemplar performance is 800 square yards per week for a crew of the same size.

References

Adrian, J., and D. Adrian. 1995. *Total productivity and quality management in construction.* Champaign, IL: Stipes Publishing, LLC.

Alarcon, L. F., and A. Serpell. 2004. *Performance measuring, benchmarking, and modeling of construction projects.* Santiago, Chile: Pontificia Universidad Catolica de Chile.

Alfeld, L. E. 1988. *Construction productivity.* New York: McGraw-Hill, Inc.

Forbes, L. 1999. *An engineering-management-based investigation of owner satisfaction, quality and performance variables in health care facilities construction.* Dissertation. Miami: University of Miami.

Forbes, L. H., and W. A. Golomski. 2001. A contemporary approach to construction quality improvement. In *The best on quality*, ed. M. N. Sinha, IAQ Book Series, 12: 185–200. Milwaukee, WI: ASQ Quality Press.

Griffis, F. H., J. V. Farr, and M. D. Morris. 2000. *Construction planning for engineers.* New York: McGraw-Hill.

Koskela, L., and G. Howell. 2002. The underlying theory of project management is obsolete. In *Proceedings of PMI research conference 2002,* eds. D. P. Slevin, D. I. Cleland, and J. K. Pinto. Seattle, WA: Project Management Institute.

Oglesby, C., Parker, H., and Howell, G. 1989. *Productivity improvement in construction.* New York: McGraw-Hill Inc.

Sumanth, D. J. 1984. *Productivity engineering and management.* New York: McGraw Hill, Inc.

Thomas, H. R., W. F. Maloney, M. W. Horner, G. R. Smith, V. K. Handa, and S. R. Sanders. 1990. Modeling construction labor productivity. *Journal of Construction Engineering and Management* 116(4): 705–726.

Bibliography

Ashford, J. 1989. *The management of quality in construction.* London: E & F.N. Spon.

Ballard, G., and G. Howell. 1994. Implementing lean construction – Stabilizing work flow. *Conference on lean construction,* September. Santiago, Chile.

Forbes, L., M. Ahmed, and S. Azhar. 2003. Productivity management, improvement, and cost reduction. In *System-based vision for strategic and creative design*, ed. F. Bontempi. Lisse, The Netherlands: A.A. Balkema Publishers, Swets and Zeitlinger, B.V.

Part B: Lean process measurement

Measuring lean construction performance

As in any process improvement endeavor, measurement is all important. In the words of one quality guru "you cannot improve what you cannot measure." Research studies have indicated that approximately 54% of project commitments are completed in the typical weekly schedule, using traditional project delivery. Lean project delivery can raise the reliability of work plans to 85% or more. Work accomplishment is recorded as a graphical plot of percent plans complete (PPC); it shows the percentage of the assigned plans; that is, commitments that were completed (fractional completion is not considered). Some lean construction practitioners refer to PPC as commitment reliability. In The Last Planner® System, teams commit to completing designated tasks in a given week; failure to complete all of those tasks represents less than 100% reliability. It must be noted that a PPC value does not give a true indication of how efficiently assignments have been carried out (Mohammed and Abdelhamid 2005). It does not measure the level of utilization of a work crew. Instead, it measures production planning effectiveness and workflow reliability.

Observations on commitment reliability

Figure 6.2 relates to an EPC project and shows a highly fluctuating level of commitment reliability (as represented by the PPC values). The mean value of reliability ranges between 44 and 48%, indicating that, on the average, fewer than half of the planned assignments are completed in a given week. This value is at or near zero on three occasions and between 80 and 82% on two other occasions in a 6-week period. In the context of lean construction, the desired outcome is to have minimal variability in the hand offs of assignments to downstream crews. Wide fluctuations in the completion of tasks upstream subjects downstream crews to the risk of scheduling labor and materials for tasks that are not in a full state of readiness. Having crews available for delayed assignments represents time consuming and costly waste.

In this case, the reasons for high variability were studied through a review of reasons for noncompletion (RNC). The most frequently occurring RNCs were found to be shortcomings in commitment/planning and inadequate prerequisite work. By addressing these issues, significant improvements were made, variability was reduced, and PPC/reliability was increased from the 45% range to over 80%. These changes accelerated the project from being significantly delayed to being on time with virtually the same resources.

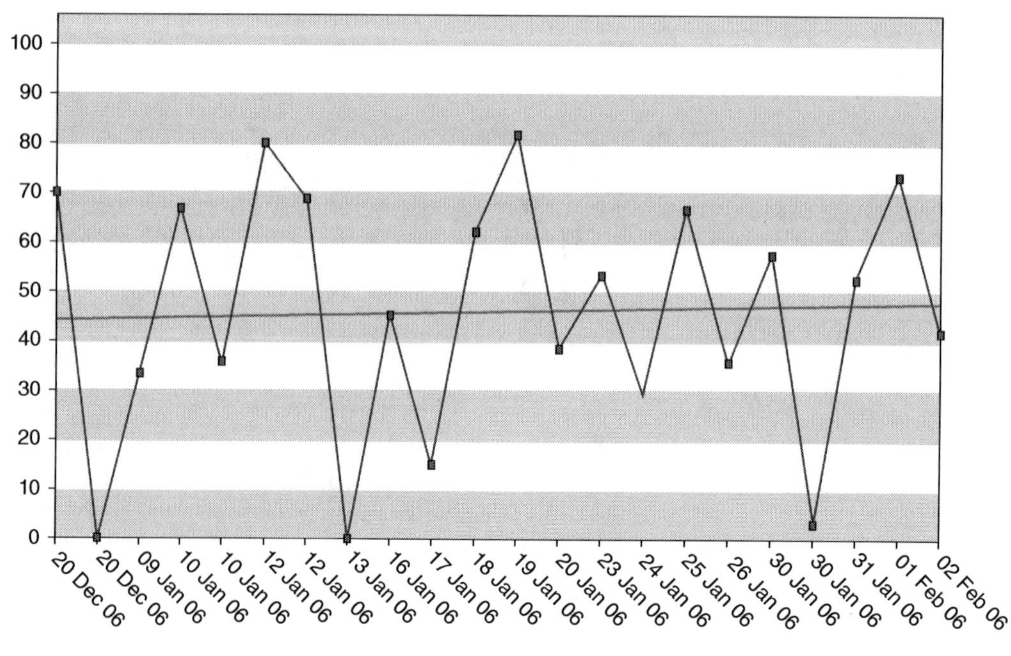

Figure 6.2 Tracking commitment reliability

Learning: reasons analysis and action

At each weekly meeting, time is devoted to learning why certain tasks were not accomplished in the previous week before creating a weekly plan for the work to be executed in the following week. Incomplete plans are studied to determine the root causes (reasons for non-completion) and why they were not completed in order to improve the effectiveness of future work plans. A Pareto chart reflects the most frequently occurring reasons. Typically bad planning and poor coordination are frequently occurring causes.

This exercise serves as an all-important learning tool. Without learning, crews are likely to simply repeat the mistakes of the past.

1 Percent plan complete (PPC) measures the degree to which work is completed as planned. It indicates the likelihood of future work accomplishment; however, the value changes as teams learn from past performance and improve future work plans. A CII-sponsored study by Ballard and Kim (2007) of pipe fitters on a major project, identified a positive correlation between PPC and productivity at the 95% confidence level. PPC is calculated by determining how many individual work tasks were initially planned, then computing a ratio of completed plans to planned tasks. There is no credit for plans that are incomplete. This measurement is also used for commitment reliability.
2 Reasons for noncompletion: These reasons are tabulated by studying incomplete plans to determine the root causes for lack of completion. These causes should be categorized in meaningful groups to facilitate identifying system-related causes. A Pareto chart is a convenient tool for ranking the reasons for noncompletion. The knowledge from the root-cause analysis enables crews to avoid obstacles in future work cycles.

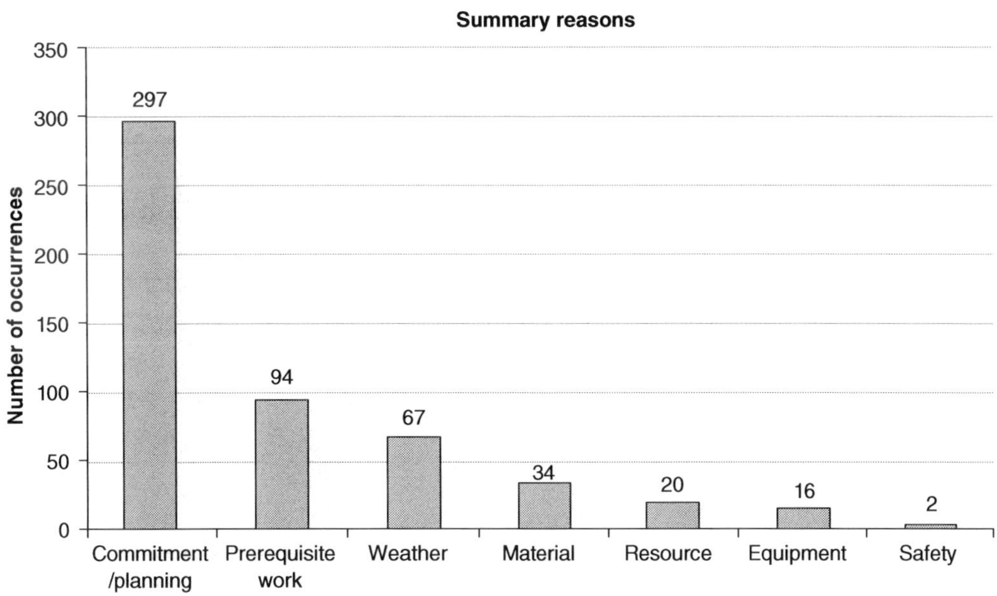

Figure 6.3 Tracking of RNC (reasons for non-completion)

Comments on the RNC

Tracking of RNC points to specific problem categories that are observed to prevent timely completion of scheduled assignments (Figure 6.2). In the example provided, commitment planning and prerequisite work were the most frequently occurring reasons for noncompletion. These categories were subjected to improved methods in the actual project with positive results. The RNC can be further investigated with five-Why analysis or a fishbone diagram.

Summary of available metrics for the Last Planner® System

Metrics are used on construction projects for monitoring and measuring the progress of work to help planners in making important decisions on time and cost improvements. There are several metrics that are currently in use for real-time control on construction projects, but researchers and practitioners are continually developing new metrics to better understand and improve project performance with the Last Planner® System (LPS).

These metrics have been developed to address the different stages of the LPS, covering both long-term and short-term planning (Ballard 2000; Chitla and Abdelhamid 2003; Mitropoulos 2005; Jang and Kim 2007; González *et al.* 2008; Hamzeh *et al.* 2008, 2015; Emdanat and Azambuja 2016). They are shown in Table 1 below, and basic descriptions are provided to explain their usage. These metrics quantify productivity, assess workflow reliability, and/or realize the progress on projects.

NOTE: Readers should refer to the description of the Last Planner® System provided in Chapters 3 and 4 to relate the metrics to the various stages in the execution of projects.

1 Percent Planned Complete (PPC)

Developed by Glenn Ballard (2000) the PPC value is obtained by calculating the number of completed tasks from the total tasks promised to be completed during a planning horizon, which is normally a week.

PPC = Did / Will

The PPC is the most commonly used metric associated with the LPS to reflect planning reliability on projects. However, perceiving the PPC metric in isolation will lead to an incomplete understanding of the overall planning process. Therefore, recent research expressed the need to complement PPC with other metrics to measure performance.

Accompanying metrics, such as the Tasks Anticipated (TA) and Tasks Made Ready (TMR), measure the success of anticipating tasks and removing constraints at the lookahead planning level.

2. Tasks Made Ready (TMR)
(Ballard 2000; Hamzeh *et al.* 2015)

$$\frac{\textit{Did}}{\textit{Can}}$$

3. Tasks Anticipated (TA)
(Ballard 2000; Hamzeh *et al.* 2015)

$$\frac{\textit{Anticipated Tasks}}{\textit{Total Tasks on WWP}}$$

TMR represents the ratio of executed tasks that were made ready by removing all their constraints. TA represents the total anticipated tasks out of the total tasks found on the Weekly Work Plan (WWP) and that continued from the lookahead week.

4. Planned Work Ready (PWR)
(Mitropoulos 2005)

$$\frac{\textbf{\textit{Work Expected to be Performed in Lookahead}}}{\textbf{\textit{Work that Should be Performed}}}$$

The Planned Work Ready (PWR) metric was introduced by Mitropoulos (2005) to assess the quality of the lookahead planning process. PWR measures the percentage of work expected to be ready as planned in the lookahead horizon, thus acting as a forecast for the schedule performance.

Moreover, Mitropoulos (2005) suggested three measures termed "deltas" to evaluate the make ready process: Delta 1 measures the constraints that are expected to be removed from the total constraints identified, Delta 2 measures the actual removed constraints, while Delta 3 looks into new identified constraints.

Delta 1 (Δ_1)
(Mitropoulos 2005)

$$\frac{\textbf{\textit{Constraints Promised to be Removed}}}{\textbf{\textit{Constraints Identified}}}$$

Delta 2 (Δ_2)
(Mitropoulos 2005)

$$\frac{\textit{Constraints Removed}}{\textit{Constraints Promised to be Removed}}$$

Delta 3 (Δ_3)
(Mitropoulos 2005)

$$\frac{\textit{New Constraints}}{\textit{Constraints Identified}}$$

Additionally, the Percent of Constraint Removal (PCR) is a metric introduced to indicate workflow predictability and is considered a leading indicator which correlates well with PPC (Jang and Kim 2007).

Percent Constraint Removal (PCR)
(Jang and Kim 2007)

$$\frac{\textbf{\textit{Ready}}}{\textbf{\textit{Can}}}$$

PCR is calculated by comparing the number of ready tasks when scheduling the WWP (Ready) to the total number of planned tasks found in the lookahead plan (Can).

Performance Factor (PF)
(Ballard and Howell 1998)

$$\frac{\textbf{\textit{Actual Labor Hours}}}{\textbf{\textit{Earned Labor Hours}}}$$

On the other hand, some metrics were introduced to relate workflow reliability to productivity. The Performance Factor (PF) above was introduced by Ballard and Howell (1998) to indirectly assess the impact of the LPS on productivity by calculating the ratio of actual labor hours to earned productivity (represented in labor hours).

Labor Utilization Factor (LUF)
(Chitla and Abdelhamid 2003)

$$\frac{\textbf{\textit{Effective Work}} + \frac{1}{4}\textbf{\textit{Essential Work}}}{\textbf{\textit{Effective}} + \textbf{\textit{Essential}} + \textbf{\textit{Not Useful}}}$$

Likewise, Chitla and Abdelhamid (2003) investigated the effective and non-effective utilization of labor. through the Labor Utilization Factor (LUF).) They found that an increase in PPC causes an increase in LUF; however, an increase in LUF does not result in an increase in PPC. Therefore, they recommended that project managers should focus mainly on PPC instead of focusing on the local production represented by LUF.

Project Productivity Index (PPI)
(González *et al.* 2008)

$$\frac{\Sigma\left(\textbf{\textit{Activity Prouctivity Indexes}}\right)}{\textbf{\textit{N}}} \times \mathbf{100}$$

Another two LPS metrics were suggested by González *et al.* (2008) as performance indicators at the activity level. The first metric above is the Project Productivity Index (PPI) calculated by finding the average of the Activity Productivity Indexes (API), with API being the ratio of the average labor productivity to the maximum labor productivity.

Planning Reliability Index (PRI)
(González *et al.* 2008)

$$\frac{\textit{Actual Progress}}{\textit{Planned Progress}} \times 100$$

The second metric is the Process Reliability index (PRI) concerned with the planning reliability at the activity level. PRI is measured by calculating the ratio of actual to planned weekly progress of a certain activity, thus overcoming the shortcoming of PPC for analyzing LPS at this level. Similarly, PPI overcomes the limitation of the PF metric by computing the maximum productivity on-site instead of the expected productivity.

Construction Flow Index (CFI)
(Sacks *et al.* 2017)

$$10\sum_{i=1}^{7} W_i P_i^{x_i}$$

Sacks *et al.* (2017) introduced the Construction Flow Index (CFI) which measures the quality of construction flow in repetitive construction projects. CFI enables project managers to improve production flow on projects by tracking and sharing this metric with trade crews. This value would be calculated and tracked with the help of project management software.

Another attempt to align short and long-term planning was performed by Emdanat and Azambuja (2016) by introducing three metrics to complement PPC, TA, and TMR. The Commitment Level (CL) metric measures the percentage of the committed required activities (critical activities) to the total required activities in a given work plan cycle.

Commitment Level (CL)
(Emdanat and Azambuja 2016)

$$\frac{\textit{Required Will}}{\textit{Should}}$$

Percent Required Complete and Ongoing (PRCO)
(Emdanat and Azambuja 2016)

$$\frac{\textit{Required Completed} + \textit{Required Ongoing on Track}}{\textit{Required Will}}$$

required activities (critical activities) to the total required activities in a given work plan cycle. Tracking CL fluctuations and relating them to the project status revealed the correlation between CL variations and the decrease in floats as a result of the increase in late paths. Another developed metric is the Percent Required Completed or Ongoing (PRCO). PRCO is the percentage of required/critical activities that are completed or expected to be completed (ongoing) on or before their promised completion dates with respect to the total required activities committed to on the work plan. The third metric is the Milestone Variance (MV) which represents the difference in days between the forecasted date to complete all remaining activities and the required date of the milestone. Emdanat and Azambuja (2016) demonstrated that teams who adjust plans to maintain high CL, PRCO, and PPC had a lower overall MV and were, therefore, more reliable.

Finally, Rizk *et al.* (2017) investigated the issue of matching load to capacity and proposed several metrics in this regard: Capacity to Load Ratio (CLR), Required Capacity Ratio (RCR), and Required Percent Complete (RPC). CLR measures the tasks completed based on the capacity available to the chosen load (tasks committed to be completed that week); the CLR can also be measured in terms of man-hrs. RCR gives the planners an indication of how many tasks were critical out of all those executed that week. Lastly, RPC depicts the percentage of critical tasks completed on a certain week. Additionally, the Weekly

Deviation (WD) is a metric, along with the three aforementioned metrics, that helps measure the extent of deviation from the WWP.

(16) Capacity to Load Ratio (CLR) (Rizk *et al.* 2017) $\dfrac{\textbf{\textit{Total Completed}}}{\textbf{\textit{WWP}}}$

(17) Required Capacity Ratio (RCR) (Rizk *et al.* 2017) $\dfrac{\textit{Required Completed}}{\textit{Total Completed}}$

(18) Required Percent Complete (RPC) (Rizk *et al.* 2017) $\dfrac{\textit{Required Completed}}{\textit{Total Required}}$

Summary remarks on common metrics:

> While these metrics equip planners with valuable input on certain aspects of planning, each metric is solely concerned with a specific planning feature and disregards other facets. This leads to an incomplete understanding of the entire project performance and the needed planning intervention actions (Emdanat and Azambuja 2016; El Samad *et al.* 2017). Additionally, many of those metrics were either inconsistently used or required data that was too difficult and time consuming to collect.

For proper management and communication among project stakeholders, the use of metrics is usually accompanied by the use of tools such as dashboards. Dashboards are substantial when multiple metrics are employed to measure and track progress on projects. Construction dashboards are thus a control tool used to facilitate communication on construction projects and to share the monitored metrics for control purposes. Such dashboards convert raw data into meaningful information regarding project performance by highlighting its weaknesses and strengths (Song *et al.* 2005; Suk *et al.* 2012). The conveyed information is used to aid planners in making responsible decisions on projects (Kerzner 2013). Furthermore, planners base such decisions on an evaluation process which considers the different metrics used and several accompanied factors. However, most metrics fall short when considering the activities' status. In fact, El Samad *et al.* (2017) demonstrated that many metrics assume that all activities are equally import when they are not. Therefore, in order to take precise decisions, planners need to have advanced metrics that reflect a deeper understanding of activities and their statuses.

References

Ballard, G. 2000. *The Last Planner® System of production control.* PhD Dissertation. University of Birmingham, UK.

Ballard, G., and G. Howell. 1998. Shielding production: Essential step in production control. *Journal of Construction Engineering and Management* 124(1): 11–17.

Ballard, G., and K. Yong-Woo. 2007. Roadmap for lean implementation at the project level. *Research Report 234-11.* Austin, TX: Construction Industry Institute.

Chitla, V. R., and T. S. Abdelhamid. 2003. Comparing process improvement initiatives based on percent plan complete and labor utilization factors. *Proceedings of the 11th annual conference of the international group for lean construction.* Virginia, USA.

El Samad, G., F. Hamzeh, and S. Emdanat. 2017. Last planner system – The need for new metrics. *Proceedings of the 25th annual conference. of the international group for lean construction.* Heraklion, Greece, 637–644.

Emdanat, S., and M. Azambuja. 2016. Aligning near and long term planning for LPS implementations: A review of existing and new metrics, *Proceedings of the 24th annual conference of the international group for lean construction.* Boston, MA, 103–112.

González, V., L. F. Alarcon, and F. Mundaca. 2008. Investigating the relationship between planning reliability and project performance. *Production Planning and Control* 19(5): 461–474.

Hamzeh, F. R., G. Ballard, and I. D. Tommelein. 2008. Improving construction workflow-the connective role of lookahead planning. *Proceedings of the 16th annual conference of the international group for lean construction*. Manchester, UK, 635–646.

Hamzeh, F. R., E. Zankoul, and C. Rouhana. 2015. How can 'Tasks Made Ready' during lookahead planning impact reliable workflow and project duration? *Construction Management and Economics* 33(4): 243–258.

Jang, J. W., and Y. W. Kim. 2007. Use of percent constraint removal to measure the make-ready process. *Proceedings of the 15th annual conference of the international group for lean construction*. East Lansing, MI, USA.

Kerzner, H. 2013. *Project management metrics, KPIs, and dashboards – A guide to measuring and monitoring project performance*, 2nd ed. Hoboken, NJ: John Wiley & Sons.

Mitropoulos, P. T. 2005. Planned work ready: A proactive metric for project control. *Proceedings of the 13th annual conference of the international group for lean construction*. Sydney, Australia, 235–242.

Mohammed, T., and T. Abdelhamid. 2005. *Understanding percent plan complete data using statistical quality control charts*. International workshop on innovations in materials and design of civil infrastructure, December 28–29, Cairo, Egypt.

Rizk, L., F. Hamzeh, and S. Emdanat. 2017. Introducing new capacity planning metrics in production planning. *Proceedings of the 25th annual conference of the international group for lean construction*. Heraklion, Greece.

Sacks, R., O. Seppänen, V. Priven, and J. Savosnick. 2017. Construction flow index: A metric of production flow quality in construction. *Construction Management and Economics* 35(1–2): 45–63.

Song, K., S. Pollalis, and F. Pena-Mora. 2005. Project dashboard: Concurrent visual representation method of project metrics on 3D building models. *Proceedings of the computing in civil engineering*. Cancun, Mexico.

Suk, S-J., B-G., Hwang, J., Dai, H. Caldas Carlos, and P. Mulva Stephen. 2012. Performance dashboard for a pharmaceutical project benchmarking program. *Journal of Construction Engineering and Management* 138(7): 864–876.

Bibliography

Shehata, M. E., and K. M. El-Gohary. 2011. Towards improving construction labor productivity and projects' performance. *Alexandria Engineering Journal* 50(4): 321–330.

Tab Tommelein, I., D. Riley, and G. Howell. 1999. Parade game: Impact of work flow variability on trade performance. *Journal of Construction Management* 125(5): 304–310.

SUMMARY OF LPS METRICS; RETRIEVED FROM HAMZEH, *ET AL.* IN CHAPTER 16

Appendix	*Metric*	*Formula*
(1)	Percent Planned Complete (PPC) (Ballard 2000)	$\frac{Did}{Will}$
(2)	Tasks Made Ready (TMR) (Ballard 2000; Hamzeh *et al.* 2015)	$\frac{Did}{Can}$
(3)	Tasks Anticipated (TA) (Ballard 2000; Hamzeh *et al.* 2015)	$\frac{Anticipated\ Tasks}{Total\ Tasks\ on\ WWP}$
(4)	Planned Work Ready (PWR) (Mitropoulos 2005)	$\frac{Work\ Expected\ to\ be\ Performed\ in\ Lookahead}{Work\ that\ Should\ be\ Performed}$
(5)	Delta 1 (Δ_1) (Mitropoulos 2005)	$\frac{Constraints\ Promised\ to\ be\ Removed}{Constraints\ Identified}$
(6)	Delta 2 (Δ_2) (Mitropoulos 2005)	$\frac{Constraints\ Removed}{Constraints\ Promised\ to\ be\ Removed}$
(7)	Delta 3 (Δ_3) (Mitropoulos 2005)	$\frac{New\ Constraints}{Constraints\ Identified}$
(8)	Percent Constraint Removal (PCR) (Jang and Kim 2007)	$\frac{Ready}{Can}$
(9)	Performance Factor (PF) (Ballard and Howell 1998)	$\frac{Actual\ Labor\ Hours}{Earned\ Labor\ Hours}$
(10)	Labor Utilization Factor (LUF) (Chitla and Abdelhamid 2003)	$\frac{Effective\ Work + \frac{1}{4} Essential\ Work}{Effective + Essential + Not\ Useful}$
(11)	Project Productivity Index (PPI) (González *et al.* 2008)	$\frac{\Sigma\left(Activity\ \mathrm{Pr}\ ouctivity\ Indexes\right)}{N} \times 100$
(12)	Planning Reliability Index (PRI) (González *et al.* 2008)	$\frac{Actual\ Progress}{Planned\ Progress} \times 100$
(13)	Construction Flow Index (CFI) (Sacks *et al.* 2017)	$10\sum_{i=1}^{7} W_i P_i^{x_i}$
(14)	Commitment Level (CL) (Emdanat and Azambuja 2016)	$\frac{Required\ Will}{Should}$

Appendix	*Metric*	*Formula*
(15)	Percent Required Complete and Ongoing (PRCO) (Emdanat and Azambuja 2016)	$\frac{\textit{Required Completed} + \textit{Required Ongoing on Track}}{\textit{Required Will}}$
(16)	Capacity to Load Ratio (CLR) (Rizk *et al.* 2017)	$\frac{\textit{Total Completed}}{\textit{WWP}}$
(17)	Required Capacity Ratio (RCR) (Rizk *et al.* 2017)	$\frac{\textit{Required Completed}}{\textit{Total Completed}}$
(18)	Required Percent Complete (RPC) (Rizk *et al.* 2017)	$\frac{\textit{Required Completed}}{\textit{Total Required}}$

7
LEAN CULTURE AND TRAINING

Setting the stage for a successful lean culture

The successful implementation of lean construction requires new knowledge and new behaviors in the project team, in effect a new culture.

A survey of 193 construction practitioners was conducted by the ENR SmartMarket Report in 2013. As shown in Figure 7.1, the survey determined that some of the most – if not *the* most – significant challenges to implementing lean on a job site stem from inadequate knowledge and understanding of lean principles and their application to construction project processes). In particular, the major challenges to implementing lean on a job site include: (1) the lack of lean knowledge, (2) the lack of sufficient support, (3) the perception that lean is too complex, (4) employee resistance, and (5) the lack of industry support.

Despite the above-mentioned challenges, the construction industry has steadily worked on lean implementation, and practitioners have noted that its benefits include improved safety, higher customer satisfaction, high quality construction, reduced schedules, reduced costs, greater profitability, and better risk management (ENR SmartMarket Report 2013). It is also clear that lean projects have a very high rate of success and that there is a close connection between lean knowledge and the success of lean projects.

Prerequisites for lean design and construction

There are several prerequisites that have to be met before a design or construction organization can successfully implement lean principles.

a. A willingness to change is essential. Lean methods are a departure from conventional methods and their adoption requires changing the behavior of people. Cultural change is the most compelling quest along with the physical transformation of an organization. One cannot force change on people – they have to be engaged so that the intrinsic satisfaction of outstanding performance will motivate them. In a lean culture, people have to be treated as the only appreciating asset. Lean enables organizations to have Responsiveness, Reliability, and Relevance.
b. A commitment to training and learning. Stakeholders at all levels need to be trained in lean techniques in order to become successful participants in lean projects. Lean implementation also requires that completed assignments have to be continually examined as a source of learning for future improvements instead of serving as sources of blame.

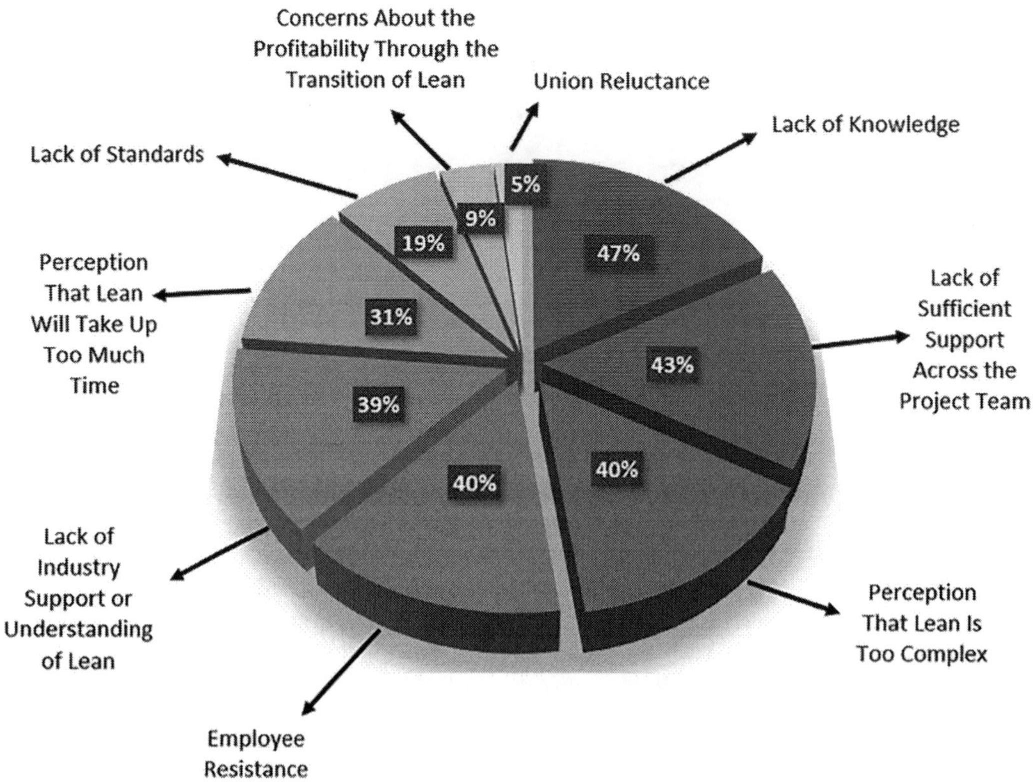

Figure 7.1 Top challenges faced in the implementation of a lean approach (according to lean practitioners)

Source: Adapted from the ENR SmartMarket Report 2013

c. A quality-oriented culture is needed for successful application of such techniques as Just-In Time (JIT), Lean Construction, and Supply Chain Management. JIT in particular, demands discipline, as there is no room for unreliable suppliers. JIT does not work in an atmosphere of suspicion, distrust, and internal competition.
d. A "Shared Vision" is essential to have all stakeholders on the same page (Macomber and Howell 2005). Promote the importance of a shared vision in which a workforce aligns itself with the direction set by a leader. This alignment is far different from carrying out orders. It is based on a sharing of beliefs and a common view of a future state that benefits everyone and makes them receptive to the changes necessary to reach that future state. In the case of lean construction, the shared vision would lead to a lean mind. A study-Action Team™ (SAT™)[1] is recommended to develop the shared vision. Members start as a reading group and focus on learning as much as possible in order to bring about change. The SAT members should be volunteers from various groups involved in construction.
e. A commitment to reducing or eliminating waste is a fundamental principle of lean construction (Polat and Ballard 2004). A commitment to improving safety is critical to lean implementation as construction accidents are rivaled only by mining. In 2006, construction accounted for 21% of all deaths and 11% of all disabling injuries/illnesses in private industry in the United States (Bureau of Labor Statistics 2006).
f. A commitment to cost and performance measures. These measures are important indicators of the impacts of lean in construction projects. Benchmarking is based on comparing performance with

other organizations (Alarcon *et al.* 2001). Measurements of percentage projects completed (PPC) and commitment reliability are essential as a foundation for continuous improvement.

g. A willingness to implement lean during the design stages. Tsao and others (Tsao *et al.* 2000; Tsao and Tommelein 2001) identified that committing more project resources to the earlier design phases improved the impact of lean techniques.
h. Collaborative relationships. Lean requires close collaboration between the parties; the standard forms of contract are adversarial in nature. Relational contracting is a transaction or contracting mechanism that apportions responsibilities and benefits of the contract fairly and transparently based on trust and partnership between the parties. It provides a more efficient and effective system for construction delivery in projects that require close collaboration for execution. Working relations between the stakeholders are improved, and improved efficiency and reduction in conflict lead to better financial returns. The most significant factors that underlie relational contracting are cooperation and dependency between the parties.
i. Information technology makes it possible to effectively manage the construction process to transform physical resources such as money, materials, manpower, and equipment. Integration is greatly improved by the sharing of information between the parties. building information modeling (BIM) is a technology that effects many economies in construction projects (see Chapter 5). It makes a reliable digital representation of the building available for design decision making, high-quality construction document production, construction planning and performance prediction, and cost estimates. Having the ability to keep information up-to-date and accessible in an integrated digital environment gives architects, engineers, builders, and owners a clear, overall vision of all their projects, as well as the ability to make informed decisions quickly.

Creating a lean culture

Keiser (2012) cited several practitioners who experienced lower costs, shorter schedules, improved safety, and better quality (Lichtig 2010; Izquierdo 2010; Angelo 2010). These practitioners closely linked the benefits with the degree of adoption of a lean culture in their organizations. The creation of a lean culture involves having stakeholders embrace the principles of *respect for people* and *continuous improvement.* Additional steps are needed to inculcate these values in the hearts and minds of people who held opposing, and long established, beliefs and attitudes.

Lichtig (2010) points to the need for a lean transformation to make a lean journey successful, and links it to the element of effective leadership. Project characteristics vary widely, as do the project teams. That creates a challenge for those who provide or support the lean function. Some companies have detoured from lean because of an inability to create a company-wide transformation (Keiser 2012).

Leaders must visibly lead the change – senior managers (in the project delivery team) must decide to commit to lean, and declare this to employees. This declaration must be followed with a steady effort to align the organization with a lean culture. The lean transformation must be organization-wide, and continuous, not sporadic. The organization must provide resources – training, facilitation, and equipment such as meeting facilities and computer systems. Initially, a consultant may be needed to fill this role, whether external or internal).

Training should be just-in-time – when people start to work on their projects they should already have had the necessary lean training, but not too long before. As observed in a UK study three key issues influence the readiness of organizations for undertaking a lean journey: lack of adequate lean awareness and training; lack of top management commitment; and culture and human attitudinal issues (Sarhan and Fox 2012).

Lean methods require new behaviors. One cannot effectively enforce change, but skillful management can lead people to derive intrinsic satisfaction from outstanding performance. A "Shared vision" places all stakeholders on the same page (Macomber and Howell 2005). The work force aligns itself with the direction set by a leader, based on a sharing of beliefs and a positive view of the future. A Study Action Team™

(SAT) is recommended to develop the shared vision. Members may start as a reading group and focus on learning as much as possible to bring about change.

An Integrated Form of Agreement (IFOA), such as used by Sutter Health on their projects, or ConsensusDocs300 or its equivalent should be provide a legally enforceable relational contract that incentivizes collaborative behavior between project team members. Standard commercial contracts are adversarial in nature, emphasizing penalties for underperformance. The IFOA seeks to align the commercial relationships of a construction project's design and construction participants as a temporary production system. Disputes are settled at the lowest possible level. It requires collaboration to bind team members as a network of commitments. It formally requires lean project activities including training, joint planning, performance measurement such as percent planned complete (PPC) and continuous learning. Completed assignments serve as a source of learning for future improvements instead of being a search for sources of blame.

Developing a successful relational contracting culture

The people who are a part of the project team need to fit in the concept of lean construction; they require special attention. Management can best enable this by:

- Employing people who are receptive to a new way of doing things. A belief in relational contracting is essential.
- Providing training and development companywide to all staff. Training industry wide would help to increase the pool of people who understand and accept the RC culture.
- A commitment by all involved companies to send their staff to workshops on relational contracting.
- Alignment of team objectives: Common vision/goals and an understanding of each individual's responsibilities and values are identified as two of the most important aspects of relational contracting. The criteria for selecting partners in RC approaches must reflect the client's business objectives and comprise both "hard" and "soft" qualities.
- Team members, along with the partners thus selected, should agree on their mutual objectives, and devise common performance appraisal plans.
- Team members should commit not to do anything that is contrary to the spirit of the relational contract. They should not impede the work of others (for example, by doing work out of sequence).
- Team members should not accept an assignment that cannot be performed correctly or safely. They should not leave the job until they know what they are planning to do the next day.
- Relational contracts should communicate management's commitment to optimizing performance at the level of the project and not push for local optimization of resources.

Organizing lean construction

The adoption of the lean construction methodology represents a paradigm shift for many owners, designers, contractors, and suppliers who have been accustomed to traditional construction practices. The success of The Last Planner® System hinges on counterintuitive actions by members of the construction supply chain to optimize the entire project instead of maximizing the efficiency and profitability of their respective portions of the work.

Greg Howell's recommendations of several years ago still hold true today. A co-founder of the Lean Construction Institute, he described. The Last Planner® System (LPS) as a disruptive technology." In effect it reveals issues and opportunities invisible to those operating in current practice. The LPS changes the way work is planned and managed on projects and in construction organizations.

Observations over the past twenty years indicate that companies implementing LPS always face these contradictions and eventually must decide to transform their company or somehow wall off those projects operating on LPS. The first projects managed on LPS can show companies what they will need to change to support wider implementation. Proponents of transformative change suggest that at this point the senior management of the respective company (in the project delivery team) must decide their future. Then they must declare this to the people in the company in an inspiring manner. The world of business has shown us many examples of companies that lost to competitive forces from others that were more agile and forward-thinking. Not that long ago, Sears was comparable to Walmart! People need a compelling reason to change; they must be able to understand their role in the change and where it will lead. It is senior management's job to say, "We are going Lean" and then tell why, what it will mean, what it will require, and what it will produce.

This declaration must be followed with a steady effort to shift the organization to the adoption of a lean culture. It isn't easy to put a company on a learning track, and people will follow their leaders as they explore the ideas and plan for implementation. Regardless of whatever structure is adopted, the focus needs to be on learning.

Glenn Ballard (1999) in "The Challenge to Change" suggests that companies should first understand themselves before rushing out to hire a consultant to make a lean transformation. Years later, it is still good advice. He cites Peter Block (Flawless Consulting) on the three possible roles for a consultant: (1) as a technical expert who tells people what to do to accomplish a specific goal; (2) as a temporary member of the organization, often for unpleasant jobs such as layoffs and reorganization; and (3) as a facilitative consultant that teaches an organization new skills. However, in the case of lean construction the people in the organization need to learn to think, see, and act in a lean mode.

Companies that understand and that have practiced self-transformation can benefit from the fresh ideas brought by a consultant. Those that have not tried to change from within may delude themselves by expecting consultants to bring about change extrinsically, when it needs to occur intrinsically. A lean consultant would benefit a company most by helping it to learn how to change. Ballard views a consultant as "always an outsider." If he or she makes the mistake of acting like management, then real change does not occur as an outsider cannot force anyone to learn new behaviors.

Ballard recommends that companies start with the LPS as it identifies improvement opportunities while freeing up resources to address them. A consultant can prepare project staff (with great diplomacy) to face unpleasant truths about their need to change their beliefs and actions in order to benefit from improved project outcomes.

Actions for change

Leading change successfully requires several necessary actions (Macomber and Howell 2005); construction project teams need to be guided in adopting new attitudes and behaviors in order to deploy lean construction. Five specific actions[2] have been identified as a minimal requirement; while adopting all five does not guarantee success, excluding any one of them will derail lean efforts.

1. Communicate the need for change in a manner that resonates with the people involved. Maintaining the status quo makes everyone vulnerable in a competitive environment. Provide specifics of how proposed changes can have concrete benefits to all involved.
2. Present stakeholders with reasonable performance standards they are expected to meet and secure their agreement. Make it clear by example that lean operates within a culture of respect. Maintain transparency with setting standards in order to project fairness, and establish on-going measurement in order to hold people accountable to their "customers." These customers may be downstream actors. Reinforce positive behaviors with sincere, public appreciation.
3. Demonstrate clearly the new behaviors instead of merely talking about them, and engage people in emulating them. Tools such as webinars are excellent for conveying knowledge and new information.
4. Measure, acknowledge, and reward new behavior. Work with people hands-on and try to see them doing something right. Point out the negative consequences of old behaviors and help to correct mistakes.
5. Help to make the changes positive for the participants. That reduces frustration and engages the workforce in responding to benefit all stakeholders, including themselves. A change agent must positively

display passion in order to convey a sense of commitment to the work force. To do less is to diminish the value of lean endeavors in the eyes of those who need to be an integral part of the change.

Leadership's role

The primary responsibility of lean leaders is:

- Ensure processes are running as designed
- Improve processes
- Empower and develop people

Leaders must visibly lead the change. The lean transformation is organizational, not just a project commitment. It needs to be continuous and not sporadic. Leaders would do well to heed the words of Aristotle:

> "We are what we repeatedly do. Excellence, then, is not an act, but a habit."
>
> Aristotle

As Deming points out in The New Economics, Management controls the system in which organizations function, and that system determines the abilities of workers to carry out their jobs. By Deming's estimates, over 85% of the failures experienced in the workplace are due to the limitations of the system, and by extension, of management's limitations! Lean requires time from people for training, conference space, equipment and supplies ranging from computers to sticky notes. The organization must provide these resources. Mentors are essential; they help people to understand change and their role in it as a positive experience. Training should be Just-In-Time; when people start work on their projects they should already have the necessary lean training, but not too long before.

Desirable leaders' attributes are to:

- Practice servant leadership,
- Teach – be the sensei,
- Be a humble learner, and
- Enable and celebrate the successes of others.

Preparing the project team

Construction involves the collaboration of a number of permanent organizations (contractors, subcontractors, suppliers, etc.) to form a temporary organization in order to execute a project. If lean construction is to be adopted, then some orientation and preparation need to be provided to the members of those organizations. A number of consultants suggest that it takes a "champion" and a firm (general contractors [GC], CM, owner) interested in investing in the process. A model project of limited risk and cost may provide a good foundation for learning; it may involve implementation of only one element of lean construction (such as The Last Planner® System). Thus, the involvement of the firm's top executive may be very limited beyond approval of a model project. A number of lean projects have evolved based on having the major training effort focused on a ranking executive responsible for an individual project, and then subsequently moving down to foremen and trades.

Experience suggests we should be wary of the idea that people can be trained in lean; some learning is required but it does not persist without leadership support. This support is usually the limiting resource. Lean facilitators need not be lean experts but they do need to be an active force for studying, learning, and thinking. Simulations such as lean-based games (see lean coaching, p. xxz) convey and reinforce very powerfully that collaborative, sharing behaviors hold mutual benefit for stakeholders in contrast to the destructive results of adversarial attitudes.

With the passage of time, an increasing number of projects have emerged that use lean principles. They run the gamut from so-called "lean-light" or "ipd-ish" projects to others that fully utilize integrated project delivery. As there are still relatively few people in the construction industry that are highly skilled in lean principles and practices, a need for lean training and coaching has grown significantly to prepare each project team.

The development of training and coaching programs

The development of formalized lean training has not been a linear process over time. When Greg Howell and Glenn Ballard founded the Lean Construction Institute in 1997 to develop and disseminate new knowledge regarding the management of work in projects (LCI 2017a), a number of companies donated company time and funding in the early days to support pilot projects based on the lean philosophy. Howell and Ballard became the driving force in the dissemination and adoption of lean knowledge in design and construction projects. The first LCI Congress was held in Denver in 1999, and it started a tradition of sharing lean construction knowledge with the OAEC community. As the construction industry cautiously began adopting lean principles in projects, some of the lessons learned were shared with the industry through the Lean Construction Institute. One significant development in the evolution of lean practice was started by the Sutter Health organization in the early 2000's as they embarked on a major construction program mandated by the California Senate to rectify major deficiencies in their health care facilities (see Chapter 9 on IPD). Concerned over the past underperformance of traditional forms of project delivery, Sutter Health took a leap of faith and adopted what was at the time a promising, yet untested, new approach to design and construction. That action proved to be a major catalyst to the adoption of lean construction, and to the many positive lessons that were learned from those projects and subsequently shared with the industry.

The roles of the Lean Construction Institute and the AGC

Both the Lean Construction Institute and the Associated General Contractors of America (AGC) have had a major impact on the development and delivery of training on the topic of lean construction. Each Annual LCI Congress since the early 2000's has routinely included training sessions; conference attendance rose from roughly two hundred in 2010 to over 1500 in 2018. Training activity was initially offered on the first day, and in response to a strong demand was increased to two days. Reportedly, over 2/3 of the 2018 attendees attended the training days as well as the Congress presentations. LCI's rapid growth is also attributed to the creation of communities of practice (CoPs) in several cities. These CoPs provide a forum for owner, Architecture, Engineering, and Construction (OAEC) members around the USA to meet and discuss the benefits of lean construction and provide guidance and training to interested organizations. Following an initiative by Cynthia Tsao in 2008, the CoPs grew in number to 31 in 2018. LCI has focused on enhancing awareness through its annual Congress and Design Forum events and has also developed interactive lean simulations that are shared with the communities of practice (CoPs) around the country. LCI also partners with the Lean Enterprise Institute (LEI) to provide additional lean training to LCI members. In addition, LCI has a collaboration with the American Institute of Architects (AIA) to offer continuing AIA education credits at each Annual Congress. While both AGC and LCI have provided training to the construction community a number of companies have developed internal resources for providing this training. Many consultants and consulting firms have mobilized to offer lean construction training and coaching.

In the interim, the Associated General Contractors of America (AGC) recognized a growing need for formalized lean training among their membership – that consisted of construction contractors and other professionals throughout the entire US. Former AGC president Chuck Greco engaged the services of Tariq Abdelhamid of Michigan State University from 2009 until 2013 under a research contract to develop a standardized lean construction certification program. Abdelhamid's initial four units subsequently expanded to comprise an interactive 50-*minute Lean 101: Foundations of Lean Construction* web-based introduction and a 35-hour in-person modular lean certification program. AGC offers the program through a network of chapters throughout the USA, but has also developed a strategic partnership with LCI to offer modules of the course at each annual LCI Congress.

Unit 1: Variation in production systems (4 hours)
Unit 2: Pull in production (4 hours)
Unit 3: Lean work structuring (4 hours)
Unit 4: The Last Planner® System (4 hours)
Unit 5: Lean supply chain and assembly (8 hours)
Unit 6: Lean design and pre-construction (4 hours)
Unit 7: Problem-solving principles and tools (7 hours)

More than 120 lean construction specialists throughout the US taught the seven units in 2019. Instructors are selected from applicants who contact AGC. Preference is given to those who have a construction background and field experience with implementing lean construction. Instructors teach the lessons in a bricks-and-mortar (versus online) setting at AGC chapters around the US, and learning sessions typically include one or two units at a time. Occasionally, AGC chapters may hold a one-week event that includes all seven units and a Lean 101 video. After completing all the required training students can take a four-hour, 150-question multiple-choice test online. Students who pass the exam are awarded a certificate as confirmation.

Training content: an overview of typical training exercises

Typical training courses include a number of games that are termed "serious games". Lean pioneer Greg Howell early on began experimenting with serious games and simulations (Tsao and Howell 2015), to illustrate lean principles to students and effectively engage them at multiple levels Serious games are distinguished from simple "gaming" in that the primary aim is educational – i.e., to *learn* through entertainment rather than to be purely entertained (Wouters *et al.* 2007). The strength of these games lies in their replication of controlled experiments; a change in the rules of each game results in a different outcome. As participants in the game observe different outcomes they can naturally arrive at a point where the association between actions and outcomes is clear. As noted by Rybkowski *et al.* (2018), in academic programs the most frequently played simulations include the Parade of Trades, the Lego® Airplane Game (or its lower cost variants – the Cup Game, Light Fixture Game, or Make-A-Card game); and Silent Squares. This implies that instructors consider the following concepts critical to Lean thought: reduction of variability (illustrated by Parade of Trades); and materials management, pull, one-piece flow, and balancing workflow (demonstrated by the Lego® Airplane Game and/or its variants). These three simulations teach the principles of pull, one-piece flow, and cross-training (Lego™ Airplane Game); impact of reduced variability (Dice Game); and importance of optimizing the whole over the parts (Silent Squares), respectively.

Silent Squares is the third most commonly played simulation. This may be due to the fact that it is an easier simulation to set up and instructors regard the demonstration of optimizing the whole over the parts as a critical lean principle to cover with students. At each session, attendees are provided with envelopes that contain large cardboard pieces that are similar to a jigsaw puzzle. Typically, there are five envelopes (numbered 1 through 5) and each piece is numbered to match its envelope. The attendees are instructed to use the pieces to complete a layout in the form of a square. They are also cautioned that no talking is allowed. They are also instructed that they may see an incomplete section in their puzzle, and that another player may have a piece that can fill the gap. But attendees are instructed not to speak, point, shrug, grunt, or make any sound. Pieces can only be passed by an offer from another player.

They can be accepted, or otherwise waved off. Essentially, they have to depend on the willingness of another player to see their need, and to fill it. Two players of five can immediately form a square, but the other three experience missing pieces. Ironically, players with complete squares have to take the risk to disrupt their own layout in order to help another player. Failure to act leads to an incomplete game. The pieces used in the game are cleverly designed so that players can exchange pieces that fill the empty spaces left with a player who has sacrificed to help another. Eventually, pieces get passed around between players until everyone has a completed square. The game teaches important lessons. Players have to discard selfish behavior in order to help others. Helping others to complete their game is similar to working together to optimize the performance of the whole team as opposed to the success of individual players.

A more recent addition to the lean construction game scene is the Target Value Design (TVD) simulation, which asks participant teams to collaboratively design and build a 2 ft. tall free-standing tower that is no more than 2 inches out of plumb, calculate its cost, and then redesign the tower, reducing the total cost by 20% (Rybkowski *et al.* 2016). Each team is provided with a kit of materials such as plastic straws, rubber bands, scotch tape, and a marshmallow (to place on the top of the tower). Each component is priced, so that the price of the completed tower can be determined.

By way of observation, this is a very engaging game; it brings out the natural competitiveness in participants while teaching powerful lessons. One version of the game has "designers" developing plans for the tower by themselves, followed by "contractors" who interpret the design as shop drawings. In the next round "designers" and "contractors" form one team to design and build a subsequent tower. The results can be dramatic – participants stretch the limits of their creativity to build a better tower than their competitors, and at a lower price. The lessons from the exercise are clear – having designers and contractors collaborate on a project is far more effective than having each discipline operate in its own silo.

An overview of the Villego®, Last Planner® simulation

This overview, reported by Gilles Bélanger is based on a half-day Villego® presentation at the LCI-Canada Congress in Montréal, Québec, Canada, by Jean-Marc Legentil and Marc-Olivier Legentil on April 30, 2019. Both presenters are co-authors of "The Toyota Kata Memory Jogger."

This simulation game is designed to represent the collaborative planning method used in the Last Planner® System. Each of the participants is assigned a task (as a subcontractor) related to building a little home with Lego® bricks (Figures 7.2 and 7.3). The goal is to discover a better way to plan a construction project using a team approach.

In Round 1 of this simulation, each subcontractor is identified with one of the 6 Lego brick colors. A general contractor and a project manager are also selected from the participants.

Each participant is identified with a color-coded badge (color as per the table above). No hats were used.

CARD CODE	SIX TRADE SUBCONTRACTORS
GR	Gray Lego parts: foundations, slab, roof
BL	Blue Lego parts
WH	White Lego parts
YL	Yellow Lego parts
GN (green)	White windows and doors
RD	Red Lego parts

Figure 7.2 The Lego building parts

Each subcontractor works with a Villego® plan that describes where the Lego bricks needs to be inserted. They plan their own work and they put their Lego bricks in place when they are called by the project manager. Lego bricks are described by color and number of studs.

Round 1 is simulated under the following constraints (not all are shown):

1 Contract duration is 10 weeks (1 min = 1 week, 10 seconds = 1 day)
2 No prefabrication allowed
3 Each defect found by clients will cost $5,000
4 Early delivery bonus of $1,000 per day
5 Liquidated Damages for late delivery of $1,000 per day
6 For safety reasons, only two subcontractor personnel are allowed on the site at a time (that is a square taped on the floor)

There are a couple of additional rules that are used to compute the project cost and the remaining profit or loss.

Round 2

In the second part, subassemblies are allowed and there is a collaborative pull planning session that includes the general contractor, the project manager and a representative of each sub-contractors.

This planning effort starts with the end. The first topic is the last thing that will be done just before completing the project and the team goes back to the first deliverable that opened the site, one step at a time.

For each task to be done and for each day, there is a Post-it note (each type of subcontractor has their own color) that shows the action item on the top section and the prerequisite at the bottom, see image below:

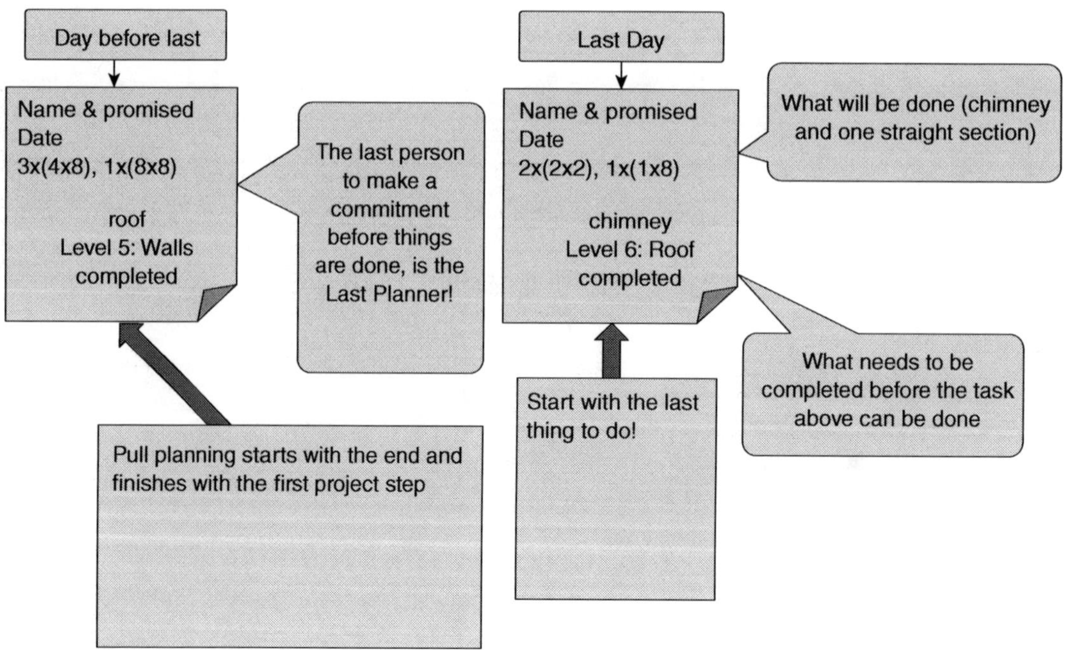

Figure 7.3 Post-it note details

The players in the Villego® game prepared pull plans as they would in a real project, working backwards from the promised completion date to the present, indicating the various building elements that have to be erected. The information required for the Post-it notes detailed in Figure 7.3. It was fascinating to see how much fun the players had with the exercise!

Lessons learned

Using the Excel spreadsheet provided by Villego, which considers the constraints and time required for building the Lego home, the team went from a loss of $2155.00 to a $400,000.00 profit. More importantly, the huge difference of emotions felt by the participants between Round 1 and Round 2 (see Table 7.1).

Many participants were interested to learn more about this collaborative planning method. This approach is useful in any environment where several specialties are working toward a common goal. Stakeholders are motivated when they can share the $ gains of a better planning method when such money is made available by the owner.

The need for coaching at project inception

Digby Christian, Director of Lean for Sutter Healthcare views coaching as on-site, on-project consultation with an experienced IFOA project manager. This is so much more effective than training which is typically offsite and divorced from the actual project.

Some practitioners think the best course of action for effective coaching would be to have a single lean development office coordinating all coaching and training. Some "all hands" work is needed, where all involved people are brought together. They may work in small groups on specific types of problems, but cross-functional teams are an absolute requirement in order to secure smooth interaction during the project. A fair amount of time – at least a few weeks, should be allowed for people to absorb new concepts. Experience suggests that people adopt lean when they make the connection with what they see, and when it influences how they explain the world. Lean simulations such as the "Airplane Production Simulation www.visionaryproducts.biz/" and "Parade of Trades" do this for some people. These simulations are typically used in a classroom setting for lean training, and are especially helpful with a mixed group of design/construction professionals. The Airplane Production Simulation involves having a team of people – often four, five, or six set up as a series of work stations in an assembly line. Lego blocks are used to assemble a number of model planes. Varied work rules clearly demonstrate the impact of the structure of the work. The Parade of Trades demonstrates the impact of reliable promising on the variability of work flow.

After becoming aware of lean construction, many firms (particularly general contractors or mechanical subcontractors) form their own internal teams to carry forward the coaching/training/facilitation to future projects. However, this approach is limited, as it benefits each firm internally, but not external stakeholders.

One coach generally starts coaching on lean implementation with a Study Action Team™ (SAT™ is the trademark of Lean Project Consulting), which is very effective in helping participants to understand lean principles and to visualize the possibilities that are available to them. The SATs are most effective with

Table 7.1 Results of the Villego simulation

ROUND 1	*ROUND 2*
Stress	Better understanding of project constraints & milestones
Annoyed by mistakes made by others	Increased trust and sense of team
Not feeling in control of the construction site	Improved quality, better delivery schedule
Empathy toward other people's problems	Feel good working with a coordinated plan of action

a cross-functional group because team members can see how others work and a relationship can begin to grow. The very act of having people in the same room can have a synergistic effect.

Cross-functional teams benefit especially with explanations of The Last Planner® System. It is helpful for the project engineers and superintendents to work together. This also applies to the project leadership of the general contractor/CM and the leadership of the owner and the architect/engineers. The LPS is built on making and keeping promises, and all team members need to internalize these principles. For small group improvement teams, cross-functional representation is also essential because these teams have a broad impact on a project, and bring a systems approach to finding solutions.

Duration of coaching

People should be coached until they are able to "swim on their own". Another approach involves, at a minimum, the coach throughout the originally assigned project. Many firms establish an internal position related to lean implementation after becoming aware of it. JE Dunn is one such company (see Chapter 10). Typical coaching engagements in the industry may last between 3 and 5 months, involving about 3 days on-site every other week but this varies widely, depending on the needs of each project. With an ambitious team, this is generally enough to set the basic habits in place. Larger organizations such as nationwide builders with many divisions are more complex and generally require a longer time. They often need additional coaching/facilitation to standardize approaches on a companywide basis. As shown in Figure 7.1, with the passage of time, a team's knowledge increases while the coach's input decreases.

One coach sees his role as building the capability of project staff, instead of doing any administrative work. He familiarizes them with spreadsheets for data capture and tracking and helps them through the daily and weekly meetings to accomplish that critical function. This lean-related role is typically carried out by a junior field superintendent.

Sustaining lean initiatives

Some lean consultants usually have a monthly follow-up, at least until procedures are firmly established; at that point the team can begin to innovate on their own. They should be able to carry out PPC measurements,

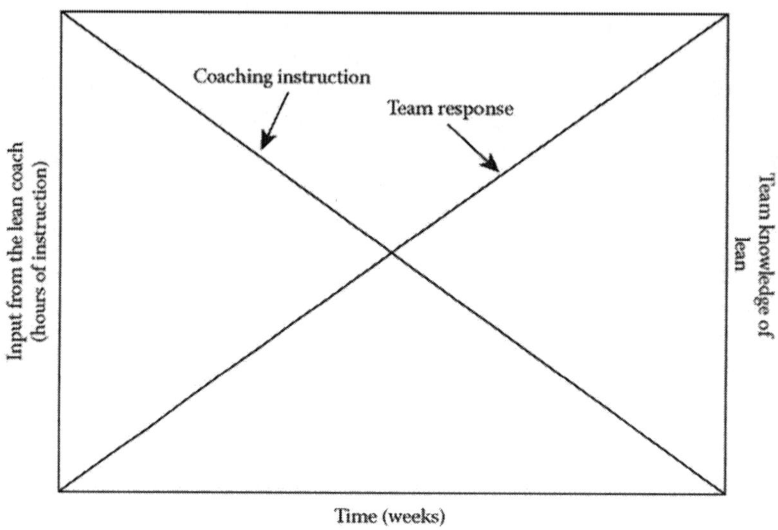

Figure 7.4 Impact of lean coaching

and to formulate improvement strategies based on them. (See Chapter 15 for more information on sustaining lean efforts.)

Qualities of a lean coach

A lean coach is a unique resource for facilitating the learning of lean principles by project teams. Successful coaches have the following attributes:

- Facilitation skills – the ability to engage a wide variety of people
- Patience
- The ability to not take things personally
- Charisma and presence are great qualities in a coach

The Toyota Way Fieldbook describes the functions of a lean coach that include both process-related activity and training activity. They include:

- Promoting the lean transformation within the organization,
- Coaching leaders at all levels,
- Teaching lean Tools and philosophy,
- Leading value stream mapping,
- Developing the lean operating system for the organization as well as metrics, and
- Leading Kaizen events.

Ed Anderson, Principal of Lean Implementation Services of Florida is also an advocate of the Socratic method. His mantra, which he attributes to Socrates, states "Ask questions to learn, ask questions to teach". He thinks that the people who have the best solutions for poor policies are the ones that are affected by them. If asked, they can share a wealth of information on how to eliminate waste and increase production and productivity.

As one former lean coach explains, people set out on a lean journey for a variety of reasons. In many cases, management sees the positive results of lean projects and wants to derive similar benefits. Some self-education occurs and a manager experiences a heightened interest in the topic; but is unsure of how to get started on a lean journey. Hiring an expert to work with the organization's team for a couple of months is a typical approach.

Lean is best described as a management philosophy for maximizing value and minimizing waste, based on the concepts of *respect for people* and *continuous improvement*. It requires fundamental change in the attitudes of the organization and this is where an expert coach can help. Three factors must be addressed in order to change the future: (1) our commitments, (2) our self-narrative or story we tell ourselves about ourselves, and (3) our action or practice. Changing any one of these three factors can yield positive results, but the change is most effective when all three have been addressed.

It is also the reason that most implementations fail. For example, changing one's actions is typically the first thing that happens in a lean implementation. Although 5S or the LPS may be initiated, some managers continue to operate in traditional ways. Their story doesn't change. They sustain the belief that the most effective way for getting work done is the traditional approach. They emphasize the use of "carrot and stick," command and control, lowest initial cost, and risk diversion strategies. These strategies contradict the lean philosophy and neutralize attempts at implementing methodologies such as the LPS, which is based on "bottom up," commitment-based planning.

Making a commitment to changing the future is another effective approach for changing the results, but also does not work completely when used alone. When people make strong declarations to change in

Figure 7.5 The three As of learning

Source: These ideas are based on original contributions by M. Horvat, originally of Lean Project Consulting

public, miraculous things happen. People's perspectives change. If a strong leader tells teams that they are going to adopt the LPS and be the best in the industry or region using this approach, the followers will make a point to familiarize themselves with the approach. Even with repeated declarations, this is typically still not enough, although it represents an integral part of the change.

The creation of a lean culture involves having stakeholders embrace the principles of *respect for people* and *continuous improvement*, but these concepts may be too abstract for most people to begin working on directly. Additional steps are needed to inculcate these values in the hearts and minds of an organization's people, and a certain amount of time has to be invested in the lean journey as it is impossible immediately change long-established beliefs and attitudes.

Activating all three of the foregoing approaches is the best method for inducing change. Leaders have to make strong declarations in public to announce their commitment to the lean philosophy. This needs to be followed up with active deployment of a lean process such as the LPS. To accomplish the transformation, stakeholders have to take an objective look at their self-narrative or how they function, and see how it differs from the lean philosophy. It is only when this type of introspection takes place that people can begin to adopt a new perspective and be receptive to the new behaviors that a lean environment requires.

All too often, managers want the benefits of lean without doing the learning themselves. That is a prescription for failure as people in an organization are quick to detect how well their leaders believe in a particular philosophy; subliminal messages convey management's sincerity even in the absence of the spoken word. If leaders do not have their hearts and minds invested in a new initiative, why should workers embrace it?

Lean coaches have been effective in building knowledge with a model termed the three As of Learning: Awareness, Acceptance, and Action. Leaders will do best with a coach that knows a lot about their desired future state; that is, *how* they wish to change and who is also expert in their industry (Figure 7.5). Beyond imparting a new curriculum, lean coaches can sometimes help simply by being available for advice while an organization is transitioning to a new operating mode. On typical construction projects, adopting lean represents a quantum leap for most construction professionals, and indicates a new way of life for practically all of the project personnel. For these two reasons a professional coach experienced in lean construction is by far the best resource to have. That individual can help to create an implementation plan and can maintain contact with the organization as lean implementation takes place.

A coach should spend time with the leaders of the company or project clarifying their goals. More importantly, he or she should make an assessment of the world that they live in, try to see situations through their eyes and spend time meeting with and becoming familiar with people all across the project.

Details of the three As

Awareness

With the project team together, a pair of production system simulations are used to give people the experience of changing the rules of how work moves through the system. The Airplane Game (Lean Zone® Production Methodology, www.visionaryproducts.biz/) shows the effect of pull scheduling, work leveling, and other concepts. The game is a hands-on simulation in which participants operate work stations along an assembly line. Through three sessions, participants explore cellular manufacturing, pull-systems, small batch flow, work in process as a manageable asset, inventory risks, cross-training, empowerment, teamwork-in-action, and many other best practices in today's manufacturing and project management/construction environments.

Traditionally, work gets done in the field by having forepersons start their crews as soon as possible. The Airplane Game puts project participants in a simulated manufacturing environment. The participants (often four, five, or six) are asked to represent an assembly line in which one worker assembles a few Lego blocks to represent an airplane section, then passes it to the next person for further assembly. One team member serves as the quality inspector. The last person should complete the airplane and set it aside as finished work.

With each round of the assembly, people are given different work rules to see how they affect the performance and output of the assembly line. In the first round people are told that they are being paid by the piece (build as much as you can – "local optimization"), that they are there to do the work and nothing else (no talking or thinking), and that to reduce set-up time they must maintain integrity of a batch size of five airplanes before moving them to the next work station. Downstream workers wait anxiously to get started while upstream workers struggle to finish a batch. In this round it is typical to see people running to get inventory, and large amounts of inventory accumulate between workstations (unbalanced line). Results are recorded on the following factors: work in process (WIP), rework, time to produce the first "good" plane, and the total number of "good" planes produced. The coach then discusses the impact of the work rules and possible improvements to the system.

The simulation is restarted in the second round, reducing the batch size to one and a queue space of one. That is, a maximum of one assembly can be in the waiting area between workstations. Some thinking and talking are allowed, a bonus is paid for team performance, and the quality inspector remains as an inspector only. Results are recorded, and the coach discusses the experience with the team.

In the third round the line is balanced, communication is completely open, people are given the authority to assist others, the inspector is put to work as an assembler, and people are made responsible for their own quality – they cannot simply pass a defective or incomplete unit down the line to another workstation.

The results are recorded. The improvement is dramatic. It is common for a team to complete five good planes in the first round and to complete 30 planes by the third with only changes to the way work moves through the system. In addition, there are interesting safety impacts. Most people say that the first round is the most unsafe.

The Parade of Trades™ (www.ce.berkeley.edu/~tommelein/parade.htm) shows the combined effects of dependence and variation in a very visual manner. A team of seven people sits around a table and represent seven building trades. A smaller team can be used, provided that each member represents a building trade.

a. Each team is provided with a die and a cup for tossing. The dies vary – one is a regular die with faces labeled 1–6. A second die has faces labeled 3 and 4, alternately. A third die may have a different numbering sequence. They are used in different rounds or by different teams to show the effects of variation induced by different dies.
b. The first player is provided with a number of tokens that represent the number of work assignments to be carried out.
c. The team has a chart subdivided into columns labeled as follows: Production Capacity, Actual Production, Lost Capacity, and Workable Backlog.

One member records values on the chart to indicate respective die rolls; each roll represents the number of people that come on the job for a week (i.e., the production capacity). That member passes tokens to the next trade, in accordance with the value of the toss, and notes that value in the production column. If the die toss indicates a value greater than the number of tokens held by that member, then the shortage is listed as lost capacity. As the game progresses tokens are passed between trades representing the building process: concrete, masonry, façade, carpentry, mechanical, electrical, and paint in that order. Passing a token is equivalent to having one trade complete its work and release that work for continuation by another trade.

If a given trade does not have enough tokens to pass on, then the duration of the project is extended. The game ends when all tokens have been passed through from the first trade to the last. Performance is tracked by noting the number of weeks elapsed, and by totaling each column, especially lost capacity and workable backlog. With more than one team, it is possible to show how the variability of the dice affects the speed at which work moves through. Typically three teams are formed and three different dice are used. One team has a standard six-sided die with the numbers being $\{1, 2, 3, 4, 5, 6\}$. The other dice that are used have the same average of 3.5, but the spread is different. Overall the dies have a mean toss of 3.5, but the range of probable values is different, and that introduces less reliability into the project. Typical roll counts are $\{1, 2, 2, 5, 5, 6\}$ and $\{2, 3, 3, 4, 4, 5\}$.

Like the airplane simulation, the results are dramatic. People estimate (correctly) that it will take approximately 10 weeks to move 35 pieces of work through the first workstation. People also generally assume that the painter, workstation 7, will complete in the 17th week. In fact, the team with the higher spread in the numbers $\{1, 2, 2, 5, 5, 6\}$ takes considerably longer to complete all the work. The concrete workstation can finish much earlier than the other teams, but bottlenecks happen with the higher likelihood of low rolls. This represents the outcome of work that is highly dependent on multiple trades that are coupled together. As is typical of the construction environment, people typically prefer speed to reliable performance. With this simulation, it is easy to see that there is a much greater improvement in throughput when work reliability is given priority over speed.

Performing these simulations with the expanded team of subcontractors, architects, engineers, designers, owners, construction managers, general contractors, project engineers, and so on brings out more conversation. Participants have a great opportunity to learn firsthand how the factors used in the simulation can be implemented and used practically in their projects. Although learning is personal, it is much more effective when it is provided by a peer group – a lean leadership group can have a much greater impact than one teacher with many students.

Acceptance

Through a powerful debriefing, a lean coach can bring out awareness of the current state of affairs in a construction organization as a good starting point. This awareness now needs to be translated into a commitment to change the managing principles in keeping with the model to change the future: commitments, story, and actions.

In initial presentations and through conversations about lean, the leadership learns to say the right things. From there, strong declarations are made about their commitment. A lean coach can continue to work with the lean leadership group throughout the engagement to help them keep in touch with their people and express the needed level of leadership during this transformation.

The lean coach can then be highly effective by introducing the leadership to lean management in a book club style group. With the shared understanding of lean, new possibilities are discussed. Simulations help participants to understand the lean manufacturing concepts of pull, kanban, waste, goal setting, and so on. This is needed to enable leaders to see the difference between current approaches and the lean approach.

Action

Learning is not useful unless it is coupled with action. An effective coach will engage participants in small group improvement activities to develop the habit of continuous improvement. Adoption of The Last Planner® System takes place throughout the beginning stage of transformation. In addition to coaching, someone familiar with lean tools can do training on techniques such as problem solving with sticky notes, group facilitation methods, visual management, five-Why analysis, and A3 thinking.

The objective of the coaching engagement is to enable participants to adopt lean management principles. Project managers typically want to adopt the LPS (they see the results), but are generally not prepared for its requirements and may not fully understand this shortcoming. Engaging a coach is the most effective way to accomplish the needed transformation; it involves gradual discovery as lean students may not be aware of how lacking they are in understanding fundamental concepts, and learning takes time and a lot of energy.

Commitment is essential – with a coach, it can take a couple of months to build the appropriate foundational habits to begin a lean transformation. Then, constant reinforcing and reflection are required to keep learning. The bigger the team is, the longer the transformation will take; the less people are ready to drop their existing habits the longer it will take as well. Shock therapy may work. By the end of a professional coaching engagement, there should be people in place to maintain the new systems and behaviors. But in reality, a successful implementation will reduce waste and result in freed capacity not additional overburdening.

Training or coaching by companies in the design/construction industry

Several organizational members of the OAEC community have developed in-house programs to meet a rapidly growing need for lean training and coaching. The following examples describe some of these efforts:

DPR Construction – team preparation for lean success

DPR Construction is a commercial general contractor and construction management firm based in Redwood City, California. The company has 27 offices throughout the United States and specializes in technically complex and sustainable projects for the advanced technology/mission critical, life sciences, healthcare, higher education and commercial office markets. DPR was #15 in ENR's 2018 list of Top 400 Contractors. DPR has been active in shaping lean thought through the pioneering involvement of Dean Reed in LCI and his interaction with Stanford and the University of California, Berkeley during lean construction's formative years.

Chris Dierks is a project executive with DPR Construction, and East Coast Lean & IPD Leader. Chris is a strong believer in workshops at the onset of a project and he views how they focus participants on what it means to develop a lean mindset. Projects start with teaching and learning and Chris plans these activities to best match the dynamics of a project by varying the duration and agenda for each workshop. A project with a shorter overall duration may need only one short workshop, while others may work best with multiple half-day or full-day workshops.

As Chris notes: We win projects at different stages, and our kick-off workshop is adjusted to that specific project. In some cases we may become involved very early – which is a best case scenario for us. Our passion for lean is important as we organize as a team and use Target Value design (TVD) on a project. We assist the owner to determine the functionality of the building, establish a budget, and enable designers to design to it. In order to maximize our success as a project team we need to decide how to become a learning organization together. Also, we need to know the right dose of knowledge that is needed, and understand how to sustain it. In our experience the best value is obtained for the project when you can pause and take the time to learn important information at the very outset. You don't want to dive into the project immediately – it is important to get people up to speed first.

Getting started:

> A kick – off workshop is an excellent starting point for getting the people in a project team together and ensuring that the project starts on a solid foundation. It provides for having everyone become familiar with each other and agree on the various processes that are necessary to carry out a project. It also serves as a vehicle for training on lean construction principles and procedures. The workshop explores a number of key questions and covers several important topics such as:
>
> - How can we develop a lean mindset?
> - What general learning do people need?

- What are the different processes involved?
- It is important to give everyone a base knowledge of lean to prepare them for working as a lean organization.
- What will provide the highest value to the entire team?
- How do we organize as a team?
- What is it that makes the most sense for this project?
- How do we establish a team for Target Value Design/Delivery (TVD)?
- How do we implement a general approach to TVD?
- We need to determine who the right people are for different TVD clusters in the team. (The PSU project diagram illustrates the composition of clusters for that project.)
- How do we become a learning organization together? Also, how do we sustain it?
- What is the right dose of knowledge to administer?
- Each month we should pause and ask – What do we need to change in moving forward?

Description of a two-day workshop agenda

Day 1

At the beginning of the workshop we take a good amount of time in getting to know team members, and to get them to talk about themselves and who they are as individuals. It is important to the team because everyone is wired differently. This is the first time that everyone is meeting as a group, and that activity is very important. It sets the tone for how the team will work during the course of the project.

The project team members are asked to stand and deliver, i.e., to get up in front of the group and share personal information such as: who they are, where they are from, their favorite food and their favorite sports team. We always ask for one unique personal fact that no one may have known before. Everyone on the team should do this – architects, engineers, inspectors, owner's representatives and so forth. Some of the questions may be changed for each project. We should have a good discussion around that, because when we learn about people we tend to care about them and we understand how they operate. It is important to a team because it leads to better understanding by everyone and better promises in the project.

With regard to owner conversations, Chris likes to have the owner stand up and explain to the team how decisions are made in the owner's organization. It helps everyone to know what details owners need to make decisions. The needs of owners' internal groups are also important. In the case of a hospital, these groups may include doctors and nurses and facility managers. This process allows the owner to explain who the key people are, and also how long it takes to make decisions depending on which stakeholders are involved. The owner's rep can also explain how the decision-making process works.

In terms of being project-specific, the owner or architect should say: "Here is our blue sky thinking. Here is why the project is happening. Here is what the program looks like. Here is where we want to go".

Other project team members should also explain the decision-making process in their own organization so that everyone understands how other parties operate and what limitations they have to work with. The team should also discuss selecting the right tool to make and document key decisions. Chris finds A3s to be very effective. This approach enables teams to move towards having measurable goals.

Goal setting: Sometimes a pre-workshop survey is administered to establish reference points for the team. A sticky note approach is also used. Team members are asked 4 or 5 questions, and they respond by writing on sticky notes and placing them on a chart. Red and green colors are often used for the notes. The details below may change depending on the type of project.

Team members are allowed 1 red dot and up to 3 green dots for each question – e.g., what is the single most important item for the Conditions of Satisfaction (CoS)? The pattern of the dots shows trends in the development of the team. One note is project-specific and starts the goal setting by answering the question "What is the one thing you do not want to see happen on this project?

Another note on goals – What do you want to accomplish, based on your knowledge?

What do you want to learn on this project? (This step helps to establish the basic CoS)

Day 2

On day 2, we do a recap of what was covered on day 1, touching on the highlights in order to reinforce the material and get everyone up to speed. In doing that we should ask – what did we miss yesterday? What do we need to talk about today?

We may be fortunate to have people who already have experience with lean, so there are topics that we may not need to cover as a team. If we do, then someone may be able to help in facilitation. It is important not to overload people with slides. It is better to stimulate them with simulations and games.

We then cover basic lean theory and use that to lead into the Last Planner® System. We typically use the Airplane Game, which shows people how the rules used in performing work affect the overall accomplishment of a project team. For example, producing in batches is subject to imbalance in the production system, with delays as some people wait on others to receive their assignments. Passing defective work down the line results in lower quality and more failed inspections. Communication and collaboration between the people at different work stations greatly increases work efficiency, and so forth. The lessons demonstrated by the simulation are excellent preparation for the DPR Pull Planning game.

With regard to decision-making conversations – we recognize that people have different needs for making decisions, and they need to be understood by all parties. Different tools exist to help teams to make and document decisions. One team may prefer an A3, another may prefer Excel.

Conditions of Satisfaction (CoS): The CoS are important and may vary by the needs of different stakeholder groups. At the end of the workshop we should have a draft of the CoS as a point of reference for the team. We should be able to say "Here is the Core team, here are some metrics that they will use.

Some people may experience a honeymoon feeling at that point.

This is a Milestone Alignment Process (MAP) – it varies from project to project and helps to establish key milestones so we can be clear on specific deliverables. For example, In the next 30 days what do we want to have? When will the first shovel hit the ground? Depending on the project the approach can be more detailed. With a longer workshop we like to establish specific milestones

Each workshop should be ended with specific action items. For each action item, there should be a responsible individual (RI), (not just a company) and a Due Date assigned

The role of work clusters in managing lean projects

A work cluster is a select small group of people drawn from different companies to work together on a project. An MEP cluster, for example, includes representatives for mechanical, electrical, and plumbing disciplines. They usually include MEP engineers, representatives from the owner and architect, each of the MEP trades, and estimators. The MEP cluster may include the trades responsible for fire protection and drywall as they typically affect each other.

Organizationally, a core team oversees the daily management activities based on the needs of the project. It ensures that clusters are formed and work collaboratively. A senior management team supports the core team, promoting team health and making key decisions that are not made by consensus.

Target Value Design/Delivery (TVD) relies heavily on the expertise of people organized as clusters. As described in Chapter 3 on design issues, when developing clusters for TVD, it is advantageous to have people with different skill sets. These people are knowledgeable about building systems and design concepts related to major building elements. This identifies the skills needed in a particular cluster. The next step would be to select the individuals who are right for these positions.

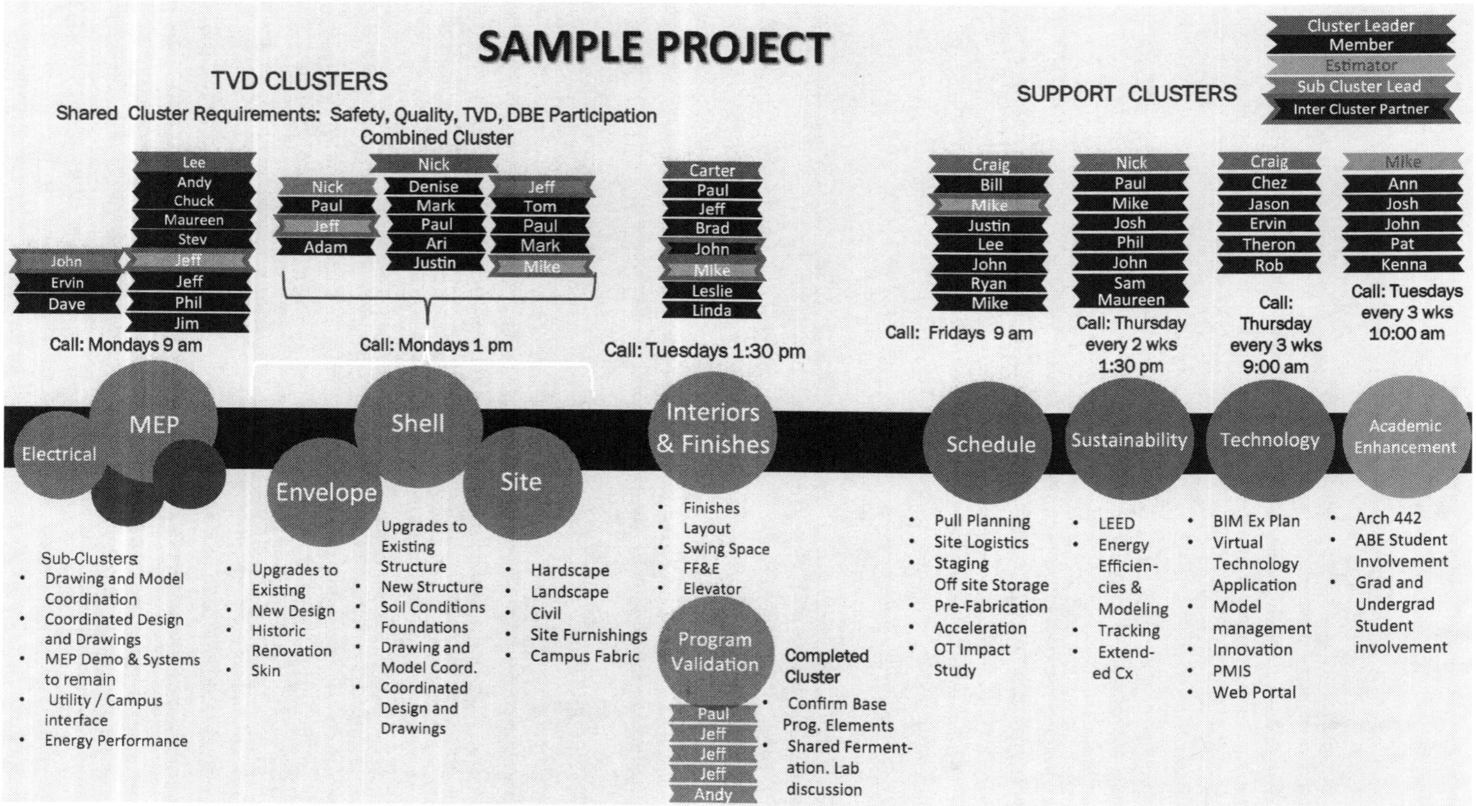

Figure 7.6 Cluster organization for a major project building renovation (Courtesy of C. Dierks/DPR Construction)

Composition of the cluster:

> There should be about seven people – typically the owner's representative, the builder, and the architect, etc. Each cluster needs a leader. He/she does not have to be an expert in the assigned area, but should be knowledgeable, understand budgets and able to be a good facilitator. That individual must keep people focused so they can come to a useful conclusion.

The owner's representative can be a facilities person in a particular discipline. People in the project team are allowed membership in a maximum of two clusters.

Examples of clusters:

> With a major university renovation project, the clusters included MEP, Shell, Schedule, Sustainability, Technology and Academic Enhancements.

The MEP cluster

The Mechanical, Electrical, and Plumbing cluster includes Sub-Clusters for: Drawing and Model Coordination; Coordinated Design and Drawings; MEP Demo & Systems to Remain; Utility/Campus interface, and Energy Performance.

The schedule cluster

This cluster includes: Pull Planning; Site Logistics; Staging; Off-site activity; Storage; Prefabrication; Acceleration; Overtime Impact Study.

The sustainability cluster

The cluster includes professionals with expertise in LEED; Energy Efficiencies & Modeling; Tracking and Extended Commissioning (Cx).

Responsibilities of cluster leaders

Cluster leaders are responsible for guiding their assigned teams toward clear and definitive decisions on the design or construction features of a project. They do not have to be experts in each specific topic, but they should have some foundational knowledge and they should be good facilitators to draw out needed ideas from the team in a collaborative manner. They are also responsible for presenting each team's decisions to the Core team (depending on the management structure used in the project).

Chris continues: At DPR we strive to have a disciplined approach to measuring and taking action in regard to how can we get better. Our process and tool we implement on projects is the Monthly Team Health Assessment. This Assessment can be facilitated by either the project's Lean Coach or another team member. We've found it should be held at the end of each month to make sure the team is following through with their commitments outlined in the project's Conditions of Satisfactions. The Monthly Team Health Assessment is the team's vehicle to drive change when needed; whether it's personnel, processes or continuous learning.

The last question requires a written answer: What is one thing you would like to see improved? (It can be anything.)

Table 7.2 IPD team health questions

KPI Key performance indicator		*IPD Team Health Questions* *Questions 1–11: Rate 1 to 5 (5 being the highest score)*
Procurement	1	The procurement process is allowing work to be executed as planned
Learning	2	The project team embraces an open, collaborative learning process.
Decision Making	3	The Supers/Foreman are empowered to make on the spot decisions in the field.
Innovation	4	I am encouraged to seek out and bring innovative ideas, solutions, and processes to the project.
BIM & Coordination	5	The field staff successfully utilizes BIM to execute the coordinated work.
Trust	6	Team members demonstrate trust and respect across all levels of the team
Visual management	7	The team successfully utilizes (including updates) visual control methods – Dashboards, A3's, Logs, Drawings, etc.
Mapping commitments	8	The Pull Planning Sessions have been effective
Schedule awareness	9	I fully understand the construction schedule for the project and how it impacts me.
Co-location	10	The trailer is a functional workspace for me
Academic enhancement	11	Team members are actively engaging in teaching classes, involving students, etc.
		Questions 12–16: Answer "Yes" or "No"
Safety	12	Do you feel safety is demonstrated as a top priority of this team?
Prefabrication	13	I am being encouraged to prefab as much as "practical."
Cost	14	Are all team members adequately managing manpower and material resources to promote a productive and efficient workflow; using the productivity log, etc.
Participation	15	Do you feel openly that you have a voice and are important member of this project team?

The assessment may be based on a survPRey or on reflections in a Big Room setting where a cross-section of inputs can be available. DPR places a high priority on having honest and open communication between team members, and determining how quickly concerns and problem questions are identified and addressed.

JE DUNN Construction's training approach

JE Dunn Construction Company, Inc. is a family-owned and employee-owned company headquartered in Kansas City, Missouri. It operates as a general building contractor and engages in the construction of projects in many industries including multiunit residential, commercial office, healthcare, education, advanced industries, manufacturing and many others. JE Dunn is #25 in Engineering New Record's (ENR) 2018 list of Top 400 Contractors.

The company embarked on a lean journey in 2012. JE Dunn had assembled a working group to study lean and IPD, and then made a serious commitment to lean. They joined LCI as a corporate member in March 2012, a clear indication of their commitment to lean. They embarked on their first IPD project in March 2012, and later they incorporated lean into the company's goals and objectives. Rebecca Snelling, a lean consultant at that time, worked with lean construction pioneers Hal Macomber and John Draper to bring JE Dunn substantial time and cost savings on several projects. JE Dunn then decided to engage Snelling in late 2012 on a permanent basis instead of a project-by-project basis. JE Dunn had a project team start working on Last Planner® System implementation with a full day of training. They then completed a phase pull planning session for all demolition work. In implementing the demolition phase pull plan, JE Dunn successfully met their objectives of early completion, no accidents, and a clean work environment. Over the next 5 years with the ongoing success of lean, the lean staff grew to 10 employees by 2018 as Snelling's role in the company was expanded to VP, National Lean Director.

The lean methodology and lean thinking have been embraced and applied as much as possible. In fact, JE Dunn has applied the methodology to projects that are traditional in nature and do not have a lean-oriented form of contract.

Although the majority of the firm's projects are design-build or CM@Risk, project teams including the trade partners, are given a full day of training that focuses on lean thinking and on the specific tools that a project will use.

Snelling attributes a significant part of the ongoing success to assigning the function of their "Lean Specialists" supporting the teams that helps keeps lean deployment on track at the project level. Depending on the size of the project, these responsibilities get transferred to different members of the project team. The Specialist helps the team learn to monitor the performance metrics and ensures that lean behaviors are continued even without a contractual obligation.

It is worthy of note that JE Dunn's high level of commitment to lean is infused in their project team as well. On each project, someone is identified who would be a good candidate for owning the Last Planner process, facilitating meetings and helping the team with execution. They try to find 2 or 3 people – the lean coach trains one, and may oversee him/her training the other Champions.

With regard to support, Snelling's team typically helps a project team with lean facilitation at the inception of each project. Rather than teaching pull planning using the DPR Pull Planning Game or Villego® (Villego 2017) for example; they first hold a training and then the Lean Specialist demonstrates the process to a superintendent or project manager (PM) on an actual project in three ways: first, by facilitating a pull plan in front of the superintendent or PM; second, by co-facilitating a pull plan with the same superintendent or PM; and third, by observing the superintendent or PM administer a pull plan on his or her own. Snelling's team offers coaching and feedback?

Training by owner organizations

Sutter Healthcare is committed to using Lean Project Delivery – especially Integrated Project delivery for its major construction projects (see Chapter 9). Since the success of lean projects became visible in the early 2000s, all projects have been lean-based, including design and construction work. One provider, Herrero Builders, through their lean champion Paulo Napolitano, created a Lean training program for its partners and clients on an IPD project.

Building on these successes, Sutter appointed Digby Christian as the Director of Lean Integrated Project Delivery. He leads a 5-day Lean and IPD training for all PMs and project controllers. The simulations used include the DPR Pull Planning Game, the Parade of Trades, and Villego®. One day per month is reserved for staff training, although it is primarily process-driven. Sutter has developed in-house resources for IPD best practices but utilizes LCI meetings and conferences for training on the Last Planner® System and Target Value Design. Sutter also often hires consultants to get new project teams started.

The majority of lean training is hands-on at the project level. In particular, Last Planner training and education are done at the start of projects, including the design and construction partners. In selecting trade partners, preference is often given to companies that have had exposure to lean practices such as the LPS. Christian's training is held once for each Project Manager, and a sequel to that program is being developed. He thinks the more integrated your team is the more likely you are to achieve your goals no matter how ambitious those goals are, and the more likely you are to achieve them with a team that is energized and happy.

James Pease is a Regional Manager with Sutter Health in Northern California. He leads a team of Project Managers who manage approximately 100 projects valued at $350M in the greater Sacramento Area. Sutter's mission is to enhance the well-being of people in the communities served through a not-for-profit commitment to compassion and excellence in healthcare services.

Pease points out that the dedication to lean from the internal construction administration structure has proven to be successful in creating a lean culture and in engaging the services of design and construction providers who are highly capable.

Choosing the right IPD Team

Paz Arroyo of DPR Construction thinks that building project owners face an important decision when assembling a project team. As she points out, when an owner commits to using an IPD approach the decision making process is going to be different than their traditional approach. An IPD approach means that they will be more collaborative and transparent with their partners, to allow for innovation. When owners need to decide which team members to choose, they have to ask many questions, such as: What are we looking for in an architecture/construction management firm? What are the factors that help to differentiate the companies? In my experience this cannot be "copy and paste" from other projects. Decision-makers need to develop their own list of relevant factors. Yes we can look at previous decisions, but each case is different. For example, safety performance may be a differentiator for choosing a general constructor if one of the companies has a higher or lower performance than the rest, but if they are all the same it will not help in differentiating. In addition, other non-traditional factors will need to be evaluated in an IPD, such as trust, evidence of collaborative behaviors in past projects, and lean IPD experience.

Many companies have established procurement procedures and treat decisions as a standardized process, rather than holding conversations to discuss concerns and develop standards (or criteria) to make assessments. As a coach, Paz guides teams by using Choosing By Advantages (CBA) to develop these conversations. When choosing an IPD team her advice is:

1. Have conversations to develop factors and criteria before interviews.
2. Make assessments and assertions of the teams, based on previous experiences and from interviewers' perceptions. This helps provide "attributes" and "advantages" for the companies in CBA language.
3. Apply the CBA Tabular method collaboratively. This means sharing assessments and the basis for them.

Making collaborative team decisions on potential IPD partners

When the owner team is open to sharing assessments about their potential IPD partners wonderful things happen. Decision makers quickly reveal different perspectives, even when looking at the same numbers! Sharing the team's experience is very relevant. Only after that conversation is the decision-making team able to share the rationale for the advantages of each company. In addition, one of the most important effects of applying CBA is avoiding going directly to a cost discussion. This means avoid choosing the cheapest option without understanding what it means for the future of the project. Decision-makers have to question how much they are willing to pay to work with a team that presents more important advantages, such as being more experienced in lean IPD. When owners have the right conversations and are willing to engage in the CBA process the result is an empowered and resolute team, eager to move forward and work with the chosen partner; understanding and willing to work with their strengths and weaknesses. That is a great start to being committed to IPD team success.

Onboarding

Bringing new team members on board is an activity that has to be well planned in order to maintain effective teams that share a common culture and a lean mindset. Onboarding not only brings new members up to speed, but ensures that the team is familiar with them and that they are enabled to fully contribute their expertise to the project at hand. The training that is typically conducted for a number of days at project inception shares a common content with everyone on the mechanics of the lean processes. Joint activity with simulations and other training activities break down barriers that may naturally occur between people that are employed by different companies, but need to work together closely to have seamless handoffs of work from one activity to the next. As new team members continue to join a project as the work progresses, the need for onboarding continues.

There should be a formal structure to onboarding – not just introductions. Presentations should be made by managers at different levels in order to create an understanding of concepts like value and waste at a practical level. The importance of collaboration should be stressed through team activity so that everyone will recognize it as the way business is done. Fundamental practices such as "going slow in order to go fast" are counter-intuitive, but must be embraced by all project team members. The importance of weekly measurement and reflection must be stressed, as the traditional practice is to avoid them.

Establishing a lean mindset

A collective lean mindset is especially important in situations where project teams have to collaborate and generate new ideas on improving project delivery. Target Value Design/delivery for example, calls for close interaction between designers and builders to formulate alternatives for lowering costs while meeting the owner's project requirements through the use of lean methods. Therefore, a common understanding of lean principles and practices is essential. It is a challenge, especially when the team members belong to different companies and have not worked together before. The desirable attributes a team are:

- An openness to new ideas and new ways of accomplishing the objectives of the project;
- A high level of trust;
- Honest and open communications, resolving issues of concern quickly, and at the lowest possible level;
- A culture of respect, which is a foundational principle of lean construction;
- A collaborative culture, where people from different companies can work together to find solutions to problems that enable the entire team to excel;
- Owners should recognize that design is not a linear process;
- Collaboration is most important – costs should not be viewed from the standpoint of initial costs;
- Designers should strive to meet the intent of the design, but should be receptive to innovative inputs from the project team. They should be receptive to new ideas to provide value – for example, project savings may potentially accommodate new owner needs if possible; and
- Contractors should respond to ideas quickly and support the design process by estimating costs and providing recommendations for constructability.

Accomplishing a lean state

Lean can be characterized in terms of objectives, principles, and methods or tools. According to Glenn Ballard, co-founder of the Lean Construction Institute, the lean ideal is to provide a custom product exactly fit for purpose, delivered instantly, and with no waste. If one attempts to conduct business or deliver projects while maximizing value and minimizing waste, then in a minimal sense that represents lean. But it would make little sense to call someone lean who claims to pursue those objectives but disregards fundamental principles. These principles are identified in the Construction Industry Institute's report (CII PT 191), which focus on the process level, or the 14 principles defined in Jeffrey Liker's *The Toyota Way*, as described above that deal with the lean enterprise. Ballard points out that the same holds true for those claiming allegiance to these principles but may not be using available methods. For example, a candidate might proclaim the principle of creating connected process flow, but fail to use pull mechanisms to release work between specialists or fail to reduce batch sizes or fail to right size work-in-progress inventories. Lastly, since lean is a never-ending journey in pursuit of perfection, it is more appropriate to ask about the rate of learning rather than the level of conformance to the ideal – remember the fable of the tortoise and the hare.

How can you tell if a project or company is lean? Ballard suggests you assess yourself by answering the following questions:

- Are you pursuing the lean ideal?
- Are you following the appropriate principles when striving for the lean ideal?
- Are you using the best methods for implementing those principles?
- How fast are you learning as a project organization?

Conditions of Satisfaction (COS) example

The Conditions of Satisfaction outline the success factors and the expected behaviors that are essential for implementation of a successful lean project. This example relates to contractors primarily. The owner's CoS would emphasize the owner's priorities. By signing this document, team members acknowledge their commit to working in a spirit of trust and collaboration for the best interests of the team and the project.

Expected behavior and commitment

As a member of this team, I understand that my focused and active participation is needed to ensure that work flows in a productive fashion for all team members. By signing this COS, I commit to the principles and practices of lean project delivery as outlined in the LPD 101 Lean Project Fundamental Guide, learning the systems and continually working in a spirit of trust, transparency and collaboration to eliminate waste and improve the performance of the team.

The team will measure success by achieving all of the following conditions of satisfaction:

- Everyone is profitable
- The project is delivered to the mutually agreed Target Cost
- The project is delivered to meet a mutually agreed schedule
- All team leaders will have excellent knowledge of their work and their scope
- RFIs will be only those where information is clearly missing in the documents or site conditions require resolution
- RFIs will be issued and answered in a timely manner to avoid becoming project constraints
- Change notices resulting from legitimate RFIs will be processed in a timely manner to avoid becoming project

Constraints

- The overall objective will be to minimize RFI's and Change Notices
- There will be exceptional teamwork
- There will be minimal rework due to poor quality work
- All schedules will be developed and executed using the Last Planner® System
- All necessary stakeholders will be involved with planning, decisions and problem solving
- Collaborate to rapidly mitigate problems, issues and constraints

As a valued member of this lean project team, I commit to working to the best of my abilities in a spirit of trust, transparency and collaboration utilizing lean project delivery practices in accordance with the above noted expected behavior.

The Big Room

The Big Room (or Obeya in Japanese) is a major project management tool for enhancing effective and timely communication. It contains highly visual charts depicting program timing, milestones and progress to date. It is an important component of project culture as its use provides critical timely information and prompt decisions that all-important to lean project delivery. It is a physical space that brings together different project team members and promotes cross functional collaboration. Ed Anderson, Principal of Lean Implementation Services of Florida views a Big Room on the construction site as an investment and an example of good foresight. He suggests a double-wide trailer provided with 4 foot by 8 foot dry-erase boards on every wall surface. It can be highly beneficial to equip a Big Room with technology such as Smart Boards that can be programmed (with computers) to keep meeting notes, building details, schedule dates and other critical information readily available for review and discussion.Video conferencing capabilities are also highly valuable as they increase the flexibility of the space by making it possible to communicate with key people who may be offsite.

Other desirable features of a big room are:

"Break-out" rooms that can be used for smaller meetings such as resolving an issue that is not central to the entire meeting

Electrical power strips to allow team members to get power for multiple devices, phones, laptops, tablets

Rest rooms

Food preparation space (kitchen with a sink)

Big Room rules of engagement

These rules help to make the Big Room an effective environment for collaboration and decision making.

- Have a Positive Attitude,
- The Site is a Safe Zone,
- Everyone is a Team Member,
- Speak so You Can be Heard,
- Listen to Others – Respect Their Opinion,
- Stay on Topic,
- No sidebar Conversations,
- Mute Cell Phones,
- Take Urgent Calls Outside,
- No Multi-Tasking,
- Stay on Time,
- Be Kind, and
- Have Fun.

Questions for discussion

1 Name at least four prerequisites for lean construction. Why are they important?
2 Explain if a lean coach is needed to begin a lean construction program and why.
3 How does the Parade of Trades simulation illustrate the impact of reliability on project duration?
4 What is a lean mindset? Explain its significance in a project and how it can be created.
5 Explain the benefits of cluster groups in a project team.

Notes

1 Study-Action Team™ (SAT™) is the intellectual property of Lean Project Consulting, Inc.
2 Macomber and Howell's five Necessary Actions for Change, Lean Project Consulting, pp. 1–2. With permission.

References

Alarcon, L. F., A. Grillo, J. Freire, and S. Diethelm. 2001. Learning from collaborative benchmarking in the construction industry. *Proceedings of the 9th annual conference of the international group for lean construction (IGLC-9)*, August 6–8. Singapore, University of Singapore.

Angelo, T. 2010. Grunau Company, presentation at the 12th Annual Lean Construction Congress. Boulder, CO, USA, October 21.

Ballard, G. 1999. The challenge to change. In *Selected articles from the lean construction chronicle*. Spring edition.

Bureau of Labor Statistics. 2006. *Fatal and non-fatal occupational statistic from 1992–2004*. Online at www.bls.gov.

ENR SmartMarket Report. 2013. *Lean Construction: Leveraging Collaboration and Advanced Practices to Increase Project Efficiency*. McGraw-Hill Construction. <https://www.leanconstruction.org/media/docs/Lean_Construction_SMR_2013.pdf>

Izquierdo, J. 2010. Grana y Montero, presentation at the 12th Annual Lean Construction Congress. Boulder, CO, USA, October 21.

Keiser, J. 2012. Leadership and Cultural Change: Necessary Components of a Lean Transformation. *Proceedings of the 20th Annual Conference for the International Group for Lean Construction*, July 18–20, San Diego, CA.

Lean Construction Institute (LCI). 2017a. <https://www.leanconstruction.org/about-us/what-is-lean-design-construction/history/>. Nov. 26. Accessed January 8 2020.

Lichtig, W. A. 2010. The Integrated Agreement for Lean Project Delivery, in *Improving Healthcare through Built Environment Infrastructure*, M. Kagioglou and P. Tzortzopoulos (eds). Oxford, UK: Wiley-Blackwell. doi: 10.1002/9781444319675.ch6

Liker, J. K. 2004. *The Toyota way*. New York: McGraw-Hill.

Macomber, H., and G. Howell. 2005. *Using study action teams to propel lean implementations*. Louisville, CO: Lean Project Consulting Inc.

The parade of trades. Online at www.ce.berkeley.edu/~tommelein/parade.htm.

Polat, G., and G. Ballard. 2004. Waste in Turkish construction: Need for lean construction techniques. *Proceedings of the 12th annual conference of the international group for lean construction (IGLC-12)*, August 3–4. Helsingor, Denmark.

Rybkowski, Z. K., and L. Forbes. 2016. 'Chapter 8: Lean Construction,' in *Handbook of Construction Management: Scope, Schedule, and Cost Control*, edited by A. Razzak Rumane. Boca Raton, FL: CRC Press, Taylor & Francis Group.

Rybkowski, Z. K., L. H. Forbes, and C. Y. Tsao. 2018. 'The evolution of lean construction education (Part 1 of 2): At US-based universities', In: Proc. 26th Annual Conference of the International Group for Lean Construction (IGLC), González, V.A. (ed.), Chennai, India, pp. 1013–1023. Available at http://www.iglc.net

Sarhan, S., and A. Fox. 2012. 'Trends and Challenges to the development of a lean culture among UK Construction Organizations'. Paper presented at the 20th Annual Conference of the International Group for Lean Construction, San Diego, CA, July 18–20.

Tsao, C.C.Y., and G. A. Howell. 2015. 'A history of the use of simulations, the parade of trades, and the pull planning technique in lean construction learning and development', White Paper, March 3.

Tsao, C. C. Y., and I. D. Tommelein. 2001. Integrated product-process development by a light fixture manufacturer. *Proceedings of the 9th annual conference of the international group for lean construction (IGLC 9)*, August 6–8. Singapore.

Tsao, C. C. Y., I. D. Tommelein, E. Swanlund, and G. A. Howell. 2000. Case study for work structuring: Installation of metal door frames. *Proceedings of the 8th annual conference of the international group for lean construction (IGLC 8)*, July 17–19. UK.

Villego® Last Planner® Simulation. 2017. <https://www.villego.com/> (Nov. 26, 2017).

Wouters, P., H. K. Tabbers, and F. Paas. 2007. 'Interactivity in video-based models', *Educational Psychology Review*, pp. 327–342. New York, NY: Springer Science+Business Media.

Bibliography

Ballard, G., and G. A. Howell. 2003. *Lean Project Management Building Research and Information* 31(2): 119–133.

Howell, G. A. 1999. What is lean construction. *Proceedings of the 7th annual conference of the international group for lean construction*. Berkeley: University of California.

Womack, J., D. Jones, and D. Roos. 1990. *The machine that changed the world*. New York: Harper Collins.

8
MANAGEMENT AND WORKER FACTORS

This chapter addresses management-worker factors in several distinct dimensions that leaders of construction organizations must consider if they wish to implement lean construction and IPD. It describes:

Part A: Management-related issues

Introducing Behavioral Change in Construction

Maslow's Hierarchy of Needs

Managing Diversity

Part B: Environmental factors

Temperature and Wind Effects on Personnel

Effects of Scheduled overtime

Part C: Performance improvement techniques

Work sampling

Stop watch time study

MTM

The Learning curve

Part D: Ergonomics

Part A: Management-related issues

Introducing behavioral change in construction

Although technology is being gradually incorporated in construction processes, the industry is still driven largely by the human factor. All levels of the supply chain are people-dependent: designers, contractors, subcontractors, and suppliers all interact to take a construction project from the owner's project requirements (OPR) to the finished product. Improving performance, therefore, secures the greatest return on investment by addressing the human element. Performance improvement initiatives in construction encompass a wide range of alternatives. These include Total Quality management, Kaizen, Six Sigma, ISO 9000, and more recently, Lean construction.

Although these initiatives vary widely in their approaches, they share one important characteristic – they require a new paradigm in which people who work in the construction industry adopt new behaviors. This need exists whether the people involved perform design work, manage construction projects or apply physical labor in construction activity.

Improving management–worker relations through the Malcolm Baldrige National Quality Award Criteria

Performance in the construction industry is linked to motivational issues in the workforce. Project success is largely attributable to the human factor. When a worker is satisfied with the job he or she is assigned to, that individual tends to be more productive and pleasant to work with as motivation levels are high. Job satisfaction among trades and forepersons is improved in projects that are well planned and where all the facilities are available to allow them to be productive. Such conditions provide an incentive for workers to increase their productivity. The success of lean construction depends heavily on having workers engaged and actively involved in improving work processes.

Benchmarking against the Criteria:

> Construction leaders can secure significant gains in performance by benchmarking their organizations against the Baldrige Criteria, even if they do not wish to apply for an award. This applies equally to all construction-related organizations – architects, engineers, general contractors, construction managers, subcontractors, and suppliers.

The Malcolm Baldrige National Quality Award is based on seven Criteria that serve as a model for organizations that aspire to be world-class (refer to Chapter 11). Candidate organizations that apply for a quality award are evaluated against the seven criteria and may succeed only if they have embodied them in their business model and strategic plan for a number of years.

Construction projects are accomplished by people, hence the greatest benefit to projects would be derived from the 5th Criterion, termed the Workforce Focus category. It is concerned with how well an organization engages its workforce, manages and develops it to utilize its full potential. It also examines how an organization creates a work environment that enables workers to maximize value to the organization.

Non-prescriptive nature of the MBNQA:

> It is important to note that the Baldrige standards are not prescriptive. They do not dictate specific means and methods for improving performance. Rather, they provide a framework for performance improvement by asking questions that point to behaviors necessary for superior performance. Furthermore, awards are based on sustained activity, not ad-hoc applications of a particular technique.

The MBNQA Criteria refer to "workforce" as the people actively involved in accomplishing the work of one's organization. In keeping with the construction environment, it includes permanent, temporary, and part-time personnel as well as any contract employees supervised by the organization.

Readers are referred to the MBNQA Criteria for further information.

Motivating workers

Many techniques can be used to motivate workers in the construction environment. They include: Goal setting, incentives, positive reinforcement and worker participation.

Worker motivation and job satisfaction are highest when a project is well planned and all the resource are available to enable everyone to be productive.

Goal setting has succeeded in improving workforce productivity in the manufacturing and service industries, productivity has increased by up to 20% with a programmatic use of goal setting and feedback. Challenging and specific goals have produced better performance than medium, easy, or do-your-best goals; or no goals. Despite the difficulty of applying goal setting in construction, productivity improvements of 10% to 20% are possible with that technique.

Goal setting should be based on schedules that are realistic and are built on work standards that are historically based. Attainable targets should be established, based on the best historical performance.

Incentives

Construction organizations generally use incentive plans based on profit sharing. One significant limitation in such plans is that market conditions and other factor beyond the organization's control may have a major impact on the profits generated in a particular firm.

Positive reinforcement

Recognition programs have been used extensively in both the manufacturing and service industries; these typically include the "employee of the month" or "of the year". More sophisticated suggestion award programs involve detailed review of employees' suggestions, and implementation if benefits can be substantiated. When suggestions are implemented the employees are recognized very openly; in some cases they receive monetary rewards as well. Such programs have proven to be very effective in improving morale. They play a very important role in promoting a sense of empowerment in the work force, especially at the field level where lean construction implementation benefits from their knowledge and input.

CHANGING MANAGEMENT ATTITUDES

The success of construction-related organizations is closely linked to the desire of the workforce to excel. There is a wide spectrum of skills and abilities in the construction industry. Designers, on one hand are highly trained architects or engineers with professional registration. Contracting organizations have formally educated managers as well as a work force that is generally vocationally trained.

Regardless of their background, all providers are human beings and their performance is influenced by the extent of their motivation. In turn, the leadership of their respective organizations sets the tone for the level of motivation experienced by workers. In particular, workers' willingness to innovate and adopt new ideas and behaviors is positively correlated with their motivation. Similarly, workers' effectiveness in providing high quality products or services to an organization's external customers is closely linked with their level of motivation. Without motivation, workers (and many managers) resist change as they fear a loss of the ability to control their lives. Dr. Edwards Deming places the responsibility for the system at the feet of management. Deming takes the view that workers at all levels wish to experience a feeling of achievement, and to take pride in their jobs. This desire is often thwarted by faulty systems – they do not allow people to fully take pride in their jobs and their work quality. As management controls the system, it is up to them to remove the barriers to good performance.

Deming was known for exhorting western managers to transform themselves to "improve and innovate the system of independent stakeholders of an organization over the long term to allow all people to experience joy in their work and pride in the outcome," and to improve and innovate the condition of society. While this advocacy may sound to some like altruistic rambling, it gets to the heart of the source of quality – the worker. The power of human resource is contained in the intrinsic motivation from the sheer joy of an endeavor, not just the extrinsic reward of a paycheck, no matter how large that may be. In his 14 point system of Profound Knowledge, Deming prescribes continuous improvement of product and service as the best route to competitiveness, to stay in business, and to provide jobs. He prescribes leadership as a substitute for management by numbers. Deming also points out that quality can only be built in, not inspected in, hence the worker is a critical component in the creation of quality construction. Traditionally, owners have deferred to the designer's best judgement and have not sought out ideas from field trades that actually execute construction work.

Top Management must recognize that performance improvement is a journey, not a destination; it involves a delicate, harmonious interplay between such disparate factors as high technology, state of the art erection systems, and employee training and empowerment programs. For the transformation to occur, Top Management must establish an organizational culture that promotes quality principles, customer satisfaction and continuous improvement as the business paradigm. Through Strategic Planning and Implementation, this paradigm must be ingrained in the organization's Vision statement, Mission statement, Guiding Principles, Broad Strategic Objectives and Specific Tactics. Managers must exert leadership by example, and ensure that the paradigm is known and understood at all organizational levels.

Lean construction and integrated project delivery are highly dependent on Top Management's commitment to lean thinking; these initiatives can succeed only if leaders "buy in" to the concept and have their enthusiasm permeate through the organization and motivate the workforce at all levels. The Strategic Plan and each company's business model should reflect a commitment to adopt and implement lean principles to satisfy both internal and external customers.

Communication is an essential component of motivation. It involves not just imparting information, but verifying that the communication is understood by the receiver. Effective leaders are able to communicate in language that is familiar to the recipient, but also understand how the message is likely to be interpreted. As the old adage goes, "meanings are in people, not just in words". Good leaders also have the ability to inspire others, motivating them to perform at a higher level. Leaders can also bring about transformation by coaching, providing feedback, and rewards.

THEORY "Y": MOVING BEYOND TRADITIONAL THINKING TO LEAN THINKING

Today's managers are generally aware of Theory "X" as described by Douglas McGregor (1957) with respect to work force attitudes. Theory X was based on three propositions:

1) Management is responsible for organizing the elements of productive enterprise in the interest of economic ends.
2) Management has to direct the efforts of the work force, motivate them, control their actions and modify their behavior to fit the organization's needs.
3) People are either passive or resistant to organizational needs and have to be persuaded, rewarded, punished or controlled.

McGregor advanced the philosophy of Theory Y that shifted from a reliance on external control of human behavior to self-control and self-direction, essentially moving from treating people as children to treating them as mature adults. Theory Y held to a more positive view of human nature – that people are not passive by nature, or resistant to organizational needs, but rather had been conditioned through negative experiences with organizations. It assumed that people have the capacity to grow and assume responsibility; management's essential task is to create the organizational conditions that enable people to self-actualize and direct their own efforts toward organizational objectives.

In the developed countries, Theory Y thinking has influenced the management of manufacturing and service organizations as these sectors have responded to the challenge of global competition. Quality management initiatives have created a cadre of workers in those sectors that have adopted a culture of "building in" quality as opposed to "inspecting in" quality. In the United States, the Malcolm Baldrige National Quality Award and state-level awards that are based on it, charge management with the responsibility of creating the conditions in the workplace that encourage workers to take the initiative in adopting quality-oriented behaviors. And yet, the construction industry as a whole has not yet broadly practiced "Theory Y' thinking. The attitude prevalent on many construction sites is still rooted in Theory X – workers are conditioned to do only what they are directed to do, with little room for initiative. It is not surprising, therefore that many work sampling studies have identified as much as 45% waste in direct labor activity.

Lean thinking: another level beyond Theory Y

Theory Y did not require that production workers be trained in management techniques, but Lean construction does. In the "Last Planner" approach, the people closest to the work, i.e., the crews, are empowered to make decisions on detailed work assignments, as they know their own capabilities as well as the need for materials and equipment resources. Lean construction requires the crews to commit to and promise to carry out specific work assignments. To do this the crews ensure that the required resources are available and that the sequence is appropriate. Weekly meetings are held to review the work accomplished for the past week and plan the work for the following week.

Very importantly, these weekly meetings emphasize learning from experience with the most recent assignments – what went well, and what did not. The purpose of the exercise is not to apportion blame for mistakes, but rather to determine how best to plan assignments to interact effectively with other trades and disciplines for optimal project results.

Lean construction requires a different culture in the work force. Traditionally, a little over 50% of the tasks planned for a given week may be accomplished in that week. Stops and starts typically occur or "making do," i.e., keeping occupied with non-critical tasks, and that approach significantly lowers work productivity. It also lowers the expectations of workers.

The concept of having work "pulled" from each crew by the downstream crew is an important lean principle, but it is counterintuitive. Tradition has conditioned workers to complete tasks as quickly as possible, advancing the accomplishment of their specific crews. Such actions may exceed the capacity of upstream trades and simply create work in process that cannot be completed without creating delays. By the same token, they may "push" their production to downstream crews faster than their capacity allows.

Lean coaching is an important activity in the implementation of Lean Construction. The Toyota Way Fieldbook describes the functions of a lean coach that include both process-related activity and training activity. They include:

- Promoting the lean transformation within the organization,
- Coaching leaders at all levels,
- Teaching lean tools and philosophy,
- Leading value stream mapping,
- Developing the lean operating system for the organization as well as metrics, and
- Leading Kaizen events.

Kaizen events, in particular, are very successful in engaging the workforce. As described in Chapter 14, these events are usually of a short duration – approximately one week – but they provide an intense focus for everyone in the organization to investigate opportunities for improvement and implement them. One contractor held a Kaizen event to clean up their main facility in order to smooth workflow and material storage activities. Rank and file workers voiced a number of comments that indicated an interest far beyond wanting to carry out minimum job requirements.

Examples:

Everyone brought a lot of quality ideas to this event.
Teams do have power (they have powerful ideas).
It was hard to get started because of the disorganized yard.
There are no dumb ideas.
We save too much stuff in our everyday activities.
Work together as a team . . . not against each other.
We all have a better understanding what is a 5S.

Clearly, construction organizations can enhance lean implementation efforts by the insight and enthusiasm that empowered workers can offer.

Maslow's hierarchy of needs

In order to effectively motivate workers and tap into the type of insight and enthusiasm described above, managers would do well by understanding Maslow's contributions and applying them to the work force.

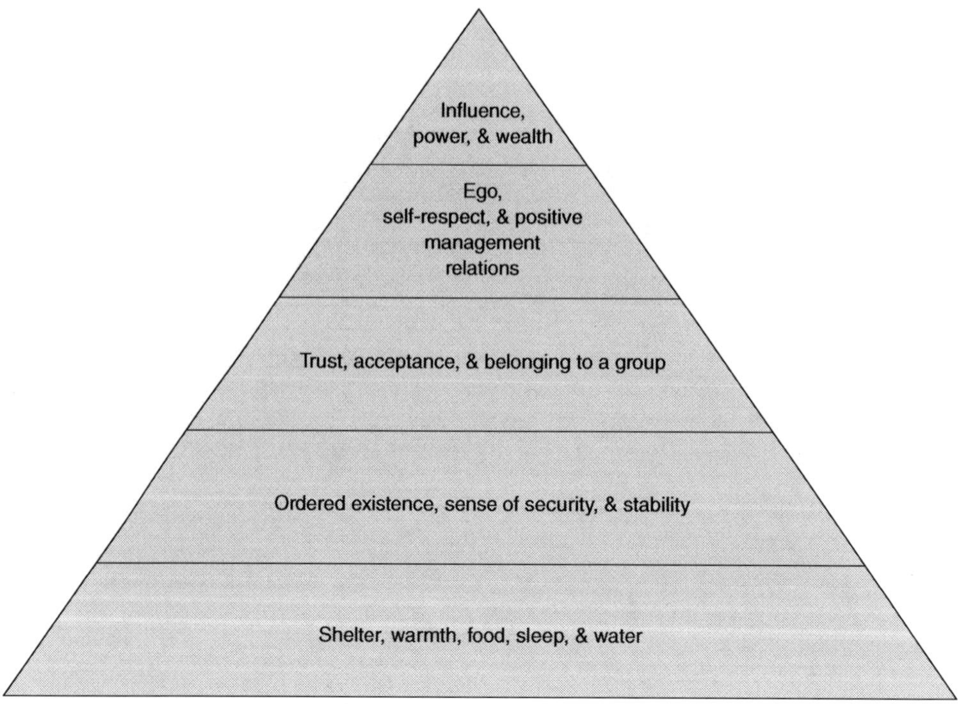

Figure 8.1 Maslow's hierarchy of needs

Abraham Maslow was born in 1908 and passed away in 1970. He is considered by many to be the father of humanistic psychology. He became interested in the concept of "self-actualization" upon meeting Kurt Goldstein in the 1940s as Goldstein was studying human mental functioning and healing from a holistic approach. After being introduced to Goldstein's work Maslow became particularly interested in studying highly-fulfilled individuals and the steps that they took to become fulfilled. In 1943 he presented at the Society for Research in Psychoanalysis and Psychosomatics which led to the publication of his paper titled "A Theory of Human Motivation"

In that publication Maslow discussed his now famous Hierarchy of Needs. He discussed how all humans are born with a basic set of needs and the desire to reach self-actualization which he described as the desire to become everything that an individual is capable of becoming. Maslow believed that human behavior is the consequence of human choices and that all people are inherently good and have unlimited potential for learning, growth, and development.

Levels of Maslow's hierarchy of needs

Maslow's Hierarchy of Needs consists of the following basic needs: physiological needs, safety needs, belongingness, ego needs and ultimately self-actualization. According to the Hierarchy the motivation to meet higher level needs emerges as lower level needs are met. It takes into account employee wages, benefits, organizational culture, and management style and how they impact the motivation and productivity of employees. Psychological needs such as food, shelter and sleep can be satisfied by wages that are paid to employees. Safety involves an ordered existence and a sense of security. In the construction industry safety needs can be met by providing job security, having defined responsibilities and offering structure. A sense of belonging can be obtained when an individual is accepted by co-workers and peers. Providing a positive work culture enables employees to develop a rapport with each other and enhances their sense of belonging in the workplace. Ego involves esteem and self-respect. In the workplace ego can be satisfied through positive management relations that make employees feel trusted and empowered. Self-actualization involves influence, power, and wealth. These needs are more difficult to satisfy as they require a management team that actively engages and motivates their employees to attain their potential. Failing to fulfill the basic needs of employees can lead to personal detriment and makes it difficult to achieve higher levels of the hierarchy.

Maslow's hierarchy of needs in the business sector

The Hierarchy of Needs can easily be applied to organizational behavior and is considered to be one of the best motivational theories utilized by managers in different business sectors. This theory is popular among business managers because it is simple and easy to explain to others. It can also help businesses to establish a learning environment that motivates employees and meets their needs.

According to Maslow, self-actualization should be the main goal of learning, and education should focus on self-development. Continual learning is essential to the success of a business as employees and managers must acquire new knowledge and skills in order to adapt to the constantly changing business environment. One of the issues that businesses run into is keeping employees motivated to be receptive to new knowledge and skills which is where using Maslow's Hierarchy of Needs can be helpful.

There are two different types of motivation: intrinsic and extrinsic. Intrinsic motivation refers to factors that make certain activities rewarding in and of themselves; it helps employees to internalize a desire to learn and tends to last longer than extrinsic motivation. Ego and self-actualization are important intrinsic motivators and have a more sustainable impact on employees than extrinsic motivators do.

Extrinsic motivation refers to rewards and reinforcement that are used by external agents to influence and direct behaviors; extrinsic motivation can be either positive (encouraging desired behavior) or negative (punishing undesired behavior). In the business sector managers tend to gravitate toward using extrinsic factors such as salary, job security, workplace conditions, promotions and termination to motivate employees to learn new skills. If used in a negative way extrinsic factors can create tension between employees and managers.

Managing diversity

This section addresses two important contexts of diversity management; first with U.S.-based construction, and second with international projects involving US managers in other countries.

Diversity may be defined as the variability in the workforce. Rasmussen (1996) states that the workforce is comprised of a variety of backgrounds, styles, perspectives, values, and beliefs as assets to the groups and organizations with which they interact. Some of its characteristics are inherited, while others are acquired. The management of diversity is essential for creating an environment in which people from different walks of life can come together as a unified team to produce high quality construction. This is sometimes very difficult to accomplish with workers of different cultural, racial, and ethnic backgrounds. An organization cannot have satisfied external customers until it has satisfied internal customers. It cannot meet, let alone exceed external customers' needs when internal customers are at war with each other, as is often the case with construction projects.

In his book *The New Economics*, Dr. W. Edwards Deming saw the organization as a system, with people representing components that work cooperatively with each other. Dr. Deming cited the importance of informal communication between people in various components of the organization, regardless of level of position. He recommended teamwork as a way of promoting joy in work, whereas he saw the competition that often occurs between people in the workplace as demoralizing to the individual and counterproductive to the organization as a whole.

Diversity management in the United States

Several events over nearly four decades have influenced the status of the diversity question today: The Civil Rights Act, Affirmative Action, Equal Employment Opportunity, Sexual Harassment, and the Americans with Disabilities Act. These events, combined with increasing immigration have drastically changed the profile of the workforce; resulting in a very heterogeneous mix of workers, far different from the way it was in 1965. The workforce of the future will become increasingly multiracial, multiethnic, and multicultural, furthering the need for diversity management. The management of diversity is about good business, not just meeting state or federal mandates; it involves optimizing the interaction between people from all walks of life.

Model of diversity

A typical model of diversity comprises two concentric circles. The inner circle represents attributes that can be seen readily, such as age, race, ethnicity, physical qualities including disabilities, and gender. Although not necessarily visible, sexual orientation is also included. These factors cannot be changed and form the basis of stereotypes. The outer circle has other factors that individuals have the ability to change, such as marital status, religious beliefs, education, income, geographic location, and work experience (Figure 8.1).

Discriminatory behavior is a fact of life, despite any hope for its eradication. Many large corporations have been the subject of lawsuits in recent years, charged by employees (or customers) with gender, racial, age, and other types of discrimination; many have settled out of court to avoid publicity. Historically, women and minorities have been underrepresented in the industry.

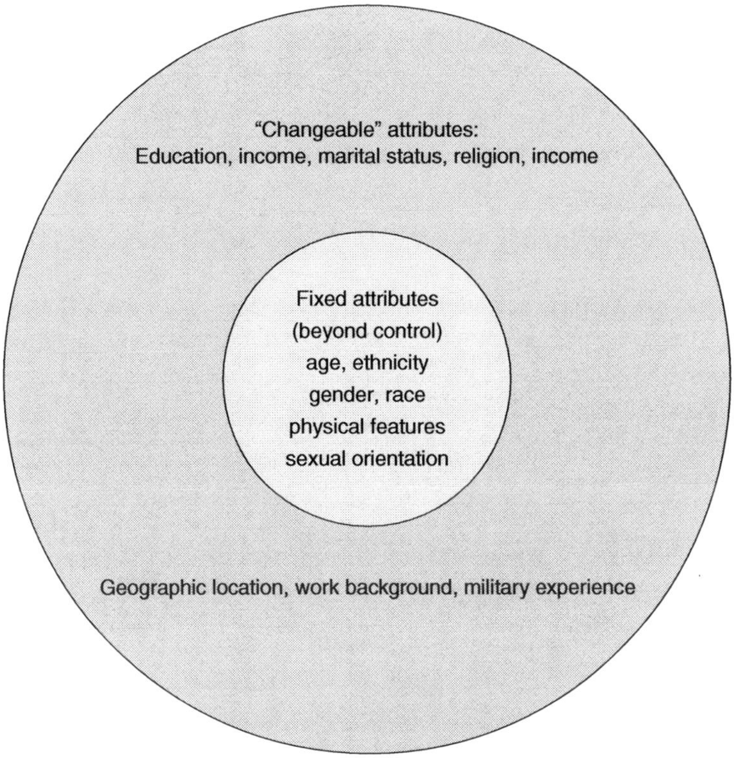

Figure 8.2 Representation of diversity factors

Construction companies do not usually interact with the general public and are driven by specific projects that may last from a few months to 1, 2, or 3 years. Discriminatory practices in their ranks scarcely reach the public view, yet their negative behaviors affect construction quality performance in subtle but tangible ways.

Lean construction offers an excellent framework for overcoming this performance gap. It depends on a cohesive team, regardless of discipline or background, focused on delivering a project with the least amount of waste (i.e., without losses due to delays, quality lapses, cost overruns, or safety incidents). This requires working seamlessly, using the "pull" discipline for total project optimization, not local optimization. Every actor in the chain must be "in the loop" for this to occur.

Women in construction

Women represent approximately 47 percent of all employed individuals but they are under-represented in the construction industry comprising only about 10 percent of industry employment. A disproportionately small number of women work in actual construction activity; in fact only about 2.5 percent of trades persons are women. At the same time, industry observers note that the rebound from the major downturn in the US economy in 2008 onwards has led to a shortage of skilled trades, and this could be a significant opportunity for women. A number of reasons explain the shortage of female workers – the industry has been perceived as male dominated, and many young women may not have considered it as a career choice.

Although attitudes have changed with the passing years, there may still be an unconscious gender bias that inhibits the support and mentoring that would enable females to fully develop their skills and talents to become high performers. The example set by women in the armed forces should dispel any misconceptions about their physical capabilities. The other skills possessed by women have been shown to be critical to an organization's bottom line – a McKinsey study has shown that companies ranking in the top quartile of executive-board diversity were 35% more likely to outperform others financially. Business professors Cristian Deszö and David Ross of the University of Maryland and Columbia University respectively estimated an increase in a firm's value of $42 million resulting from gender diversity. At the field level, and in the context of lean construction, it would appear that women have attributes that greatly support lean construction – Women's sensitivity, intuition and emotional intelligence equip them (more than many men) to both listen to the ideas of others and use that to enable better decision-making. This quality makes them very effective at problem solving.

Women prefer to do every job with fairness, and are very valuable on a team.

Women are well equipped to create and maintain a healthy balance between work and life. Interested parties should consider contacting the National Association for Women in Construction (NAWIC), a US organization that promotes activities and resources that help women to succeed in the construction industry.

Diversity in international/overseas projects

These projects are a very important market sector; at one point, 50% of the value of projects carried out by US companies has been due to overseas projects, but there is growing competition from other countries. A typical project could be the construction of a hotel in the Cayman Islands with an American construction manager, a British architectural firm, a Trinidadian engineering firm and a Venezuelan contractor, using Latin American and Caribbean workers. Choudbury and Haque (2001) of Texas A&M University investigated 35 US contractors working on international projects in Africa, Asia, Europe, Latin America, and the Middle East. The study findings pointed to the importance of cross-cultural training to project success; critical factors were the physical environment, language, economic environment, labor conditions, and social environment. The nature of sociocultural interaction appears to have a major impact on the productivity of project personnel; mistakes in this area can be costly, indeed.

The management of staff with multiple languages makes integration especially challenging because of the difficulty in translating heavily nuanced and complex technical concepts. The use of bilingual project managers appears to be the best strategy. A thorough understanding of the abilities and availability of foreign labor forces is critical in order to have realistic expectations of their production. Also importantly, the cross-cultural training of project staff should include cognitive studies of the host country population and their social and cultural norms. Swedish contractors have excelled in Saudi Arabia because of extensive pre-project training on social values.

Decision making in some countries is far different from the way it is done in the United States. Whereas we have become accustomed to shallow or flat organizations and staff can collaborate openly and informally in making decisions, other cultures may approach the process in a far more complex manner, based on their power-distance index. Countries with a high index have a small middle class, high social stratification, and high levels of privilege for the powerful, as exemplified by such countries as India, Kuwait, Saudi Arabia, Nigeria, The Philippines, Mexico, Venezuela, and Panama, to name a few. Low index countries include the United States, Canada, Great Britain, Australia, Israel, Germany, and Costa Rica. In high index countries, the decisions tend to be more centralized and less democratic. These patterns are crucial in the conduct of negotiations. US managers need to be patient and respect that approach.

The host country's customs may be very unfamiliar, yet US managers must observe them. For example, Americans tend to use first names freely, even in formal communication. Some cultures take umbrage at this perceived overfamiliarity.

Attitudes to time may diverge widely throughout the world. North Americans and Northern Europeans treat time as a very precious commodity; they observe monochronic time, which involves a linear, orderly time progression. According to Seelye and Seelye-James (1995), polychronic time is the norm especially in many agrarian countries. Time is treated far more flexibly as an objective, not a precise commitment. Several tasks are pursued simultaneously, adapting to changing circumstances. An American project manager needs to understand this in order to manage successfully in other countries.

Taking action on diversity management

Leaders should take the trouble and time to understand the potential of diversity; it goes far beyond hiring a portion of the workforce from "other groups" simply because it is said to be a good practice. To be meaningful, diversity needs to be applied across the organizational spectrum. Unfortunately, many organizations that desire to have a diverse workforce actually have a number of women and minorities in clerical or low-level professional ranks. That situation is a beginning, and leaders should build on it to tap people's potential by assembling the best team based on talent regardless of the individual's background.

Be sure to implement plans for mainstreaming women – both new hires and existing staff to optimize their integration into the organization. NAWIC may have ideas that can help this transition.

Effective diversity management

A design or construction organization should set up a formal structure for managing diversity, so that it does not get taken for granted; the degree of complexity should be based on the organization's size. Several major US corporations have established formal programs with diversity criteria integrated into their strategic plan, with associated measurement systems. Texas Instruments and Kraft Foods/ Philip Morris Companies are well-established examples that have the important feature of measurement systems. As the old quality adage goes, "what gets measured gets done." Without measurement, the success (or failure) of a diversity management program may be impossible to determine. Some companies have appointed high level positions such as Chief Diversity Officer to guide their efforts to address this issue.

The following recommendations should help companies to be effective:

- Communicate a commitment to staff about providing an environment that is conducive to diversity. This fosters cooperation, inclusiveness of all staff, and recognizes value of knowledge, talent, and skills from all sources.
- Diversity/sensitivity training for all staff is an absolute necessity. Role-playing is an indispensable component of that training.
- Managers should remember the fact that no ethnic group has a monopoly on natural mental ability or leadership. They should communicate belief in dignity and respect for all, identify and remove barriers to minority groups in the organization, eliminate judgmental attitudes, and learn to listen.
- Managers should practice employee development in order to make diversity successful: Establish and communicate performance expectations, use appropriate motivational approaches, provide working conditions they appreciate, provide meaningful feedback, both positive and negative, but respect the individual's dignity. Use appropriate rewards and recognition, preferably things that the employee values, not just a routine award.
- For international projects, train management staff on the social and cultural values of the respective host country. Use bilingual, bicultural staff to interface where possible.
- Develop conflict resolution procedures that allow a win-win outcome. Differences in viewpoints and cultural factors must be respected and addressed.

- Recognize that the management of diversity is an ongoing long-term process. Like all quality endeavors, a culture of continuous improvement is the best prescription for a design or construction organization to ascend to world-class performance.

Everyday practices to promote harmony

Managers and staff employees can use the following ideas alike to improve harmony in the workplace, especially in design-related organizations.

Take a "vacation" without leaving home. We vacation in distant lands in order to see other cultures and broaden our knowledge of the world. Why not take advantage of a talk with a coworker who may be from a particular geographic region; it costs far less than an airline ticket and hotel stay, and can even equip you to better enjoy your visit if you decide to go.

Try to learn at least a few phrases of another language. View it as preparation for future opportunities. Students of international business pay universities handsomely for the privilege of learning about other cultures while they learn about marketing, finance and project management, and often go on to lucrative assignments with firms that have foreign affiliates. Interactions with coworkers of other backgrounds cost nothing, and at the very least increase the quality of teamwork in your organization. Better teamwork improves organizational performance and leads to greater competitiveness.

Nurture employees' potential. It is a good business practice to hire the best people. It is also good business to recognize the potential of individuals and provide them with an opportunity to grow. In turn, people who have grown with an organization often develop a high level of loyalty to it, but even more importantly, are willing to take ownership to provide customers with service beyond their expectations.

Actively create a mentoring culture. Mentoring is a critical factor in retaining minority professionals. While they may be hired on the basis of their training and skills, their advancement and longevity in an organization depends heavily on their ability to gain mission-critical experience and credibility. It is important to note that minority employees do not necessarily need to be mentored by other minorities; they simply need someone with more experience to show an interest in their welfare, provide some career guidance, and make them feel like a valued part of an organization.

Practice sensitivity. Managers in the construction arena should avoid appearing to "talk down" to people from less developed countries. People in host countries do not want to be made to feel their culture is inferior; they often regard external influences as irrelevant. Company representatives in a host country should be careful not to flaunt the fact that their compensation is higher than that of locals. This information may simply alienate others and destroy any hope of having a cohesive team.

Construction professionals should recognize that unwitting actions may offend others. Some managers from developed countries occasionally put up their feet to rest their shoes on desks or other furniture, sometimes facing their audience. Some cultures are deeply troubled by this action.

Ultimately, a project team can only excel if its people are able to take pride and joy in their work, regardless of their background. Those people must work together harmoniously, as a system, and treat each other as important internal suppliers and customers without reservation. Effective diversity management helps to accomplish that end.

Part B: Environmental factors

Temperature effects on personnel

Arousal theory explains the impact on a worker's productivity caused by the demands of a task and the physical environment. The physical environment creates physical stimuli that, in turn, create arousal in the human body, particularly heat and noise. The body regulates its equilibrium through thermal exchanges with the surrounding environment (i.e., evaporation, conduction, radiation, and convection).

If the approximate level of a worker's productivity is estimated, then that information can improve the accuracy of resource management activities. Variability in workforce performance can be compensated for by basing resource allocation on a forecast of seasonal weather conditions. Four major parameters can affect human performance: ambient temperature, relative humidity, wind speed, and radiant temperature.

Workers are most effective when working in an environment where temperatures vary between 50 and 70° F. In this temperature range the human body gives up heat by evaporation from the skin and by respiration, without the need for much sweating. As the temperature and humidity rise, the body is less able to lose heat, and its stress level rises as well. This results in lower levels of productivity. There are a number of thermal comfort indices:

1 Effective temperature,
2 Wet bulb globe temperature,
3 Operative temperature,
4 Heat stress index, and
5 Standard effective temperature.

Normal body temperature is approximately 98.6° F (Table 8.1). There are a number of thermal and humidity effects that impact workers.

High temperatures

- Continued exposure to heat and humidity is a common cause of heat cramps, heat exhaustion, or heat stroke.
- Heat evaporation is caused by excessive perspiring, because of a hot environment or strenuous physical exertion.
- Heat stroke is much more serious than heat cramps or exhaustion.

Table 8.1 ASHRAE seven-point scale

ASHRAE Scale	*Sensation*	*Physiological*
3	Hot	Profuse sweating
2	Warm	Sweating
1	Slightly warm	Slight sweating, vasodilation
0	Neutral	Neutrality
–1	Slightly cool	Vasoconstriction
–2	Cool	Slow body cooling
–3	Cool	Shivering

Source: ASHRAE Fundamentals Handbook (SI), 8.1–8.28, 1997

High temperature and performance

- Stresses generated by high temperature degrade performance.
- The degree to which temperature affects performance is a topic of constant debate.
- Five categories that determine the effects of heat:

  1 Intensity of heat,
  2 Duration of exposure,
  3 Tasks involved,
  4 Persons performing the tasks, and
  5 Presence of other stresses.

- High humidity can cause severe stresses, especially at high temperatures.

Effects of cold

- Chilblain: a relatively mild form of tissue damage.
- Local itching and swelling.
- Wet cold syndrome: trench foot and immersion foot. These result from exposures below 53°F for several days.
- Frostbite: results from prolonged and severe vasoconstriction at temperatures below 32°F.

Effects of wind on personnel

- High and low temperature variations can all cause damage directly or generate other conditions that can result in damage.
- Wind impacts body temperature by removing heat from the skin surface. In effect for a given ambient temperature an increase in wind speed creates a lower "effective" temperature.

The National Electrical Contractors Association (NECA) sponsored a study in 2004 to examine the effects of extreme combinations of temperature on electrical labor productivity. The principal researcher was Awad S. Hanna, PhD, PE. A number of tables and charts were compiled to document the interaction of temperature, humidity, and wind, as may be encountered in field situations. NECA recommends that the measurement data compiled in these tables and charts be used in factoring of estimates, job planning and scheduling, and in estimating change orders related to extensions of time. Tables 13.6 and 13.7 were adapted from the findings of the NECA study. The general findings from the research were:

1 Worker efficiency of 100% can be achieved only when the temperature is between 40 and 70°F, and the relative humidity is below 80%.
2 While humidity of any percentage is not a significant factor in the 30–80°F range, it plays a very important role at elevated temperatures.
3 In extremely cold conditions, temperature is far more significant than humidity. Regardless of humidity, an effective temperature of –20°F or lower may justify a journeyman to stop working.
4 High humidity at colder temperatures lowers productivity but the effects may be reduced significantly by wearing appropriate clothing and by providing protection from the wind.

They cite the benefits of weather prediction to project planning and execution as:

a Time savings
b Improved job safety

c More efficient handling of materials
d Improved labor planning
e Other economies

The NECA points out that the shifting of job tasks from one season to another may have significant impacts on environmental conditions, and on the productivity of workers as a result. A quantification of these changes in productivity can serve as justification for adjustments in contract budgets and schedules. Productivity estimate involves determining an effective temperature from Table 8.2 based on wind speed and actual temperature. The effective temperature is then used in Table 8.3 to estimate productivity.

Impact on lean construction

Weekly work plans are based on pulling activities or work assignments from a lookahead schedule. As described in Chapter 4, eligible assignments are the ones that have no current constraints, and that have resources available and assigned. Workers are a major component of those resources, hence their capabilities must be known with some certainty in order for a foreman and a last planner to accept a crew's commitment to having a promised activity or task completed. Adverse weather conditions can significantly impact workers' productivity. Tables 8.2 and 8.3 can provide fair approximations of productivity for job planning purposes. Example:

- Air temperature: 30°F
- Wind speed: 10 mph
- Relative humidity: 70%

A simple 2-stage estimation is required. Table 8.2 reflects the relationship between relative humidity and perceived temperature. From Table 8.3, the effective temperature equals 21°F. Determining productivity from Table 8.4 at an effective temperature of 20° and 70% R.H, productivity equals 93%. At an effective

Table 8.2 Effective temperature vs. wind speed

Wind Speed MPH	*Actual Temperature in – Degrees Fahrenheit (Reference temperatures are displayed horizontally in the row below)*														
	40	*35*	*30*	*25*	*20*	*15*	*10*	*5*	*0*	*−5*	*−10*	*−15*	*−20*	*−25*	*−30*
Calm	40	35	30	25	20	15	10	5	0	−5	−10	−15	−10	−25	−30
5	34	31	25	19	13	7	1	−5	−11	−16	−22	−28	−15	−40	−46
10	32	27	21	15	9	3	−4	−10	−16	−22	−28	−35	−33	−47	−53
15	32	25	19	13	6	0	−7	−13	−19	−26	−32	−39	−45	−51	−58
20	30	24	17	11	4	−2	−9	−15	−22	−29	−35	−42	−53	−55	−61
25	29	23	16	9	3	−4	−11	−17	−24	−31	−37	−44	−59	−58	−64
30	28	22	15	8	1	−5	−12	−19	−26	−33	−39	−46	−63	−60	−67
35	28	21	14	7	0	−7	−14	−21	−27	−34	−41	−48	−67	−62	−69
40	27	20	13	6	−1	−8	−15	−22	−29	−36	−43	−50	−69	−64	−71
45	26	19	12	5	−2	−9	−16	−23	−30	−37	−44	−51	−59	−65	−72
50	26	19	12	4	−3	−10	−17	−24	−31	−38	−45	−52	−63	−67	−74
55	25	18	11	4	−3	−11	−18	−25	−32	−39	−46	−54	−67	−68	−75
60	25	17	10	3	−4	−11	−19	−26	−33	−40	−48	−55	−69	−69	−76

Source: Adapted from Table 2.2, NECA study: The Effect of Temperature on Productivity, 2004

Table 8.3 Effect of Temperature and Humidity on Productivity

RH.%	*Journeyman Productivity Percentages at Various Environmental Conditions*												
90	56	71	82	89	93	96	98	98	96	93	84	57	0
80	57	73	84	91	95	98	100	100	98	95	87	68	15
70	59	75	86	93	97	99	100	100	99	97	90	76	50
60	60	76	87	94	98	100	100	100	100	98	93	80	57
50	61	77	88	94	98	100	100	100	100	99	94	82	60
40	62	78	88	94	98	100	100	100	100	99	94	84	63
30	62	78	88	94	98	100	100	100	100	99	93	83	62
20	62	78	88	94	98	100	100	100	100	99	93	82	61
	−10	0	10	20	30	40	50	60	70	80	90	100	110

Effective Temperature (F); Effective temperatures are in the row immediately above

Source: Adapted from Table 2.1, NECA study: The Effect of Temperature on Productivity, 2004

Note: Relative humidity is listed vertically in the left-most column. Worker productivity is estimated by navigating with the effective temperature and the relative humidity.

Table 8.4 Relationship of hours worked to productivity

1	*2*	*3*	*4*	*5*	*6*	7	*8*
50-Hour Overtime Work Weeks	*Productivity Rate*		*Actual Hour Output for 50-Hour Week*	*Hour Gain Over 40-Hour Week*	*Hour Loss Due to Productivity Drop*	*Premium Hours*	*Hour Cost of Overtime Operation (at 2×)*
	40-Hour Week	*50-Hour Week*					
0–1–2	1.0	.926	46.3	6.3	3.7	10.0	13.7
2–3–4		.90	45.0	5.0	5.0	10.0	15.0
4–5–6		.87	43.5	3.5	6.5	10.0	16.5
6–7–8		.8	40.0	0.0	10.0	10.0	20.0
8–9–10		.752	37.6	−2.4	12.4	10.0	22.4
>10		.75	37.5	−2.5	12.5	10.0	22.5

temperature of 30° and 70% R.H, productivity equals 97%. Interpolating between these values yields a value of 93.4% approximately. A worker productivity level of 93.4% would have to be considered when sizing the work crews for the project.

The NECA points out that the reference standards from the study provide basic data for planning and cost estimation, but that individual contractors will need to apply additional job factors for the specific workers and projects involved.

Test your understanding

Assume that a carpenter can erect 160 square feet concrete wall forms per eight-hour day under ideal environmental conditions. The budgeted labor rate for a carpenter is $15 per hour. Unexpectedly, weather conditions change such that wind velocity is 25 mph, the temperature drops to 30°F, and the humidity is 65%. Calculate the number of square feet of forming the worker will place under these conditions. Use the tables supplied.

Managing the impacts of scheduled overtime on productivity

The Business Roundtable (BRT) conducted a study in 1974 on the impact of scheduled overtime operation on construction projects with respect to worker productivity and labor costs. The BRT repeated the study in 1980 and obtained similar results. The BRT study identified disproportionate increases in project costs to accomplish construction work when it was done with scheduled overtime, as compared with a nominal 40-hour week. This premium became larger as overtime was continued for several weeks; the BRT found that a schedule of 60 or more hours per week would potentially delay a project if it were continued for more than two months with the same crew size. Table 8.4 shows raising scheduled hours from 40 to 50 results in lower productivity for the entire 50-hour period.

More recent studies by other researchers obtained similar results to the BRT's findings. Thomas and Raynar (1997) observed losses of efficiency of 10–15% for 50- and 60-hour work weeks. For this analysis, disruptions in three categories – resource deficiencies, rework, and management deficiencies – were analyzed. The analyses showed that the disruption frequency, which is the number of disruptions per 100 work hours, worsened as more days per week were worked. This led to the conclusion that losses of efficiency are caused by the inability to provide materials, tools, equipment, and information. Hanna *et al.* (2005) observed behaviors similar to those identified in the BRT study. Mayo *et al.* (2001) included travel time in their analysis and observed that longer work days could be more productive. That special circumstance, however, does not apply to most construction projects.

Implications for lean construction

As indicated in Table 8.4, hours worked beyond the nominal 40-hour shift may result in lower productivity. More total labor hours are utilized to carry out scheduled tasks than would be needed with a 5-day, 40-hour shift. The lost hours as shown in the table are in effect waste, representing the very antithesis of lean construction. These losses may even be compounded by difficulties in keeping crews supplied with needed materials and instructions. Careful attention should be directed to planning work activities and monitoring accomplishment, at least on a weekly basis, to avoid bottlenecks and maximize crew performance. The methodology of The Last Planner® System combined with daily huddles can be effective in optimizing work output.

Part C: Performance improvement techniques

Lean construction seeks to improve construction performance primarily by integrating the actions of the stakeholders in the supply chain. This approach can be made even more effective by applying a number of analytical techniques that have been adapted to the construction environment. In a traditional environment, these techniques can provide significant benefits at the activity level. In a lean project environment, they can be even more beneficial by supporting project-wide optimization.

Performance improvement requires a frame of reference in order to be effective. It provides that reference point by determining the current state of key processes in order to measure the impact of subsequent improvement strategies. It provides the information needed for process control and makes it possible to establish challenging yet practical goals. These actions support the implementation of competitive business strategies. Although some construction managers recognize the importance of performance measurement, it has not been widely implemented in a formal sense. These managers often base their decisions on intuition and common sense, as well as on a few broad financial measures that are no longer adequate in today's competitive environment.

The above mentioned techniques, therefore, need to be applied in the context of performance measurement. Unless the current state of a process is measured it cannot be fully understood. If it is not fully understood, then the impact of performance improvement techniques cannot be properly evaluated.

As described below, work sampling, stop watch time study and Methods Time Measurement (MTM) may be used to increase construction performance.

Work measurement techniques have been traditionally used by industrial engineers in the manufacturing and service industries. They include:

- Work sampling
- Stop watch time study
- MTM

Construction professionals can also use work measurement techniques to increase construction productivity. Whereas standard work times are often used by the manufacturing and service industries, these standards need to be reviewed and updated for the construction industry. These standards can be tailored further to specific projects to reflect the logistics of the work site and to adjust the standards to represent methods improvement. The more accurate the information that is available on work standards, the better construction managers can conduct the preplanning of projects and exert control over the costs and schedules of these projects. Many construction standards need to be reengineered to reflect the use of technology in work processes.

Work sampling

Work sampling is a useful method for conducting work measurement studies. It can estimate the proportions of the total time spent on a task in terms of its various components. It is particularly useful in distinguishing between direct and indirect work in situations where stopwatch time study is not practical or appropriate. Work sampling involves taking a small portion or sample of the occurrences in an activity being studied in order to make inferences about the frequency of occurrences in the overall activity. In construction, for example, certain tasks have high variability – installing drywall panels may involve taking an increasing number of steps to take them from a stack to the point of installation. Also, the panels are often not installed as whole sheets – workers may need to cut holes in the panels for electrical outlets and other devices. On the other hand, it is useful to know what percentage of the time workers are actively working instead of being idle, and work sampling can provide that information.

Work sampling's origins lie in its application as ratio delay studies by L. H. C. Tippett in the British textile industry (Tippett and Vincent 1953). Ralph Barnes (1980, 2009) also documented the work sampling technique. It is also known as activity or occurrence sampling.

A process is observed at intervals such that each activity has an equal chance of being observed; the probability of an occurrence is based on the binomial distribution. For convenience, the normal distribution is used, especially when the number of observations is large.

Typically, the variable p is used to represent the proportion of an occurrence; that is, the proportion of a sample that will have a particular feature or characteristic.

Proportion $p = x/n$ where x is the number of samples that exhibit a particular characteristic, and n is the size of the sample.

Accuracy is expressed by the limit of error. It refers to the range in which the value of p will lie. It is the percentage of variation above and below the sampling estimate where the true value will lie, for a given confidence interval.

Absolute accuracy (A) $= p \cdot r$ where p is the proportion and r is the relative accuracy.

If the proportion $p = 40\%$, and relative accuracy r is 5%, then the desired absolute accuracy

$$A = 0.4 \times 0.05 = 0.02 \text{ i.e., } (2\%).$$

The true value of p will lie in the range

40% ± (40 × 0.05) = 40% + 2%, $0% – 2% → 42% to 38%

Confidence interval

The confidence interval refers to the dependability of an estimation based on a sample. It denotes the long-term probability that the ratio p will lie within the limits of accuracy. A 95% confidence interval implies that 95 of 100 samples would contain the true value of p. The higher the value of the confidence interval, the greater the number of samples that are required for a given level of accuracy.

In most construction situations, a confidence interval of 95% and a limit of error or accuracy of 5% are generally adequate. The percentage of occurrences is also usually between 40 and 60%.

With:

Confidence interval estimate: $P \pm Z \cdot \sqrt{\frac{p(1-p)}{n}}$

Table 8.5 shows the number of standard deviations above and below the mean of a normal distribution that represents the population sample. A confidence level of 95% corresponds to 1.96 standard deviations (approximately two) above and below the mean.

Similarly, a confidence interval of 68% corresponds to one standard deviation on either side of the mean.

Number of samples required

In order to conduct a sampling study, an observer needs to predetermine the confidence interval and desired limit of error. The number of samples required may be determined from the equation below. n = number of samples, z = number of standard deviations, s = limit of error, and p = category proportion.

$$n = Z^2 \frac{p(1-n)}{S^2}$$

If a number of observations n is selected first, then the resulting limit of error S may be determined from the equation

$$S = Z \cdot \sqrt{\frac{p\ (1-p\_}{n}}$$

An example is shown here of a preliminary estimate from a work sampling study with a percentage occurrence p of 40%.

Table 8.5 Confidence levels and corresponding "Z" values of standard deviation

Confidence Level	*Z Values (Number of Standard Deviations)*
68%	+ −1.00
90%	+ −1.64
95%	+ −1.96
99.73	+ −3.0

Table 8.6 Examples of methods-time measurement application data

Activity	*TMU*	
Reach 2"	4	1 TMU = 0.00001 hr.
Grasp (simple)	2	= 0.0006 minute
Turn	6	= 0.036 second
Regrasp	6	
Look (eye time)	10	
Leg motion	10	
Kneel on one knee	35	
Arise	35	
Total	108	

Note: All times include a 15% allowance
Normal time = 108 TMU × 0.036 seconds/TMU = 3.89 seconds.
In calculating the standard time, provision must be made for breaks.
If personal allowances are 25% then standard time = 3.89 × 1.25 = 4.86 seconds.

Confidence level desired = 95%
Accuracy desired = 5%.

Find the number of samples needed to meet those requirements (see Table 8.6).
Number of standard deviations z = 1.96 from Table 8.6

p = 0.4, s = 0.05

Substituting: $n = (1.96)^2 \cdot 0.4(1–0.4)/(0.05)^2 = 3.842 \cdot 0.24/0.0025 = 369$ samples Check your understanding:

1 How many samples would be needed for a 90% confidence interval?
2 At the 90% confidence interval, what accuracy would be obtained with 721 samples?

Work sampling procedure

- Establish the purpose of the sampling study
- Clarify the categories of the activities to be measured
- Estimate percentage (p) with a trial study
- Select the confidence intervals and the accuracy desired
- Calculate the number of observations needed
- Schedule observations based on randomized observation times
- Develop data collection forms for sampling
- Collect the data
- Summarize the results

Duration

The duration of the study will be determined based on the total number of observations needed and the number that can reasonably be taken each day. If 384 observations are needed and 40 can be taken per day, the number of days required = 384/40 = 9.6 days. For planning purposes, 10 days would be required for collecting the required number of samples.

Randomization

The validity of work sampling is based on the assumption that each component of an activity has an equal chance of being observed. The observations are required to be random, unbiased, and independent. The use of a random number table provides a convenient method for structuring a study so that the required number of observations can be planned for, within the time frame allotted. Each observation should be made at a time determined by chance such that no bias is introduced by having the subjects anticipate the observations. The observation times can be selected from random number tables or from electronic random number generators.

Stopwatch time study

The stopwatch time study is based on the successive observation of a repetitious task to measure the time taken for each of its components. The purpose of such a study is to determine a reasonable and predictable time for the performance of a task. This is termed "normal time." In turn, this time is adjusted to create a "standard time," and serves as a mechanism for determining the staffing required to accomplish a given task or set of tasks.

The greatest benefit of stopwatch time is derived when it is based on a "best method" or "best practice." It would be pointless to dedicate extensive effort to developing a standard based on a method that is not the most effective one. Before embarking on the time study, considerable attention should be given to clarifying a best method that is agreed upon by all concerned. If workers "buy in" to the method, then the time standards that are based on it are much more likely to be accepted.

Allowances

Workers develop fatigue by performing a task repeatedly. They also need to take breaks for personal time, delays, and other routine activities that are not a part of the task itself. Standard time is determined by adjusting normal time to include allowances. A typical stopwatch time study is structured as follows:

1 Document the method used for the activity in detail. The work cycle should be subdivided into discrete steps or elements.
2 Measure and record elemental times. Observe multiple cycles of the same activity to meet accuracy requirements, based on the variability of the elemental times.
3 Rate the effort exerted by the subject in order to establish a fair base for the elemental times. The observer must have the experience and expertise necessary to determine if a subject is working at a faster or slower than normal pace. The standard should be based on a so-called average worker, working at a normal pace that can be sustained. The observer may rate individual elements of the activity separately, but should provide an overall effort rating for the entire activity.
4 Calculate normal time. This is done by multiplying the observed time by the performance rating. For example: mean observed time = 2.0 minutes per piece; effort rating = 115%; and normal time = 2.0 × 1.15 = 2.3 minutes per piece.
5 Calculate allowances. For example, the allowance for personal time, delays, and other activities may be 25%.
6 Calculate standard time. Standard time = normal time × allowance factor =
7 2.3 minutes × 1.25 = 2.875 minutes per piece.

Methods time measurement

Methods time measurement (MTM) can be used to develop engineered standards. The MTM is based on the concept that a method must first be developed, elemental steps defined, and standard times developed. The standard must be based on the average times necessary for trained experienced workers to perform tasks of prescribed quality levels, based on acceptable trade practices. This approach is most practical with repetitious tasks of a relatively short duration.

In the MTM system, operations are subdivided into tasks; tasks are further reduced to individual body movements such as reaching, grasping, applying pressure, positioning, turning, and disengaging. Other movements include eye travel; focus; and body, leg, or foot motions. Each body movement is subdivided into individual actions, such as reaching two inches, grasp, apply pressure, turn, and so on. Each action is assigned a standard time stated in time measurement units (TMU; Table 8.6).

In applying the MTM system (or any other standardized measurement system) it cannot be overemphasized that an appropriate method must first be established that can be successfully applied by the average, trained worker at definitive quality levels. The effect of the learning curve should also be considered when establishing work standards to ensure that repetition does not render the task times excessively long. Further details on MTM can be obtained from the current edition of the *Handbook of Industrial Engineering*, J. Salvendy, editor.

The learning curve

A learning curve is the phenomenon demonstrated by the progressive reduction in the time taken by an individual, or by a team to perform a task or a set of tasks on a repetitious basis. The individuals performing the task or project become more proficient with each repetition; the observed improvement serves as a motivator and a learning tool resulting in successively shorter performance times. The learning curve is represented by an equation of the form

$$T_n = T_1 \cdot n^{(-a)}$$

where Tn is the time for the nth cycle, T1 is the time for the first cycle, N is the number of cycles and a is the constant representing the learning rate. This equation produces a hyperbolic curve. (Readers, please note that this generalized formula is not effective in solving problems. Instead, the methods shown below should be applied.)

In order to determine the learning rate of a given activity, the time study may be applied to a worker who is performing the task. For example, masons installing concrete blocks to form a wall would be timed as they perform successive iterations of the process. The learning curve can be applied to construction projects. It can be highly relevant in repetitious projects such as housing construction, but the success of this application requires the IE to understand that interruptions to the construction process limit its use. Examples of such interruptions include prolonged shutdowns or over holidays.

Construction tasks are often varied and nonrepetitive, so the user has to apply the concept very judiciously. On-site managers who understand the learning curve rates for different types of tasks can improve work performance by selecting alternative work methods, especially with less experienced crafts persons. There are three distinct phases:

1 When construction crews are familiarizing themselves with a process.
2 When a routine is learned so that coordination is improved.
3 A deliberate and continuing effort to improve with successive iterations of the process.

Learning curves for construction typically fall in the 70–90% range.

The curve in Figure 8.3 represents a project involving the installation of a number of generator units. The expert's estimate for carrying out this work was 11,000 man hours. The contractor's bid was lower – 7200 hours per unit. It is unlikely that a bid based on the expert's estimate would have been successful. The use of the learning curve allowed the contractor to complete the project at an even lower level of man-hours – 5900 hours per unit. By using the benefit of the learning curve, the contractor was able to reduce the labor hours for the project by 1200 × 8 = 9600 hours over eight installations. This savings could translate directly to an increased profit margin (Figure 8.3).

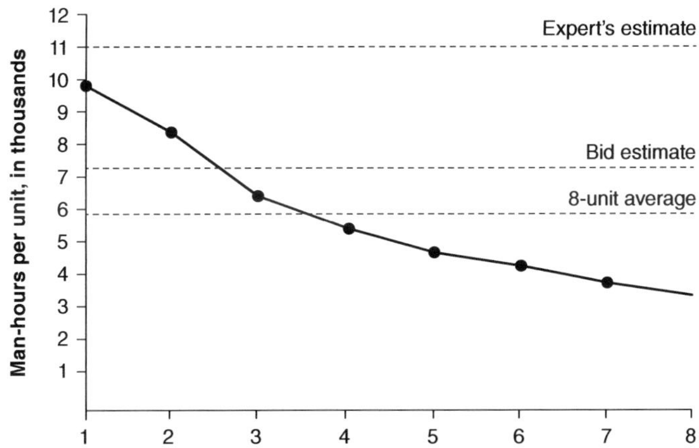

Figure 8.3 An arithmetic plot of worker hours expended for installing successive generator units

Source: With permission of G.A. Howell

Example: learning curve calculations

Calculations are done with the ratios that are shown below:

The following equation displays the relationship between the involved variables:

$$r/s = (j/i)^{\text{Ùn}}$$

Where i	=	unit sequence of the earlier unit (r hours or dollars)
j	=	unit sequence of the later unit (s hours or dollars)
Learning rate	=	$2^{-n} = 1/2^{\text{Ùn}}$ [2 raised to the power – n]

Where n is the exponent that describes the variation. It is used for calculating the learning rate.

An Example:

A construction crew is carrying out a repetitious task. The first cycle takes the crew 5 hours to complete. The third cycle takes the crew 4 hours. Calculate the learning rate.

Applying values to the equation:

i	=	1 (first cycle), j = 3 (third cycle)
r	=	5 hours, s = 4 hours
r/s	=	$(j/i)^{\text{Ùn}}$ Then $5/4 = (3/1)^{\text{Ùn}}$
Substituting, 1.25	=	$3^{\text{Ùn}}$
Using logs: log (1.25)	=	n x log 3
Solving: n	=	0.203
Learning rate	=	$2^{-n} = 1/2^{\text{Ùn}}$ [i.e., 2 raised to the power – n]
Substituting for n, the learning rate	=	0.868 = 86.8%

Another example:

If the learning rate remains the same, how long should it take to complete the 4th cycle of the task? Substituting the given values

i	=	1 (the first cycle)
j	=	4 (the fourth cycle)
r	=	5 (5 hours)
s	=	?

From above,

learning rate	=	86.8%, n = 0.203		
r/s	=	$(j/i)^{Ùn}$		
5/s	=	$(4)^{Ù}0.203$	=	1.3205
Cross multiplying, s	=	5/1.3205	=	3.79 hours

The fourth cycle takes 3.79 hours.

Questions for discussion

1 How can the development of standard times for repetitious tasks benefit the construction process?

2 A repetitious task takes 3 hours for the second cycle, and 4 hours for the fourth cycle. Calculate the percentage learning rate.

3 The learning curve: The first unit takes 5 hours. The third unit takes 4 hours. Determine the learning curve expressed as a percentage. If the learning curve remains the same, how long would the fifth unit take? If the learning curve were 80%, how long would the third unit take?

Part D: Ergonomics

Improving worker performance: Ergonomics-based strategies

Note: The purpose of this section is to provide an overview of Ergonomics and its impact on construction processes. Readers should consult with specialized sources for further information. Some of these sources may be available from the references and the bibliography listing at the end of the chapter.

Ergonomics is a practical discipline that deals with the design and improvement of productivity and safety in the workplace. Ergonomics involves studies of human perceptions, motions, workstations, machines, products, and work environments. The main methodology of ergonomics involves the mutual adaptation of the components of human-machine-environment systems by means of human-centered design of machines in production systems.

The application of the principles of ergonomics in the construction environment has many benefits. Physical work input is a major component of the construction process. Building components are typically large, heavy, and cumbersome. Building blocks (CMUs) and bricks are used in large quantities and are installed by hand. In the course of a day a bricklayer may handle large quantities of bricks or building blocks; this handling imposes a physical workload on the worker. Reinforcing steel is also used in the construction process and its weight is physically stressful as well.

Construction involves the use of tools and equipment in tight and awkward spaces, requiring workers to assume uncomfortable postures. Especially in cases where such aids are used for prolonged periods of time, workers' effectiveness and capacity to work with high levels of concentration is reduced. These situations often lead to on-the-job injuries and worker health impacts. Workers cannot be expected to "build in" quality in constructed facilities if they are subjected to excessive physical stress caused by tools and equipment that are difficult to use.

Reducing musculoskeletal stresses

Musculoskeletal disorders (MSDs) are different from other occupational injuries and illnesses because they develop over a long period of time. Musculoskeletal diseases (mostly low back problems) represent 27% of non-fatal injuries and illnesses in construction. In 2001 disorders resulting in lost work days were:

Overexertion while lifting	45%
Overexertion without lifting	32%
Bending/twisting	17%
Repetitive motion	7%

Overexertion injuries are mostly observed in the following trades: roofing/siding/sheet metal, masonry, carpentry, and mechanical trades (US Department of Labor 2001).

Work-related musculoskeletal disorders (WMSDs) account for 34% of all lost workday injuries and illnesses according to the Bureau of Labor Statistics (BLS) and account for between $15 billion and $20 billion in worker's compensation costs in the United States each year. These disorders have a critical impact on the construction industry as workers with severe injury may experience permanent disability that renders them unable to perform their job tasks in keeping with standard output levels.

The WMSDs include: carpal tunnel syndrome, tenosynovitis, tension neck syndrome, and lower back pain. Carpal tunnel syndrome accounts for more days away from work than any other workplace injury. All body parts involving joints, muscles, tendons, nerves, and blood vessels are susceptible to WMSD. Overuse disorders include: strains, which are injuries to muscles or tendons; inflammation, as a protective response to limit bacterial invasion; and sprains involve the stretching, tearing, or pulling of a ligament.

Carpal Tunnel Syndrome (CTS)

CTS is one of the most common and well-known WMSDs. It receives its name from the eight bones in the wrist, called carpals. The disease occurs with the gradual swelling of the tendon sheaths that surround each of the tendons. CTS symptoms are:

- Painful tingling in one or both hands
- Finger numbness
- Fingers are described as being swollen
- A weakened grip

There are non-occupational factors that contribute to CTS. They include: systematic diseases such as rheumatoid arthritis, acromegaly, gout, diabetes, myxoedema, ganglion formation, and certain forms of cancer. Congenital defects may also lead to CTS. It can be caused by acute trauma to the median nerve inside the carpal tunnel that can may result from a blow to the wrist, a laceration, or a burn. Women tend to have more susceptibility to CTS due to their generally smaller body frame.

Risk factors for WMSD

The National Institute for Occupational Safety and Health (NIOSH) attributes WMSDs to a number of causes:

- Repetitive, forceful, or prolonged exertions of the hands
- Frequent or heavy lifting
- Pushing, pulling, or carrying of heavy objects
- Prolonged awkward postures
- Vibration

There is a relationship between workplace factors and MSD development. Some joints display significant susceptibility to work-related injury. They are: the neck and neck/shoulder area, the back, and the hand/wrist area. It is not clear that elbow injuries are generally work-related.

Back disorders may be associated with both occupational and non-occupational factors. Back injuries account for almost 20% of all injuries and illnesses in the workplace, and the prevalence in the general population has been estimated at 70%. The costs of these injuries in the United States range from $20 to $50 billion annually, hence its impacts on the economy are significant. At the organization level such injuries exact high costs by both reducing workforce availability and inflate insurance costs.

Evaluating and addressing MSDs

There are seven steps for evaluating and addressing MSDs in the workplace. Strategies are described in this section below for addressing their effects. The steps are:

- Be able to identify signs of WMSDs
- Set the stage for action
- Train personnel in order to build in house expertise
- Gather and examine evidence concerning WMSDs
- Identify effective controls for WMSD-risky tasks, and evaluate these controls to determine effectiveness
- Implement health care management
- Maintain a proactive ergonomics program

Work system design

According to Albers *et al.* (2005), tools and their selection influence work postures and work stress. In order to reduce worker stress and improve productivity, work system design should anticipate the postures required to accomplish the respective tasks. Using electrical work as an example, the work system comprises:

- The work task itself (e.g., high level cable tray installation)
- The available equipment (e.g., scaffolding)
- Tools, power or hand such as electric drills
- Design processes that affect these components

Modular and architectural innovations

Modular and architectural innovations can have an impact on the construction process and worker activities. Modular innovations can be implemented by one trade with minimal effect on others. They reduce exposure to one or more risk factors. Architectural innovations change the nature of the activity for a number of trades. For example, a hammer drill is often used to drill holes and embed concrete inserts for the hangers for piping, cable trays, or other devices. Instead, inserts may be cast into concrete beams by inserting them in the form work used for pouring concrete to form the beams. One trade would be relieved of the drudgery of drilling holes in beams to set the concrete inserts after the building structure has been erected. On the other hand, another trade would have the responsibility to place the inserts in the concrete form work.

NIOSH recommended best practices for work tasks

In the absence of ergonomic standards for US construction, NIOSH assembled a number of best practices from a 2003 stakeholder meeting (Albers *et al.* 2005). These stakeholders included mechanical/electrical contractors/installers and trades; 39 industry representatives, 17 construction ergonomics researchers from government and academia, and four ergonomics consultants with construct experience. Common tasks were (a) drill holes and shoot fasteners, (b) place and install systems, and (c) lift and carry materials and equipment (Table 8.7).

Table 8.7 Suggested ergonomics intervention to reduce musculoskeletal loading during drilling holes and shooting fasteners

Problem/ Musculoskeletal Load	*Electrical*	*Piping*	*Sheet Metal HVAC*
Install anchors for hanging systems in concrete		Specify embedded concrete inserts (i.e., Unistrut, Anvil, etc., or beam clamps, caddy clips)	Specify embedded concrete inserts (i.e., Unistrut, Anvil, etc., or beam clamps, caddy clips) Place minimum number of hangers required
Operate rotary-impact hammer drill	Use clutch-driven drill to control torque Use sharp bits Use low vibration tools Use vibration damping gloves	Use clutch-driven drill to control torque Use sharp drill bits Use vibration damping gloves Implement a tool and bit maintenance program	Use drill with side arm to control torque Use low vibration tools Identify, purchase tools on performance criteria
Operate powered actuated tool overhead		Use extension pole and remote triggering	Use extension pole and remote triggering Use drill bit extender
Drill overhead	Support head with neck pillow Work on platform or scaffold	Use drill bit extender Use suspension and balance system for tool	Work on powered lift or scaffold
Work overhead	Work on powered lift or scaffold	Work on powered lift or scaffold	Work on powered lift or scaffold
Standing on concrete	Stand on antifatigue mat	Stand on antifatigue mat or use shoe inserts	
Working at floor level			Use knee pads

Source: Adapted from Albers, J., Estill, C., and MacDonald, L., *Applied Ergonomics*, 36, 428–39, 2005. With permission from Elsevier

Voluntary ergonomics guidelines developed by OSHA

Design for construction

Design processes should consider the construction process in terms of safety, health, and ergonomics as design decisions are being made (BNA 2002). The proximity of pipes, ventilation ducts, and cable trays can be a problem for construction workers. Workers are forced to work in very confined spaces, and to assume postures that result in very high musculoskeletal stresses. The inclusion of fastening anchors and hanging devices in the building structure can have a significant impact on the complexity and risk factor exposure of construction workers. The configuration of building elements that place more emphasis on mechanized transportation and handling rather than on manual methods can also have a significant impact on construction processes. It can also impact the speed of construction and the rate of productivity.

Interventions

Some interventions are within the reach of contractors and some are not. The use of embedded concrete inserts or beam clamps are determined during the design phase of the project, and are considered to be engineering interventions. Administrative controls are within the reach of contractors for reducing musculoskeletal loading on workers:

- Worker training programs in body mechanics
- Job rotation and microbreaks
- Physical conditioning (i.e., stretching programs)
- Prejob planning and hazard analysis
- Optimizing communication and coordination between contractors and trades

Strategies for manual material handling

- Observe weight limits – require weight labeling
- Optimize building site organization and housekeeping (this is complicated by having multiple contractors)
- Utilize lifting programs – training programs can significantly lower the occurrence of low back pain

Six recommended lifting methods are:

- Straight back, bent knees lift (most effective lifting method)
- The hip-flex lift
- The kinetic lift
- The dynamic lift
- The stooped lift
- The pelvic-tilt lift

Back injury prevention programs have three general parts. Pre-employment and job-placement screening, training and workplace design. Back belts have been proven to be effective in preventing back injuries related to manual holding jobs. However, a disadvantage of back belts is a risk of homeostasis, in which a person believes that he/she can lift more than they are capable of when wearing the belt.

Ergonomics applications in structural ironwork

Forde and Buchholz (2004) observed higher rates of musculoskeletal injuries and disorders in construction workers than workers in other industries. In studying this disparity, they determined that each construction trade and task represents a unique situation and proposed that the identification and application of prevention measures, tools and work conditions is best derived from trade and task specific studies. Their findings suggest that this approach is the most likely to minimize the incidence of construction trades work-related musculoskeletal disorders (WRMSDs).

By way of illustration, Forde and Buchholz (2004) studied construction ironworkers to identify mitigating measures in that group. Construction ironwork refers to outdoor work (not shop fabrication) as four specialties; the erection of structural steel (structural ironwork, SIW), placement of reinforcing bars (rebars; reinforcing ironwork RIW), ornamental ironwork (OIW), and machinery moving and rigging (MMRIW). Previous studies determined that construction ironwork involves the lifting, carrying, and manipulating of heavy loads; maintaining awkward postures in cramped quarters; working with arms overhead for extended periods; using heavy, vibrating pneumatic tools; and extensive outdoor exposure in temperature and weather extremes. Forde and Buchholz (2004) made the following observations and recommendations on the various categories of ironwork.

Reinforcing ironwork

The preparation of reinforcement cages and tying of rebars was seen to cause non-neutral trunk postures up to 50% of the time. The handling of heavy loads (50 pounds or greater) was observed to occur for 1.9 hours of an eight-hour shift, representing significant long-term risk. Forde *et al.* identified a need to improve the design of hand tools used for securing rebars. Such redesign would reduce non-neutral hand/wrist postures such as flexion, extension, and radial and ulnar deviation. These postures put construction workers at risk of repetitive motion injuries.

Drywall hanging methods

Drywall lifting and hanging are extensively conducted in both residential and commercial building construction; drywall board has become the standard for interior wall panels. It is the standard for surfacing residential ceilings. Workers are required to handle heavy and bulky drywall sheets and assume and maintain awkward postures in the course of performing installation work. These activities often cause muscle fatigue and lead to a loss of balance; studies have identified drywall lifting and hanging tasks as causing more fall-related injuries than any other tasks. Pan *et al.* (2000) studied 60 construction workers to identify the methods resulting in the least postural stability during drywall lifting and hanging tasks.

The subjects' instability was measured using a piezoelectric-type force platform. Subjects' propensity for loss of balance was described by two postural sway variables (sway length and sway area) and three instability indices (PSB, SAR, and WRTI). The study results indicated that the respective lifting and hanging methods had significant effects on two postural-sway variables and the three postural instability indices. The recommended methods were:

- Lifting drywall sheets horizontally with both hands positioned on the top of the drywall causes the least postural sway and instability.
- Hanging drywall horizontally on ceilings produces less postural sway and instability than vertically.

References

Albers, J., C. Estill, and L. MacDonald. 2005. Identification of ergonomics interventions used to reduce musculo-skeletal loading for building installation tasks. *Applied Ergonomics* 36: 428–439.

ASHRAE. 1997. ASHRAE fundamentals handbook (SI). 8.1–8.28. In *American society of heating, refrigerating and air-conditioning engineers, Inc*, 1791 Tullie Circle, Atlanta, GA 30392.

Barnes, R. 1980, 2009. *Motion and time study design and measurement of work*, 7th ed. New York: Wiley.

BNA. 2002. Ergonomics: OSHA's plan for reducing ergonomic injuries. Occupational safety and health reporter. *Bureau of National Affairs* 32(15): 358.

Choudbury, I., and M. Haque. 2001. A study of cross-cultural training in international construction using GLM procedure and neural network approach. *ICCM conference proceedings*. Singapore.

COBRA. 2001. *Conference organised by RICS research foundation*. National Institute for Occupational Safety and Health, January 15, 2004. Glasgow, UK.

Forde, M., and B. Buchholz. 2004. Task content and physical ergonomic risk construction ironwork. *International Journal of Industrial Ergonomics* 34(4): 319–333.

Hanna, A. S., C. S. Taylor, and K. T. Sullivan. 2005. Impact of extended overtime on construction labor productivity. *Journal of Construction Engineering and Management* 131(6): 734–739.

Mayo, R., K. Knutson, G. Barras, and J. Pineda. 2001. Improving construction productivity with scheduled overtime. *Royal Institution of Chartered Surveyors Conference* 53(2): 897–900.

Pan, C. S., S. Chou, D. Long, J. Zwiener, and P. Skidmore. 2000. Postural stability during simulated drywall lifting and hanging tasks. *Proceedings of the 44th triennial congress of international ergonomics association, and the 44th annual meeting of the human factors and ergonomics association*. Ergonomics for the New Millennium, 679–682.

Rasmussen, T. 1996. *The ASTD trainer's sourcebook: Diversity*. New York: McGraw Hill.

Seelye, H., and A. Seelye-James. 1995. *Culture clash: Managing in a multi-cultural world*. Wausau, WI: NTC Books.

Thomas, H. R., and K. A. Raynar. 1997. The effects of scheduled overtime on labor productivity: A quantitative analysis. *Journal of Construction Engineering and Management* 123(2): 181–188.

Tippett, L. H. C., and P. D. Vincent. 1953. Statistical investigations of labour productivity in cotton spinning. *Journal of the Royal Statistical Society* 116: 256–271.

U.S. Department of Labor. 2001. *Accident statistics for the construction industry*. Online at www.bls.gov/data/.

Bibliography

Aghazadeh, F., and A. Mital. 1987. Injuries due to hand tools. *Applied Ergonomics* 18(4): 273–278.

Albers, J., C. Estill, and L. MacDonald. 2004. Identification of ergonomics interventions used to reduce musculo-skeletal loading for building installation tasks. *Applied Ergonomics* 36: 428–439.

Asfahal, R. 2004. *Industrial safety and health management*. Upper Saddle River, NJ: Prentice Hall.

Auliciems, A., and S.V. Szokolay. 1997. *Thermal comfort*. Brisbane: Dept. of Architecture, University of Queensland.

Bjelle, A., M. Hagberg, and G. Michaelsson. 1979. Clinical and ergonomic factors in prolonged shoulder pain among industrial. *Work, Environment and Health* 5: 205–210.

Burati, Jr., J. L., and T. H. Oswald. 1993. Implementing TQM in engineering and construction. *Journal of Management in Engineering, ASCE* 9(4): 456–470.

deCatanzaro, D. A. 1999. *Motivation and emotion: Evolutionary, physiological, developmental, and social perspectives*. Upper Saddle River, NJ: Prentice Hall.

Fanger, P. O. 1970. *Thermal comfort*. Copenhagen: Danish Technical Press.

Forde, M., and B. Buchholz. 2004. Task content and physical ergonomic risk construction ironwork. *International Journal of Industrial Ergonomics* 34(4, October): 319–333.

Kubal, M. T. 1994. *Engineered quality in construction*. New York: McGraw-Hill.

McGregor, D. M. 1957. "The Human Side of Enterprise." In *Adventure in Thought and Action*. Proceedings of the Fifth Anniversary Convocation of the School of Industrial Management. Cambridge, MA: Massachusetts Institute of Technology, 9 April.

Mohamed, S., and K. Srinavin. 2001. Impact of climatic environment on construction productivity. *Proceedings of the third international conference on construction project management*. Singapore, 120–129.

NIOSH. 1997. *Elements of ergonomic programs*. US Department of Health and Human Services, Public Health Service, Centers for Disease Control and Prevention. National Institute for Occupational Safety and Health, Cincinnati, OH, DHHS (NIOSH) Publication No. 97-117.

Oglesby, C., H. W. Parker, and G. A. Howell. 1989. *Productivity improvement in construction*. New York: McGraw Hill Inc.

Pan, Christopher S., Sharon Chou, Daniel Long, Joyce Zwiener, and Paige Skidmore. 2000. Postural Stability during simulated drywall lifting and hanging tasks, *Proceedings of the XIVth triennial congress of international ergonomics association, and the 44th annual meeting of the human factors and ergonomics association, 'Ergonomics for the New Millennium,* 679–682.

Rosecrance, J., T. Cook, and C. Zimmermann. 1996. Work-related Musculoskeletal disorders among construction workers in the pipe trades. *Work* 7: 13–20.

Schaffer, R., and H. Thomson. 1998. Successful change programs begin with results. In *Harvard Business Review on Change.* Watertown, MA: HBS Press, 189–213.

Sillanpaa, J., A. Lappalainen, M. Kaukiainen, and P. Laippala. 1999. Decreasing the physical workload of construction work with the use of four auxiliary handling devices. *International Journal of Industrial Ergonomics* 24(2): 211–222.

9
THE INTEGRATED PROJECT DELIVERY (IPD) METHOD

*Samir Emdanat, Digby Christian, and Lincoln H. Forbes**

This chapter introduces applications of the integrated project delivery (IPD) concept, starting with early examples and continuing with more recent developments within the Sutter Health organization. It devotes a major section to a description of the conditions and methods that are needed for IPD to be implemented and managed successfully. Much of the narrative presented reflects developments in IPD based on lessons learned from the Sutter Health projects. Readers are encouraged to review these conditions carefully in order to have a reliable foundation for embarking on IPD.

IPD was developed to improve innovation by moving money across boundaries (Alarcon *et al.* 2013). The Lean Construction Institute (LCI) defines IPD as a project delivery approach that integrates people, systems, business structures and practices into a process that collaboratively harnesses the talents and insights of all participants to reduce waste and optimize efficiency through all phases of the project from early design through project handover. The three contractual components of IPD include Organization structure, Lean Operating Systems and Commercial Terms.

The IPD concept has proven to be highly successful since its development in the early 2000s. The Westbrook Air Conditioning and Plumbing project in Orlando is described as an early example of the IPD method, based on the creative modification of a small design-build project and resulting in significant time and cost savings. The project is described in detail in Section B of the chapter from a historical perspective. As CEO of Westbrook, Owen Matthews wished to utilize a relational contract approach with his project team, that was separate and apart from the design-build contract with the owner. The general contractor (Westbrook) the designer, and several subcontractors were designated as primary team members (PTMs) and were bound together legally in a relational contract that apportioned risk equitably. This temporary organization was established for the duration of the project, and was based on a commitment to implementing lean practices. The project was completed early, and cost savings of 10% were derived on a project cost of $6 million. The savings were shared by a pre-arranged formula between the PTMs and the owner. This successful project is recognized by the Lean Construction Institute as a pioneer example of the IPD process.

The Sutter Health organization in particular took a leap of faith by adopting lean construction to deliver a major construction program with an aggressive schedule mandated by the State of California. Lean construction was a promising, yet not guaranteed, methodology at that time, but the urgency for action propelled Sutter's executive management towards adopting lean as a calculated risk. Out of those early forays into lean construction practices emerged an evolving model of IPD that has been highly successful over many years since that time. The capital projects program has been highly successful and was delivered under budget, ahead of schedule, and without compromising any of the projects' operational objectives as evidenced by the outcomes of:

- 20 projects completed ($4.39 billion in total),
- Just 1 failure (slightly over budget), and
- 95% success rate compared to an industry average of 28%.

Sutter Health attributes many of these successes to the implementation of lean construction, particularly the IPD method. They are all the more significant because the method of project delivery was new to Sutter, and in fact new to the construction industry. There is good reason to believe that they would not have been achieved with traditional project delivery.

Origins of IPD

The IPD contracting method has been steadily gaining traction in the construction industry as the need for supply chain collaboration continues to increase. IPD provides a contracting framework that encourages teamwork and overcomes arbitrarily placed barriers between owners, designers, and builders. It allows project participants to collaborate and innovate by identifying new ways to organize themselves, manage workflow, and accelerate the design-to-construction cycle.

IPD contracts recognize that projects are complex and dynamic. They are designed to promote harmonious relations between the parties and reward positive and collaborative team behaviors that deliver projects in ways that meet their original goals. Early successes resulted in the gradual improvements to the approach and its adoption in the United States and Canada by various professional groups and owners, both private and public. It continues to gain popularity, and several studies show that it is resulting in less risk and more reliable outcomes when compared to other delivery methods.

Relational vs. transactional contracts

In contrast with IPD contracts, the most common contracting methods in the AEC industry have been transactional in nature and are usually established between two parties, in which one party owes a deliverable to the other (a.k.a. the transaction). Examples are provided in Chapter 1 such as Design-bid-build, design-build, CM-at Risk, and others. Those contracts require the parties to have clearly defined roles and responsibilities. They also obligate the supplying party to hold the risk of an incomplete deliverable. As such, those contracts assume perfection where perfection does not exist. For example, the owner's agreement with the designer requires designers to deliver 100% complete contract documents (100% CDs) so that contractors can provide detailed resource-loaded schedules as the basis for their contract with the owner. The architects' agreements with consultants as well as the general contractors' agreements with their subcontractors would have similar expectations. Over the years, this resulted in highly developed legal and risk management strategies to mitigate the risks of this type of contracting for each of the individual parties involved. These strategies provide protection for many stakeholders, ranging from the owner to the last trade contractor or design consultant. These contracts encourage behaviors that are inwardly focused as each party seeks to protect its interests rather than doing what is required for the overall success of a project.

Relational contracts are different. They are designed to align the interests of the parties around common goals rather than individual interests and to allow them to manage risk rather than shifting it from one party to the other. The parties gain more profits when a project is delivered in keeping with its targets, and they are at risk when the project overall fails to achieve its objectives. Relational contracts align well with lean construction principles, which advocate that projects should be managed collaboratively as production systems.

Disadvantages of traditional contracting methods

Traditional construction contracts are adversarial in nature – they typically include penalties for under performance or nonperformance by each party in a project. Owners contract with a general contractor (GC) or construction manager (CM). In turn, these parties contract with subcontractors for different disciplines. In Matthews and Howell (2005), the authors point to four systemic problems with traditional contracting that can be addressed with a relational contracting approach.

Design Ideas often Lack Field Input: Designers are focused on the form of the finished product such as a pipe section, a room, or a control system, and not so much on the method of how it gets produced. Constructability reviews during the late phases of design seek to address that problem, but with limited success because of the difficulty in making changes at that stage.

Cooperation and Innovation are Inhibited: The traditional contract structure inhibits coordination, discourages cooperation and innovation, often rewarding some of the parties for optimizing their performance at the expense of others. For example, communications and electrical subcontractors must complete their installation of raceways and outlet boxes before a drywall contractor installs the drywall that covers and surrounds them. If the drywall is installed first, then expensive corrective work is needed. Drywall installers are focused on meeting a schedule that is the foundation for their progress payments, and may actually be unaware of the schedules of other disciplines.

Planning Systems are Not Coordinated: Subcontractors in traditional contracts have a legal obligation to the GC or CM, but often no clear-cut responsibility to the other subcontractors. Consequently, their planning is guided by the master schedule, with little consideration of each other's schedules. The GC/CM often is not concerned with the detailed planning of its subcontractors, resulting in many unexpected events as field conditions change.

Self-Preservation Is the Subcontractors' Mantra: There are many examples of subcontractors being tied to a fixed price contract with the prime contractor, with little protection against unknown site conditions and rising material prices. Subcontractors may try to optimize their performance with little concern for the overall project, which they see as the prime contractor's responsibility.

Overview of relational contracting

An understanding of relational contracting principles is essential for a successful deployment of lean construction methods, including IPD. It is a contracting mechanism that apportions the responsibilities and benefits of the contract fairly and transparently, based on trust and partnership between the parties. Relational contracting provides a more efficient and effective system for construction delivery in projects that require close collaboration for execution. The relationship between the parties in that situation transcends the exchange of goods and services and displays the attributes of a community with shared values and trust-based interaction.

Relational contracting enables the parties to work together for mutual benefit, and reduce risk instead of shifting it to others. A successful outcome benefits all stakeholders, especially the client, and minimizes the need for troubled participants to improve their financial position by reducing quality and cost. Working relationships between the stakeholders are improved, resulting in more efficient work processes and less conflict. A spirit of teamwork leads to better project outcomes, including relatively higher financial returns. Consequently, the most significant factors that underlie relational contracting are cooperation and dependency between the parties for mutual benefit.

Benefits of relational contracting

- The ability to work on a face-to-face basis. The collaborative approach eliminates the need to formally document everything. Direct discussion leads to faster decision making. Job satisfaction is higher due to a more pleasant working environment.

- More innovative solutions become possible, and quality is improved as more people come together to discuss problem issues.
- An elevated level of trust in the contractor's competence and trustworthiness relieves inspectors and others of the chore of being on the spot at all times to see every detail of the work. When such a situation arises, quality costs are reduced as the contractor can be trusted to carry out the job correctly; that is, "do it right the first time."
- Better, faster solutions are possible that lead to savings for the contractor as well as the owner/client.
- More harmonious relations free up the parties to focus on work issues and not contractual issues.
- The shared use of Information and Communications Technology (ICT), avoids duplication by project team members and significantly reduces information costs while improving response time.
- There is an atmosphere of goodwill that helps to resolve situations where one party fails.
- Disputes are resolved early.

Characteristics of IPD agreements

For consistency, the term integrated project delivery agreement (IPDA) will be used throughout unless a more specific term is needed when discussing the evolution of the IPD concept. IPDAs are forms of contracts that govern the relations among the parties involved with the design and delivery of a project including as a minimum the owner, the designer, and the builder; hence the term 3-party agreements was used initially. Additionally, IPDAs can involve several other parties (such as engineering consultants and trade contractors) as it will be illustrated later in the chapter.

There are certain commonalities between all versions of IPDAs. Those will be discussed first as they are the foundation for defining the relations among the parties involved in project delivery in the construction industry.

1. Multiple signatories to a single agreement

An IPDA is a multi-party agreement with a shared risk/reward mechanism for the non-owner signatories. As a minimum, the signatories to this single agreement are the owner, the designer, and the builder. It can include more signatories such as other trade contractors and design and specialty consultants. Regardless of the number of its parties, the IPDA must contain a mechanism that rewards the non-owner signatories for meeting the project objectives and punishes them for failing to meet the overall objectives. In contrast, under most other contract types the signatories are rewarded or punished based on how well they execute their company's particular contracted scope. Signatories of non-IPDA contracts may also be allowed compensation due to the errors of other parties. The classic example of this is the builder being able to submit a change order to the owner for errors in the drawings provided by the designer. IPDAs are different. The signatories collectively own the risk associated with such errors, and they are rewarded collectively for their ability to make an overall project successful.

Depending on the project type, the owner may select the designer and then the general contractor in collaboration with the designer. In other cases, the owner may choose the general contractor first and then choose the designer collaboratively with the contractor. Alternatively, the designer and the general contractor may self-organize as a response to a request for qualifications from the owner. In all those cases, the selection is value-based where "value" does not mean "cost". Once a team is identified, representatives from the owner, designer, and general contractor jointly and collaboratively manage the project. They select the remaining members of the team including engineers, technical consultants, key subcontractors, and suppliers based on the specific needs of the project. The additional IPD team members may also be included in the IPDA as risk-reward members either directly under the main agreement or they may sign a sub-agreement with the general contractor or the designer. In those cases, their sub-agreement terms would include pass-through provisions to ensure alignment with the overall language of the IPDA.

2. Early engagement of an integrated team

The selection of teams under IPDA is fundamentally different from other forms of contract. As this delivery model requires a different set of skills, typically, one or more of the signatories will already be experienced with this way of working, or will have a strong reputation for their experience, creativity, willingness to collaborate, and demonstrated the application of lean principles.

Teams are selected sooner than with other contract models to ensure they are able to maximize their ability to influence positive project outcomes. Specifically, major trade companies as well as the general contractor are typically brought on board during the early phases of design to help with constructability and feasibility analysis. At the latest, they must be selected before the final overall project budget is set so that they can participate actively in the discussions that lead to finalizing it.

3. Team alignment on the project's goals

Though the exact methodology and timing can vary in different IPDAs most of the signatories are brought on board during what is termed the validation phase of the project. They are brought on with the understanding that once the validation is complete, and if the project is then funded and authorized to proceed, they will be selected to deliver the project. They will then be expected to design and build the project in accordance with the scope, budget, and schedule developed during the validation phase.

The most important part of validation is the creation of a clearly stated project definition and associated conditions of satisfaction that are produced as by the collaboration between the owner and the selected team. Typically those will include:

- A description of the operational purpose of the building.
- A space program required to deliver on that operational purpose.
- A proposed budget to design, build and open the facility to house that program.
- A proposed schedule to design, build and open the facility to house that program.
- Any other key performance metrics that are to be met by the facility.
- Any potential risks, and whether the budget and schedule do or do not include provisions for those risks. Where they do not, some assessment of the likely cost and schedule impacts is provided as a risk that the owner would manage outside of the IPDA.

It is worth noting that goals should be measurable within the duration of the IPDA. Owners often have longer term operational goals that can only be measured years after a facility opens. In those cases, a proxy goal is written into the IPDA. For example if the facility is being built to increase production capacity by 25%, and the owner believes that the space program that is set for the IPDA would achieve that long-term goal, then the team is at risk for building the space program and not for the ability of the owner to increase production capacity by 25% after the building opens.

As the non-owner signatories are part of the process that establishes the goals for the project this creates transparency across the team as to what the goals are, and creates buy-in and accountability to deliver on these goals.

4. Shared risk/reward mechanism

The IPDA signatories' risk and reward is tied to their collective ability to deliver all project goals as stated in the project definition and the conditions of satisfaction. There are many ways to achieve this, but typical characteristics across IPDAs are:

1) The signatories are transparent about their costs (actual costs, overhead, and expected profit) and review and challenge each other on those costs. Owners typically have the right to verify this information through a third party audit of the signatories' financial data.

2) Generally, the total cost of work, absent of profit, is guaranteed to be paid by the owner. Some more recent IPDAs have required partial payment of the overhead at risk.
3) The total cost of work usually contains a contingency amount that the team carries to offset the risk of the project and to fund opportunities that will help in meeting their goals. Various IPDAs handle contingency allocation differently, and it is generally contextual based on the needs of the project. Typically the team will receive a share of unused contingency at the end of a project.
4) The non-owner signatories of the IPDA allocate a substantial percentage (often 100%) of their profit at risk.
5) Overhead is handled in many different ways. Sometimes it is guaranteed, sometimes it is fixed as a lump sum at project commencement. In some instances a portion of it is also put at risk.
6) The owner has the risk of paying the non-signatories' costs when the at-risk portions of the budget are consumed until the project is completed.

The illustration above represents a dynamic risk/reward model that makes everyone accountable for the success of an overall project. The most important distinction that needs to be made from the owner's perspective is that the IPDA does not include a guaranteed maximum price concept. Instead, at the end of validation, the team produces what is known as the Estimated Maximum Price (EMP). This distinction is fundamental as it is what defines the owner's risk under the IPDA. The owner's risk starts during execution, when the total cost of work including all the at-risk portions of the non-owner signatories scopes are consumed and exceeds the EMP.

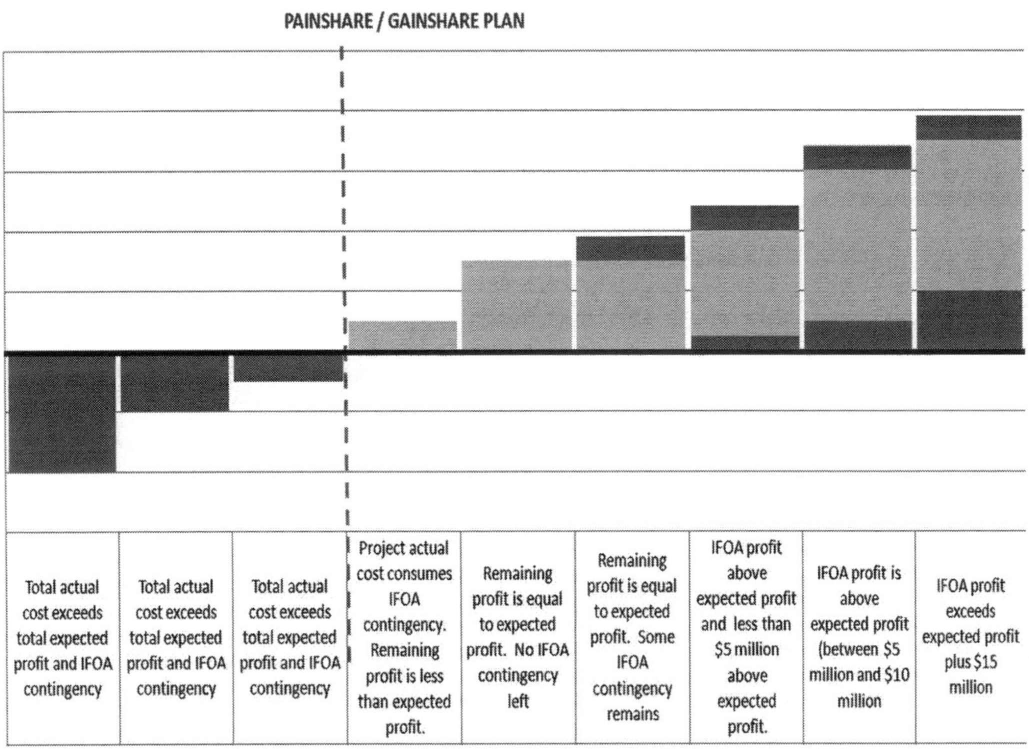

Figure 9.1 The shared risk and shared reward concept in IPDA showing the various scenarios of how risk and reward are allocated among the parties

Source: Courtesy of Sutter Health

IPDA signatories thus have a comprehensive and dynamic model at their disposal to manage overall project risk. They have control over how they spend the project budget to manage risk without the arbitrary contractual transactions of non-IPDAs.

The sharing of savings between the owner and the team varies from one IPDA to the next, and from project to project based on negotiation. The team can receive as little as 20% to as much as 100% of the savings. IPDAs may also contain tiers of savings whereby the team gets a different percentage depending on how much is saved. Often there is a cap to the maximum savings that can be obtained.

5. *The distributed governance model*

IPDAs describe a governance model wherein (in return for putting their profit at risk), companies get to be represented on the project leadership team which also includes representatives of the owner. This leadership team has different names in different IPDAs but some examples are core group, core team and project leadership team (PLT). This team has the primary responsibility for most strategic decisions on a project. Typically the IPDA requires decisions to be unanimous, though there are exceptions to the rule and those would be defined in the documentation of the IPDA.

6. *Limited grounds for change orders*

It is highly beneficial if the team has been together since early in the design phase as they collectively hold the responsibility for ensuring that the design meets the goals of the owner and that it can be built for the agreed budget and within the agreed schedule. Consequently, very few change orders are permitted under the IPDA. The team bears responsibility for several critical activities, i.e., errors or omissions in the design or coordination; errors or omissions in the estimate; errors or omissions in legal compliance, constructability, and the "inspectability" of the installation. Change orders are typically limited to those caused by the

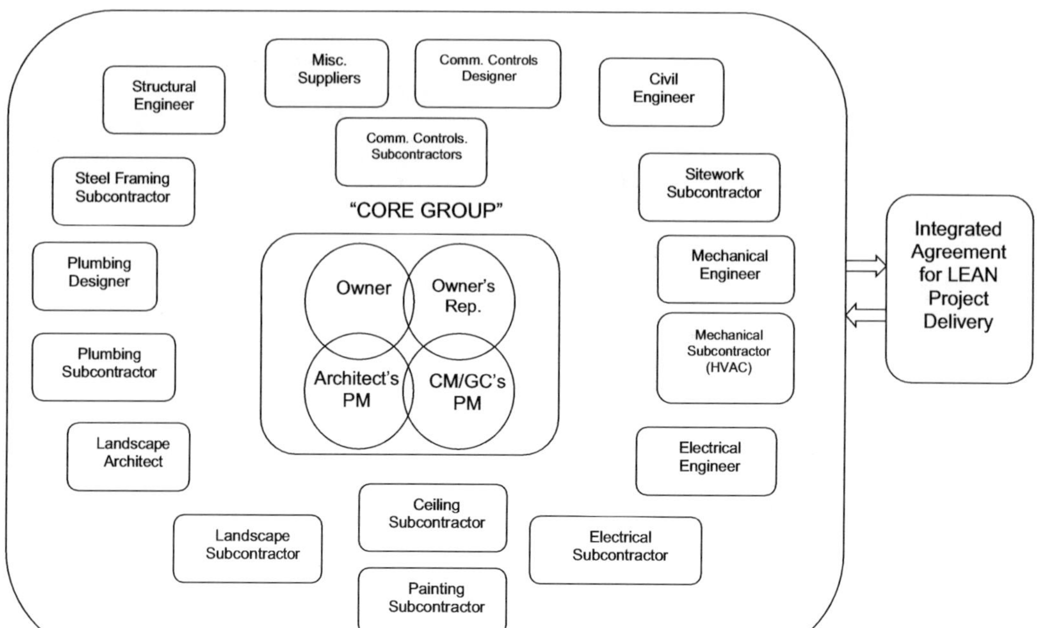

Figure 9.2 A generalized representation of a typical IPD project team

owner's changing requirements, by changes in the law, or by external events. These events must be reasonably considered to be outside the team's control, but within the contingency that is controlled by the owner, outside the scope of the IPD. Whenever such unforeseen conditions occur, the owner can modify the IPD scope accordingly from the owner-controlled contingency.

Sutter Health's IPD approach

History

Sutter Health is a not-for-profit community-focused service-oriented healthcare network that operates in Northern California. Their vision is to lead the transformation of healthcare to achieve the highest levels of quality, access and affordability. The network has 50,000 employees, 5,000 physicians, 30 hospitals, and over a hundred general and specialty clinics and laboratories. It serves a patient population of approximately 3.5 million that is one of the most demographically diverse in the United States. In support of this it spends an average of $500 million annually on capital projects.

While several factors, organizations, and individuals influenced the development of IPD concepts, Sutter Health in California played a foundational role as it sought to improve its methods for securing the design and construction work for its healthcare facilities, reduce project cost overruns, and improve schedule reliability.

The Northridge earthquake in January 1994 caused significant damage to several hospitals, and the California legislature passed Senate Bill 1953 that same year requiring all healthcare providers in that state to upgrade or replace their acute-care facilities (e.g., hospitals) so that they would withstand major earthquakes and remain operational afterwards. In 1994 the deadline for compliance with the new law was extended to 2008; subsequently the deadline was extended to 2013 and tiers of compliance were introduced later with deadlines of 2020 and 2030. In the mid-1990s Sutter Health estimated that it was going to need to spend $5.5Billion over several years to comply with this legislation. Sutter Health engaged the firm of Lean Project Consulting Inc. to guide the adoption of lean construction. McDonough Holland & Allen PC, Attorneys at Law, provided legal services especially with developing an early version of the Integrated Form of Agreement (IFOA) that was an essential framework for its projects.

Prior to the use of the IFOA, the first two major projects in the replacement program did not go well, and both were ultimately delivered over budget and late. Sutter knew that if this pattern repeated itself over the whole replacement program, it would create a significant risk to the organization as a whole. It began to investigate how to mitigate the risk and found that there were alternative approaches to design and construction that were said to outperform the conventional delivery model.

Among the early influences on Sutter's approach were the efforts of the Lean Construction Institute's (LCI) founders Greg Howell and Glenn Ballard as they explored how production management, lean thinking, and supply chain collaboration could improve project delivery by reducing variation. These principles were embodied in the Last Planner System and the Lean Project Delivery System (Ballard 2000). At the time Sutter started its exploration LCI advocated improving collaboration through the formation of cross-functional teams, with the use of concurrent engineering including the involvement of specialty contractors during the design phase.

In March 2004, Sutter organized a Lean Summit for its design and construction partners in Concord, CA in which they shared their vision of collaborative project delivery. The summit concluded with the development of what came to be known as The Five Big Ideas, namely:

- Increase relatedness,
- Collaborate, really collaborate,
- Optimize the whole,
- Tightly couple action with learning, and
- Projects as networks of commitments.

In the projects that followed, the participating partners and Sutter Health signed the Five Big Ideas document declaring their intent to use these concepts in project delivery.

Projects began to be implemented with this vision in mind, but no change was introduced in the contracting method. The results were mixed, but certainly not transformative. At regular progress meetings, the message back to Sutter was mainly: "You are asking us to change our behaviors, but the contracts themselves do not align with that request."

Because of this, in 2004, William (Will) A. Lichtig took on the challenge and came up with a modified GMP contract that included much of the intent around the Five Big Ideas. At that time, Will was a construction attorney with McDonough, Holland & Allen, P.C. and he acted as Sutter Health's external legal counsel for design and construction contracts.

However, again the results of using the modified GMP contracts were still not transformative. Therefore, in 2005, Will wrote the first version of Sutter's IPDA, known as the integrated form of agreement or IFOA. (Lichtig, 2005, 2008) It was a tri-party agreement that was signed by Sutter Health, the architect and the builder. The agreement described the goals clearly, regarding the clinical program, the opening date, and the budget, but it did not contain a risk/reward mechanism. It was not until the Sutter Medical Center Castro Valley project (SMCCV) began in 2007 that Sutter fully developed their IFOA contract.

Characteristics of Sutter's IFOA contract (current procedures)

Multiple signatories

The Castro Valley IFOA started with three signatories and quickly expanded to include 11 direct signatories. On subsequent Sutter IFOA projects, there have been anywhere between 3 and 17 at-risk signatories, but most have been three-party agreements in which additional at-risk partners worked through the architect or the general contractor. In such cases, Sutter requires that the language of the agreement signed by the architect and the builder with Sutter is passed through into the contracts they (the architect and builder) sign with the other at-risk team member companies.

1. Early engagement of an integrated team

Before Sutter Health funds a project that it knows will cost more than $10 million, it funds a validation effort to confirm the goals, scope, budget and risks of a project, and then puts that before the board for funding approval. To initiate the validation effort, Sutter assembles an integrated team of designers and builders. Sutter retains the right to not use members of this team for the approved project, but usually the team is retained and awarded the project once it is approved. Across the 23 Sutter Health IFOA projects (at the time of this writing), only in rare cases have team members involved in the validation study not been retained for the full project.

2. Team alignment on the project goals

Any project in excess of $10 Million must have a Validation Study. This study turns Sutter's goals for a proposed project into a functional narrative that describes how different spaces will be used, and a space program that shows how much of each clinical function is required to meet the operational goals. It includes a proposed budget and schedule. It also lists all the assumptions that the validation effort has made and the risks to the scope, budget, and schedule.

Sutter selects an architect and builder as an integrated team to carry out the validation study. Typically the cost of such studies has been in the region of 0.5 to 2% of the total project budget. The team also includes specialty design, consultant and trade partners where justified by project complexity and risk warrant such.

Typical consulting and preconstruction services contracts are used for the Validation Study, and teams earn their expected overhead and profits. However, a mandatory condition of being selected to be on the validation team is the company's agreement to the terms and conditions of the Sutter IFOA without exception, so if the project is approved, and the EMP is then finalized, they would have already agreed to sign the IFOA.

Having the team involved in the process that creates the Validation Study for the project, creates transparency and prepares the project for future success.

3. Shared risk/reward mechanism

The Sutter IFOA states that the requested program, along with any other listed built-environment goals must be built for the Estimated Maximum Price (EMP) and to have 100% of the proposed space ready for use by the date specified in the agreement. The IFOA does not offer a goal by goal option for earning incentives. It requires that all goals be fully realized before the team can earn incentives.

The Expected Maximum Price is calculated as follows:

EMP = Team's Actual Costs + Team's Overhead + Team's Profit + Team's Contingency

The final number is also supported by historic data as well as by market analysis of similar and comparable projects.

Actual costs

For at-risk trade partners, these costs include the agreed hourly rates, excluding overhead and profit for every person on the project as determined by an external audit of each company's financial records. For at-risk design partners, the situation is the same except that overhead is considered part of the designers' actual cost and as such is included in their hourly rates. This is because design companies have an entirely different business model from builders wherein overhead pays for a substantial part of their overall cost structure. In the healthcare design and construction market in which Sutter operates; overhead rates for designers are in the range of 10 to 20 times that of trades.

Actual costs also include the costs of materials reimbursable expenses, rental rates, and sub-contract buy-outs as verified by back-up paperwork. The IFOA guarantees that Sutter Health will reimburse the actual costs incurred by at-risk signatories. This means that if the total project cost during execution exceeds the EMP, then Sutter continues to pay the at-risk partners' actual costs. The downside for the team is that in this scenario they will have eroded their contingency and profit and so will not make any money.

Overhead

This is the agreed level of overhead for each company for non-project specific costs required to support the project as determined by an external audit of each company's financial records, typically going back over the previous two to three years. As noted above, for at-risk design partners this audited calculation is included in their hourly rates. For the at-risk trade partners their expected overhead is then fixed as a lump sum at the time of setting the EMP. The at-risk trade partner team members are then paid their lump-sum overhead on a pro-rata basis through the life of the project.

While the audit typically sets the overhead if there is disagreement over whether certain categories of cost are considered overhead such issues are resolved on a case-by-case basis.

Profit

Profit is what each at-risk team member company expects to make if the project is delivered exactly per the formula that establishes the EMP.The profit amount is proposed by each team member and typically would align with industry benchmark data or is based on historical data.The total amount of profit establishes the profit pool for the project and 100% of this total is considered to be at risk. In the event, the actual costs rise so high that they entirely erode the Contingency they next will start to erode the profit.

When the EMP is set, each partner's fixed profit expectation is converted to a percentage of the total profit pool. Thus they can only make more profit by increasing the size of the total profit pool, and they can only do so by collaborating to reduce overhead, actual costs, or both without compromising the project goals. For example if there were just an architect and a general contractor and the architect's profit amount in the EMP figure was $300,000 and the builder's profit was $700,000 these would be converted to 30% and 70% respectively of a total profit pool of $1,000,000. From that point on the architect has a contractual entitlement to 30% of the total profit, and the builder has 70%.

Contingency

The contingency is a sum of money controlled by the team to manage risks and fund opportunities. It varies by project type and context as agreed between the team and the owner. It has varied between approximately 3 and 7% on the 23 IFOAs used so far at Sutter Health. An IFOA contract for a low-complexity new clinic would be at the lower end, while one for a replacement hospital would be at the higher end.

When the EMP is set, the contingency is converted to a dollar figure. Typically, it appears as a line item within the builder's estimate but is money available to the full team of signatories.

The contract allows the team to share in a portion of any unused contingency at project completion.

The risk/reward deal

As previously mentioned, Sutter's projects are managed against the EMP using the formula previously mentioned. The team's risk starts after the contingency is exhausted. Once the contingency is exhausted, the team's profit pool is used to pay for the overrun. Sutter's risk starts after both the team's contingency and the team's profit pool are exhausted. At that point, Sutter pays the team's actual costs, without limit, until the project is complete. Note, this actual cost overrun that Sutter guarantees does not include overhead for the at-risk trade-partners as that remains at its original lump sum value.

Rewards are available based on meeting overall project goals and conditions of satisfaction. The team must achieve all the goals. For example, the project must open on time. This is important, as the team is not allowed to compromise the goals, and thus protect their profits by failing to respond to a project projecting to finish late. The team is required to track progress against the goals and report regularly and visually on the trends to achieving those goals. If after meeting all the goals there is some contingency left, a share of that is added to the team's profit pool. Additionally, if after achieving all the goals the team's actual costs are less than in the EMP calculation (above) then a share of that savings is added to the team's profit pool. At the end of the project, each team member company receives it's agreed on the percentage share of the remaining profit pool.

The sharing of savings between the owner and the team varies from one IFOA to the next. The sharing is based on how aggressive Sutter feels the EMP number is. Less aggressive EMP numbers have a lower share of the savings for the non-Sutter team members, while more aggressive numbers have a higher share. Sometimes tiers are established whereby as more savings are achieved the percentage that goes to the team increases or decreases. There is always a cap, wherein, once that level of savings is achieved 100% of additional savings reverts to Sutter. This cap is still a substantial figure, and to date, no project has exceeded such a cap. Where an EMP is thought to be exceptionally conservative, there may be an initial tier where all the savings go to Sutter.

An important note about behavior under this risk/reward deal

Even though it was discussed previously in the IPDA section, it is important enough to repeat here how the team behaviors change under an IFOA as compared to transactional agreements through an example. Under the IFOA model, the designer can increase their profit by lowering the builder's costs, and the builder can improve their profit by reducing the designer's costs. This is a transformatively different incentive environment for participating companies. It typically takes players new to this contract some time to adjust their behaviors and to take advantage of the opportunity. So much of the team behavior in the industry is driven by local optimization of one's own costs of performing one's own scope of work.

Under the IFOA, if a partner Company A "innovates" in a silo in a way that reduces their own costs, but this increases the costs of its fellow partners by more than that then the total cost of the project has increased, which decreases the size of the team's profit pool, which decreases the dollar value of Company A's share of that team profit pool. I.e. Company A reduced its cost but in doing so reduced its profits.

Conversely, if partner Company A innovates in a collaborative, team-focused way that actually increases Company A's costs, but by doing this they lower the costs of its partners by more than that, then the total cost of the project has decreased, which increases the size of the team's profit pool, which increases the dollar value of Company A's share of that team profit pool. Therefore, Company A increased its cost but in doing so also increased its profits.

It can take a long time for individual players to understand this fully, as it flies in the face of decades of ingrained behavior driven by contracting models that are based on self-interest.

5. Distributed governance model

The governing body in the IFOA is the core group as shown in the generalized representation above. The membership of the core group varies from one IFOA to the next, but the general intent is that a member of the core group represents each at-risk partner. In projects with an extraordinarily large number of at-risk signatories, decision-making can become unwieldy; the preferred arrangement is to have Sutter Health, the architect, the general contractor, one trade partner who represents all the at-risk trade partners, and one design partner who represents all the at-risk design partners. Sutter typically has two members on the core group. One member is the Sutter Health project manager, and the other is an operational manager or representative thereof from the part of Sutter Health for whom the project is being built. The core group members must each be able to make decisions on behalf of the organization they represent.

The IFOA requires all core group decisions to be unanimous. There are very few cases where this is not the case. One of which is, and for obvious reasons, the decision to remove a company from the project – this decision requires a majority decision with Sutter being a member of the majority.

The core group makes the decisions with regard to the sub-contracts that are issued and all strategic issues that affect the overall processes and direction of the project. The core group meetings are run by Sutter's project manager. The larger projects also require a Senior Executive Team (SET) that meets quarterly to oversee the performance of the project and the core group. The executive team makes final decisions in the event that the core group has been unable to achieve the required unanimity which is extremely rare.

6. Limited grounds for change orders

The IFOA only allows change orders for

1) Owner changes.
2) Changes in the law that could not reasonably have been anticipated.
3) Decisions by the authorities having jurisdiction that are unreasonable and could not reasonably have been anticipated.

4) Discovered conditions that could not reasonably have been anticipated.
5) Nothing Else.

Note: Errors and omissions in the work of the design team is not an allowed grounds for a change order. This eliminates, by number, nearly every change order under traditional contract approaches. The purpose of this is to incentivize the team to ensure the work of the design team is fully complete before the project proceeds into the construction phase. It has been Sutter's experience that this has transformed the quality of the work during the design phase of its IFOA projects.

7. Dispute resolution

The integrated form of agreement requires the core group to resolve disputes rather than the architect. If the group is unsuccessful, the senior management executives from the group try to resolve the dispute in question. If their efforts are unsuccessful, an independent expert may be brought in to have closure in an unbiased manner and without animosity.

8. Everything else that is different in Sutter's IPD approach

When Sutter changed its contract to that described above, it did so to allow its project teams to behave in a way that supports overall project success. What Sutter did not anticipate is that over the next eleven years almost everything else on its projects would change. This resulted in a transition from siloed (or self-interested) contracting with contradictory and misaligned goals to an integrated contract with a single set of goals that all key participants needed to achieve in order to be successful.

Specific requirements are described below that are common to all Sutter IFOA-contracted projects; they create the environment for integrated teams to be successful. The IFOA simply sets the framework for collaboration to happen but it does not by itself create the success. A project with an IFOA but without these processes is very likely to still fail. In particular, reference *Validation: See "3. Team Alignment on the Project Goals" in section "2. Characteristics of Sutter's IFOA"*

- Big Room Co-Location: All Sutter projects are required to use some form of co-location where the owner, designers, and contractors share a common space to collaborate on the design, planning, and management of the project. On the larger projects, several hundreds of team members would leave their home office and co-locate for an extended period of time (years), and on others, they may col-locate for 1 day a week or perhaps 2 days every 2 weeks.
- The Last Planner® System (LPS): which embodies the management by commitment concept from the 5 Big Ideas. LPS is used through all phases of the work from entitlements through to commissioning, fit-up, and move-in. It is a highly effective framework that allows an integrated team to organize its work effectively and reliably.
- Target Value Design (TVD): During Validation and in the early phase of the project post-approval this process uses the budget as an input to design, instead of having it as an output of design. A satisfactory design is one that delivers Sutter's operational goals and is informed by the integrated ability to innovate in ways that allow the project to be designed, built, and opened for the agreed level of funding in the agreed upon period.
- BIM for Risk Management. Sutter focuses its teams on aligning on what the BIM is for, what it will include, when to include it, and at what level of detail. Then it has its teams use LPS to create the workflows to create and manage that content. Within these workflows, there is a subset of tasks that describe all the decisions that Sutter needs to provide to the team.

- Design for Fabrication: this is to reinforce the concept of design definition being more than design intent. Many project teams new to the IFOA enter the process thinking the primary purpose of design is to create a paper (or a 3D model) package to be signed and approved by the Authority Having Jurisdiction (AHJ) and represents the designer's intent. They then feel it is the responsibility of the contractors to translate that intent into reality. Sutter has its teams primarily think of design as a deliverable to a complex web of downstream trade customers who will use it to procure, fabricate, install and build the design. Sutter gets its project teams to think of what goes to the AHJ as a point-in-time extraction from this "design for fabrication" workflow.
- Risk and Opportunity Tracking: the integrated team meets, typically, weekly to review all the project's risks and opportunities, assess the impact and likelihood of such, and work to eliminate, mitigate or manage the risks, and take advantage of opportunities. The aggregate financial risk is added to the teams projected costs, which incentivize the team to manage the risks and to implement the opportunities because both affect the size of the team's profit.
- Cost-to-Complete Forecasting: Sutter has found that companies are highly variable in their competency in forecasting the cost of the work that is not yet in place. Team members share how they do it and collectively influence the weakest to get better. Under the IFOA if a team member does a poor job, the whole team feels the impact.
- Takt Planning and Location Based Management: On the more recent projects takt planning has been implemented to increase the speed of construction and create a predictable flow of work for the trade partners. The approach has proved effective at managing the overall supply chain for the project and to create clear execution strategies and production targets for all involved.
- Dash boarding and Visual Management: All project goals must be tracked, updated and reviewed by the whole team. On Sutter IFOA projects all key leaders of the integrated team companies know each week or month whether the project milestones are on track and whether the project is on budget.

Since 2006, Sutter has completed 23 IFOA contracted projects totaling over $4.1 billion. They were delivered on time, on budget, and without scope compromises with only one exception where a five year project was a few months late and a few percent over budget. Currently, Sutter has another $500 million in IFOA projects in progress and does not intend to move away from the contract model and delivery processes.

The following sections will illustrate some of the breakthrough innovations that were achieved for the first time and which are now commonly used on many Sutter projects as standard practice. They will be highlighted through case studies from various notable projects over the past several years.

Sutter Health Eden Medical Center (SHEMC)

During the design and construction phase of this project, it was referred to as the Sutter Medical Center at Castro Valley (SMCCV) or the Eden Medical Center. In this chapter we will use its current name as the Eden Medical Center project.

The two primary authors of this chapter Digby Christian and Samir Emdanat collaborated for the first time on this project to combine their knowledge and experience in maximizing the potential of the supply chain collaboration enabled by the IFOA. From the start, the project was envisioned to be a "laboratory for learning." This wording was chosen intentionally to encourage the team to push the limits of possibility, explore opportunities, assess emerging technologies, and implement appropriate solutions to meet project goals as efficiently as possible. The project has influenced the direction of IPD for various owners and in various industries. It has been subjected to various studies; however, none had gone in depth to illustrate how breakthrough change actually took place on the project. The following detailed examples will explain these innovations.

Figure 9.3 An overview of the final 3D model of the SHEMC project showing all the facility systems

Source: Composed by Samir Emdanat, Courtesy of Sutter Health

SHEMC is a $309 million 230,000 square foot replacement hospital, which includes 130 private beds, a level 2 trauma center, seven operating rooms, and 24 ER stations. The project initially included an additional $11 million for the demolition of the existing building. This was split during validation and set aside as a separate project. Design started in mid-2008 and the project opened on December 1, 2012 ahead of schedule and under budget, meeting all of its conditions of satisfaction. This was despite the fact that when the team signed the IFOA the project was over budget and had no funds remaining in the contingency and had already used part of their profit pool.

The project was the first to formally introduce the concept of risk and reward as a component of the IFOA. It started as a three-party agreement, which quickly evolved into an agreement between 11-parties to promote more transparency and direct communication. The original project team included Sutter Health and nine partner companies. Namely: the Devenney Group as architect, DPR Construction as general contractor, TMAD Taylor & Gaines (*now IMEG Corp*) as structural engineer, Capital Engineering as mechanical engineer, The Engineering Enterprise as electrical engineer, JW McClenahan as plumbing contractor, Superior Air Handling (*now Harris Companies*) as mechanical contractor, Morrow-Meadows as electrical subcontractor, and Transbay Fire Protection as fire protection contractor.

As the project started, and at the request of Sutter Health, Ghafari Associates' Management Services team (*now vPlanner*) joined the team initially as a lean and BIM consultant and then became the 11th at-risk partner assuming the role of the overall project integrator and lean facilitator. Ghafari was invited due to their successes in implementing 3D-enabled lean thinking for streamlining automotive facility design and construction that had received national attention at the time the project was starting.

One of the most significant challenges on any IPD project is the transformation of the behavior of the various players and the alignment of their interests and those of their organizations to benefit the project. Each participating company came to the project with its preferred way of doing things, its preferred tools, and workflows that while effective internally, may not benefit the project as a whole. Initial efforts focused

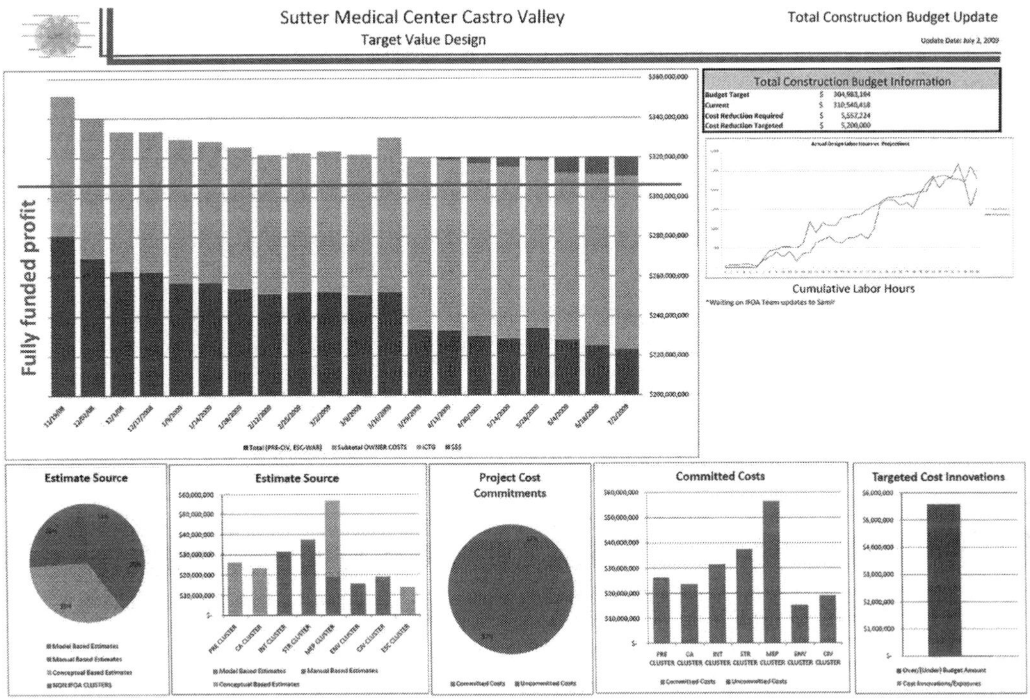

Figure 9.4 The visual budget reporting dashboard used by the team throughout the project to report on budget and trends related to the Target Value Design process

Source: Courtesy of Sutter Health

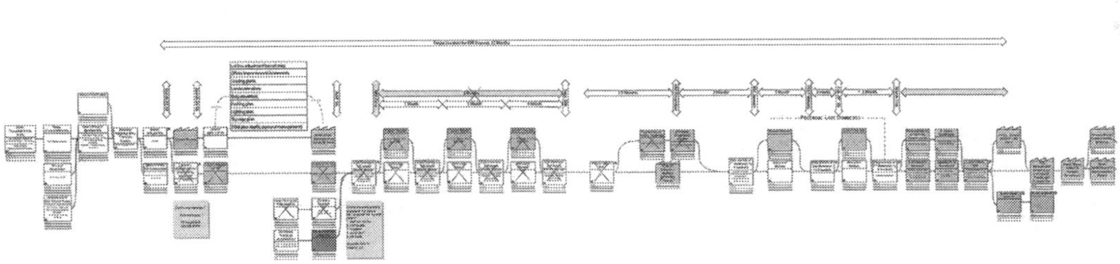

Figure 9.5 A process map of the Environmental Impact Review (EIR) process

Source: Courtesy of Sutter Health

Note: The workflow was developed in collaboration with the 25 EIR consultants and designers in the design team. The map was used to identify opportunities to improve the current state of the EIR process and reduce the risk to the ongoing design efforts.

on clarity of goals and on process and information flow mapping to objectively and visually map out how the various organizations would produce and share information to collaborate on the project and achieve the goals. This was fundamental as it provided an objective way to identify risks and opportunities as well as evaluate options for streamlining the process measured by its net impact on achieving the goal. Decision-making concerning which changes were to be implemented was highly process focused and goal focused. Before a decision to implement a change was made, the team mapped existing processes and information flows to fully understand the current situation. This resulted in improved alignment and allowed the team to weigh various improvement alternatives against the current state, and, to implement changes to the workflow if they would have a significant overall benefit.

In many cases, the introduced workflow changes would involve investing in upfront efforts and new technologies that would result in significant downstream gains. Examples of the team's achievement have been documented in various cases studies. Two examples are provided to show how the team changed the conventional workflow to achieve breakthrough results. The examples show the advantages of the IPD framework over other contractual methods when it comes to implementing transformative changes that affect the transactions between the parties.

- Management of the Steel Design to Fabrication Workflows
- Management of the Hangers Design to Fabrication Workflows

The focus of this overview will be on the processes themselves, how the opportunities were identified, and how the team implemented new processes that challenged established industry norms. The project team co-located once every other week for two days at the Big Room to review the design, update the pull plans, and identify work sequences to achieve the overall project goals. In parallel, they also planned the approach for phased OSHPD approvals and pull planned in detail the environmental review processes. This was a significant commitment in time and was met initially with resistance from various ranking individuals on the team. The fear was that the schedule was too aggressive and based on the team's past experience there was no time to spare and so design work needed to start as soon as possible. Sutter stepped in at that point and made planning a requirement before proceeding with design.

The initial pull plans served several functions including the management of risk and the identification of key decision points for the project. Another important function of those initial high-level pull plans was the alignment of the team's thinking around an objective way of achieving the respective goals. The plan

Figure 9.6 A pull planning meeting where the MEP team is reviewing their part of the plan in support of the structural steel needs

Source: Courtesy of Sutter Health

became the physical manifestation of how the various designers would sequence their decision making to deliver only what was required to achieve the objective of the early release of the steel structure with 100% certainty it would not change. At the same time they had to keep their eyes on the remaining scope items. The pull plan represented sequences of process steps, not deliverable dates, and as a result, it served as the current state for the various improvements that followed.

Information flow, tools, and means of exchanging information including preferred tools were overlaid over the initial pull plans and evaluated. This identified risks from using incompatible tools for representing the parts of the project that were interdependent. Obvious decisions were quickly made to keep all MEP systems for example, on the same platform that the trade contractors intended to use for fabrication and some of the engineers and trade contractors had to retool to align with the overall project requirements.

Pull planning was not done in abstraction, the pull plans informed the level of detail required by each of the disciplines represented in the 3D models. A collaborative multidisciplinary review of the in-progress design using the 3D model was held at each big room meeting. The planning informed the progress of the design, and the review of the design informed the planning. This pattern of collaborative in-model reviews was repeated throughout the project. It started bi-weekly during the early phases of design, then became weekly for detailed design, and daily for detailed coordination among the trades. Ghafari successfully applies this practice in the manufacturing industry and it was equally successful for managing the progress of design for the hospital by reducing negative iteration.(DEFINE) At the end of the OSHPD deliverables, the project team realized $1.2 million in labor savings related to the pre-construction efforts. That rigorous approach is now the standard on all of Sutter's large hospital facilities, and was implemented throughout the design and construction phases of the recently completed Van Ness and Geary Campus and Mission Bernal Campus hospital projects. These two projects are discussed later in this chapter.

Improving the steel design to fabrication workflow

For any steel structure, the locations and sizes of the steel elements are fundamental for managing risk as they are needed for the accurate representation of the systems that follow. However, this is not merely a question

Figure 9.7 The first 3D-based design review which focused on showstopper decisions regarding steel layout and the overall concept of MEP routing through the building

Source: Courtesy of Sutter Health

Note: This was before modeling any of the systems.

of accuracy. The timing and availability of the information is equally important. The initial pull planning efforts focused on the start of steel installation, the placement of the steel mill order, and the information required from all the other disciplines to begin steel construction. The goal was to identify an approach for releasing the steel and foundations with 100% certainty that the information would not change once the steel was released for fabrication. It is worth noting that the design and systems coordination of the facility interiors was schematic at best at the time the steel and foundations design was finalized and that the design processes of the interiors would continue for well over a year after the steel construction start date.

The structural engineers used Revit BIM software during design, and the fabricator (Herrick Steel) used XSteel. Since the detailer's model was the one that would be used to fabricate the steel, the goal was to transition to the detailer's model as soon as it was available. Once the first digital exchange between the engineer and the fabricator was completed, and the first 3D steel detailing model was available, it was compared to the engineering model and made the basis for future coordination. The structural engineer agreed to inform the detailers of any changes to the steel locations or sizes, so that the fabrication model was kept up to date.

The idea of eliminating the steel 2D shop drawings from the workflow and replacing them with the approval of the 3D detailing model was explored from the start as it was just made possible by Ghafari on several of its manufacturing projects in the Midwest with significant schedule savings. The workflow approach had not been attempted in California and surely not for an OSHPD hospital project. The team both on the engineering and detailing side had not done this before, and were opposed to the idea, thinking it would be too risky. Action was delayed for several months as the engineers developed the permit drawings using Revit.

In the months leading up to the start of steel detailing, it became clear that the rest of the design coordination would not be adequate to guarantee against possible changes to the steel design after it was released for fabrication. The sense of urgency was heightened by the length of time usually needed to detail the steel, prepare and submit the steel shop drawing. Even more time would be needed for an engineering review and of the shop drawings, incorporate the changes in the steel before fabrication. It would also be

Figure 9.8 A process mapping meeting to identify the overall process steps from the start of detailing to release for fabrication for the structural steel system and to identify opportunities for improvement

Source: Courtesy of Sutter Health

difficult to maintain the milestone date for the start of steel installation. A detailed step-by-step process map was created for the 2D approval process (the current state) and it was compared to a detailed step by step process of the 3D-based approval process (the future state). The analysis identified significant advantages for the 3D approval process over the 2D process in terms of the accuracy of the information and the timing of the reviews. However, it was the identification of a possible 6 months schedule reduction in the detailing, approval, and fabrication of steel that made the switch to the modified process an imperative. The start of the steel installation date did not change, but the time savings were critical for the other design disciplines to complete their work in time to fully inform the detailing of the steel and complete the layout of the systems.

Once the decision was made to use the 3D process, things moved very quickly. The future state became the current state with full buy-in from the various team members including the original skeptics. Implementation

Figure 9.9 A weeklong meeting at the engineer's office, attended by the steel detailers to train the engineers on the new 3D-based submittal and approval process

Source: Courtesy of Sutter Health

Note: On the left is the training session, on the right is the model being reviewed for comments or approval.

plans were developed that included detailing standards, review standards, and training for the engineers on how to approve the detailing model. As a result, the 3D detailing model became the preferred deliverable and replaced the cost 2D drawing submittals. This required the engineers to purchase licenses of XSteel, and required the detailers to assist with the training of the engineers on the new processes. The cost was covered by the overall project budget once the core group approved the modified process.

There were several adjustments to the process during the review of the first few steel sequences. Most of them related to the level of review that all other team members were expected to perform. For example, because of the 3D nature of the deliverable, the rest of the disciplines had an opportunity to perform a comprehensive parallel review of their work against the steel fabrication model. They were able to assess how their work could be impacted by those elements of the steel that could only be known after the detailing stage. Other process steps would eliminate clashes with MEP systems, steel plates, and large slab openings; coordination with skin, coordination with the edges of slab, brace locations and corridors, as well as stair coordination. Eventually, the steel design model was completely replaced with the 3D steel detailing model.

The results were very positive. The entire structure was erected with fewer than 130 RFIs relating to steel design. The steel detailer reported that they had not seen a project of this size installed so quickly and with so few changes to the steel after fabrication. The steel installation was completed ahead of schedule, and because the fabricator was hired under a cost-plus contract, the IPD team realized $1.5 million in savings related to the steel due to labor savings and reduced rework. However, the real benefits of getting the steel right were yet to be realized. The accurate steel model served as the foundation for all the detailed design coordination that followed with the building skin, the elevators, and all the services throughout the building. It significantly improved the quality of the coordination of all of the facility systems. Herrick has since refined this workflow and used as their standard review process on several projects that followed. According to Bob Hazelton, President, it is now their preferred way of managing the steel information flow.

The system hangers workflow

The location and exact sizing of the steel structure was the start of the routing and coordination of the building systems. The placement of the systems hangers indicated the readiness for installation. The trade partners on the project debated about new technologies that would allow the placement of system hangers from above the metal deck as opposed to under it. This led to ongoing debate on using inserts vs. drilling for the hangers. Also discussed at almost every Big Room meeting was "can vs. core" system for the conduit and pipe penetrations through the concrete floor slabs.

For readers who are unfamiliar with the concept, the support hangers for mechanical systems that run overhead in ceiling spaces can be installed by either drilling for their attachment anchors from below the floor slab after the concrete is poured on the floor above, or their attachment anchors can be punched through the metal support deck from above before the concrete floor slab is poured (thus the name inserts). There are many reasons why inserting is better than drilling. It is safer, more accurate, faster, and far less expensive. Drilling requires renting expensive lifts, more labor to install, and is a slower process. The main constraint on the insertion process is the timing – all hanger attachment points have to be identified before the floor slabs are poured. Slabs need to be poured approximately six months before the majority of the mechanical systems go to fabrication.

Naturally, all trade partners in the project decided to use the insertion process as their default practice. If they failed to locate the hangers ahead of the concrete pours then drilling would be the second choice. Provisions were made in the budget for drilling attachment there were doubts about the method that would be used for the hangers. As the team proceeded with coordination, parallel efforts where underway to map out

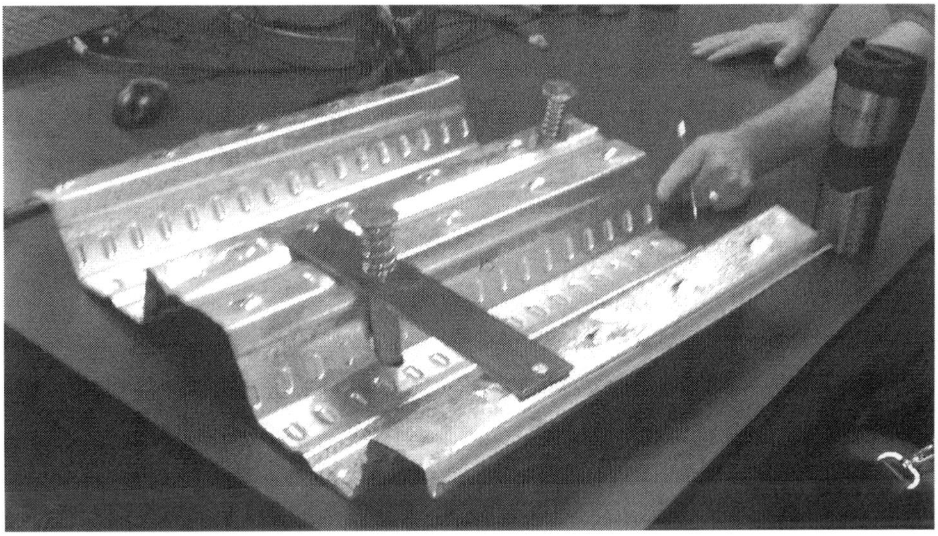

Figure 9.10 A mockup of the ISAT blue banger hanger system selected by the team to study how to improve the accuracy of the placement of the point load when the hanger needs to be on the slanted side of the metal deck

Source: Courtesy of Sutter Health

Figure 9.11 A pull planning meeting to identify the standard workflow steps from the sign off on the model to the start of the inserting process in the field

Source: Courtesy of Sutter Health

the submittals and approvals processes from the coordinated 3D models to install systems in the field. This approach ensured the compatibility of the team's model-centric approach with the levels of documentation required to obtain OSHPD permits. It allowed for field inspections to proceed unhindered and included the following goals that were set for the insert process:

- All system hangers were to be inserted before concrete was poured (approximately 30,000 hangers).
- Trade partners were required to procure the same type of insert from the same supplier.

Several months after coordination started, it was time to consider how the information from the models would be finalized in the form of layout drawings. The only method that the team knew from experience was based on the sequence of the pull plan:

- Prior to field installation, structural approval was required for the final hanger locations.
- Once coordination is completed and hangers are located each trade contractor produces a 2D CAD layout that identifies their hanger locations and associated load conditions, then submits it to the structural engineer.
- The structural engineer consolidates and reviews the information and provides comments regarding how the placement of the hangers could affect the loading of the slab.
- Trade partners would re-coordinate some of the systems based on the engineer's comments to be ready for inserting before the concrete pour date.

If the team was able to complete the steps before the pour, then they would insert; otherwise they would drill. The structural engineers needed about 10 days to review each concrete pour area, and the trade partners needed about 10 days to produce each drawing for the review. If there were changes, then an additional 5 to 10 days were needed. Obviously, the process was inefficient, open-ended, susceptible to error, and unacceptable from a risk management point of view. Additionally, there was simply not enough time between the completion of coordination and the timing of the slab pours. The process itself needed to be re-engineered if the team expected to insert all the system hangers before the concrete was poured. The next step for the integration team was to facilitate the development of an improved process and convince the team to adopt it.

The elements of the solution began to emerge when the engineering review and approval processes were reviewed in detail with the structural engineers. An improvement event was scheduled, and all participants were required to bring exhibits from previous projects that illustrated their work products.

At the meeting, the engineers described their work steps as follows: they took the CAD drawings provided by the trade contractors and inserted symbols (halos) that represented the load conditions of the systems as identified by the detailers. On visual inspection of the drawings, intersections of the symbols indicated overloading of the slab. Wherever this occurred, the hangers had to be relocated. This immediately presented a significant improvement opportunity. Rather than having the structural engineer identify the loading condition after the completion of coordination, making the load condition symbol a part of the hanger element itself provided a better solution. The information about the load was known to the detailer at the time of placement. With this simple change to the process, several lean principles were implemented (built in quality for example) and significant non-value added steps were eliminated, such as the need for the detailed structural review itself. The structural review criteria were combined with the coordination process itself. Figure 9.11 below shows an example of the resulting work product as a top view of the coordination model. The yellow and blue circles represent the load conditions and their diameter represents the load influence on the slab as determined by the structural engineer. As long as visual inspection did not indicate intersections between the circles, then it was assumed that the involved slab was not overloaded. The process was then reviewed with the field inspectors and accepted as the way forward.

Similar to structural steel, this simple change in the process resulted in significant gains. It saved the team approximately 30 days of non-value added time per slab pour which was about the time required to complete the coordination of the hangers ahead of each pour. The MEP team then streamlined the conversion process from 3D models to field installation. Within one day of declaring a pour area as fully coordinated, the crews would begin installing their hangers, and within a few more days the slab would be ready for rebar placement and a concrete pour. After a few slab pours, the inserting crews were able to reduce the cycle time of the inserting process itself. In the end, over 95% of the system hangers were installed as coordinated, and several months later they were used as installed to support the systems – an industry first.

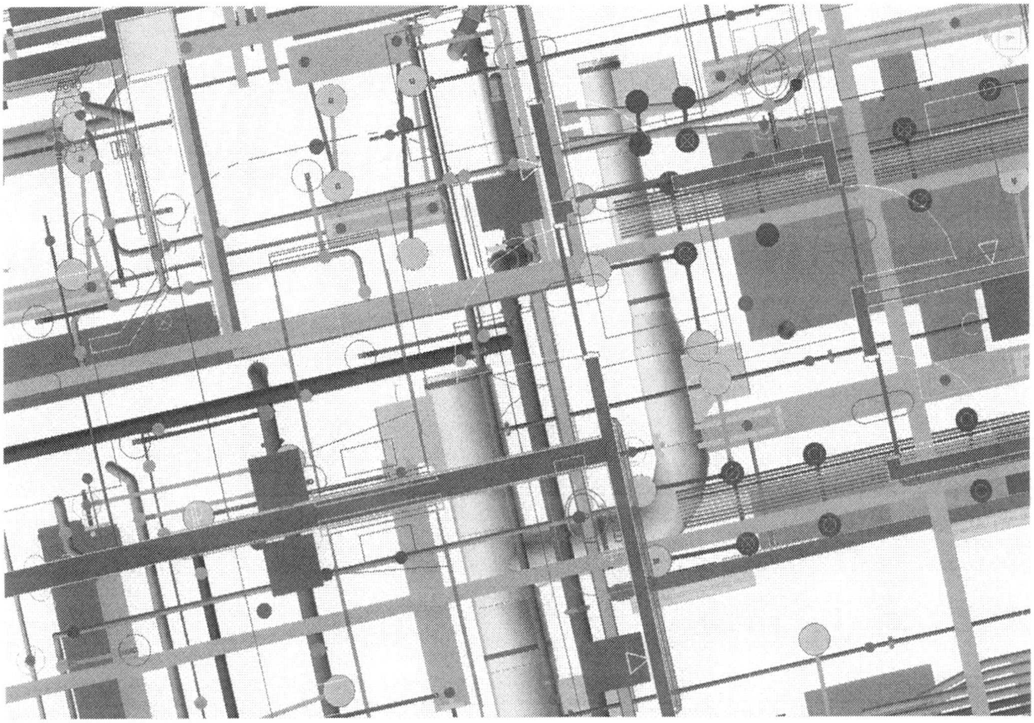

Figure 9.12 The work product as a top view of the 3D model showing fully coordinated systems after the detailed clash reviews

Source: Courtesy of Sutter Health

Note: The circle colors and sizes represent the influence of the load each hanger caries on the slab. Intersecting circles represent an overloading condition. (No clashes are indicated.)

Figure 9.13 The MEP crews on the metal deck locating the inserts using x,y,z coordinates directly from the 3D model

Source: Courtesy of Sutter Health

Incidentally, the detailed workflow analysis revealed another hidden risk to the project related to the edge-of-slab coordination. No one on the team had the full responsibility for the coordination of the edge of the slab and no element in the model existed that is the fully coordinated slab. The architect identified large penetrations such as stair openings and elevator shafts. The structural details showed the slab depressions and edge details but not the actual conditions. The steel detailing model showed the bent plates, the plumbing and the electrical models identified locations where vertical systems penetrated the slab but not the actual slab openings. In order to mitigate risk, a special team worked for several weeks to identify all the scope gaps, design workflows and review processes.

Summary

The foregoing examples were a few of the innovations implemented on the Eden Medical Center project. There were many other examples of innovation in the design-to-fabrication value stream on the project. The positive outcomes and lessons learned illustrate how true collaboration can happen on IPD projects as compared to non-collaborative models. The project took a record 4.5 years from the start of design to receiving its first patient (30% faster than similar projects). It halved the time from the start of design to start of construction for comparable projects. It was delivered within budget (the project started approximately $60 million over budget) and six weeks ahead of schedule with less than 0.5% scope variance from the original program. Rework was reduced by more than 60% across the trades, and trade productivity was roughly 8% higher than the estimates by the trade contractors at the start of the

Figure 9.14 A view of the SVCC project during construction from the roof of the SHEMC building

Source: Courtesy of Sutter Health

project. There were 4,955 state inspections conducted, and they indicated a first-time pass rate of 96%. The project demonstrated that IPD produces results and it set a high bar for the projects that followed.

The Castro Valley Care Center (CVCC)

During the design and construction phases of this project, it was known as the Sutter Health Castro Valley Clinic (SHCVC) however here we will reference it by its current name. The Castro Valley Care Center (CVCC) project is the continuation of the main hospital project. This 92,000 square foot care center is a building attached to the main hospital. The $41 million program included 74 exam rooms, a 6-station urgent care unit, labs, and an imaging center. The schedule goal for the project was to have it open at the same time as the main hospital. Similar to other Sutter medical office buildings, a developer built the Core and Shell (C&S), and Sutter built the interior. The C&S contract was a design-build contract with a GMP. DPR construction was selected as the GC and the Devenney Group was the design architect. Sutter selected NBBJ as the architect for the interior construction together with DPR as the GC under an Integrated Form of Agreement (IFOA). New trade contractors were selected for the clinic, and they performed under both the design-build and the IFOA scopes of work.

Setting up the project with separate design and build teams on the Core and Shell and TI (interior?) presented inherent difficulties. With two different design teams working on the design of the same building at the same time, the risk to coordination between the two teams was high. To mitigate this risk, the teams planned the design work together as one project schedule. All the best practices from managing the main hospital were implemented on this project. The entire team co-located weekly either in-person or virtually

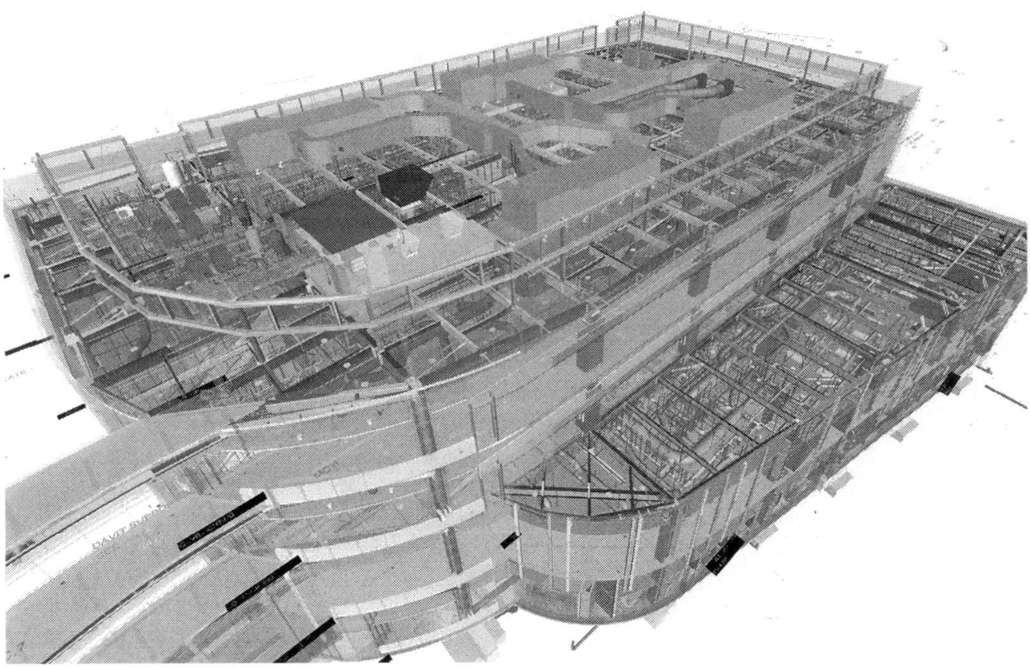

Figure 9.15 A view of the final 3D model for the SVCC project showing all systems

Source: Courtesy of Sutter Health

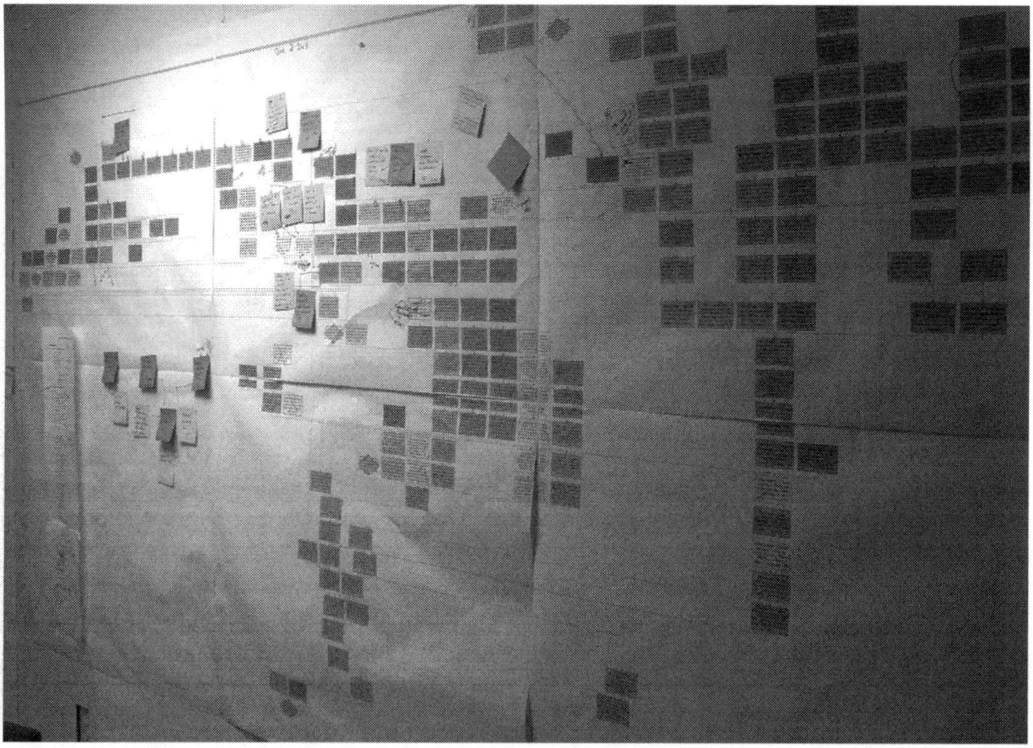

Figure 9.16 An image showing the outcome of the make ready planning weekly review meeting of the pull plan

Source: Courtesy of Sutter Health

at the project big room for 1 day a week during design and then daily during construction. A single file share site was used by all involved for all project documents and 3D models. The Last Planner® system was used to manage both design and construction planning. The team met weekly to plan and to review the design in depth using the 3D models and all open design issues were tracked directly in the model until resolved.

However, things did not start this way and the project could have had a different outcome had Sutter not intervened. At the start some of the clinic project team members that had just completed the adjacent hospital feared that implementing the same rigorous processes including BIM on the new clinic would be both unnecessary and overly costly. At Sutter's request, Ghafari held a 2 hour workshop to assess the risks of not using BIM and lean for the clinic project. The workshop was attended by both architects, the general contractor and the newly selected trade contractors.

The discussion focused on the overall risks to the projects, processes and information flows between all the parties when BIM and lean are implemented, as compared to conventional approaches. This allowed most of the fears to be discussed openly. The workshop also explored the design for fabrication potential and efficiencies of model-centric processes that minimize the use of 2D information. At the end of the workshop the entire team chose to implement the same model-centric design to fabrication workflows as well as the rigorous planning approach used with the hospital project, without increasing the overall project cost. It is worth noting that situations similar to this one exist on almost every IPD project where teams push back on innovation based on a past negative experience or fear of change. Teams that take this resistance at face value without a serious attempt to analyze the workflow end up missing valuable opportunities.

The project started with the big room agenda setup as previously mentioned with a weekly pattern of onsite meetings for 1 day that involved all parties. During that day, the team explored opportunities to improve the workflow and the risks to the project. They pull planned both design and construction, and conducted an "all hands" design review using the 3D model. The project did not start with a search for breakthrough change; instead it started having standard best practices as the baseline to reduce variation, and improvement followed naturally.

Framed wall panels

The unique innovation on this project was came from the use of prefabrication. The project team used the coordinated 3D model to pre-assemble most of the interior walls at a nearby warehouse facility. The pre-assembled walls included all framing, backing, in-wall electrical and low voltage systems, and in-wall plumbing (see Figure 9.17).

In order to release the framed walls to fabrication, each wall in the building had to be reviewed with all disciplines and also with the operations team using the 3D model for final approval. This was the signal to release the wall for fabrication. Views from the models were extracted and reviewed for appropriate placement of outlets, plumbing, and casework. Once the team signed off on a batch of walls, they were released to the warehouse for spooling and fabrication.

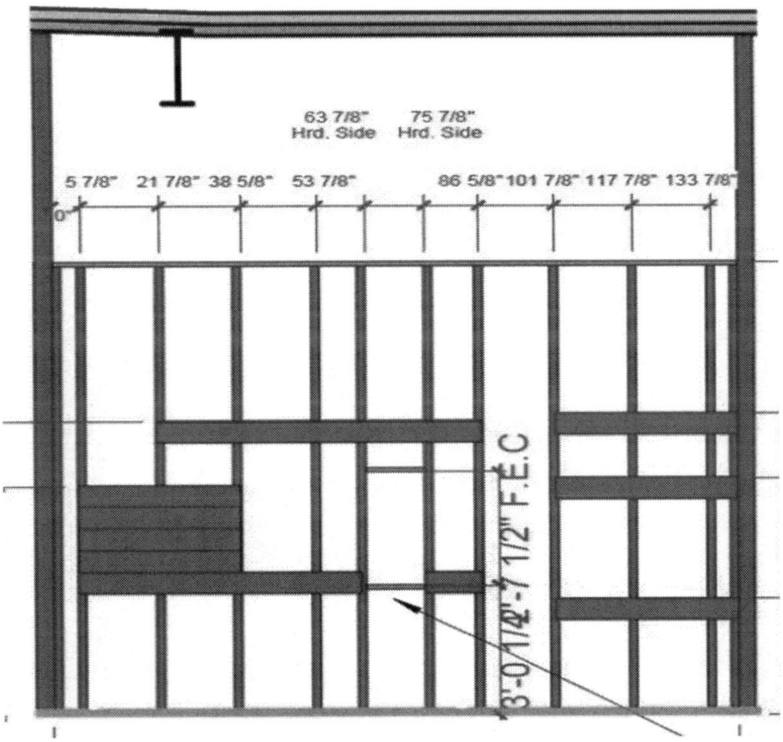

Figure 9.17 Final wall panel elevation sheet containing information from the various models including the casework, outlets, plumbing, and framing

Source: Courtesy of Sutter Health

Figure 9.18 The model to installation process illustrated by actual workflow steps

Source: Courtesy of DPR Construction and Sutter Health

Note: The coordinated model was used to produce the spool sheets needed for fabrication and pre-assembly offsite. In parallel, overhead installation progressed uninterrupted by framing. When the panels were ready, they were delivered and installed in a fraction of the time required for a "stick build" process.

Trade workers from framing, electrical, low voltage, and plumbing used the same coordinated models to fabricate and pre-assemble their systems. In parallel, the overhead construction proceeded uncoupled from the wall framing. Once overhead systems were completed on a floor, the prefabricated walls were delivered and installed in approximately 4 days. This was a significant acceleration from the prior time frame of 15 days per floor.

The clinic opened ahead of its contracted date, on budget and meeting all of the required goals set at the start of design. The total design to construction time was 21 months for both the C&S and the TI projects. Design started in January 2011, and the project was completed in October 2012. Most of the walls (about 90%) on the third and fourth floors were pre-assembled offsite. Whereas the original schedule for on-site framing was 40 days per floor, the actual onsite framing duration was four days per floor, leaving the floor ready for final inspection and drywall.

Because of the accuracy of the information and the use of offsite pre-assembly, on-site productivity increased, quality improved, and overall onsite material waste was reduced. The framing contractor's productivity doubled compared to the on-site construction baseline; there was less than 1% material waste compared to 15–20% when stick building the same types of wall. Decoupling the framing and the overhead construction resulted in the more efficient installation of the overhead systems and less trade crowding on the floor. It also eliminated the onsite in-wall construction phase.

The two projects above were selected in this chapter because they illustrate a significant point. Many of the team members involved had a steep learning curve and made significant process adjustments in mid-stream. Many of the companies involved had to retool and purchase new software or hardware. Their teams had to learn and implement new processes. Moreover, their last planners® had to meet weekly to plan and review the work for an entire day. Yet they were able to complete those two projects on budget, ahead of schedule, without compromising the project goals set forth at the start of design. This is how IPD is different than other delivery models as the above would have been disastrous to any project where the contracts are set based on predetermined transactions between the parties that set the rules for how they will interact with each other. IPD allows teams to identify waste across the supply chain, while they are working on the project, and to be able to take corrective action. These projects set the stage well for the projects that followed later.

The journey since 2012

At the end of 2012 Sutter had two challenges; on one hand they needed to re-start their largest project formerly known the Cathedral Hill Hospital. On the other hand they needed to spread the lessons learned and best practices to the rest of the organization in a way that was sustainable for managing dozens of smaller projects. The Cathedral Hill Hospital was split into two projects in San Francisco; one at the new Van Ness at Geary Campus location and the other at the existing Mission Bernal Campus.

As the two projects started, Sutter assembled a team of its most senior managers to be in charge of them. Under to overall leadership of Larry Kollerer, Digby Christian initially acted as the program manager for

Figure 9.19 A view of the Van Ness and Geary (VNG) hospital

Source: Courtesy of Sutter Health and Amanda Barnes, Southland Industries

Figure 9.20 A view of the Mission Bernal (MBC) Campus hospital

Source: Courtesy of Sutter Health and Amanda Barnes, Southland Industries

both projects; Panos Lampsas who managed the Castro Valley project with Digby became the project manager for the VNGC (Van Ness and Geary) project and later, Tom Minard joined them to manage the MBC project just after finishing Sutter's other successful IPD hospital at Santa Rosa.

The VNGC project had already been fully designed in the previous cycle (iteration) but the restart removed substantial scope from it and added substantial scope to the MBC project instead. The program for the new VNGC project included 740.000 sq. ft of program space, approximately 250K sq. ft. of underground parking space, and 12 stories of above grade construction, including a full service acute care facility with an emergency department. At the end of the validation phase, the overall project budget was set at $2.1 billion. Within the IFOA there was $1.0 billion for Construction, $160 million for Design, with $40 million for Contingency. As the project started in 2013, the team was undergoing a complete assessment of the existing design and followed with a significant redesign effort.

In addition to Sutter's representatives, the VNGC project had 16 at-risk partner companies including HerreroBoldt (HB) as the General Contractor, SmithGroupJJR as the architect, Custom Engineered Openings (Curtain Wall), California Drywall (Interior Framing and Drywall), Charles Pankow Builders (Concrete), Degenkolb Engineers (Structural Engineers), GFE NorCal (Acoustical Ceilings), Harrison Drywall (Exterior Framing and Plaster), ISEC (Door, Frame, and Hardware), RLH Fire Protection (Fire Protection), Rosendin Electrical (Electrical), RTKL Associates (Medical Equipment Consultant), Silverman & Light (Electrical Engineering), Southland Industries (HVAC and Plumbing Systems), Ted Jacob Engineering Group (Mechanical Engineering), and Vantage Technology Consulting (Technology Consultant).

At MBC, besides Sutter, the MBC project had 12 at-risk partner companies including HB as the GC, Harrison Drywall (Exterior Framing and Plaster), Charles Pankow Builders (Structural Concrete), RLH

Fire Protection (Fire Sprinkler), Southland Industries (HVAC and Plumbing), Rosendin Electric (Electrical), SmithGroupJJR (Architect), Boulder Associates (Architect), Degenkolb Engineers (Structural Engineer), Silverman & Light (Electrical Engineer), Vantage Technology Consulting (Technology Consultant), and RTKL Associates Inc. (Medical Equipment Planner).

The MBC project started with a new program for the replacement hospital. The 7-story building included 215,000 sq. ft. of program space, 10 ICUs, an imaging department with an MRI unit, 5 operating rooms, an emergency department, and 16 exam rooms. The total project budget at the end of validation was set to $540 million. The target opening date for the VNGC project was set to end of February 2019. The target completion date for the MBC project was set to October 2019.

As the projects restarted, Sutter asked the vPlanner team (*formerly Ghafari Management Services*) to be part of the Sutter management team to help in implementing the lessons learned and best practices from the previous projects. The team joined the project initially as an observer of the current state for several weeks, and then provided discussions and recommendations at the core group level on both projects to begin implementing workflow improvements.

Early improvements included the requirement for the teams to implement the concept of a single-source of truth for all the various information sources on the project. For example, task teams were formed to implement the same document management system (ProjectWise) which had been used successfully on the Castro Valley projects. The same filing structure was used to organize the files for both projects including the models, CAD files, and other related documents. The teams were required to use the system as the source document (and not the copy). This subtle change promoted transparency and trust that everyone was working on the most current documents. The vPlanner production management platform was selected as the primary system for managing all pull planning, make ready planning, and commitment management. That decision provided a single source for managing the work streams and prioritizing work plan activities from the start of design to installation and final testing.

From a team organizational perspective, the project teams from design and construction were physically collocated in the same Big Room space. They had organized themselves around the idea of work clusters and functional teams but they did not have a weekly event where all hands would meet. The team was required to set aside a day each week where all the cluster leads and production managers both from design and from the trade partners would meet in the big room. The agenda for the day varied but in general, it

Figure 9.21 VNGC Weekly Big Room Meeting which started in 2013 and continued as a full day meeting every Tuesday until project completion

Figure 9.22 Weekly make ready planning for design at the VNGC project

included three main elements each week. A time to plan, a time to review the design, and a time to have breakout sessions for priorities identified by the team.

The **big room day** for each project would start with project headlines, followed by a review of commitments remaining from the previous week, and a review of the reliability metrics and emerging late paths as identified in the vPlanner system. This was followed by an-all hands review of the in-progress design using the 3D model. Designers were required to describe their ongoing efforts with the use of the 3D model. The team gave high-level constructability feedback. Any issues that were in an open status were tracked in the model. This practice worked well on earlier projects and reinforced the idea of a model-centric design to fabrication process. It also helped the senior leaders in the team who were not very familiar with 3D modeling as well the inspectors of record to contribute to the design reviews without waiting for 2D drawings to be produced. At the end of the model review, breakout meetings were held to re-plan any late paths in the pull plans and resolve any critical issues that the team identified during the reviews. All team members were required to come back at the end of the day and make commitments for the following week. The process started in 2013 and was continued every week throughout the duration of both projects. The big room meetings were initially led by the vPlanner team then the responsibilities were transitioned to members of the project teams once the team reached a steady state and enough leaders were coached on how to lead the sessions. This served as the framework for managing the big room to create visibility and consistency to all the moving parts that impacted each project as a whole.

Many breakthrough innovations were introduced in the VNGC and MBC projects; like the Castro Valley project they have been subject to various academic studies, industry presentations, and publications. A few examples are discussed, that had a significant impact on the projects.

Production strategies using takt

The BOLDT team was just completing a project for Sutter Health in Sacramento where they had successfully implemented takt to improve crew flow. They were also in the process of implementing a comprehensive production strategy using takt in collaboration with the trade partners for all aspects of work on the two projects. The technical details of the effort can be found in the various references cited at the end of this chapter. Here the focus will be the overall effect of this clear execution strategy on design and the supply chain. The clear delineation of the project phase plans, the locations where and when the work would take place, and the ability to clearly define conditions of satisfaction for completing a work area helped organize the upfront efforts of detailed coordination and design decisions. The OSHPD permitting strategy was revised in light of the production strategy. The priorities for the phased review process and goals for the completion of the design were better articulated to the design teams as they developed the memorandum of understanding with OSHPD.

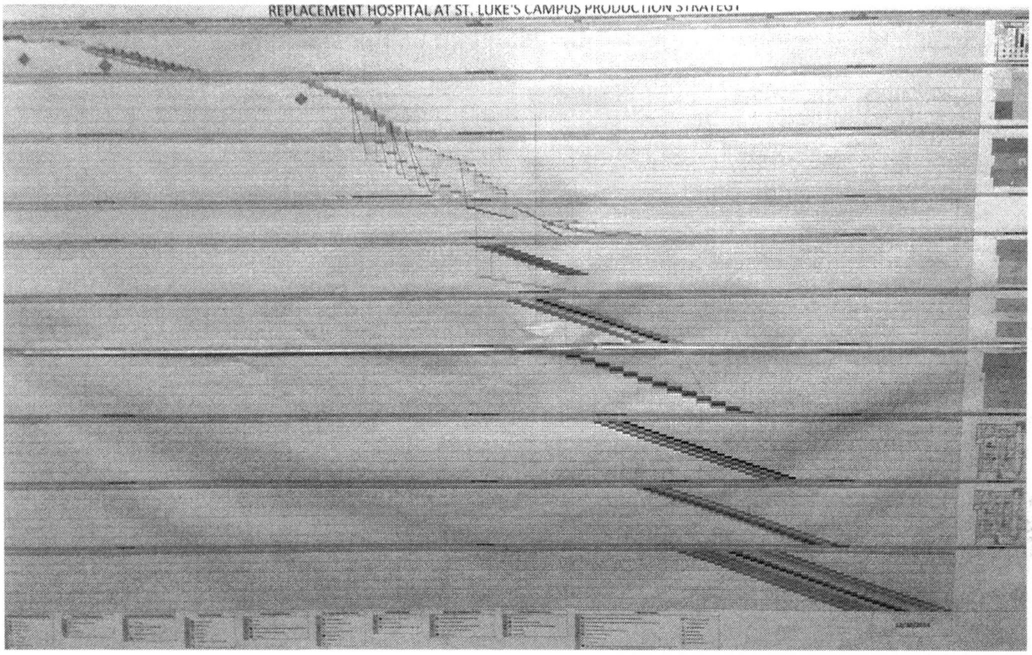

Figure 9.23 An early version of the MBC project production strategy using takt

Figure 9.24 A weekly make ready planning meeting at the MBC project big room to review the lookahead based on the takt plan

The clarity of the takt plans gave the coordination and detailing team clear goals for their resource requirements to keep up with production flow and the overall pull planning process gave immediate visibility and priority to design issues and decisions that would impact production flow. In preparation for fabrication and construction and due to lack on onsite laydown space, the team identified the need for an offsite assembly area and selected a common space for all offsite trade work at Treasure Island. The trades used this space to pre-assemble their work and stage their just-in-time deliveries to the nearby construction site. The space was adapted over time for the evolving project needs and was later used for staging the owner's medical equipment. The costs were shared from the overall project budget.

Electronic OSHPD review and approval

The execution strategy and takt planning analysis on the VNGC project identified that if the team was to hold the target completion date and take into account the time required for permit approvals, then they would have to be working on about 7 floors at a time moving through 24,000 sq. ft of work each month. At the start of the project an opportunity presented itself to streamline the OSHPD approval process and utilize an electronic deliverable to OSHPD instead of submitting thousands of paper drawings. A task team was quickly assembled to try to improve the timing of the reviews. The idea of an electronic review had appeal for various reasons, it would save time both for the team and for OSHPD, it would create the space required to review a project of this size at the OSHPD office, it would save significant printing costs, and would improve how review comments were tracked and resolved. The constraint became apparent that while the idea had merit, OSHPD lacked the technical resources (hardware, network, and staff capability) to perform electronic reviews. The team with Sutter's support proposed a modified process to OSHPD and OSHPD accepted that the team instead of submitting paper drawings in the conventional way, that they would submit their sets both electronically and with a supplemental hard copy. The solution was simple, at the time, the large touch screen all-in-one computers were just being introduced and 4G networks were highly reliable. The deliverable to OSHPD was defined as a set of Touch Screen computers, a printer, and associated 4G networking. The project team committed to support the setup at the OSHPD office and teach the reviewers how to use the system.

The touch screen displays were about the same size of a half-size plots and allowed the reviewers to zoom, pan, and have a comprehensive look at the floor plans during code review vs. having to piece together dozens of 2D sheets. The design team introduced new sheets that showed the entire floor plan rather than just a section of the floor plan. Additionally, the design team was able to color code the systems to make it easier for a reviewer to identify them. Leading to the review, the project team collaborated with Bluebeam to improve the review tools and to be able to track and review the status of the comments.

While the reception varied by reviewer as they adapted to the new workflow, the process improved the flow of comments back to the design team. The team was able to respond to comments in smaller batches. They were able to spot and begin to resolve critical reviewer comments almost immediately which cut the response time the typical 18 months to approximately 10 months total duration for more than 6,000 drawings.

The MBC project opened in August 2018 and the VNGC project Opened on March 2nd, 2019. Both projects were delivered within budget and schedule; they have also not compromised Sutter's overall goals that were set forth at the start of the design process back in 2013.

Summary of Sutter observations

Sutter Health uses its IFOA contract on the vast majority of its capital spending (approximately 90%). By doing this it has created a business environment for its design and construction partners that allows them to self-innovate, i.e. to create new processes and solutions to the wide range of problems that typically emerge on every project. It has taken the major lessons learned on 23 projects over eleven years, and $4 billion dollars of spending to create a delivery model that is markedly different from the norms of the industry as a whole. The way Sutter Health's teams plan, execute and track their work is different. Their attitude and their approach to problem-solving, goal setting and working as a cohesive, integrated team to achieve common goals is different. Together this creates an innovative, challenging, supportive work environment in which skilled professionals from entitlements, through design and construction to commissioning can do their very best work.

Sutter has benefited from its steady message to its supply chain, stability of it management structure, and consistency of the approach. Other organizations that attempted IPD were not as lucky and some despite initial breakthrough successes defaulted back to conventional contracting methods after the first set back. Our hope in writing this chapter is to encourage continuity and persistence in applying lean to construction projects as the IPDA process does not guarantee success but it significantly improves the chances for success.

The Westbrook IPD project: section B

The Westbrook IPD project is presented in this chapter as a historical reference to an early adoption of the collaborative approaches used in IPD projects.

Owen Matthews and Greg Howell (2005) defined IPD as a relational contracting approach that aligns project objectives with the interests of key participants, creating a temporary organization able to apply the principles and practices of the Lean Project Delivery system.

Their paper was based on the creative modification of a small design-build project, resulting in significant time and cost savings. Westbrook Air Conditioning and Plumbing of Orlando, Florida, won a design-build contract in December of 2003 to design and construct a central chilled water plant housed in a steel framed building. The capacity of the plant was 12,000 tons of chilled water, starting initially at 3000 tons. The project comprised a main utility building housing chiller equipment, supported by steel columns with large diameter piping supported by hangers from the roof structure. A GMP of $6 million was established with the help of value engineering. As CEO of Westbrook, Owen Matthews wished to utilize a relational contract approach with his project team, that was separate and apart from the design-build contract in order to have more effective teamwork and better outcomes in delivering the project. This project is recognized by the Lean Construction Institute as a pioneer example of the IPD process.

As indicated in Figure 9.25 the following steps were taken with the project team

1 Creating a business entity (i.e., the IPD team that incorporates designers, contractors, and subcontractors). A contractual agreement bound the members of the business entity together. Westbrook successfully applied the concept to a design-build team, holding it together by a single "pact." It is represented by the dotted line in the diagram,

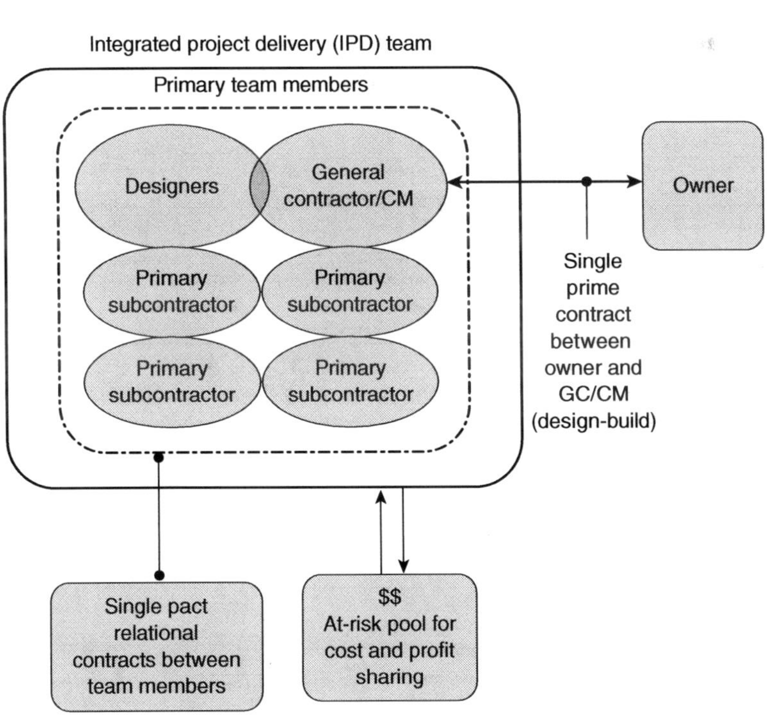

Figure 9.25 The integrated project delivery team

2 Having the parties serve as business entities in their own right and held together through a relational contract or integrated form of agreement (IFOA). Some of the parties established a classic transactional contract with the client and suppliers.

Westbrook's project, as a single business entity contracted with the owner, and the entity's project team comprised the architect/engineer, the GC, and the primary subcontractors. These parties are termed primary team members (PTMs). There were other subcontractors that were not PTMs. The client has a single prime contract with the IPD team. This contract may be of the traditional type such as design-build, design-bid-build, and so on that defines scope, design and specification requirements, cost, and schedule. PTM members may display their cohesiveness with the IPD team by wearing the same hardhats and the same logo. They operate under the direction of one general superintendent whom they authorize to seek the best overall methods to get the work done.

The term IPD was originally developed and trademarked by Westbrook Air Conditioning and Plumbing of Orlando, FL. Owen Matthews, CEO, developed the concept to have more effective teamwork in Westbrook's projects and better outcomes as a result.

The fundamental principle of IPD is the close collaboration of a team that is focused on optimizing the entire project as opposed to seeking the self-interest of their respective organizations.

As was configured for Westbrook's project, a single business entity contracts with the owner, and the entity's project team comprises the architect/engineer, the GC, and the primary subcontractors. These parties are termed primary team members (PTMs). There may be other subcontractors that are not PTMs. The client has a single prime contract with the IPD team. This contract may be of the traditional type such as design-build, design-bid-build, and so on that defines scope, design and specification requirements, cost, and schedule. However, a design-build contract has the distinct advantage of promoting collaboration between designers and contractors. PTM members may display their cohesiveness with the IPD team by wearing the same hardhats and the same logo. They operate under the direction of one general superintendent whom they authorize to seek the best overall methods to get the work done.

Westbrook and their IPD team established two principles for the project: all PTMs were responsible for all provisions of the prime contract with the client and the PTMs shared the risk and profit for total project performance.

One primary team member had a prime contract with the owner/client. The PTM with the prime contract and the other PTMs entered into a single pact; they jointly and severally bound themselves to each other and to carrying out all the conditions, terms, and requirements of the prime contract. The price contracted between the owner and one PTM was the contract price for the IPD team. This scope included all the work required to complete the project.

The PTMs agreed to share the costs incurred in the job and also to share the profits, using a formula that reflected their participation in the project. There was no accounting by these team members to determine who was over or under budget as that would represent a backward step to the traditional approach of "every man for himself." PTMs agreed to open their books for the project to other PTMs and the owner. Each PTM obtained a certificate of insurance based on the prime contract's requirements.

Other key provisions in the agreement consisted of the pact between the PTMs ended when the project was completed and all terms and conditions of the prime contract had been fulfilled, including warranty obligations. At that point, the profits were to be determined, adjusting for any cost overruns that may have occurred. The resulting profits would be distributed to the PTMs based on the previously agreed-upon formula.

Teamwork was practiced by innovation and optimization. Team members were expected to shift work and costs between different companies and trades to improve performance and reduce overall costs. Team members agreed to promptly advise others if there was any impediment that would delay their work. This would enable the team to make necessary logistical adjustments to meet the requirements of the prime contract. Team members agreed to support each other through thick and thin. If a member ran into difficulty, the others would seek to help instead of alienating or even terminating that member. They treated honest mistakes as a "part of the job" and focused on making sure that the overall job was not jeopardized. The fact that this situation deviated so much from traditional construction practice was a testament to the strength of the relational bond between team members. Team members viewed each other as having integrity, character, and competency and who deserved support if they made honest mistakes.

Cultural factors in IPD implementation

Matthews and Howell point to barriers that had to be overcome before PTM members could "buy in" to the IPD concept. Prior to joining the project team, the respective subcontractors were accustomed to traditional project delivery methods; they focused on optimizing their own performance and did not feel responsible for overall project outcomes. The IPD principle of shared responsibility is a major departure from the norm; if one PTM makes an error the cost is shared between the entire team. It reduces the profit in the overall project. Similarly, cost savings are shared between team members and with the client; these savings could be considerable, depending on the extent of the collaboration on a project.

One important factor in the cultural change on this project was the PTMs' exposure to training on the principles of lean construction. They became aware of the benefits of lean systems design, and that local optimization would diminish project-level optimization. A team spirit, open communication, and mutual respect were seen as essential ingredients for IPD to succeed. The top executives of the PTMs met twice monthly for breakfast and fellowship, and this interaction fostered a spirit of trust and mutual respect. It also enabled them to discuss and refine the details of the IPD process.

The team created an organization for the duration of the project that was best suited to its needs. This team included a project executive, director of design services, project manager, project superintendent, manager of ICT. The most appropriate team members were selected for these positions. As they performed their functional duties their respective companies were reimbursed for their time and these costs were charged to the project as direct job costs. This also fostered these leaders' identification with the team instead of with their respective companies. Each PTM company was paid monthly to cover its direct costs incurred. At the end of the project, gross profits were determined and paid to the PTMs based on a prearranged formula.

Project logistics of the IPD team

During the design process, the IPD team had the benefit of the entire spectrum of skills – architects, engineers, sub-consultants, the GC, subcontractors, and so on. That they functioned as members of a common organization contributed greatly to full collaboration between all involved disciplines. Value engineering efforts benefited from the collaboration between all these stakeholders. Constructability was also significantly enhanced as product and process issues were reconciled. Very importantly, a GMP reflected the creativity and innovation provided by this collaboration and was therefore more realistic than would be derived with a standard bid process.

Team members looked for duplication of construction-related activities and avoided it in their cost structure (e.g., rental equipment and other resources were shared by the team and were essentially charged once on the project). This contrasted with traditional construction where several different trades might each rent an air compressor and add its cost to the project.

The IPD team decided to aim for a zero accident rate in the course of the project. Extensive safety compliance procedures were established, with the superintendent having authority over all site safety issues of all trades. The team shared the cost of these safety measures.

The IPD team adopted the concept of "spending more to save more." The mechanical engineer laid out the equipment room (a major part of the project) in detail using object-based 3-D. This layout was

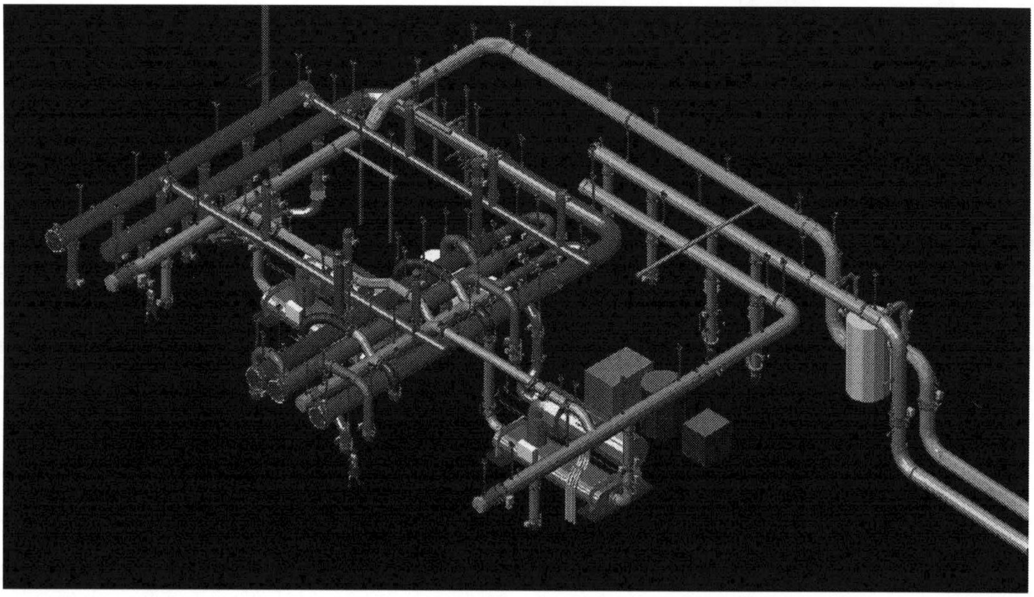

Figure 9.26 Three-dimensional view of piping in the equipment room

Source: From Matthews, O., and G. Howell, *Lean Construction Journal*, 2(1), 46–61, 2005. With permission

immediately usable by the fabrication shop for producing sections of large bore pipe. The engineering cost was increased, but was more than offset by the savings in shop drawing preparation and installation time.

The PTM organization participated in various team-building exercises to promote communication and commonality of interest. Team members typically met twice monthly for breakfast or other social activity and often used the occasion for training on lean concepts to improve their knowledge base. Disputes occurred far less frequently than in traditional contract structures, largely because team members were aligned with a common goal in mind.

Project specifics

The project comprised a main utility building housing chiller equipment, supported by steel columns with large diameter piping supported by hangers from the roof structure (Figure 9.26). A GMP of $6 million was established with the help of value engineering. The team took several actions that led to job completion much faster than in conventional projects. They recognized the importance of an early finish and cooperated closely to obtain that result.

The steel erector was included in the process and integrated the fabrication of structural steel columns with the construction schedule. Innovative changes were made to the layout of steel columns to accommodate equipment installation. The mechanical contractor used 3-D imaging to identify desired equipment locations, and collaborated with the structural engineers to make the needed changes. Such rapid design changes would not likely have occurred in conventional projects. The team realized that, with an aggressive construction schedule, the location of underground utilities was of paramount importance. Through weekly team design meetings, the column footers were positioned to accommodate electrical ducts with elbows for the main service.

Another design innovation involved deviating from a power duct system that would have been installed overhead with expensive hangers, and would have taken considerably longer and been more costly. The team devised a method for the underground installation instead, which involved having the GC wash aggregate into a trench around the duct with fire hoses. Prefabrication enabled major savings to be obtained in the project. Through the use of object-based 3-D CAD, the routing of the piping was designed and displayed in 3-D, facilitating integration of the hangers with the supporting roof structure. Figure 9.24 shows the 3-D CAD representation of the piping layout. This technology also provided details for fabrication off-site. The main headers were 30" and 24" in diameter; the piping was trucked in from the fabricators on a just-in-time basis for installation. The installation itself was simplified to the level of simply bolting components together without the need for field fabrication.

Overall, two 1500 ton chillers were mounted in place and interconnected with piping, all within a period of 10 calendar days. This was a remarkable accomplishment that very probably would not have been accomplished with traditional construction methods at a comparable cost. The schedule was as follows:

Contract award	12/30/03
Design development complete	1/26/04
Demolition complete	1/7/04
Permit Issued	4/14/04
Work start date on site	5/4/04
Plant complete, operational	7/28/04

The effectiveness of the IPD methodology was proven by the fact the team obtained savings of 10% (i.e., $600,000; the final contract price was $5,400,000).

Observations on the IPD process

The IPD process was seen as a clever solution to tough organization and contracting problems in the contemporary market (Matthews and Howell 2005). As the contracting incentives and contractual rules reward cooperation and encourage innovation, project transactional costs are reduced and disputes are unlikely to occur – either within a team, or between a team and a client.

Questions for discussion

1 What is Relational Contracting and how is it different from other contracting methods used in the construction industry?
2 Discuss the objectives and the major factors in building a successful integrated project delivery team.
3 Briefly explain the major steps involved in implementing relational contracting and making it successful.
4 What are the Five Big Ideas?
5 How do they impact project delivery?
6 How are disputes resolved in Integrated Project Delivery?
7 Are there any specific advantages to having a Risk Pool for savings and extra costs in integrated project delivery contracts?

Note

* Samir Emdanat is the Director of Management Services for the vPlanner Division of Ghafari Associates LLC, of Dearborn, MI, USA. Samir is a well-known advocate of integrated project delivery and lean principles. At Ghafari he applies his passion for collaborative methods and technologies to help owners improve their capital project delivery systems and maximize value throughout the design and construction processes.

Digby Christiaan is the Director of Integrated Lean Project Delivery at Sutter Health, a not-for-profit community-based health care and hospital system headquartered in Sacramento, CA, that serves over one hundred communities.

He trains, mentors, supports and oversees a large staff on Lean Integrated Project Delivery implementation across Sutter's highest risk projects. His projects have been ground-breaking examples of lean success with major time and cost savings. Digby also serves on the national board of the Lean Construction Institute (LCI).

Lincoln Forbes, Ph.D., PE, CM-Lean is a specialist in Lean Project Delivery and lean management systems. He has over 35 years' experience in quality/performance improvement as well as facilities design and construction management support. He is a member of The Lean Construction Institute (LCI), and the International Group for Lean Construction (IGLC). He is also a Fellow of the Institute of Industrial and Systems Engineers, (IISE).

References

Alarcon, L., H. Mesa, and G. Howell. 2013. 'Characterization of Lean Project Delivery', paper presented at the *21st Annual Conference of the IGLC,* Fortaleza, Brazil, 31 July–2 August.

Ballard, G. 2000. *The Last Planner® System of production control.* PhD Thesis. Faculty of Engineering. School of Civil Engineering. Birmingham, AL: The University of Birmingham.

Lichtig, W. A. 2005. Sutter Health: Developing a contracting model to support lean project delivery. *Lean Construction Journal* 2(1): 105–112.

Lichtig, W. A. 2008. Sutter health. *Integrated agreement for lean project delivery between owner, architect and CM/CG.*

Matthews, O., and G. Howell. 2005. Integrated project delivery an example of relational contracting. *Lean Construction Journal* 2(1): 46–61.

Bibliography

Cheung, Y. K. F., and S. Rowlinson. 2005. Relational contracting: The way forward or just a brand name? In *ICCEM 2005 conference: Globalization and collaboration in construction,* October 16–19. Seoul, Korea.

College, B. 2005. Relational contracting – Creating value beyond the project. *Lean Construction Journal* 2(1): 30–45.

Howell, G. A. 1999. What is lean construction? *Proceedings of the 7th annual conference of the international group for lean construction.* Berkeley: University of California.

Howell, G. A. 2000. White paper for Berkeley/Stanford CE & M workshop. *Proceedings of the construction engineering and management workshop.* Palo Alto, CA: Stanford University.

Koskela, L. 1992. *Application of the new production philosophy to construction.* Technical Report #72. Stanford University: CIFE.

Lichtig, W. A. 2006. The integrated agreement for lean project delivery. *Construction Lawyer, American Bar Association* 26(3).

Macomber, H., and G. Howell. 2005. *Using study action teams to propel lean implementations.* Louisville, CO: Lean Project Consulting Inc.

Mauck, R., W. Lichtig, D. Christian, and J. Darrington. 2009. Integrated project delivery: Different outcomes, different rules. *The 48th annual meeting of invited attorneys.* Chevy Chase, MD: Victor O. Schinnerer & Company.

Rowlinson, S., and Y. K. F. Cheung. 2004. Relational contracting, culture and globalisation. *International symposium of CIB W107/TG23 joint symposium on globalization and construction,* November 17–19. AIT, Bangkok.

10

LEAN CONSTRUCTION APPLICATIONS

The previous chapter described IPD in which different design and construction organizations work together in a highly integrated manner to deliver a project. This chapter describes companies that applied lean concepts to their projects without the benefit of an integrated agreement. A tenet of lean construction is its application to an entire project, involving all related parties. Its very essence is the optimization of the project, as opposed to the local optimization of a particular trade or subcontractor. Given the fact that lean construction has yet to be instituted as a standard requirement in construction projects, there are many instances in which a contractor or subcontractor who has embarked on the lean journey will be engaged in a project where the prime contract is not based on lean principles. Nevertheless, a significant benefit can still accrue to the overall project, and to other stakeholders, including the owner. Not to be overlooked is the fact that lean successes at the contractor level can set a powerful example for other project participants. Contractors who practice the lean philosophy are well equipped to deliver their work on time, within budget, with high quality, and with few or no safety incidents. They are also inclined to be good business partners as their modus operandi is based on high schedule reliability, clear communications, clean site habits, and predictable pricing.

The following examples illustrate lean applications by four companies in the construction industry:

Kinetic Construction Limited of Canada
JE Dunn Construction Company
Grunau Mechanical Inc.
MACVAL Associates, LLC.

Case 1: The Kinetic Construction Ltd. story (Contributed by Trevor Coyle, CM-Lean, Lean Implementation Manager)

Although the concept of Lean and its application to Construction has been around for more than 20 years in the United States; Western Canada has only recently become aware of its benefits and a small. but growing number of contractors have started adopting the principles. Kinetic Construction Ltd. is a prime example of these contractors. The Lean Construction Institute of Canada was established in 2015 and officially kicked off with its inaugural national conference in March of 2016. The conference took place in Calgary, Alberta and hosted about 125 people. Speakers were brought in from international organizations with Lean experience both within and outside of Canada. Local attendees for the most part were hearing the term Lean for the very first time. The conference showcased the philosophy of collaboration and principles of flow with the greatest emphasis placed on the benefits of The Last Planner® System.

Kinetic Construction Ltd. is a mid-sized General Contractor – Construction Manager operating out of Victoria, British Columbia (BC) since 1984. Kinetic has three branches on the west coast of British Columbia; One in the greater Vancouver area and two on Vancouver Island, with their head office located in Victoria. They directly employ 150 staff and manage a revenue of $100 million/year, specializing in commercial, institutional and light industrial projects ranging from $500,000 to $35 million. Tom Plumb, President and CEO of Kinetic was appointed to the CEO position in 2016, taking over for the company's founder. As the new leader of Kinetic, Tom felt obliged to seek out progressive solutions that he could apply within the organization but also that would have a positive impact on the local industry as well. Tom had long been dissatisfied with what he saw as the same problems repeating from one project to the next. In his view the industry was not only declining in performance, but was fundamentally broken.

A CEO's revelation

Tom received an invitation to the LCI-C's 2016 Inaugural conference through the local construction association. During the first day he was introduced to the philosophy of Lean and participated in a Lego airplane simulation. The simulation had a profound impact on Tom and over the three days he was there, the metric that stood out the most was the steady increase in participant satisfaction as they learned basic pull, one-piece flow and just-in-time concepts and continually improved through collaboration and PDCA during the simulations and games. Tom had a revelation. His vision of completing projects collaboratively and more effectively, was not only possible, there was a framework available that like-minded people in our industry could learn and buy into. Tom made a commitment for Kinetic to become a lean organization, and subsequently appointed Trevor Coyle to guide the transition as the Lean Integration Manager.

The lean journey

One of Kinetic's early Lean Initiatives was to use The Last Planner® System on a $40M, 16-story high-rise located in Victoria, Canada. The project had experienced significant delays coming out of the ground and was three months behind schedule. These delays and the process of resolving them caused considerable tension between the contractor and client and in-turn the trade to trade tension was high. The decision was made to deploy LPS on the project to stabilize schedule over-runs and build cohesion within the team and with the client.

The contractor acquired the help of a Lean consulting firm to hold a two-day LPS (Villego) Bootcamp prior to implementing LPS on the project. During the bootcamp the project team identified their Conditions of Satisfaction – This was a very important step for the group and was often reflected on although not always adhered to.

The results obtained from this early foray into lean construction were mixed. But important lessons were learned about the lean methodology and the logistics of implementing it. It is generally accepted within Kinetic that this project experiment was a positive endeavor. The team collaboration improved significantly over time, the project schedule stabilized and the final contract completion date was achieved without the end date slipping for the last year of the project. The Last Planner® System was unable to reach its full capacity due to lack of experience but acted as a major eye opener for the need and potential of Lean Project Delivery – This experience was the catalyst for Kinetic deciding to fully integrate Lean into our corporate culture.

The Brookes project

In late 2016 Kinetic Construction was approached by a prospective client that was considering them for a new location of an international private school campus. The client was aware of the Lean initiatives Kinetic was involved with and interested in the opportunities they presented for the project. The Brookes project had a highly ambitious schedule and a very constrained site which made it a good candidate for applying the lean philosophy and collaboration techniques that Kinetic had been learning over the prior year. The project started with a collaborative workshop to allow the trades to assist with final design inputs – a modified version of Target Value Design (TVD), The Last Planner® System (LPS) for onsite production control, just-in-time (JIT) material supply and both on and off-site prefabrication methods for the wooden wall and floor assemblies. The development of the project quickly surpassed all projects in the vicinity and became the talk of the town among local Architecture, Engineering, and Construction (AEC) professionals. The use of lean techniques enabled the project team to complete what was expected to be a seventeen month project in just thirteen months.

Project description

The proposed project was located on Vancouver Island BC, on a 2.9-acre property. The project scope was comprised of three large structures; A six-story – 93 Unit/40,000 square foot Student Dorm Facility; A two story 24,000 Ft2 Academic Classroom and Cafeteria facility and 5,300 square foot gym and fitness facility. The site straddled an environmentally sensitive Streamside Protection and Enhancement area (SPEA). The SPEA supports fish and wildlife habitat for several months of the year and is heavily regulated by the Canadian Federal Fisheries Act.

The general layout of the project can be described as having the main parking and drop off area located on the east edge of the property and the three major structures located on the west with two areas separated by the large unusable SPEA with Concrete and Timber frame foot-bridge structure to act as the main access between the two.

Project Site:

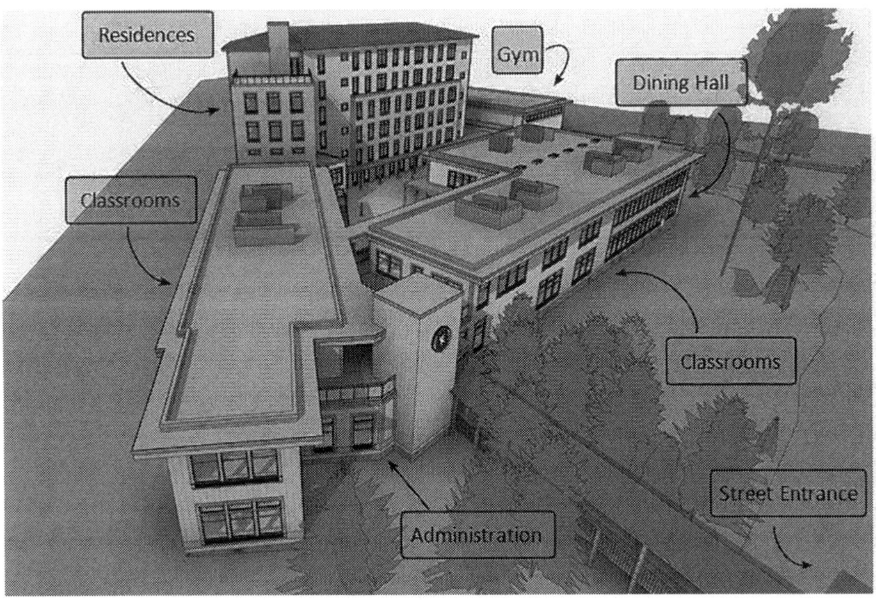

Figure 10.1 A view of the Brookes project

The proposal was to evaluate whether the project could be completed within 13 months, and be partially operational for the September 2018 school year. The complexity of the scope and timeline, and the unique challenges of the site, compounded by current market conditions that were starving projects of skilled labor, would traditionally have led to a discussion with the client about setting more realistic expectations. Perhaps to allow for an additional four-month construction period after the design was complete. Although Kinetic had limited exposure to Lean as a project delivery model, their past experience had validated enough of the concepts to give them the confidence to agree to the project. This was a historical turning point for the company.

Pre-project

During the initial project meetings, the Kinetic team and the owner's design team gathered to identify the challenges and clarify the project goals.

Initial Goals:

- A finite budget
- Academic wing and gym available for use during 2018/2019 school year
- Residential wing to follow early spring 2019
- Incorporate as much green space as possible
- Minimize design time and construction time challenges identified
- Budget – not negotiable
- Timeline –
    - reduced design period (not nearly enough time to do TVD properly)
    - to start in June and be handed over for occupancy early August 2018
- Multiple permits required
- Multiple occupancies
- Completion of residence building on occupied site
- RAR zone
- Extremely tight space on site
- Market conditions/resources

It was agreed that the team would take a lean approach to both design and construction – with an emphasis on the team approach, and the following expectations were set:

Design

- Target value design – design to specific budgets
- Contractor input on a more frequent basis – initially civil, structural, mechanical, electrical with other team members added as needed
- Eliminate the need for redesign
- Maximize innovative solutions

Construction

- Eliminate waste in all processes
- Create workflow to avoid workers waiting for work and work waiting for workers
- Use of pull planning and last planner scheduling-all trades
- Targeted – measurable production levels

- Reduce the need for excessive design input during the construction phase
- A happy – collaborative site

A select group of 6 contractors that would represent the core team was invited to a two-day Lean workshop. The group was asked to participate at their own cost with the intention that those who were serious about the project would see the value of the Lean approach and eventually enter a contract to assist with completion of the design and build the project. The business case for Lean was presented on the first day along with the principles of Target Value Design.

One attendee was heard commenting that the idea of slowing down to go faster was long over-due in our industry and that the conversation was finally making sense to them.

The owner's representative presented the group with the intent for the project and communicated the background and prestige surrounding their investment. He defined his idea of value for the project and was open-book about the funds that were available and the restrictions surrounding them.

He addressed the group with his primary concerns which were as follows:

- Market Escalation
- Regulatory Approvals
- Site Access
- Environmental regulations
- Financial Limitations and budget

Conditions of Satisfaction:

The team collected to develop the Conditions of Satisfaction for the project and the following ten points were agreed to and recorded:

- Zero injuries
- Transparency in all project activity
- Deliver what we promise
- Act with integrity
- As a team, we will not allow scope gaps
- Act respectfully to everyone
- On budget, on/ahead of schedule/meet owner value
- Commit to lean culture
- No rework of Design and Construction
- Environment

At the time of the workshop, the design drawings were at 30% and the budget had been developed to a class D estimate. The information was reviewed with the team on the second day and the disciplines were divided into groups and invited to discuss design gaps and ideas that would bring us closer to cost certainty. The concept of Target Value Design was new to Kinetic and although this was a very late stage to undertake that approach, the team felt that it was still worth the time and effort to collaborate and try to influence the final design outcome as much as possible.

Workshop team discussion highlights

- Phasing the project
    - Initiate critical path early -Start a project with the largest building – the Dorm, to ensure that trade presence is on-site and continues to the other building rather than having smaller crews

onsite and ramping up the operation at the end – it will be more difficult to get manpower commitments later, because of labor market shortages.
  - Imperative to coordinate exterior works and plot to accommodate continuous access for siders.

- Framing

  - Framer – Will prefabricate walls offsite and floor assemblies onsite; install with a crane.
  - Virtual mock-up of framing required to solve all prefabrication issues.
  - Extend walls to the underside of sheathing along parallel walls to reduce assembly components and install time.
  - Fire Blocking/FRR alternatives to be considered to reduce assembly components and install time.
  - Resolution of mechanical exhaust in the dorm – it is key to resolve and accommodate framing.
  - Discuss logistics surrounding joist direction – Shorter lengths and faster, easier assembly: Will work with mechanical crafts.
  - The building does not have large windows or decks typically used to load material – Proposed solution: Frame in extra structural headers each floor to allow wide, temporary access doors. Structural engineer to confirm location with Framing contractor.

- Sequencing of building construction

  - Gym structure 'last' and erected quickly. To allow for truck turnaround and lay down during construction.
  - Start dorm first to get envelope in fall before weather impacts and this will increase trades onsite early.
  - Importance of framer "flow" efficiencies to allow subsequent trades to follow.

- Basement (eliminate)

  - Is it possible to relocate sub-level basement mechanical area to expedite critical path activities?
  - Mechanical could be at penthouse-structural spans accommodate.
  - Eliminating the basement will avoid pumping issues and cost.
  - Code implications of a "seven- story" design with mechanical systems on the roof should be reviewed.
  - Will eliminate tanking and floodplain issues.
  - High impact solution – Team to investigate and confirm asap.

- Main Access Foot-Bridge

  - Sequencing to access SPEA window (8 weeks in summer).
  - Divide Substructure (2017) and Superstructure 2018.
  - Environmental access to be confirmed asap.

- Wall Cladding

  - Will select a system that is fast and can be installed during wet months.
  - A system that is less impacted by weather MUST BE durable (owner requirements).
  - Building skin to apply vertical and complete "Finish to Finish" with roof framing.

- M&E Distribution (Dorm)

  - Provide dropped ceilings at water closets and corridors.
  - Set up distribution to leverage horizontal plenums/shafts.

    - Centralized panels and comm. Each floor for simple distribution (shorter runs).
    - Main electrical room and ground floor.
- Shuffle the program to accommodate (electrical outside of basement).

The discussion among the groups was focused on finding efficiencies and cost savings around the current design proposal. The team identified at an early stage that the sequencing of the site would lead to the greatest cost savings given its constraints. The Last Planner® System was already intended to be used but it was communicated to the group – who had no prior knowledge of it, that this would be the mechanism to combat the perceived challenges and constraints. Kinetic would be responsible for training and supporting team members on the program. The remaining discussion was opened to the trades to offer their ideas on product substitutions that they had the experience and the most certainty with. They were evaluated for compatibility with the associated assemblies as well as availability and simplicity of installation without decreasing the value of the final product.

Design document handling process

Conversations to reduce the waste of passing documents back and forth to clarify intent resulted in the agreement that shop drawings would be produced based on thorough discussions around intent and the shop drawings would ultimately influence and in some cases become the IFC document of record. Given the time constraints, the team also agreed that the use of The Last Planner® System would enable them to produce Just-in-time drawings.

Risk management

The team's approach to risk management on the project was to evaluate the areas of perceived risk and develop a risk register with the group to determine the degree of risk and the actions that should be taken to manage it.

Risk Matrix Example:

RISK MATRIX

LIKELIHOOD		IMPACT				
		Negligable 1	Minor 2	Significant 3	Major 4	Severe 5
Very Likely	5	Low	Medium	High	Extreme	Extreme
Likely	4	Very Low	Low	Medium	High	Extreme
Possible	3	Very Low	Low	Medium	High	High
Unlikely	2	Very Low	Low	Low	Medium	High
Very Unlikely	1	Very Low	Very Low	Low	Medium	Medium

Figure 10.2 The Risk Matrix

The scoring matrix shown identifies the correlation between the likelihood of a perceived risk materializing, and the degree of impact it could potentially have to the project. This tool allows all team members to provide input and vote to provide a shared perspective. The person with the highest stake in the area of risk will have the most weight in determining the risk score.

Risk Register Example:

RISK REGISTER SCORE SHEET

Rank each item in each of the 4 columns. First 3 col'ms are based on IMPACT of Cost, Schedule,and Quality. The 4th column is PROBABILITY of this risk occurring.
Rank from 1-5. 1 being very low, 5 being very high

IMPACT						PROB'LTY	DESCRIPTION
COST	RISK	SCHED	RISK	QUAL	RISK		
5	EXTREME	5	EXTREME	1	VERY LOW	3,5	Won't achieve complete project schedule of August 2018 completion
5	MEDIUM	5	MEDIUM	3	LOW	1	Owner's FF&E requirements unclear
3	MEDIUM	5	EXTREME	1	VERY LOW	4	Not being granted permits when required
3	LOW	3	LOW	1	VERY LOW	2	Scope creep due to unclear Municipal Development expectations
4	MEDIUM	3	LOW	2	VERY LOW	1	Lack of fully documented design approaches & cost could result in missed scope, exceding expected costs &/or legal disputes
4	MEDIUM	4	MEDIUM	3	LOW	1	Misalignment of subtrade scopes of work
4	MEDIUM	3	LOW	3	LOW	1	Insufficient detailing
2	VERY LOW	5	MEDIUM	2	VERY LOW	1	Timely submittals/shop drawings/reviews/RFIs/etc.
5	MEDIUM	5	MEDIUM	5	MEDIUM	1	Risk of using unknown trades with potentially inadequate performance
5	HIGH	5	HIGH	2	LOW	3	Material delivery and installation not on time
3	MEDIUM	2	LOW	1	VERY LOW	2	Site congestion
5	MEDIUM	5	MEDIUM	5	MEDIUM	1	Risk of ineffective communication, insufficient communication, or lack of collaborative communication impacting project coordination/schedule/reputation
5	HIGH	5	HIGH	1	VERY LOW	2	Team members don't cooperate to meet the schedule
5	MEDIUM	5	MEDIUM	1	VERY LOW	1	Crane access and logistics tight

Figure 10.3 The Risk Register above illustrates four14teen of the 64 items identified during this process and the perceived risks associated with them in three categories: cost, schedule, and quality.

Risk register highlights

1 The achievement of the schedule was known to be the most challenging aspect of the project since it is, in fact, the effect of the subsequently listed causes. Interestingly, the team did not associate the risk to the quality of an expedited schedule to the degree that it transpired. The speed of the project ultimately affected the quality outcomes to a considerable degree and lead to significant re-work for the trades.

2 Lack of fully documented design approaches and cost – Exceeding expected costs and leading to disputes, was identified but ranked very low in probability. In actuality, this did become a problem for the project team and the probability was underestimated. There were some scope gaps and escalations that occasionally lead to fractured team perform-ance. Both in managing the deviations from what was expected and in production.

3 The risk of ineffective, insufficient or collaborative communication and the impact on the project coordination, schedule, and reputations of the contractors was identified as high risk to cost, schedule, and quality and given a low probability of occurring. These challenges did occur on the project, as they do with many projects – schedule, cost, and quality were also affected.

Although there is much opportunity to explore the factors leading to these outcomes, the most notable learning opportunity from these examples is with the low numbers in the probability section. In several cases, the expectation of low probability was underestimated. This identifies a heightened sense of confidence in the ability to avoid these risks which most likely attributed to the Lean methods proposed. The Lean workshop introduced the benefits of collaboration and production management to be expected from a Lean operating system on a project. The message used to sell the team on the new Lean approach was effective in gaining their buy-in to try something new. This is considered a positive outcome for the company and will be repeated during future applications, however, the limited experience the team had about how to interpret and deploy Lean techniques, led to a misplaced perception of how to manage risk for a project with the complexity of this one. This perception likely impacted early opportunities to implement more effective avoidance measures. Participants were heard saying "Yes, but with Lean and LPS the probability will be less". And although a fair assumption with a mature LPD model and experienced team, many factors in the approach of an inexperienced team can influence the ability to get the basic expected outcomes of Lean.

Summary of outcomes from the project Kick-off

- A commitment to complete 100% of the project within 13 months – A new, expedited master schedule would be developed using collaborative methods.
- The concrete basement would be deleted, and the mechanical moved to a rooftop penthouse – This solved many challenges brought up by the contractors and allowed site production to commence while the design is being completed.
- The framing contractor would produce design schematics and initiate prefabrication of wall panels and engineered floor systems. Pre-manufactured panels were to be built in Vancouver and brought to Vancouver-Island across the strait of Georgia (70miles) by way of Ferry or Barges. – An unprecedented strategy that had never been done or even considered by Kinetic before.
- Several options for material alternatives that would be reviewed for conformance.
- Clarity by all of owner value.
- Understanding and buy-in of the lean approach.
- General confidence and excitement from the team to take on the challenge.

Kinetic invited the project team to their head office to develop the pull-plan for the early pre-mobilization phases. Pulling backward from the milestone of "Slab on Grade Complete" the owner and design team, the Project Managers from the Civil, Mechanical, Electrical and General Contractor's established the sequence of activities and supporting design activities required to mobilize the project.

Although Kinetic had some experience with LPS when dealing with the Last Planners onsite, it was difficult for them to facilitate the design team, who felt they could not contribute without being given specific targets to produce the design. This was a reasonable perception at the time and led to the decision to excuse them from the meeting to gain more clarity but also eventually lead to the belief that the tool was more for the site than it was meant for the entire project team.

Figure 10.4 Early phase planning: Last Planner® System

At the time, Kinetic's team was unsure whether that assumption had merit, having little exposure to the opportunities of collaborating design and production, did not want to apply too much pressure until it was clear to them when and how the design team could get the most value out of LPS. The lack of experience and decision to try and figure things out as they went, along with the lack of clarity on how they could proceed with the project right away – lead to some division among the team early in the project but eventually repaired itself as the project picked up speed.

A valuable learning outcome for this the just-in-time drawing approach was that design constraints and production management influence each other equally. Once the project was underway, the trades and contractors were of the mind that the timing of production should be the prime influence on design and that is where the team breakdowns occurred. It is the responsibility of the people planning the work to understand and communicate their requirements with enough time for the design decisions to develop. Understanding the parameters of design development eventually allowed contractors to retard their work plans and fill in the production time with work they knew they could do. The Last Planner® System provided the framework for the degree of understanding required to communicate each others needs proactively.

The project mobilized, and a 40-foot, temporary site office was set up as the new boards were installed using standard closet door hardware and the new Last Planners were oriented on the system. During the first pull plan in Kinetics' boardroom, it was mostly the Project Managers in attendance to represent their respective firms. When the site space was set up, the Foremen were invited to participate for the first time. The concepts were new to them, but the same approach was taken as the first session – Learn by doing. This approach led to some confusion about the nuances of the system and made the task of facilitating more difficult than necessary but was repeated from the first workshop in the absence of having a standardized training system available and with the belief that it shouldn't take too long to get the hang of it.

Figure 10.5 Planning activity onsite

Kinetics' learning outcome from this was that it is important to develop and validate their own internal training program for their Project Superintendents and that they would evaluate proficiency prior to assigning them a project they intended to use the system on. There are many lost opportunities starting out slow on a project that hamper the ability to prove the system early. Proving the system early is imperative to the Last Planners buy-in because it requires a large time investment up-front to be most effective.

An early observation about having the foreman in the room was that their perspective on task durations was nearly always less than the project managers. Sometimes by more than a week or two. The actual duration tended to be somewhere in between the two but more relative to the foreman's perception. It is easy to see that the foreman, who are the last planners, had a much more grounded perspective of how flow is achieved during execution than the Project Managers, although Project Managers provided a valuable perspective on the interconnections of off-site activity and behind the scenes processes that cause delays. Such as information handling and decision making. It is reasonable to expect that both should be involved as often as possible, if not always.

Prioritizing information constraints

Immediately following the site mobilization, it was apparent that the Civil team didn't have the information they needed from the designers to proceed as planned and questions regarding site details left the team at a standstill. Assumptions were made early on and the questions from the site required considerable effort to coordinate answers to. This led to a period of limbo and daily reorganizing of tasks to the point where PPC was nearing 0%. Team health and buy-in were affected by this and this set the pace for the early part of the project. This early experience led to a silo-effect of the design team who felt that in order to keep up with the rapidly changing demands from the site, their time was best spent focusing on producing documents as quickly as possible. Because the site production varied so much – more questions were generated about information to proceed in other areas, which eventually led to more requests for the design team to respond to.

The Last Planners identified the need to have design consultants understand these rapidly changing requirements of the site activity better, so developed a means to organize and communicate them effectively. They were introduced to an information Kanban system that allowed the onsite project team to identify their highest priority items for each consultant and modify the priorities as site conditions changed. This endeavor provided clarity and focus and helped to stabilize the early activity with minimal time investment and reduced work stoppage.

The image above shows the simple Kanban board, hand drawn, on 4'x8' whiteboard and posted in the Site Office. The columns on the left allow the team to record the information constraints and the post-it notes have an identification number correlating to the list on the left. The horizontal swim lanes are organized by Discipline and progress from left to right – indicating the upcoming requirements and their importance identified by a Cold-Warm-Hot scale in the backlog. Once items are required to be acted upon, they are moved into the "Focus" column to the right. There is a Primary and Secondary column to illustrate the most urgent priorities of the teams in the field. When a task is complete, they are moved to the right and stay in the completed column until the team can discuss the Plus/Delta's of how they were handled and why they became urgent.

There is a yellow dot sticker beside the title on the Kanban board shown previously; The dot correlates to the post-it tags on the six-week look ahead. When a constraint is associated with a tag on the six-week lookahead, (Shown previously) a yellow dot is placed on the tag and the corresponding information constraint number is written on the dot. This visual cue is an example of Poka Yoke (Mistake Proofing) and allows all Last Planners to see the task with the associated constraint, progress toward the planned execution date. This eliminates the chance that issues will be forgotten or overlooked by the team, making them more reliable to each other.

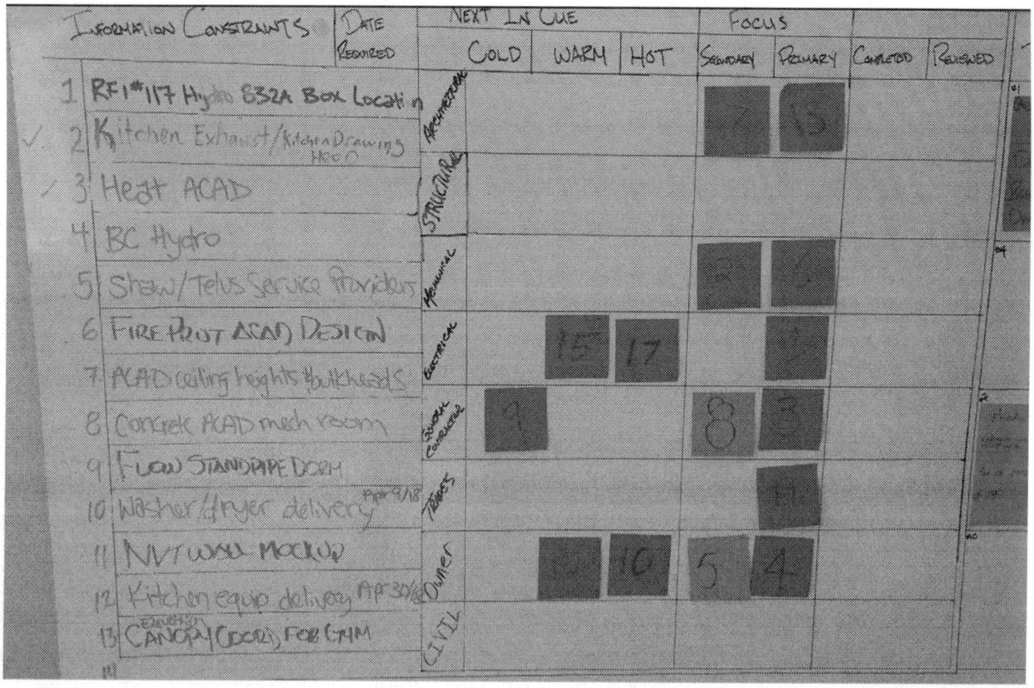

Figure 10.6 The Kanban board

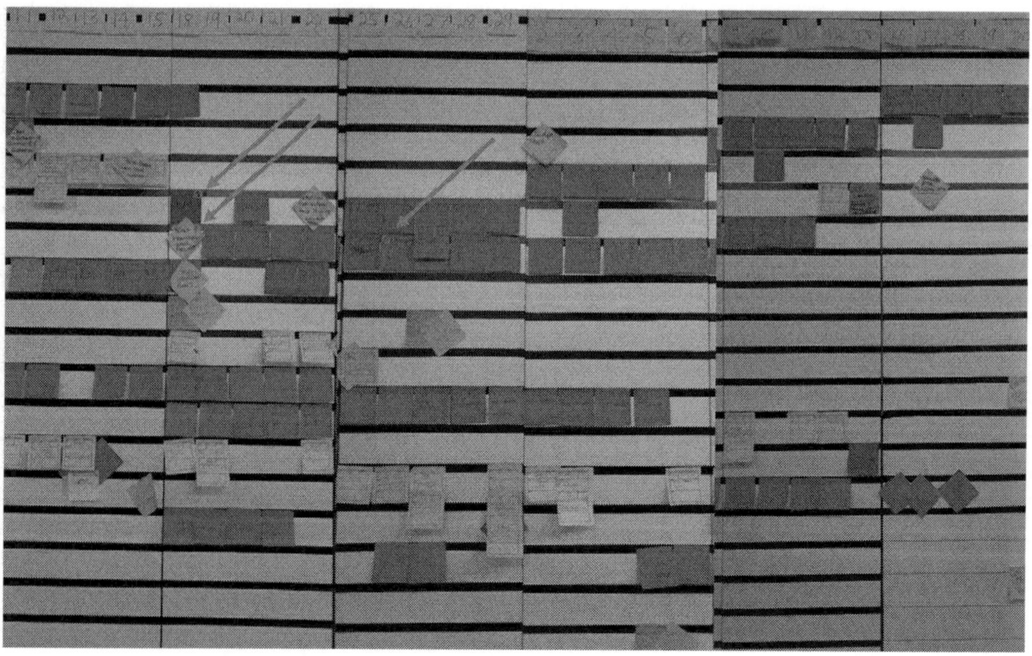

Figure 10.7 6 Week lookahead board

Weekly work plan example with Tasks Certified:

> Visual management is accomplished with the tracking of the metric – Tasks Made Ready (TMR) with dot stickers or a bingo dabber. Tasks are made ready when the task certifier declares that he/she has the Tools, Information, Material, Manpower, Equipment, Safety and Space (TIMMESS) planned and accounted for with all constraints removed. The tasks that are declared ready are stamped or stickered and then advanced to the weekly work plan. The objective is to have all constraints removed within the two-week window prior to execution in the promise period. This requires diligence and is difficult to enforce without full buy-in. Kinetic was not able to gain traction with this on this project. The weekly work plan shown above breaks down work into four-hour time blocks. The layout of the swimlanes is organized by trade, whereas it was decided that the six-week lookahead would be organized by area swim lanes to better illustrate the flow of work and people.
>
> (Shown in Figure 10.8 in the second big-room location of the project with freestanding slide boards)

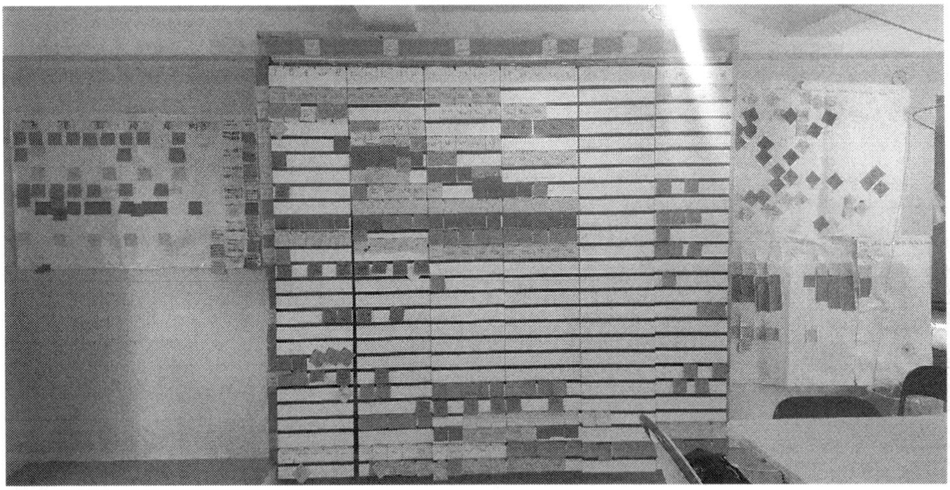

Figure 10.8 Rolling boards

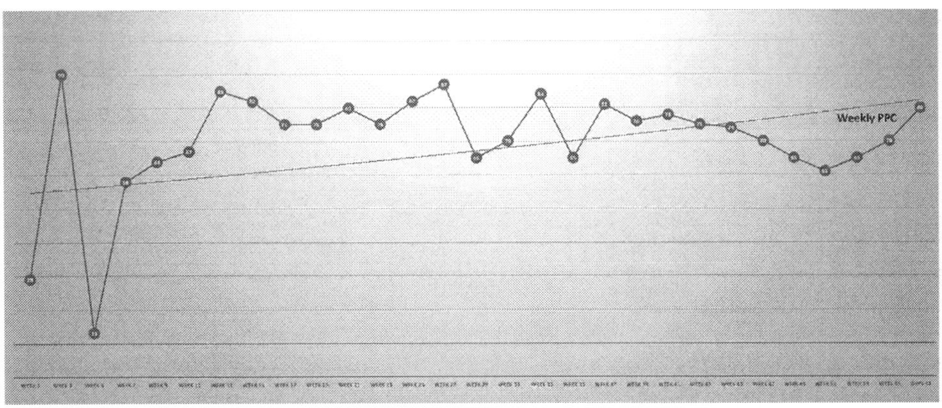

Figure 10.9 PPC Chart – including a trend line

As shown in the figures above, the rolled PPC for the project experienced some dramatic fluctuations early on. These can be attributed to the just-in-time drawing approach and the lack of clarity on what tasks were fully prepared for the team to work on. This was also a result of defining the work on the six-week look ahead, at the time of mobilization rather than during weeks prior. As the project neared the framing stage in week 12 the project began to increase PPC and became more predictable during the installation of prefabricated wall and floor systems. The nature of the work was repetitious and there was a high degree of effort going into the coordination of deliveries and flow of materials coming to the site. The following decline and then relative stabilization in the following weeks can be associated to the increase of activity and tasks happening upon completing the main structure and although it shows a decrease in PPC, it actually represents higher productivity when compared to the number of trades on-site and tasks being executed.

The project experienced some major challenges in weeks 29 and 25 – Both of these weeks show a decline in PPC and when reviewed with the Kinetic team onsite, are attributed to the difficulty of maintaining the integrity of the Last Planner® System and its benefits during major moves of the office and planning space. When planning efforts were put aside to allow the team time to re-establish in a new space, some critical errors were made in coordinating handoffs and led to a ripple effect of rework and other hand-offs being missed. As the project neared the final months, the increase in activity led the team towards a reactive business as usual approach and drew them away from the use of the Last Planner® System. This is discussed in the area below – LPS at closeout.

While the team is learning to define and estimate their tasks properly. There are also very typical reasons to record in the absence of an experienced facilitator to help them see beyond the idea that somebody else was the reason they did not deliver on time.

The interesting takeaway from this is the relatively low number of Qualified Staff Available as a reason for variance. The market conditions during the life of this project were definitely challenging and it was identified early on that this was expected to be a major constraint for the project.

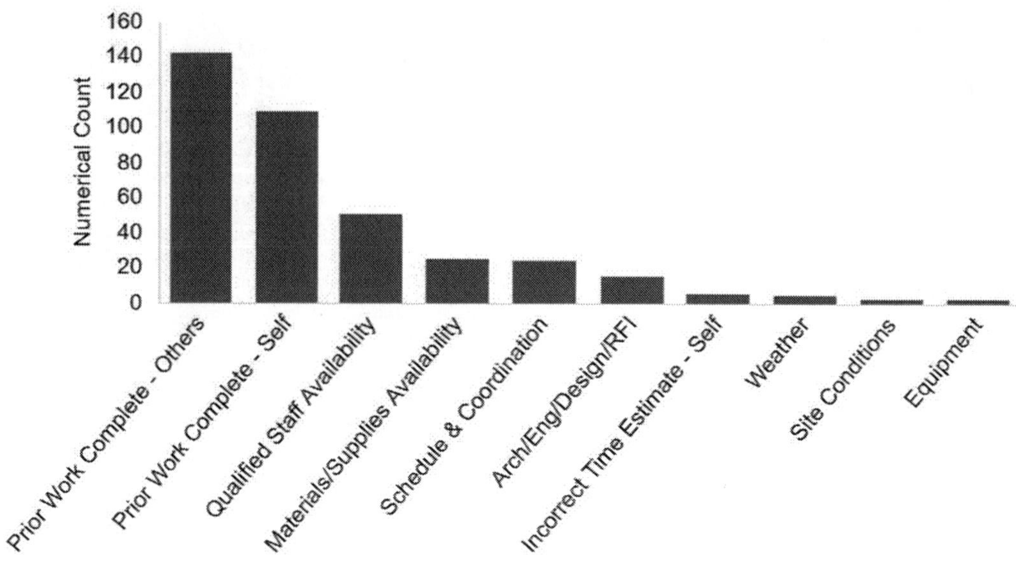

Figure 10.10 Reasons for Variance

Last Planner® System – self evaluation

A lesson learned early-on is that without a comprehensive understanding of what the system in full swing was capable of, it is not fair to expect anyone to see the value of the up-front effort required. Often trades would feel that their understanding of one element of the Last Planner meant that they understood the system and did not require further coaching. Kinetic developed a visual system to show the team where they were and what they could expect from the system as a means to keep them engaged.

the degree of expected satisfaction as it correlates to the defined outcomes. The color scale acts as an Andon (Help display system) and tells the team what level of point along their journey.

The self-evaluation board Conditions of Satisfaction, listed from top to bottom, are as follows and is meant to be achieved from the bottom up so each statement exceeds or encompasses the previous one. It enables contractors to see how they are performing, what they can expect to get by participating in the program. Contractors place a post-it note with their name and phone number on it in the second column, next to the statement they most identify with in the fourth column titled "Conditions of Satisfaction" (CoS). The color scale acts as an Andon (Help display system) and tells the team what level of support to expect at a given point along their journey.

Table 10.1 Self evaluation board

Very satisfied	The LPS has created a team environment that that delivers a safety and overall trade performance on this project consistently reliable flow of work, reduces stress and improves
	The LPS has created a team environment that that delivers a consistently reliable flow of work reduces stress and improves safety and overall trade performance on this project
	The LPS is our preferred method of production management and we rely on it completely to deliver this project
	My trade partners regularly deliver on their promises and consistently have all the resources and information available to perform their work
Mostly satisfied	I regularly deliver on my promises and have all the resources and information available to perform my work. I have a high and accurate PPC and am able to predict the work of other trades.
	I always participate in daily and weekly meetings as well as phase planning (Pull) sessions. My phase planning is accurate and feeds into the SWLA. I use the SWLA to identify constraints and make my work ready by screening my promises
	I have a good understanding of the five stages of the LPS and their respective functions and how they benefit me and the project team
Somewhat satisfied	I always participate in the regular planning sessions and rely on the LPS to schedule my work. I maintain a predictable workflow and clearly understand the WWP, PPC, and SWLA
	All of my work is organized on the six week look ahead and it has helped me to reduce constraints and improve my predictability. I see the value in this process
	I participate in all weekly planning sessions and understand the importance of my attendance to the team and am starting to experience some improvements to my daily workflow
Mostly unsatisfied	I participate in all weekly planning sessions and understand the importance of my attendance to the team but am not getting much value yet
	I attend weekly planning sessions when I can and understand the potential benefits but am not seeing much of the value
	I attend planning sessions when I am available but have not seen any returns on my investment of time
Completely unsatisfied	I do not see the value for my company's participation in the last planner system and do not wish to attend the planning sessions

The Last Planner® System at closeout

During the last couple of months, the project fell into a crisis mode – with the majority of trades feeling that their crew efforts were best spent reacting to the daily demands to keep up with their scope, the attendance to the planning sessions declined. This had a negative impact on production and the team realized that they could no longer see the bigger picture or provide any certainty as to whether they could finish on time.

As the occupancy date approached, they were responsible to provide a "Go/No-Go" decision to confirm with the owner. With more than two hundred students on the way from various locations around the world, the backup plan was to have them housed at another facility 20 miles away and bussed in to attend classes while construction finished. The uncertainty of the outstanding issues made the decision impossible to answer. The team onsite had become reactive, putting out whatever fire burned the brightest with little proactive visibility into what each other was trying to achieve.

They realized that their best option was to take a step back from the work and put as much effort as they could into coordinating the closeout with the Last Planner® System. The operations onsite had become far too complex for anyone to predict the outcome so Kinetic responded by providing a full-time lean facilitator to support the planning and coordination efforts for the remainder of the project.

The first step was to set up a new planning space (Figure 10.11) in an area that would be accessible post-occupancy, should they exceed the planned completion date, and large enough to accommodate every trade on site. Kinetic met with the client to warn them of the potential risk and requested they release the Gym and Mezzanine space from the operating program with the promise that once they had fully

Figure 10.11 The closeout Big Room

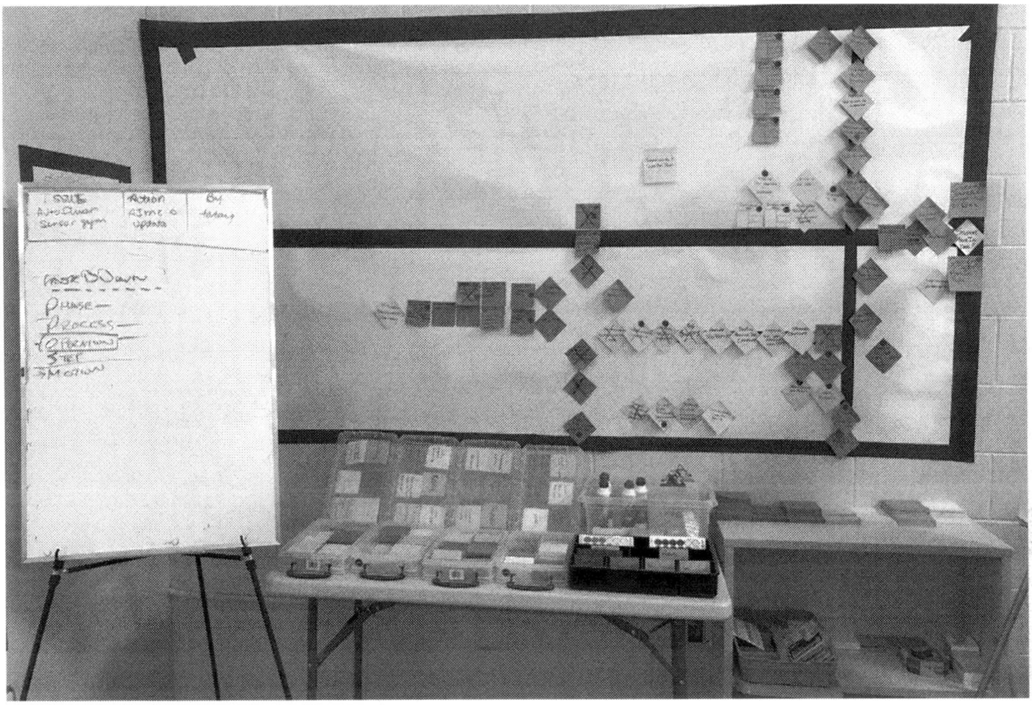

Figure 10.12 Example of closeout document pull plan

established the remaining phases of work, they would provide them with an accurate commitment and a platform and facilitator to support the coordination of their move-in activities as well. The client, feeling overwhelmed by the amount of activity still happening onsite, agreed to the plan and made arrangements for the students to temporarily attend a facility near the site.:

The project superintendent and lean facilitator set out to define the critical network of activities required to produce the final closeout document package. A pull plan of the critical documents was created (above) and subsequently linked (Via colored dot stickers and corresponding numbers) to the activities on the phase plan (below) required to achieve occupancy.

The activities supporting the closeout documentation were established and the contractors were redirected to focus all efforts on these tasks or their supporting predecessors.

With this in place, the client and their coordination team were invited into the planning room to coordinate their timelines and the construction activity was re-prioritized to support their commitments. This was a remarkable experience for the client and the construction team, who received an immediate sense of clarity and understanding of what had to be accomplished to achieve occupancy. The team negotiated their approach to eventually compress the plan enough and confirmed to the owner that the move in for the students was achievable.

Manager's reflection on the use of LPS

Although Kinetic and the project team underestimated the support required to manage the Last Planner® System at different times during the project, it is clear that it's deployment and the collaboration efforts it supports were the most fundamental reason for the success of the project. The regular communication of the daily huddle and designer, owner, contractor collaboration paid the most dividends to the team.

Kinetic estimates that the Last Planner® System was only ever operationalized to about 40% of its potential at any given time, the people who bought-in felt that they were truly experiencing a collaboration effort unlike any they had experienced before. The majority believed that the continuous communication and willingness of others to adapt to the changes made this project uniquely effective given the abundance of challenges. The trades who experienced this said that they did not want to do future projects any other way.

Project post mortem

After the completion of the project – 35 members of the project team were invited to a one day, Value Innovation Kaizen to analyze the performance of the project and narrow in on opportunities for being more successful in the future.

The team broke the project down by phase and broke into four key focus groups to review the plus/delta's and develop themes for opportunity by doing an affinity analysis. Each focus group was facilitated by a greenbelt-certified member of Kinetics' internal Lean committee and the themes were laid out using value stream mapping techniques. Root causes were identified, and countermeasures were developed among the groups and then reviewed on a Kano graph to illustrate the value achieved through the post-mortem effort.

Each team was given time to present their findings to the group and have an open discussion at the end of the session.

Table 10.2 Plus/Delta Highlights

Plus	*Delta*
Solution oriented team worked well	Conflicting service layouts caused re-work
Speed of construction	Supply chain coordination – Late equipment
Trade to trade communication – best ever	Unclear instructions on changes caused re-work
Good selection of systems – quick install	Just in time design was challenging
Scheduling and Last Planner Coordination	Some trades just doing what's best for them

Table 10.3 Issues and Solutions highlights

Issue	*Root Cause*	*Countermeasures*
Lack of clarity on the scope, project priorities and expectations.	Decision-making protocol was not defined well or early enough.	• Incorporate "Key-Decision" cover sheet into meeting minutes template and review bi-weekly. • Define decision and contractual hierarchy • Use Last Planner® System to identify major impact decisions pre-project.
Poor understanding of Lean processes by most.	Team underestimated the complexity of the project – Did not acquire enough experienced help.	• More comprehensive boot camp to kick off the project – All trades. • Hold smaller boot camps at key milestones and mid-project onboarding. • Use the same team on another project.
Multiple iterations of design and revisions took too much time – caused rework.	Time constraints led to moving forward on assumptions – an inadequate process in place.	• Establish stop gates and validation steps in the process early. • Use LPS to connect design processes to field execution dates.

Project outcome

The outcome for the project was that the school was able to open in time for fall classes due to a 25% reduction in delivery time. This gained the client an additional year of revenue and the total cost of construction by the time it was complete, represented a savings of $30 per square foot below current market value.

The Lean approach was a major opportunity for all of those involved and the high-profile nature and visibility served a catalyst for new ways of thinking about how we do construction locally. The project also gained the attention of Award magazine who published the story of this highly ambitious and collaborative achievement.

Below is an excerpt from the Award Magazine's December article that outlines the project outcomes from the perspective of the client and consultant team:

Site clearing for the school started in July 2017 and critical construction was completed on schedule in August 2018, just in time for the new school year. "The structural systems for the school were conventional. However, the project delivery methodology was unique for Herold Engineering," comments Kate Ulmer, principal structural engineer at Herold Engineering. The project followed Lean principles from design through construction completion. Chong at Integral Group adds that the kind of Co-operative relationships between designer and contractor that this methodology produces, also promotes a focus on real functional requirements, a consistent understanding of design intent, shared risk, and minimizing the tendency to take unnecessarily conservative approaches.

"A lot of emphasis was put on having a collaborative team approach to construction in an effort to deliver the project to the client on schedule and on a budget," notes Jeff Halpenny from AES Engineering. The effectiveness of such an approach is confirmed by the results. "For example, Herold Engineering worked with a framing contractor who utilized a prefab approach that allowed the six-story wood frame structure of the residence building to be constructed in a matter of weeks, instead of the typical several months," says Ulmer.

A better workflow approach for the envelope saved time as well, beginning when the structure was on the fourth floor rather than waiting until the roof was completed to start. Just-in-time delivery of the windows allowed the workflow to continue without waiting for a large inventory of windows to be stored on a site that had little storage space.

"We started construction without the completed design and without any subcontractor values established." Remembers Padraig Vaughn, project manager at Kinetic Construction. "The project team set out to collaborate from the beginning, and whole not all subcontractors were involved at this stage, the collaborative culture that was established early remained intact as we awarded the rest."

Kinetic Construction was able to deliver a project that should have taken 17 months but was completed just 13 months instead, on time and on budget. "Staying on schedule is rarer these days and the project was able to do so by using a lean construction tool called the Last Planner® System to deliver value to your customers who are building owners, consultants, trade partners, and our own employees." Adds Tom Plumb, president, and CEO of Kinetic Construction. This approach allowed the $25M school to start providing exceptional educational opportunities for 300 local and international students without delaying the start of term-1.

Case 2: JE Dunn – construction project successes with lean construction

The following cases are provided courtesy of Ms. Rebecca Snelling, VP, and National Lean Director for JE Dunn Construction Company, Inc. The organization is a family-owned and employee-owned company headquartered in Kansas City, Missouri. It operates as a general building contractor and engages in the construction of projects in many industries including multi-unit residential, commercial office, healthcare, education, advanced industries, manufacturing and many others. JE Dunn is #25 in Engineering New Record's (ENR) 2018 list of Top 400 Contractors.

The company embarked on a lean journey in 2012. JE Dunn had assembled a working group to study lean and IPD, and then made a serious commitment to lean. They joined LCI as a corporate member in March 2012, a clear indication of their commitment to lean. Rebecca Snelling, a lean consultant at that time, joined the company, and has become a major influence in helping to institute lean practices. The lean staff grew to 10 employees by 2018.

The lean methodology and lean thinking have been embraced and applied as much as possible. In fact, JE Dunn has applied the methodology to projects that are traditional in nature and do not have a lean-oriented form of contract.

Although the majority of the firm's projects are design-build or CM@Risk, project teams including the trade partners, are given a full day of training that focuses on lean thinking and on the specific tools that a project will use.

Snelling attributes a significant part of the ongoing success to assigning the function of their "Lean Specialists" supporting the teams that helps keeps lean deployment on track at the project level. Depending on the size of the project, these responsibilities get transferred to different members of the project team. The Specialist helps the team learn to monitor the performance metrics and ensures that lean behaviors are continued even without a contractual obligation.

With regard to support, Snelling's team typically helps a project team with lean facilitation at the inception of each project.

Lean training is passed on to project staff through a practice-based approach. Superintendents and project managers are encouraged to learn lean techniques such as pull planning by demonstrating it to others.

The following two projects the Skyloft and Aspen West are examples of JE Dunn's lean successes.

Example 1

Construction project successes with lean construction

As described by Ms. Rebecca Snelling, VP, JE Dunn completed two related student housing projects linked to the University of Texas at Austin in late 2018; Project Skyloft and Aspen West Campus.

Project Skyloft is an 18 story apartment building of 426, 592 square feet, budgeted at $71.3 million as a design-build project. It was designed with post tensioned concrete flat slab framing with 3–1/2 levels of parking below grade. The architects were Johnson Trube and Associates, and the owner was STG/Mark Hart.

Aspen West Campus is a 17 story student housing building of 304, 275 square feet with a 4 and a half level underground parking garage. The architect was GFF architects and the owner was Aspen Heights. The project was delivered as Construction Management at Risk (CMR) with a budget of $48.5 million. Both sites are within two city blocks of each other.

JE Dunn was the constructor for both facilities, but each one had a different project team. Although different designers were involved, some trade partners participated in both projects. Consequently, they set up a single jobsite office within easy reach of each site. This decision provided synergy to both projects, such as providing co-location that allowed the benefit of better communication and learning.

Even though neither project was awarded with a lean-oriented contract it was decided to apply lean construction initiatives such as the Last Planner® System (LPS) and building information modeling (BIM) in order to manage them more effectively. JE Dunn was committed to introducing lean concepts to the entire team and to providing similar training to the type used on lean projects; their thinking was that showing the team the benefits of lean behaviors and the LPS would secure their "buy-in". Also, sharing knowledge between project teams could improve performance and reduce costs.

Specific actions to support the lean concept included: JE Dunn provided a lean specialist for training, coaching, and support. The Regional President of the company had established an expectation that all

Figure 10.13 The Skyloft project

project teams would use the Last Planner® System. The lean specialist provided JE Dunn's standard training on the LPS with a 4-hour session that included the Parade of Trades simulation. Various parts of the process champion role were assigned to different team members in order to maintain the continuity of the lean processes.

A BIM Coordinator was appointed to manage the Mechanical, Electrical, and Plumbing (MEP) and fire protection trades. BIM coordination was used to enable a) the coordination of MEP fire protection disciplines b) clash detection of designed systems, allowing prompt correction and avoidance of field rework,

Figure 10.14 The Aspen West project. (Photo by Brian Mihealsick – courtesy of JE Dunn Construction)

and c) enabling trade partners to better plan fabricated systems offsite. Weekly BIM coordination meetings were held that improved work planning and avoided physical clashes between various building elements.

Improved quality oversight was promoted by using the same trade partners for several scopes of work; on each project each trade partner was awarded the same package. A package discount of $10,000 to $50,000 was obtained as a result. Safety oversight was improved by having the project managers of each site walk together on a weekly basis and complete a checklist of all safety issues.

The LPS provided several benefits – a) Reverse phase scheduling (pull planning) that improved work schedules b) Avoidance of construction issues by prompt action and c) Daily huddles to reinforce the work that needed to be completed. The foreman and the project superintendent conducted this activity each morning for fifteen minutes or less and it was very effective.

Pull planning sessions were conducted approximately 2 to 4 weeks before the required due dates. JE Dunn guided the trade partners' team members to serve as "Last Planners". The work covered by pull planning included:

Concrete, rebar, and formwork
Mechanical, electrical, and plumbing rough-in
Framing and drywall
Finishes and trim out for MEP
Exterior skin.

Project challenges

A number of unexpected challenges were encountered, such as Buda limestone at a depth of 40 feet that had to be removed with an excavator (because of its historical significance). The Skyloft site also had water at 13 feet below grade, and the site was very tight with respect to the close proximity of adjacent properties. Dewatering was required in order to have the site ready for the construction phase, including pier drilling for the foundation.

At the Aspen West campus, Buda limestone was also encountered, and its removal proved to be even more problematic. The project schedule did not allow for significant deviations or time extensions, so activities were re-sequenced to make that possible.

The benefits of the plan:

On the Skyloft project, activities that were to be repeated on successive floors were studied carefully, so that the involved trades could sign off and plan for subsequent floors. The BIM model was coordinated to optimize the use of exposed concrete; sleeves and slab openings were positioned at ideal locations.

The BIM 360 system:

At the Aspen West project the BIM360 system kept track of punch list items and emailed trade partners repeatedly on any items that remained open. Following the architect's punch list walk, both project teams did a "pre-punch walk" and documented to the architect by email how each deficiency had been corrected. This significantly reduced the need for "back walks" by the architect and reduced costs to the owner.

Project outcomes

The Skyloft project was completed one month earlier than expected on 8/1/2018 and was completed under budget. Substantial completion was granted on 7/3/18.

The Aspen West project saved two months on the schedule and was also under budget. It was completed on 8/1/2018, and substantial completion was granted on 6/22/2018.

At the end of the construction phase, several important lessons were learned:

Daily huddles are highly beneficial – They involve the foreman and the superintendent for only 15 mins each morning.

Hiring facility managers early is a smart strategy – it allows for a smoother handover

Weekly BIM coordination meetings with MEP coordinator are greatly effective. Trade partners must be required to participate.

Aggressive management of change orders is an important strategy; there were no change orders on Skyloft, and only for unforeseen changes at Aspen West.

In summary, JE Dunn Construction attributes these project successes to their lean practices, including the Last Planner® System. They also note that these benefits were obtained even though the contracts used were not based on lean project delivery.

Case 3: The Grunau Company Inc's lean journey

The Grunau Company is headquartered in Oak Creek, Wisconsin, with offices in Indianapolis, Pittsburgh, Orlando, and Boardman, Ohio. Grunau designs, installs and services mechanical systems for HVAC, plumbing, fire protection and specialized electrical systems. The company is also active in specialty metal fabrication under the name Grunau Metals. Four generations have led the company, starting with Paul J. Grunau, the original founder in 1920. He was followed by Paul E. Grunau, Gary Grunau and Paul W. Grunau, who purchased the company in 1999.

Grunau refocused its business strategy in recent years to concentrate on building long-term relationships with customers. Their goal is to truly understand a customer's business, and to get involved earlier in a project's development process. According to Ted Angelo, Executive Vice President, Grunau firmly believes in running their business with an orientation toward building high trust relationships. They have embraced lean principles to accomplish their strategic goals.

Since April 2003, Grunau Company has been developing business practices adopted from lean production theories. These theories seek to minimize waste and maximize value by creating better workflow, thereby increasing productivity. When used consistently, these philosophies, which were first implemented in the manufacturing industry, save time, materials and costs.

With regard to Grunau's employees, their commitment to lean practices continues to be sustained, even after the many years that have elapsed since the company's foray into the lean philosophy and practices. According to Ted Angelo, lean practices are "the way they do business". A company newsletter supports a recognition program by highlighting individual accomplishments.

Lean initiative 1. Application of the Last Planner® System (LPS) technique

Grunau has used the Last Planner® System to identify the tasks that need to be completed each week. On simple projects, the foreman serves as the last planner and may keep the schedule by hand. The foreman is a working part of the crew. In Grunau's view pull planning gets the foreman involved in the field to assist other trades.

On big jobs, the foreman is the last planner, but Grunau has a project engineer who works closely with that person. The GC superintendent often leads the group. The project engineer attends the weekly meetings and provides input on where the trades are and what they WILL do (i.e., their work commitment) the following week. This approach helps Grunau to make reliable promises, on work they are scheduled to perform, to other parties that are involved in a given project. Such reliable promises are an essential component of lean construction. Grunau uses software to track their progress.

Grunau uses the project's Master Schedule to create a lookahead schedule. In some cases, there may be a project lookahead schedule, and Grunau uses it to create their own lookahead schedule. As a sub, Grunau pulls weekly assignments from the lookahead schedule. Grunau has done this with five trades managed by the company – plumbing, HVAC, pipefitting, sprinkler installation, and sheet metal. They use an Excel spreadsheet to track 1 week and 6 week work assignments pulled from the GC's Master Schedule. It also shows percent plans complete (PPC). Grunau uses a computerized system – a Last Planner program for tracking work assignments with regard to time, labor hour, and material requirements for each construction task. The Last Planner program was upgraded with 5 pre-programmed macros that greatly reduce the steps involved in building a work plan for the subsequent week.

Lean initiative 2

a) Use of building information modeling (BIM)

On big jobs, Grunau uses BIM to keep track of the routing of various systems to ensure there are no clashes between various lines such as pipes and ducts.

Grunau gets a file on a CD in Navis Works from the A/E, and it is used to show various lines for building systems that they are scheduled to install. This enables Conflicts/clashes can be detected.

b) Human Relations: Daily huddle

The Daily Huddle is a Pre-Task planning exercise for discussing what is needed for each job for a given day. The Huddle helps to reduce accidents, eliminate waste and improve efficiency. It supports the Lean Mission: To continually examine processes to provide greater value to customers without waste.

The team members write notes on 3x5 cards to promote discussion, and a volunteer is selected each time to read questions from the card.

A typical list of questions for the huddle are as follows:

1 What are our Productivity Goals for today?
2 Do we have what we need?
3 Are there any obstacles to achieving our goals?
4 Does anyone see a Better Way?
5 Were there any Near Misses Yesterday?
6 What Safety Hazards should we be Aware of Today?

Examples of answers to typical Daily Huddle Questions

1 What are our Productivity Goals for Today?

300 Feet of pipe, boxes hung, fixtures set, heads roughed in.

2 Do we have what we need?

Yes, all materials, tools, information to get the job done TODAY.

3 Any Obstacles to Achieving our Goals?

Is another trade in our way? Is the area not completely open to us? Can we move around area?

4 Does anyone see a Better Way?

Is the routing of piping/duct OK, or is there a better way? How about the material handling – any better way? Share a better way to complete a task.

If a crew member develops a way to eliminate waste, inform a Lean Team Steering Committee member so we can share with others in the company through a future newsletter.

5 Were there any Near Misses Yesterday?

This creates safety awareness for team members Someone else may be subject to same situation.

6 What Safety Hazards should we be Aware of Today?

There is going to be a lot of crane activity on the west side of the building. The conditions out there are slippery. Pay special attention to where you are walking.

Remember to Sustain, we must keep it fresh since we are reviewing on a daily basis.

Observed benefits of Daily Huddles:

The Daily Huddles have changed workers' behavior. Even though the work for a given week may have been already planned, workers increasingly go through a thought process before taking action on a project.

Lean initiative 3

a) Space reorganization

Grunau's Oak Creek facility was housed in a 22,00 sq. ft. building. Management recognized the need to prepare for expansion, yet they opted to optimize their existing space so they could use the available space more effectively.

The CEO directed the Grunau team to reorganize the existing space, using a 5S approach and leave unused space for up to 20% growth in staff and operations. The team was also directed not to touch the exterior walls of the facility, but rather to make changes to the use of the interior space only. The underlying purpose of the transformation was to "Reduce storage, increase efficiency, and eliminate waste".

This new requirement was very difficult to accept for many employees, especially for those who had worked with Grunau for several years and held supervisory positions and titles. Paul Grunau himself set the tone by announcing that he would not have a private office. Eight "quiet" rooms were to be established that would allow staff to hold meetings, etc. These were modeled after the "big room" (obeya) concept. The work spaces make it easy for employees to collaborate on the planning of projects, as a CAD operator can talk with a foreman about the configuration of a piping layout and enable an estimator to price a job accurately at the same time.

Also, space was allocated that would facilitate personal business such as family matters, medical appointments, etc.

b) Electronic filing

Previously, Grunau's service department had organized the storage of large numbers of three ring binders that contained Operating & Maintenance (O&M) manuals from completed jobs. These binders had been indexed with their respective storage locations in a large bookshelf system. More than 500 such binders were eventually stored in Service Department cubicles, with the overflow accommodated along a building wall.

Electronic filing was seen to represent a significant opportunity for space saving.

All the O&M manuals were scanned electronically and placed in computer files, with CDs as backup copies. The storage space that was occupied by the shelving system has been regained and modified to provide employee work space. Two small CD cabinets have replaced the extensive shelving system that monopolized a fairly large storage area.

Time savings: O&M data can now be retrieved through each employee's computer, using project reference information. There are considerable time savings in eliminating the need to physically walk to the storage cabinets and locate the respective binders. (Estimated $ savings?)

c) Vendor supply team

Employees studied the supply chain for materials and took steps to improve it.

In the previous state, materials would be purchased from supply houses and stored at Grunau's facilities. In the current state, they have a purchase agreement based on delivering materials when needed, where needed. It includes having a driver do the delivery for each trip, but it provides a major saving, estimated at $70,000 annually.

Case 4: MACVAL – applying concurrent engineering and lean methods in small construction projects

This case study discusses how concurrent engineering and lean production techniques can improve overall performance of municipal construction projects. Although the available projects have not been lean-based, the company identifies opportunities for application of concurrent engineering and lean methods. MACVAL Associates, LLC is a small engineering and construction firm based in Dallas, Texas. About 80% of the firm's business is municipal projects obtained primarily through competitive bidding. Concurrent engineering and lean production techniques have proven valuable during the construction phase.

One of the obstacles to improved performance in municipal projects is government regulation due to bidding requirements. Because of this, projects are typically developed and brought forward in serial steps. This serial procedure is not always an efficient method with respect to the total time required for delivery of a project because input from the prime contractor during the design process does not usually occur.

Once a project enters the construction phase, the contractor's work is generally the same regardless of the contract form or delivery method selected. The same activities are necessary to get the project built: the subcontractor's contracts must be executed, the project staffed, a detailed schedule of values and work schedules developed, and the job site assembled. In a typical construction organization, these activities are accomplished through a serial iteration process; one disadvantage of this approach is that errors are not usually detected until after upstream activities are completed, thus resulting generally in poor project performance.

Concurrent engineering approach

MACVAL uses a concurrent engineering approach during the construction phase to help improve project performance. Here, a team that consists of project stakeholders develops each construction process needed to deliver the project concurrently. Stakeholders are included early in the construction planning process. The team identifies customer value by analyzing all activities required for project delivery. The entire process is optimized from the customer's viewpoint in order to increase value and reduce waste. MACVAL's lean production technique includes concurrent construction process planning, optimization of worker utilization and capabilities, continuous focus on the customer, development of mutual trust relationships, continuous challenge of existing methods, decision making at the construction site, flexible scheduling, and "Lean-O-Mobile." MACVAL uses overtime wisely to reduce expenditures while advancing the project schedule. Using a paved concrete surface that has not cured as an example, domestic animals or some residents typically leave ugly prints on the freshly placed concrete. When this happens, rework is quite expensive. To minimize possible rework, the firm not only places concrete during low-pedestrian times, but also has someone stay at the job site until the concrete cures. It is much less expensive to pay three hours of a worker's overtime versus having the crew tear out the damaged concrete and reconstruct the pavement. This simple action has resulted in a 90% reduction in rework costs.

Through experience, the company has learned the importance of stability in dealing with subcontractors. It developed a plan to form alliances with preferred subcontractors; those selected receive a percentage of the savings resulting from quality and improvement ideas, and early completion of projects. This approach results in a stable workforce and reduces variation in similar projects.

The company also turned a $7,000.00 moving truck into a Lean-O-Mobile. The truck started out as a simple storage unit on the job site. After applying 5S principles, the truck was organized with shelves and proper storage locations. The truck enables frequent material delivery in small quantities, allows the team to gain better control of material and portable equipment/tools, reduces the distance traveled to storage and staging areas, and supports minimum inventory required to keep the construction operation running. This is the company's version of JIT (Just-In-Time).

Below is an example of MACVAL's $1.18 million-dollar project completed in 2017 that benefited from concurrent engineering and lean production techniques.

Project Name:	Commerce Place/Tibbets Drive/Scenic Hills Drive/Red Oak Lane Water System Improvements
Project owner:	City of Bedford, Tarrant County, Texas
Bid Reference Number:	SWIFT-16–02
Project Description:	Construction of water main improvements.

Completed project results

- Quality: There were no items on the punch list.
- Cost: Utilized 79% of the budget resulting in 21% savings for the taxpayer.
- Schedule: Utilized 80% of the contract schedule to complete the project.
- Safety: There were no safety incidents.

Systems-based bid strategy for new projects

MACVAL uses a systems thinking approach to bidding on new projects. By analyzing past projects, MACVAL compared the scope of work at completion with the scope originally defined in the project documents. This difference was generally due to poor information quality in the project documents. A review of 33 completed projects revealed that 20% of the information in the bid documents was not relevant. Some design professionals simply copied and pasted data from old project documents to new ones resulting in poor data quality. Understanding this allowed MACVAL to use lean/systems thinking approach to structure their bids most competitively. Table 10.4 illustrates a sampling of frequently recurring bid items for typical projects. For example, Item 607A had a completion rate of 153%. This item accounted for 5% of the cost of a typical project, as shown in the extreme right column titled Typical % Dollar Volume. By comparison, item 790A had a similar completion rate, but only represented 1.5% of project cost.

Table 10.4 Summary of observed trends from 33 completed projects

Item No.	*Description*	*Completion Rate*	*Typical % Dollar Volume*
201	Remove concrete. Pave drive and apron	245%	3.50%
504	Asphalt concrete. Fine grade surface. Course	162%	5.00%
607A	Bermuda/St. Augustine grass sodding	153%	5.00%
623	Sidewalk lug	246%	3.20%
790A	Adjust traffic pullbox	156%	1.50%
1225A	Inlet protection for existing inlet	200%	3.80%
		Total →	**22.00%**

Use of A4 Chart for Status Reporting

Example from Street Reconstruction Operation

G

Owner: Al Attah, P.E.

Problem Statement

- Hit gas line three times at different stations and elevations during excavation operation. Construction was suspended until the root cause of the problem was identified.

NOTE: Use of Traffic Light Coding Symbols for Problem Resolution – *see upper right corner of chart.*

- Problem with unknown solution is coded with large **red dot**.
- Problem with possible solution but not final solution is coded large **yellow dot**.
- Problem with definite solution is coded large **green dot**.

Problem Resolution Method

- Formed problem solving team from people who were knowledgeable in the specific construction processes
- Initiated brainstorming session
- Used Cause and Effect (C&A) diagram to determine the root cause of the problem.
- Identified the following major cause categories on the C&A diagram: people, equipment, method and engineering plans.
- Using dispersion analysis technique, the team placed individual causes within each major cause category. Then asked of individual cause, why did the cause occur? The question was repeated for the next level of detail until the team ran out of causes.

Concern

Could not determine immediate cause though the following actions were taken:

- Held pre-construction meeting with key stakeholders to review the project and identify all potential issues before resuming construction operation
- Located underground utilities per established procedure
- Assigned competent professionals/operators to the excavation operation
- Utilized appropriate excavation equipment
- Followed all safety procedures and established methods
- Excavated per cut sheet supplied by the surveyor

Findings/Actions Taken

- After going through the major causes in detail, the root cause of the problem was identified as misinterpretation of an engineering note by the surveyor. The surveyor interpreted an 8-inch cement treated sub-grade as an 8-inch concrete pavement. With this in mind, the surveyor provided the equipment operator with a cut sheet showing that extra 8 inches of dirt would be removed instead of treating the dirt with cement in place prior to paving the street with 8-inch concrete.
- Corrective measures were taken and lessons learned were communicated to the design engineer and other key stakeholders. Similar engineering plans were reviewed and corrected to prevent future problems.

Figure 10.15 The use of the A4 chart for status reporting

A4 Chart

The A4 Chart (Figure 10.15) is a tool MACVAL uses to highlight recurring project problems. The problem statement is in the upper left quadrant. The concern is in the lower left quadrant. In this case gas lines were repeatedly hit during excavation, and construction was halted. Proper precautions had been taken as described therein. The problem resolution method is in the upper right quadrant; a problem-solving team was formed to find the root cause of the problem. The team jointly prepared a cause and effect diagram based on the categories of people, equipment, methods, and engineering plans. Possible causes were assigned to these respective branches using dispersion analysis. The Why-Why method was used to identify possible root causes. The findings and actions taken are in the lower right quadrant. Finally, the lessons learned were implemented through changes to other engineering drawings used on this project and similar projects.

End note

The firm's founder, Dr. Aloysius A. Attah, P.E., is a trained civil engineer and industrial & systems engineer. He is also an engineering faculty member at the University of North Texas. Prior to founding MACVAL, he was in the aerospace and manufacturing industries for 17 years. In the subsequent years he has been involved in the construction of municipal engineering projects and has successfully applied lean and concurrent/systems engineering techniques to these projects.

Discussion questions

1 Describe some of the benefits that Kinetic Ltd. derived through their lean initiatives.
2 What motivated Kinetic Ltd. to embark on the lean journey?
3 Describe some of JE Dunn's process improvements that were driven by lean initiatives.
4 Explain how the Grunau Company uses The Last Planner® System.
5 How was MACVAL able to price its services so competitively in public sector projects?

Bibliography

Angelo, T. 2009. *Lean Case.* The Grunau Company Inc. The description of Grunau's adoption of lean construction, including application of the last planner technique, daily huddle, space reorganization (5S) and supply chain improvements. Self-published notes, March 30, 2009, pp. 1–5. With permission.

Ballard, G., and G. A. Howell. 2003. Lean project management. *Building Research and Information* 31(2): 119–133.

Howell, G. A. 1999. What is lean construction. *Proceedings of the 7th annual conference of the international group for lean construction.* Berkeley, CA, USA.

Womack, J., D. Jones, and D. Roos. 1990. *The machine that changed the world.* New York, NY: HarperCollins.

11

QUALITY MANAGEMENT IN CONSTRUCTION AND SIX SIGMA

A complement to lean construction

This chapter addresses several topics that relate to the improvement of quality in construction, learning from projects and using the lessons learned to make improvements. The topics discussed complement the adoption and deployment of lean construction. Built-in quality is a component of lean construction. While lean practices enable project teams to work cost effectively and with efficient hand offs from one discipline to another, quality assurance and commissioning are initiatives that ensure that completed facilities perform as expected and without deficiencies. The adoption of total quality management or other quality management systems prepares an organization for subsequent adoption of lean construction. Many of the tools and techniques that are applied to total quality management (TQM) and Six Sigma are used by practitioners of lean construction.

SPECIAL NOTE: It must be recognized that lean construction and TQM are not in competition with each other. For one, TQM is a long-term philosophy that provides a foundation for a company's strategic plan; it may take several years to be instituted. It is difficult to implement TQM at the project level because of the many disciplines and companies that are involved in a typical construction project. While TQM aligns a company's workers with that company's priorities, it is much more difficult to extend that influence to the entire project team.

With lean construction it has been possible to align a team to the objectives of a project; that is a much narrower scope than is encountered with TQM. Lean has been successfully taught to project teams and deployed in a single project cycle, even without prior experience.

But Lean and TQM can be a powerful combination. It is with that thought process in mind that this chapter is provided.

Part A: Total Quality Management

Overview

The term *quality* can mean different things to different people, but needs a common definition if the people in an organization can hope to pursue it successfully. This is especially important in the construction industry where there is much variation from one project to another. Dr. Joseph Juran's definition has been interpreted as "fitness for use," "fitness for intended use," "conformance to requirements," and "conformance to specifications." Dr. Edwards Deming took the view that: "Quality is defined by the customer; the customer wants products and services that, throughout their lives, meet customers' needs and expectations at a cost that represents value." Deming also added that the quality of a company's output cannot be better than the quality determined at the top (i.e., by an organization's leaders).

Total quality management (TQM) is an approach to doing business that attempts to maximize the competitiveness of an organization through the continual improvement of the quality of its products, services, people, processes, and environments (Goetsch and Davis 2016). The principles of TQM create the foundation for developing an organization's system for planning, controlling, and improving quality.

Dr. Armand Feigenbaum played a major role in the origination of the total quality movement; his landmark text of 1951 was titled *Total Quality Control* and has influenced the quality movement. He defines quality systems as a method of managing organizations to achieve higher customer satisfaction, lower overall costs, higher profits, and greater employee effectiveness and satisfaction. Although organizations have adopted a wide variety of quality improvement programs, these programs are based on the concepts advocated by the total quality pioneers. In addition to Dr. Feigenbaum, the most highly acknowledged pioneers are W. Edwards Deming, Joseph M. Juran, and Philip B. Crosby. A number of Japanese experts including Kaoru Ishikawa and Shigeo Shingo were also major contributors to the quality improvement philosophy. Dr. Deming has emerged as the most influential and durable proponent of quality management in the United States and is best known for the Deming Cycle, his Fourteen Points, and the Seven Deadly Diseases.

TQM is based on involving everyone in an organization in an integrated effort toward improved performance at each level (Goetsch and Davis 2016). It integrates fundamental management techniques and improvement efforts in a disciplined approach toward continual process improvement. TQM emphasizes the understanding of variation, the importance of measurement, the role of the customer, and the commitment and involvement of employees at all levels of an organization. Total quality is driven by an organizational strategy and unity of purpose, an internal and external customer focus, obsession with quality, scientifically based decision making and problem solving, continuous process improvement, long-term commitment, teamwork, employee involvement and empowerment, and education and training.

Quality management systems

The American Society For Quality (ASQ) defines a quality management system (QMS) as a mechanism for managing and continuously improving core processes to "achieve maximum customer satisfaction at the lowest possible cost to the organization." It applies and synthesizes standards, methods, and tools to attain strategically important goals. In the construction project environment, quality management may be defined as the process required to ensure that a project's outcome will satisfy the needs for which it was undertaken. TQM also promotes excellence in customer satisfaction through the total involvement and dedication of each individual who is in any way a part of that product/process.

History of quality in construction

Quality has always been with us since recorded history – the code of Hammurabi ensured that a builder would have to provide his clients with good quality. If a house collapsed and hurt its occupants because of inferior quality, the builder was required to be put to death. Fortunately, the twentieth century brought a kinder, gentler approach to quality.

Historically, the Japanese were among the first to apply quality improvement approaches in construction on a large scale, although they did not embrace this concept until the oil crisis of 1973. Previously, they thought that the construction industry was inappropriate for the application of total quality control (TQC), because of the inherent variability in projects, and the difficulty in defining "acceptable quality." Takenaka Komuten Company, the sixth largest in Japan, had their formerly impeccable safety and quality image tarnished by the failure of a sheet piling system in Okinawa, in 1975, and embarked on a quality control (QC) program. They were followed by Shimizu Construction Company, the second largest in

Japan, which established a QC program in 1976, and by Kajima Corporation, the third largest, in 1978. Subsequently, several US companies adopted TQC programs and the more familiar TQM programs used by US manufacturers.

In 1992, the Construction Industry Institute (CII) published *Guidelines for Implementing Total Quality Management in the Engineering and Construction Industry*. Their research studies confirmed that TQM resulted in improved customer satisfaction, reduced cycle times, documented cost savings, and more satisfied and productive work forces (Burati and Oswald 1992).

While total quality approaches have been highly beneficial to the manufacturing and service industries, they have had limited application in the construction environment. The construction industry has been heavily steeped in the traditional ways of executing projects, and its constituents – designers and constructors – have been reluctant to make a necessary cultural and behavioral change to adopt total quality approaches. One reason for this unwillingness has been the perception that TQM is for manufacturing only (Chase and Federle 1992). The other major factor inhibiting the implementation of TQM in the construction industry has been the notion that TQM is costly and requires a long-time period for implementation. One aspect of TQM that has frustrated the construction industry the most has been "measurement" (Hayden 1996).

Many construction companies in the United States, Singapore, UK, and other European countries have been using TQM successfully for a number of years. They have been reaping rich rewards in improved client, consultant, and supplier relations. They have also experienced a reduced "cost of quality," on time and within budget project completions, and well-informed and highly motivated employees. On the other hand, many US companies have tried to adopt TQM programs and subsequently abandoned them. Studies indicate the failures can be attributed to a misunderstanding of TQM or to faulty implementation. In many cases top management and senior management are generally preoccupied with short-term, project by project profitability and not with long-term, quality-based strategies.

Attainment of acceptable levels of quality in the construction industry has long been a problem. Great expenditures of time, money, and resources – both human and material – are wasted each year because of inefficient or non-existent quality management procedures. The QC procedures that work effectively in a mass production industry have not been considered suitable for the construction industry. Typical explanations point to the uniqueness of construction projects and the lack of a clear and uniform standard for evaluating overall construction quality. Thus, there is great potential and challenge for quality improvement in the construction industry.

Instead, the construction industry has become increasingly reliant on burdensome specifications, which seldom say exactly what the owner intends them to say. This has led the owners to shift more of the project risks to the contractors (Ahmed and Aoieong 1998). The net outcome is that the construction industry has been bogged down with paperwork and defensive posturing; project participants generally tend to have a hostile attitude toward each other. TQM can help to reverse this trend.

Traditionally, inspection has been one of the key components of a quality assurance/quality control system in the construction industry. Regarding inspection, Deming's view is that "routine 100% inspection is the same thing as planning for defects – acknowledgement that the process cannot make the product correctly, or that the specifications made no sense in the first place. Quality comes not from inspection, but from improvement of the process" (1982). This does not mean that inspection ceases. Instead, it means that more effort is put into preventing errors and deficiencies.

Benefits of TQM

From the viewpoint of the individual company, the strategic implications of TQM include:

- Survival in an increasingly competitive world,
- Better service to its customers,

- Enhancement of the organization's "shareholder value,"
- Improvement of the overall quality and safety of our facilities,
- Reduced project durations and costs, and
- Better utilization of the talents of its people.

Quality management is a critical component in the successful management of construction projects; critical success factors include the support of senior management, appropriate leadership styles, cultivation of employee's enthusiasm and participation, and open communication and feedback. The most appropriate strategy for effective QMS in the construction industry is to establish effective management practices.

Quality costs

In order to quantify the benefits of TQM, quality must be measurable. Although there are numerous tools for measuring quality, the "cost of quality," or "quality costs," as advocated by Crosby (1984) and Juran (1988) represent an important indicator. Oberlender (1993) summarized quality costs as follows:

> Quality costs consist of the cost of prevention, the cost of appraisal, and the cost of failure. Prevention costs are those resulting from quality activities used to avoid deviations or errors, while appraisal costs consist of costs incurred from quality activities used to determine whether a product, process or service conforms to established requirements. Failure costs are those resulting from not meeting the requirements and can be divided into two aspects. Internal failure costs are the costs incurred on the project site due to scrap, rework, failure analysis, re-inspection, supplier error, or price reduction due to nonconformance. External failure costs are costs that are incurred once the project is in the hands of the client. These include costs for adjustments of complaints, repairs, handling and replacement of rejected material, workmanship, correction of errors, and litigation costs.

Principles of TQM

As reflected in Figure 11.1, five core principles are embodied in TQM (Harris 1995). These principles can be achieved in an organization with the aid of seven basic supporting elements. These elements are embodied in the criteria of the Malcolm Baldrige National Quality Award. Since its inception in 1987, the MBNQA has promoted national competitiveness through demonstrated quality-based achievements. Several outstanding companies have won the award such as: Federal Express, Texas Instruments Inc.,

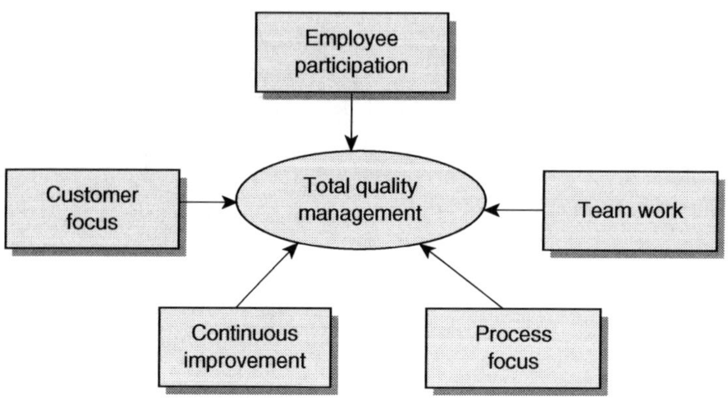

Figure 11.1 Principles of total quality management

Ritz-Carlton Hotel Co., Xerox Corporation, and Merrill Lynch Credit Corporation. It is worthy of note that no construction organizations have yet won the Baldrige Award, although many are undoubtedly adopting quality management techniques as first steps toward that coveted goal.

The Baldrige Criteria are built on seven categories of beliefs and behaviors that are typical of high-performance organizations:

1 Leadership: Senior management must lead this effort by example, by applying the tools and language, by requiring the use of data, and by recognizing those who successfully apply the concepts ofTQM.
2 Strategic planning: Senior managers may require support to bring about the change necessary to implement a quality strategy.
3 Customer and market focus.
4 Information and analysis:The use of data becomes paramount in installing a quality process.To set the stage for the use of data, external; customer satisfaction must be measured to determine the extent to which customers perceive that their needs are being met.
5 Human resource focus: Communications in a quality environment may need to be addressed differently in order to communicate to all employees a sincere commitment to change. Reward and recognition to teams and individuals who successfully apply the quality process so that the rest of the organization will know what is expected.
6 Process management.
7 Organizational results.

Organizations are scored out of 1000 points on these categories. Customer and market focus and customer-focused results account for 210 points and are therefore crucial to world-class performance.These categories address how successful organizations listen and learn with customers' key requirements and describe the methods used to satisfy customers, increase repeat business, and generate positive referrals.They evaluate complaint management, its analysis for improvement, determination of customer satisfaction/dissatisfaction, and benchmarking satisfaction with competitors.

Characteristics of the construction industry

Construction work is carried out in the form of a project. Projects are becoming progressively larger and more complex in terms of physical size and cost. Currently, the execution of a project requires the management of scarce resources; manpower, material, money, and machines to be managed throughout the life of the project – from conception to completion.The projects have six distinctive objectives to be managed: scope, organization, quality, cost, time, and safety (Figure 11.2). Construction work requires different trades and knowledge, but the management, scheduling, and control of those projects utilize the same tools and techniques, and are subject to constraints of time, cost, and quality.There are also unique characteristics of projects, which differ from routine operations.

TQM principles

There are many approaches available for improving quality:Total Quality Management, ISO 9001–2000, Six Sigma, the Malcolm Baldrige Criteria, and in lean production the European Quality Award.These approaches are based to varying degrees on the principles established by many quality experts such as Dr. W. Edwards Deming, Dr. J.M. Juran, Philip Crosby, Dr. Armand Feigenbaum, Dr. Kaoru Ishikawa, and Dr.Walter Shewhart. Dr. Deming in particular, has proven to be the most enduring because of his view of quality as a societal issue. Deming is known for the contribution of 14 Points and Seven Deadly Diseases, as well as several other initiatives.

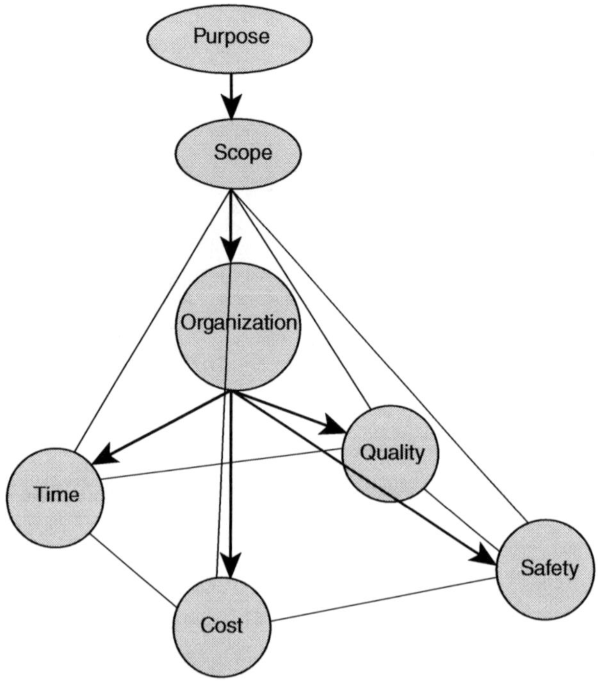

Figure 11.2 Six objectives of a construction project

A mathematical physicist by training, Deming was a major force in guiding the Japanese manufacturing effort to pre-eminence, after World War II. Humbled by their defeat, the Japanese were desperate to rise out of the economic doldrums, and eagerly grasped Dr. Deming's philosophy. Japanese manufacturers distinguished themselves with outstanding quality products such as autos produced by Toyota. The Japanese bestowed high honors on him, recognizing him with the Deming Prize. Much later, after 1980, America finally recognized Dr. Deming's genius, and reached out for his philosophy, in order to turn the tide against the losses in the US economy that resulted from poor quality and low productivity.

Dr. Deming formulated a management philosophy which he termed the "System of Profound Knowledge". A very important part of his philosophy is the concept that an organization is not just comprised of components, or departments, but is a system that can be compared with a living organism. The System of Profound Knowledge is based on three purposes and four assumptions.

Explanation of the system of profound knowledge

The purpose of his theory is threefold:

"To transform western leaders so they will:

1 Improve and innovate the system over the long term to allow all people to take joy in their work. This will restore the individual's intrinsic motivation, i.e. motivation from within, which is limitless and lends a long-lasting feeling of self-confidence and pride.
2 Optimize the system – not just its components – so that over the long term, everybody wins. The system includes customers, suppliers, employees, stockholders, and competitors.

3 Improve and innovate the conditions of society that impede the above purposes, such as ranking individuals, assigning grades, and MBO (management by objectives). Society includes systems at local, regional, and national levels, such as the economy, education, the environment, and public health and well-being."

In Deming's view there is a great need for this transformation in the prevailing western style of management as exemplified by the superior results obtained by organizations that practice Japanese methods of management. American companies have lost most of their market share especially in the automotive and electronic industries. Many American organizations have difficulty in motivating people to take joy in their work, especially with the familiar layoffs in the workplace. Many companies tend to be focused on the short term, but Deming was a strong proponent of long-term thinking.

Deming's influence can be clearly seen in the principles and practices of lean construction: his theory of management creates a situation in which everybody wins, by means of the following assumptions: To manage by seeking to improve the process that creates results, not just managing for results only.

1 To improve a process so that it meets an organization's needs better, one needs to study the underlying cause and effect systems of the process, not just to observe a "snapshot" of the process at a specific point in time; that can be misleading about what actions lead to specific results.
2 To manage by using a mixture of intrinsic and extrinsic motivation. Intrinsic motivation comes by workers' joy in doing challenging interesting work, but human nature also responds to extrinsic motivation, such as money or recognition.
3 To manage to facilitate cooperation, not competition. Cooperation between individuals or organizational units improves the functioning of the overall system that they comprise. Competition can become self-serving and compromise the success of the system.
4 To optimize the system (the enterprise) for the benefit of all stakeholders. In this scenario, everyone succeeds and benefits, and therefore contributes optimally.

Deming's 14 points

The 14 Points describe the conditions that are needed to promote a quality environment. They are summarized in the context of construction as follows:

1 Develop a program of constancy of purpose of improvement of product and service. The organization's strategic plan should include a commitment to improving products and services. Persistence is necessary for quality initiatives to bear fruit in design or construction activities.
2 Adopt this new program and philosophy. Quality needs to be ingrained in the organization's strategic plan – both its mission and vision.
3 Stop depending on inspection to achieve quality – build in quality from the start. Employees need to adopt a quality philosophy to produce to high standards, without relying on inspections.
4 Stop awarding contracts on the basis of low bids. Quality costs may be high with suppliers that have low-initial costs. Low bid awards may lead to failed inspections, low quality, and subpar performance. Suppliers should demonstrate capable quality systems.
5 Improve continuously and forever the system of production and service. Design and construction organizations should commit to continuous improvement.
6 Institute training on the job. Employees at all levels need to be trained in the best methods for delivering construction-related services. This includes executives, managers, technicians, supervisors/foremen, and workers.
7 Institute leadership. Top management has to lead with quality endeavors and set an example for their employees. Unless top management displays commitment, others will not take quality seriously.

8 Drive out fear so everyone may work efficiently. Construction workers are unwilling to volunteer ideas for improvement in an atmosphere of intimidation. Without their input, inefficient processes are likely to persist.
9 Eliminate barriers between departments so that people can work as a team. Reward teamwork instead of individual accomplishments.
10 Eliminate slogans, targets, and targets for the workforce – they create adversarial relationships.
11 Eliminate quotas and management by objectives. Institute work standards that reflect quality, cost, schedule, and safety.
12 Remove barriers that rob people of pride of workmanship. Provide workers with needed supervision, information, and resources to enable them to do the best job possible.
13 Establish rigorous programs of education and self-improvement. Commit the training and support systems necessary to equip everyone to develop and contribute to a quality environment.
14 Make the transformation everyone's job.

Deming's seven deadly diseases

These "diseases" are the antithesis of a quality environment:

1 Lack of constancy of purpose,
2 Emphasis on short-term profits,
3 Evaluation by performance, merit rating, or annual review of performance,
4 The mobility of management,
5 Running a company on visible figures alone,
6 Excessive medical costs, and
7 Excessive costs of warranties fueled by lawyers that work on a contingency fee.

These "diseases" are considered to be the antithesis of quality.

Philip Crosby

Philip H. Crosby (1928–2001) is considered one of the quality management gurus. He is recognized for creating the Zero Defects Process. This process was introduced by Crosby in 1979 and emphasizes customer satisfaction as a measure of quality control. Crosby also taught methods for eliminating the causes of defects before a defective product reached the customer. The Crosby process can help an organization by providing a quality management culture. Lean and Six Sigma can be effective "tools," but a tool becomes more beneficial when it is put to work regularly. To do that a quality "culture" is required to encourage (or insist) that everyone participates in this important process. Among his clients were General Motors, Chrysler, Motorola, Xerox and many hospitals. Crosby's first book, Quality is Free, has been credited with playing a large part in beginning the quality revolution in the United States and Europe. He published a total of 13 books, including The Absolutes of Leadership in 1996 and Quality and Me, an autobiography filled with lessons from life published in 1999.

Crosby's basic principles

– Doing It Right the First Time (DIRFT)
– Zero Defects Day
– Four Absolutes of Quality
– Prevention Process

- Quality Vaccine
- Six C's

The absolutes of quality management

- The First Absolute: The definition of quality is conformance to requirements
- The Second Absolute: The system of quality is prevention
- The Third Absolute: The performance standard is zero defects
- The Final Absolute: The measurement of quality is the price of non-conformance

Crosby's 14 steps to quality improvement

1. Management Commitment: Make it clear that management is committed to quality.
2. Quality Improvement Teams: Form Quality Improvement Teams with senior representatives from each department.
3. Measure Processes: Measure processes to determine where current and potential quality problems lie.
4. Cost of Quality: Evaluate the cost of quality and explain its use as a management tool.
5. Quality Awareness: Raise the quality awareness and personal concern of all employees.
6. Correct Problems: Take actions to correct problems identified through previous steps.
7. Monitor Progress: Establish progress monitoring for the improvement process.
8. Train Supervisors: Train them to actively carry out their part of the quality improvement program.
9. Zero Defects Day: Hold a Zero Defects Day to reaffirm management's commitment.
10. Establish Improvement Goals: Encourage individuals to establish improvement goals for themselves and their group.
11. Remove Fear: Encourage employees to tell management about obstacles to improving quality.
12. Recognize: Recognize and appreciate those who participate.
13. Quality Councils: Establish Quality Councils to communicate on a regular basis.
14. Repeat the Cycle: Do it all over again to emphasize that the quality improvement process never ends.

Crosby's quality vaccine

Integrity: Quality must be taken seriously throughout the entire organization, from the highest levels to the lowest. The company's future will be judged by the quality it delivers.

Systems: The right measures and systems are necessary for quality costs, performance, education, improvement, review, and customer satisfaction.

Communication: Communication is a very important factor in an organization. It is required to communicate the specifications, requirements and improvement opportunities of the organization. Listening to customers and operatives intently and incorporating feedback will give the organization an edge over the competition.

Operations: A culture of improvement should be the norm in any organization, and the process should be solid.

Policies: Policies that are implemented should be consistent and clear throughout the organization.

Six C's are as follows:

Comprehension: The importance of understanding what is meant by quality
Commitment: The stage when top management establish a quality policy
Competence: Developing and education and training plan

Communication: All efforts must be documented, and success stories published
Correction: Prevention and performance
Continuance: The process must become a way of life in the organization

The "1/10/100" rule – Crosby (1979) considered that the impact of rework at different stages of a project could be defined with the "1/10/100" rule. Applied to construction, changes made during the predesign phase would have an impact of $1 to the project, $10 during the design phase, and up to $100 after the beginning of construction. If rework is deferred until after project completion, the cost to implement changes not picked up previously could be as high as $1000!

Kaoru Ishikawa

The cause and effect diagram is the brainchild of Kaoru Ishikawa, who pioneered quality management processes in the Kawasaki shipyards, and in the process, became one of the founding fathers of modern management. The cause and effect diagram is used to explore all the potential or real causes (or inputs) that result in a single effect (or output). Causes are arranged according to their level of importance or detail, resulting in a depiction of relationships and hierarchy of events. This can help to search for root causes, identify areas where there may be problems, and compare the relative importance of different causes.

Armand Feigenbaum

Dr. Armand Feigenbaum played a major role in the origination of the total quality movement – his landmark text of 1951 was titled "Total Quality Control" and has influenced the quality movement. He defines quality systems as a method of managing organizations to achieve higher customer satisfaction, lower overall costs, higher profits, and greater employee effectiveness and satisfaction. Although organizations have adopted a wide variety of quality improvement programs, these programs are based on the concepts advocated by the total quality pioneers.

Joseph Juran

Dr. Joseph Juran is known for several quality contributions: Three Basic Steps to Progress, Ten Steps to Quality Improvement, and The Quality Trilogy. Juran's quality trilogy has three components: namely; (1) quality planning, (2) QC, and (3) quality improvement. This trilogy of quality processes provides a successful framework for achieving quality objectives. The processes must occur in an environment of inspirational leadership and the practices must be strongly supportive of quality. A brief description of the Juran's quality trilogy is given below.

1 Quality planning

   a. Determine who the customers are
   b. Identify customers' needs
   c. Develop products with features that respond to customer needs
   d. Develop systems and processes that allow the organization to produce these features
   e. Deploy the plans to operational levels

2 Quality control

   a. Assess actual quality performance
   b. Compare performance with goals
   c. Act on differences between performance and goals

3 Quality improvement

a. Develop the infrastructure necessary to make annual quality improvements.
b. Identify specific areas in need of improvement and implement improvement projects.
c. Establish a project team with responsibility for completing each improvement project.
d. Provide teams with what they need to be able to diagnose problems to determine root causes, develop situations, and establish control that will maintain gains made.

Customer focus

In the TQM philosophy, total customer satisfaction is the goal of an entire system, and a pervasive customer focus is what gets us there. The function of the construction industry is to provide customers with facilities that meet their needs. For a company to remain in business this service must be provided at a competitive cost. TQM is a management philosophy that effectively determines the needs of the customer and provides the framework, environment, and culture for meeting those needs at the lowest possible cost. By ensuring quality at each stage in the construction process and thereby minimizing costly rework – as well as other costs – the quality of the final products should satisfy the final customer.

By definition, customers may be either internal or external. The external customer is the consumer or client; that is, the end user of the products or services being offered. An internal customer is a second process or department within the organization, which depends on the product of the first. For example, the products are plans and specifications for designers, the customers are the owner, and the contractor is responsible for the construction. For the contractor, the product is the completed facility, and the customer is the final user of the facility. There are also customers within the construction organization. These internal customers receive products and information from other groups of individuals within their organization. Thus, satisfying the needs of these internal customers is an essential part of the process of supplying the final external customer with a quality product.

Every party in a process has three roles: supplier, processor, and customer. Juran defined this as the triple role concept. These three roles are carried out at every level of the construction process. The designer is a customer of the owner. The designer produces the design and supplies plans and specifications to the contractor. Thus, the contractor is the designer's customer, who uses the designer's plan and specifications to carry out the construction process and supplies the completed facility to the owner. The owner supplies the requirements to the designer, receives the facility from the contractor, and is responsible for the facilities operation (Burati and Oswald 1993). This clearly illustrates that construction is a process, and that TQM principles that have been applied to other processes are potentially adaptable to the construction industry (Figure 9.3).

Process improvement

A process is a way of getting things done. A process consists of the tasks, procedures, and policies necessary to carry out an internal or external customer need. According to the TQM philosophy, if the process is correct so will the end result (product). Thus the organization should work to improve the process so as to improve the end product or service.

Three different approaches have emerged for improving the efficiency or effectiveness of a process. Continuous improvement is an approach used on an on-going basis for incremental gains. Benchmarking should be used periodically, and reengineering can be launched occasionally to achieve dramatic breakthrough.

By focusing on a process by measurement and analysis, it can be improved by influencing five Ms of the process; namely, man (people), machine, material, method, and measurement. A sixth factor is that of the operating environment. A strong emphasis on process improvement centers on the measurement of variation, the control of variation, and the knowledge of variation to seek improvement. This analysis is referred

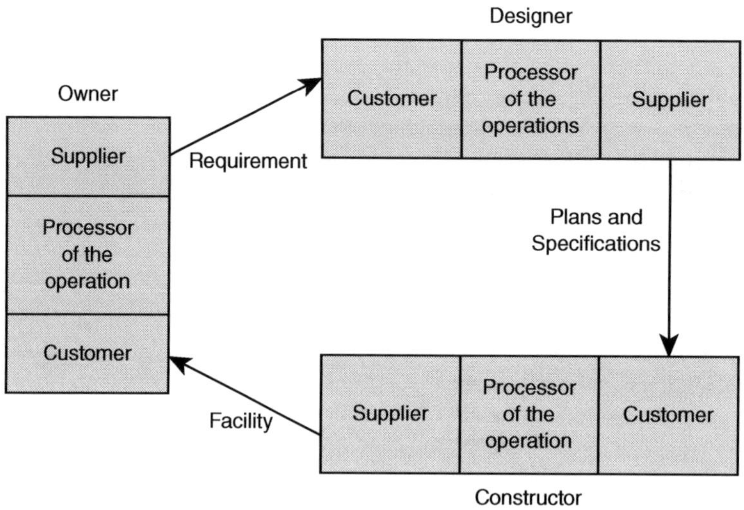

Figure 11.3 Juran's triple role concept applied to construction

to as statistical process control or statistical analysis. It is at the center of process improvement. The objective of measuring the variation in a process is to learn how to control the variation and also how to improve the process by viewing variation as a tool for improvement. The analysis of the positive side (good performance or quality) of the variation of process is referred to as a "breakthrough improvement" or "breakthrough management," which is another key component of TQM.

Continuous improvement

The goal of continuous improvement is common to many managerial theories; however, what differentiates TQM is that it specifies a specific step-by-step process to achieve it. The steps are as follows:

1 Identify the process,
2 Organize a multidisciplinary team to study the process and recommend improvements,
3 Define areas where data are needed,
4 Collect data on the process,
5 Analyze the collected data and brainstorm for improvement,
6 Determine recommendations and methods of implementation,
7 Implement the recommendations outlined in step 6,
8 Collect new data on the process after the proposed changes have been implemented to verify their effectiveness, and
9 Circle back to step five and again analyze the data and brainstorm for further improvement.

The nine-step cycle emphasizes: focusing the progress, measuring the process, brainstorming for improvement, verification, and re-measurement. These four elements are further illustrated in Deming's Plan-Do-Check-Action (PDCA) diagram shown in Figure 9.4. The PDCA diagram stresses removing the root cause of problems and continually establishing and revising new standards or goals (Deming 1986).

Under TQM, management in the construction industry has two functions: (1) to maintain and improve current methods and procedures through process control and (2) to direct efforts to achieve, through

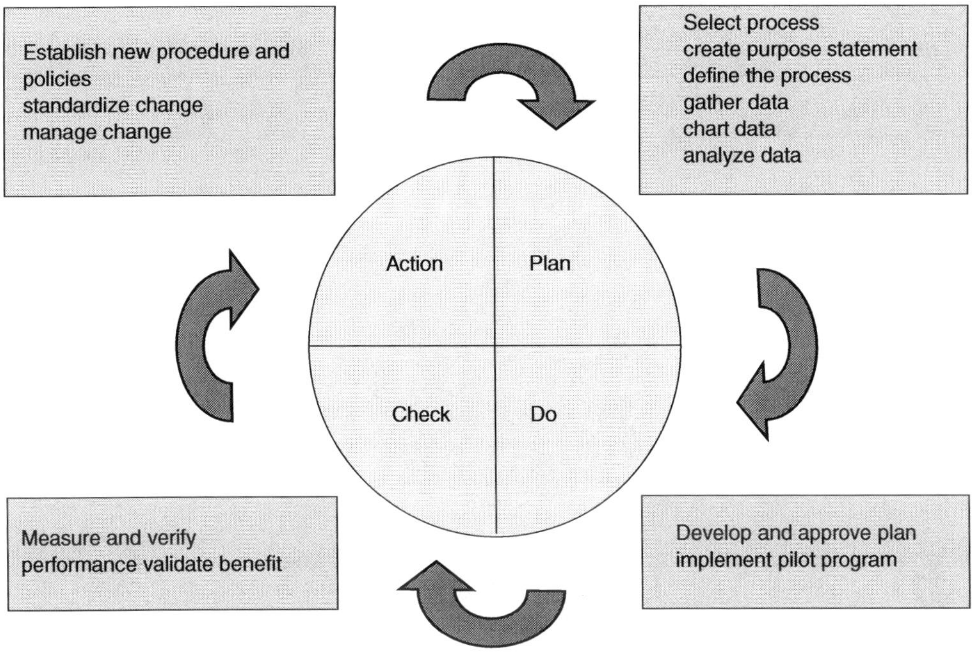

Figure 11.4 The PDCA diagram

innovation, major technological advances in construction processes. The incremental improvement of the process is achieved through process improvement and control. In every construction organization there are major processes by which all the work is accomplished. However, there are innumerable parts in the construction process. Through the use of flow diagrams, every process can be broken down into stages. Within each stage, input changes to output, and the methods and procedures directing the change of state (i.e., the construction procedures) can be constantly improved to better satisfy the customer at the next stage. During each stage the employees should communicate closely with their supplier and customer to optimize the work process for that stage. This requires each employee to recognize their place in the process and their respective supplier and customer.

Quality improvement techniques

(Note: Chapter 14 describes some of these and other tools and techniques in greater detail)
Total quality management mainly demands a process of continued improvement aimed at reducing variability. An organization wishing to support and develop such a process needs to use quality management tools and techniques. It is prudent to start with the more simple tools and techniques: The check sheet, the checklist, the histogram, the Pareto diagram, the cause and effect diagram (Fishbone diagram), the scatter diagram, and the flowchart. They are described briefly below. Readers may refer to Chapter 14 to obtain more detailed information on the use of some of these tools as well as others.

Check sheet

Used to record events, or nonevents (nonconformances). They can also include information such as the position where the event occurred and any known causes. They are usually prepared in advance and are completed by those who are carrying out the operations or monitoring their progress. The value of a check sheet can be retrospective analysis, so they help with problem identification and problem solving.

Checklists

A list of things that need to be done and when completed are checked off. It can be used in the auditing of quality assurance and to follow the steps in a particular process.

Histogram

Provides a graphical representation of the individual measured values in a data set according to the frequency of occurrence. It helps users to visualize the distribution of data and can display the amount of variation within a process.

Pareto analysis

A technique employed to prioritize the problems so that attention is initially focused on those having the greatest effect. It was discovered by an Italian economist, named Vilfredo Pareto, who observed how the vast majority of wealth (80%) was owned by relatively few of the population (20%). As a generalized rule for considering solutions to problems, Pareto analysis aims to identify the critical 20% of causes and to solve them as a priority. (See Chapter 14 for an example of a Pareto chart.)

Cause-and-effect diagram (fishbone diagram)

The diagram, which was developed by Kaoru Ishikawa, is useful in breaking down the major causes of a particular problem. The shape of the diagram looks like the skeleton of a fish. This is because a process often has a multitude of tasks running into it, any one of which may be a cause. If a problem occurs, it will have an effect on the process, so it will be necessary to consider the whole multitude of tasks when searching for a solution.

Scatter diagram

The relationship of two variables can be plotted in the scatter diagrams. They are easy to complete, and an obvious linear pattern reveals a strong correlation between the variables.

Flowcharts

A chart providing a diagrammatic picture using a set of symbols. They are used to show all the steps or stages in a process project or sequence of events. A flowchart assists in documenting and describing a process so that it can be examined and improved. Analyzing the data collected on a flowchart can help to uncover irregularities and potential problem points.

Barriers to the implementation of TQM

Employees generally show resistance to the introduction of TQM for a host of reasons, which include fear of the unknown, perceived loss of control, the "it may cost more" syndrome, and an unwillingness to take "ownership" and be committed to change. Other barriers include:

- Perceived threats to foreperson and project manager roles,
- Disinterest at the site level,

- Lack of understanding of what TQM was, particularly on-site,
- Geographically dispersed sites,
- Fear of job losses,
- Inadequate training,
- Plan not clearly defined,
- Employee skepticism, and
- Resistance to data collection (e.g., rework costs, nonconformances, material waste, etc.).

TQM efforts should be implemented to increase productivity of the organization (quantity of performance); increase quality (decrease error and defect rate); increase effectiveness of all efforts; increase efficiency (decrease time requirements while increasing productivity).

TQM implementation

TQM deployment structure

A formal structure is necessary in order for TQM to be implemented effectively in an organization. Such a structure is even more important in the construction environment where stakeholders are preoccupied with meeting schedules and responding to daily events. TQM calls for persistent analysis, evaluation, and learning; but this may be nigh impossible to accomplish by people in their traditional roles.

The proposed structure is aligned with the principles embodied in the Malcolm Baldrige award criteria and facilitates the deployment of activities necessary to secure the high level of quality in relation to the design and construction processes. It must be pointed out that the intent of this structure is not to compete with other organizations for the award; rather the purpose is to take advantage of a well-integrated structure that connects the behaviors of stakeholders with the criteria that have proven to be successful for many years.

The proposed TQM structure (Figure 9.6) shows the relationships between the functions that are essential for its deployment. Top management is the driving force behind quality endeavors. They supervise the steering committee, and this committee supervises the project teams. A consultant provides the expertise needed to initiate an effective TQM program. This function may be temporary if its scope is limited to getting the program established. TQM facilitators performs many functions that are critical to the deployment of quality-related activities and serves as a multiskilled resource for the project teams, the steering committee, and top management. The Malcolm Baldrige Criteria gives direction to all quality-related activities.

Quality implementation

There are three distinct phases to quality deployment: preparation, planning, and execution.

The process starts with a firm's top management embracing the idea of incorporating quality as a value of the organization. They need to be committed enough to generate enthusiasm in the company's employees; everyone should wish to produce quality workmanship, by "doing it right the first time." top management appoints a steering committee to oversee quality efforts. The members should be advocates of quality and placed at an adequate level in the organization where they can make the decisions needed for a program to be effective.

Limitations of TQM deployment

In traditional manufacturing and service industries, providers have had excellent results with TQM. They have worked closely with suppliers to deliver products and services to high levels of customer satisfaction.

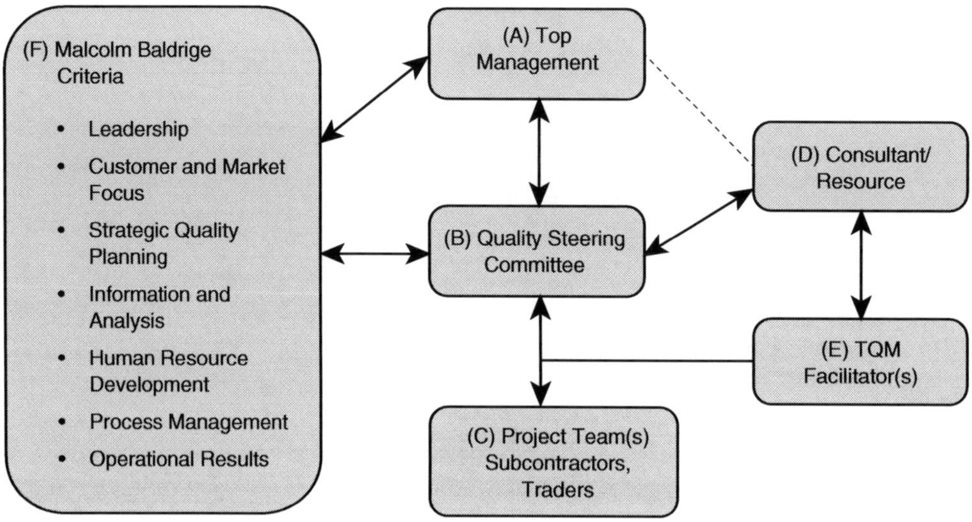

Figure 11.5 Proposed TQM deployment structure

They are well equipped to optimize the generation of those products and services, most of which are discrete and/or transaction-driven. A manufacturer takes components and subassemblies from a number of suppliers, and can optimize the consolidation of these items to generate a finished product. A service operation such as a garage is responsible in total for the service delivered and can optimize service delivery as well.

Construction, on the other hand, is a protracted process that involves interaction between many providers. While a single provider can optimize its portion of the work by improving its internal operations, project-wide optimization requires the close collaboration of other providers. The difficulty in establishing that collaboration is the Achilles heel of most traditional project delivery methods. The effect the entire team of top management leadership, therefore, may be limited to a specific firm that wishes to improve quality. It may have limited impact on the desires of another provider of construction (or design) services that is a member of the project delivery team. For example, a general contractor may embrace the TQM philosophy, but cannot reasonably expect a mechanical subcontractor to embrace these concepts and demonstrate genuine commitment to its employees with the degree of credibility needed to engage them. It is for these reasons that lean construction is able to engage the project team to "global optimization" where the entire team excels, as opposed to the "local optimization" experienced by subcontractors or work units that seek to maximize their individual performance.

Quality improvement concepts

A mission statement should exemplify an organization's focus on exceeding customers' expectations with products of high value and quality, and motivating/empowering its employees.

The organizations that have won the Malcolm Baldrige award have all shared the common attribute of having leaders that were committed to quality and demonstrated it by example through their daily activities. Such leaders were active participants or members of quality councils that were set up to promote quality endeavors. They guided and led quality improvement efforts, established and reviewed performance measures, kept quality as a major topic of all meetings and/or on-going reviews.

Measurement is critical to ensuring that an organization is meeting its goals and expectations. These goals/expectations should be delineated in a strategic plan developed with the participation of all stakeholders to ensure that they are an integral part of the quality endeavor. For quality performance to be achieved there must be specific, meaningful measures of performance based on key drivers. The strategic plan should also delineate the responsibilities of specific individuals and assign time frames for accomplishment.

Key drivers include customer satisfaction, operational performance, financial performance, employee satisfaction, and community service. By way of an example, the indicators of customer satisfaction are: customer satisfaction index, referral rates, percentage of on-time delivery, defects detected by the builder at delivery, defects determined by homeowner inspection, warranty call rates, warranty costs, and response time on warranty calls.

Quality strategies

1 Establish top management priorities

- Visionary
- Set aggressive goals, and
- Walk the talk.

First, top management of the organization must establish that total quality is the highest priority of the organization. Executives must provide a clear and reasonable vision, set aggressive goals for the organization and each unit, and most importantly demonstrate their commitment to TQM through their actions.

2 Initiate cultural change

- Paradigm change,
- Credibility development, and
- Time.

Second, the culture of the organization must be changed so that everyone and every process embrace the concept of TQM. The organization must change its paradigm to adapt to a customer focus emphasis where everything done in the organization is aligned with exceeding customer expectations. This becomes an ongoing way of life for the organization, continually improving and adapting. As part of the culture change, credibility among the employees must be built through rewarding positive steps toward the vision of TQM. The organization must also allow time for the change to occur and be indoctrinated into the everyday aspects of the organization.

3 Establish small teams – provide overall goals

- Define quality,
- Identify customers wants, and
- Measure and change.

Third, small teams need to be developed throughout the organization to define quality, identify customer wants, and measure progress and quality. These teams will be responsible for creating their own goals, given the organization's overall goals.

4 Execute change and continuous improvement

Fourth, change and continuous improvement must be implemented, monitored, and adjusted based on an analysis of the measurements.

A study of TQM in the United States (Florida) construction industry

Syed M. Ahmed and Salman Azhar conducted a 2002 study of the Florida construction industry; the study observations are presented as guidelines for successful TQM implementation in the industry. The main objectives of this study were to (1) investigate the adoption and implementation of TQM in the construction industry; (2) determine the processes ("what to measure") that are most suitable and appropriate for measurement during the construction project life-cycle; and (3) develop a model ("how to measure") for the measurement and evaluation of the quality performances of the construction processes identified in step 2 above as a tool for continuous improvement. When it comes to measuring work processes, the construction industry does not enjoy a good reputation. The problem, however, can be attributed to the nature of the industry, which lacks solid data gathering and experiences wide fluctuations in productivity.

Phase I

The first stage of this study identified the prevailing implementation and adoption of TQM principles in the construction industry through an in-depth questionnaire. The questionnaire was divided into six parts; namely, their knowledge of TQM, their perception of quality, the data acquisition methods used by them, quality in their organization, the degree of training provided to their employees toward TQM, and the obstacles faced by them in implementing TQM in their businesses. The conclusions from these six sections are briefly discussed below. (Observations are reported in real time, for the sake of convenience)

Knowledge of TQM

The results show that the majority of the contractors agreed that if a contractor satisfies clients, the profits would increase in the long run. They feel that TQM will work very well in their organizations, and that it will be beneficial to all stakeholders. They are, however, not aware of any implementation programs. Most of them feel that TQM is a philosophy used to improve cost estimating and warranty claims. This shows their lack of knowledge about the TQM and the potential benefits in implementing it in their organizations.

Perception of quality

The analysis indicates that the majority of the contractors perceive quality as a competitive advantage next to elimination of defects. They feel that product/service quality is very important for them in gaining customer satisfaction because it ultimately translates to higher profits. They feel that customer satisfaction is their main goal. Interestingly however, when asked to rank quality, safety, time, cost, and scope in the order of importance, they ranked scope and cost as the more important considerations followed by timeliness, safety, and quality.

Data acquisition method

The results suggest that most of the companies in the study collect data to measure performance. They address problems by assigning an individual to solve them. On the other hand, 52% of the companies have a system for capturing customer suggestions, but only 28% measure customer satisfaction through questionnaires. Customer suggestions are gathered by the number of complaints or other methods in 20% of the companies. In most cases, the suppliers and the subcontractor are rated (52%) and when defects in services are identified, they are required to pay for or correct them.

Quality in their organization

Although only about 50% of the contractors surveyed have a clear definition of quality in their organizations, 86% are aware of the importance of quality. A majority of the respondents say that they do not have a formal quality improvement program (QIP) in place. Those that do, however, have the full support of their top management. They use a mix of QC, TQM, and ISO 9000 principles in their QIP. Demanding customers, CEO commitment, and competitive pressures are identified as the key reasons for implementing the quality improvement programs. The main objectives of the QIP are employee involvement followed by increasing productivity and cost reduction. There were 40% of the contractors that felt the quality of their products and services improved after implementing such a program.

Training

In the majority of the companies, employees are not given formal training in TQM or other quality improvement programs. Of these companies, 44% reported that managerial/supervisory staff had undergone quality improvement training while 29% of the companies provided training on quality management philosophies to non-managerial and non-technical staff. Training programs mostly emphasize customer satisfaction as a primary goal followed by teamwork and communication.

Barriers to implementing total quality management

The following list gives a breakdown from the most important obstacle to the least important one.

1 Difficulty in changing behavior and attitudes,
2 Lack of expertise/resources in TQM,
3 Lack of employee commitment/understanding,
4 Lack of education and training to drive the improvement process,
5 Schedule and cost treated as the main priorities,
6 Emphasis on short-term objectives,
7 Tendency to cure symptoms rather than getting to the root cause of a problem, and
8 Too many documents are required (poor documentation).

It is easy to infer from the above that although TQM has been a magic word in the construction industry for the past few years, methods and techniques to implement the quality management program still need to be developed. The basic problem attributed to a lack of expertise or resources for implementing quality improvement programs is the difficulty in assessing what to measure and how to measure it – particularly the intangible aspects of quality. Without measurement, the notion of continuous improvement is hard to follow.

To address the issue of measurement, an attempt is made to determine the client satisfaction index. This provides a direct reference point from where quality improvement steps can be initiated in the construction industry. Various possibilities of client satisfaction were listed and rated. This gives us a measure or an index of client satisfaction or dissatisfaction that is discussed in Phase II of the study.

Phase II

In the second phase of this study, a second questionnaire focused on the customer was created to identify the processes for improvement. From the analysis of this questionnaire, the client satisfaction index was developed that lists the major causes of dissatisfaction for the clients. They are:

- Lack of attention to client priorities,
- Poor planning,
- Poor scheduling,
- Inadequacy in processing change orders, and
- Poor delivery schedules and methods.

Next, a third questionnaire was developed using the cause and effect diagram to identify the subcauses for the main reasons for client dissatisfaction. This was presented to contractors and their feedback was sought through structured interviews.

The major subcauses for the lack of attention to client priorities were:

- Lack of personnel training
- Lack of quality and cost control
- Inadequacy in contractor-subcontractor coordination
- Lack of conformance to specifications

The main subcauses for poor planning in the construction industry were:

- Low quality of material and workmanship,
- Poor management of change orders,
- Poor cash flow analysis,
- Construction underestimation, and
- Poor equipment management.

The sub-causes for poor scheduling were:

- Incomplete design,
- Lack of site condition supervision,
- Inadequacy in project management coordination, and
- Improper network model selection.

The main factors for the inadequacy in processing change orders were:

- Changes in design by clients,
- Errors in construction design,
- Defective materials/equipment, and
- Weather delays.

The subcauses for the poor delivery schedules and methods in the construction industry were:

- Ambiguity in methods
- Change orders from procurement department
- The availability of materials and equipment

Phase III

This analysis led to the final phase of the research study where the most important identified subcauses by the contractors were presented to the same owners who had participated in questionnaire #2. In questionnaire

#4 they were asked to indicate a measure of their satisfaction if the contractors made improvements to the identified subcauses. We call this the improvement index. Following are areas that would increase customer satisfaction the most if the contractors improved them.

- Construction underestimation
- Conformance to specifications
- Project management coordination
- Design changes by clients
- Change orders from a procurement department

Customer satisfaction can be greatly enhanced by improving construction estimating, conformance to specifications, project management coordination, and by reducing design changes by clients, and change orders from the procurement department. The above areas were identified after analyzing the results of the last questionnaire in which we had asked the clients to identify the most important processes that need improvement.

This process can be repeated until different areas of improvement are identified through another cycle of the development of the client satisfaction and improvement index. The key to understanding is that clients are now a moving target – their expectations and requirements are constantly changing. To keep up with their ever-changing goals, contractors need to have a system of identifying, measuring, and continuously improving their tangible and intangible products and services in place.

Given the limited size of the sample, it was not possible to confirm that the causes and sub-causes of client satisfaction or dissatisfaction identified in the study were statistically significant. The objective of this study was to develop and demonstrate how a system of continuous improvement can be put in place by measuring different processes.

Recommendations for improvement

Success in TQM deployment is very much driven by the adage: "What gets measured gets done." It is of paramount importance to develop metrics related to key performance indicators (KPIs) and to report ongoing measurement to all parties – top management and workers alike. Exception reports may be used to flag non-conformances. Top management should be attuned to having corrective measures implemented to address anomalies.

Factors to be measured

The following factors are examples of measurements to be applied in the application of TQM.

Designer performance:

Number of design errors
Processing time for shop drawings
Number of RFIs submitted
Response time for RFIs
Customer satisfaction ratings

Contractor performance:

Accidents, days without an accident
Number of warranty calls

Number of punch list items
Number of failed inspections
Percentages of delay in schedule
Customer satisfaction ratings
Estimated cost versus actual cost – by trade/discipline

Best practices

The following TQM-based best practices are recommended for improving performance by building customer loyalty.

- Develop a clear strategy for enhancing customer loyalty, emphasizing activities that represent high value. Resource limitations often make it impossible to be all things to all customers. Responsiveness and proficiency of customer service operations may need to be weighed against the quality and timeliness of delivery, for example.
- Capture the voice of the customer (VOC). Have customers define the criteria for quality, price, and value.
- Develop measurement systems to collect and analyze customer feedback – information on won/lost customers or projects is critical. Track problem resolution and the percentage of highly satisfied and highly dissatisfied customers.
- Base measurable performance goals on the VOC so those performance thresholds can be compared with customers' needs.
- Establish a clear understanding of market segments in order to deliver value appropriately to each segment. Public sector customers have needs that differ from private sector companies – housing developers and office building investors have different needs as well. Value does not mean the same thing to each of these customers.
- Establish competitive benchmarks – know how well your competitors are delivering value to their customers. You may increase the satisfaction of your customers, but lose them to a competitor who can satisfy them even more.
- Maintain your company's image by such "soft" issues as commitment, partnership, and integrity. If a crisis occurs that jeopardizes the value delivered, responsiveness and communication with customers can keep their loyalty intact.
- Do root cause analysis to see where problems originate; institute corrective action.
- Discard the mistaken notion that construction is a project specific proposition; remember that each satisfied customer provides a building block for the value and corporate image that forge long-term business success.

Above all, embrace change in order to adopt the new paradigm; like the dinosaur, those who do not change are doomed to extinction.

Summary for Part A

Total quality management (TQM) is an approach to doing business that attempts to maximize the competitiveness of an organization through the continual improvement of the quality of its products, services, people, processes, and environments. There is no simple, comprehensive definition of quality. Deming's view is that the customer is the most important part of the production line. Quality should be aimed at the needs of the customer, both present and future. By ensuring quality at each stage in the construction process, and thereby minimizing costly rework, as well as other costs, the quality of the final products should satisfy the final customer.

Part B: Six Sigma in construction

Definition of Six Sigma

Six Sigma management may be described as the "relentless and rigorous pursuit of the reduction of variation in all critical processes" (Gitlow 2006). It is an organizational imperative to create an order of magnitude change in processes. The Six Sigma approach has two versions. One version is a managerial initiative for measurably improving processes to better meet customer requirements; the other version includes another approach based on statistical tools and techniques that reduce product defects with a target of 3.4 defects per million opportunities (DPMO). Both versions are based on tools and techniques that were previously developed through Shewhart, Deming, Juran, Ishikawa, and others. Six Sigma management originated in the high-volume manufacturing environment with an emphasis on measuring and reducing defects. Some construction applications may involve repetitious activities that may lend themselves to this numerical target. As much of construction is non-repetitious, it can benefit from the more general process improvement approach.

Six Sigma is not focused on quality although quality generally improves with its application as a result of reducing the variation in target processes. As opposed to TQM, it is not concerned with changing the culture of an enterprise, but this culture tends to change as Six Sigma thinking permeates it. It facilitates "breakthrough" business improvement as opposed to gradual or incremental improvement. Six Sigma projects typically yield process performance improvements in the vicinity of 30–60% in 6 months or less, with attendant improvements in financial and business performance.

According to the Six Sigma Academy, Black Belts save companies approximately $230,000 per project and can complete four to six projects per year. General Electric, one of the most successful companies implementing Six Sigma, has estimated benefits on the order of $10 billion during the first 5 years of implementation. In 1995, GE first began Six Sigma after Motorola and Allied Signal blazed the Six Sigma trail. Since then, many organizations worldwide have discovered the far reaching benefits of Six Sigma.

Whereas Quality Management Systems such as TQM and ISO 9001:2000 generally require sustained effort for several years, this period of time is longer than a typical construction project cycle and limits application to the construction environment. Six Sigma projects have a much shorter cycle, making the approach more adaptable to the construction environment.

History of Six Sigma

Prior to 1978, Motorola was primarily a manufacturer of car radios and television sets. Bill Smith, a senior engineer and scientist in Motorola's communications division, introduced/developed the Six Sigma concept. He brought these ideas to CEO Bob Galvin who was deeply concerned with the quality problems that had been reported by Motorola's customers. Galvin set a target to reduce manufacturing defects by a factor of 10 in the first 5 years. By 1987, Motorola had reduced its rate of defects by that factor, then revised these improvement criteria to a factor of 10 in 2 years. Subsequently, they established a target of a hundredfold in 4 years and 3.4 defects per million opportunities (DPMO) in 5 years.

Smith and Galvin went on to establish Six Sigma as a primary component of Motorola's culture in order to provide customers with outstanding products. By the late 1980s, Six Sigma had been expanded to reducing concept to market cycle time. It also became the primary means to reduce field failures and increase customer satisfaction. Motorola's Six Sigma methodology evolved – continuous improvement became a main goal of the organization. In 1988, Motorola was awarded the MBNQ Award in its first year. The award is administered through the US Department of Commerce and recognizes US organizations for excellence in quality and overall performance.

Between 1989 and 1993, Motorola was joined by ABB and Kodak to support the Six Sigma Institute. During the mid-1990s, General Electric and Allied Signal implemented Six Sigma with outstanding, well-publicized results. A large number of other organizations have adopted the Six Sigma philosophy in recent years. They include American Express, Ford, Dupont, Dow Chemical and 3M. The Six Sigma philosophy has expanded beyond the original concept of 3.4 DPMO. It has become a business strategy and methodology for improving process performance and customer satisfaction.

Six Sigma benefits

Motorola views the Six Sigma methodology as a proven way of accomplishing transformational change within an organization. It serves as a business improvement process that aligns an organization closely with customer requirements, applying tools that seek near perfection. Organizations that have employed the Six Sigma methodology have achieved significant benefits. Typical examples are:

- Better process flows,
- Fewer defects,
- Higher productivity,
- Greater customer satisfaction (both internal and external),
- Shorter cycle time,
- Faster delivery time,
- Reduced work in process (WIP),
- Lower inventory levels,
- Greater capacity,
- Less waste,
- Higher quality and reliability,
- Lower unit costs, and
- Greater profitability.

Stakeholders typically experience the following benefits:

- Greater customer satisfaction with products and services,
- Better employee morale,
- Greater stability for suppliers, and
- Stockholder satisfaction with lower costs and increased profits.

Process basics (voice of the process)

A process is a collection of interacting components that transform inputs into outputs toward a common aim, called a mission statement (Gitlow 2006). It is management's job to optimize the entire process toward its aim, directing resources as necessary to accomplish its aim, even if a particular department or function has to suboptimize. The important thing is for the entire organization to secure maximum profit. This poses a particular challenge for the construction environment as the organization in that situation comprises separate actors (such as subcontractors) who each seek to maximize their individual profits.

Statistically, the term Six Sigma refers to a process in which the range between the mean of a process quality measurement and the nearest specification limit is at least six times the standard deviation of the process. Six Sigma seeks to center the process on the target and reduce process variation. Table 9.1 cross-references the number of standard deviations and the corresponding defect occurrence. (Note: the values

Table 11.1 Sigma limits vs. defect occurrence

Specific Limit	*Percentage of Accuracy*	*Defects (Per Million Opportunities)*
1 sigma	30.23	697700
2 sigma	69.13	308700
3 sigma	93.32	66810
4 sigma	99.3790	6210
5 sigma	99.97670	233
6 sigma	99.999660	3.4

indicated assume a process shift of 1.5 standard deviations.) The Six Sigma concept is built on the assumption that:

- Continuous efforts to achieve stable and predictable process results (i.e., reduce process variation) are of vital importance to business success.
- Manufacturing and business processes have characteristics that can be measured, analyzed, improved, and controlled.
- Achieving sustained quality improvement requires commitment from the entire organization, particularly from top-level management.

Features that set Six Sigma apart from previous quality improvement initiatives include:

- A clear focus on achieving measurable and quantifiable financial returns from any Six Sigma project.
- An increased emphasis on strong and passionate management leadership and support.
- A special infrastructure of "Champions," "Master Black Belts," "Black Belts," and so on to lead and implement the Six Sigma approach.
- A clear commitment to making decisions on the basis of verifiable data, rather than assumptions and guesswork.

Methods

Six Sigma has two key methods: define, measure, analyze, improve, control (DMAIC) and define, measure, analyze, design, verify (DMADV), both inspired by Deming's PDCA cycle. The DMAIC process is an improvement system for existing processes falling below specification and looking for incremental improvement. The DMADV process is an improvement system used to develop new processes or products at Six Sigma quality levels. It can also be employed if a current process requires more than just incremental improvement. In the construction context, Six Sigma quality levels are not generally attainable due to the lack of repetition in work tasks. The DMAIC sub-methodology may be more appropriate (Figure 11.7).

DMAIC steps

Define: It is the first step in our six sigma approach. DMAIC first asks leaders to define core processes. The critical to quality (CTQ) characteristics must be identified to determine what is important to customers. It is important to define the selected project scope, expectations, resources, and timelines. The rationale for the project must be established, relative to other potential projects. The first efforts at process documentation are often at a general level. Additional work is often required to adequately understand and correctly document the processes.

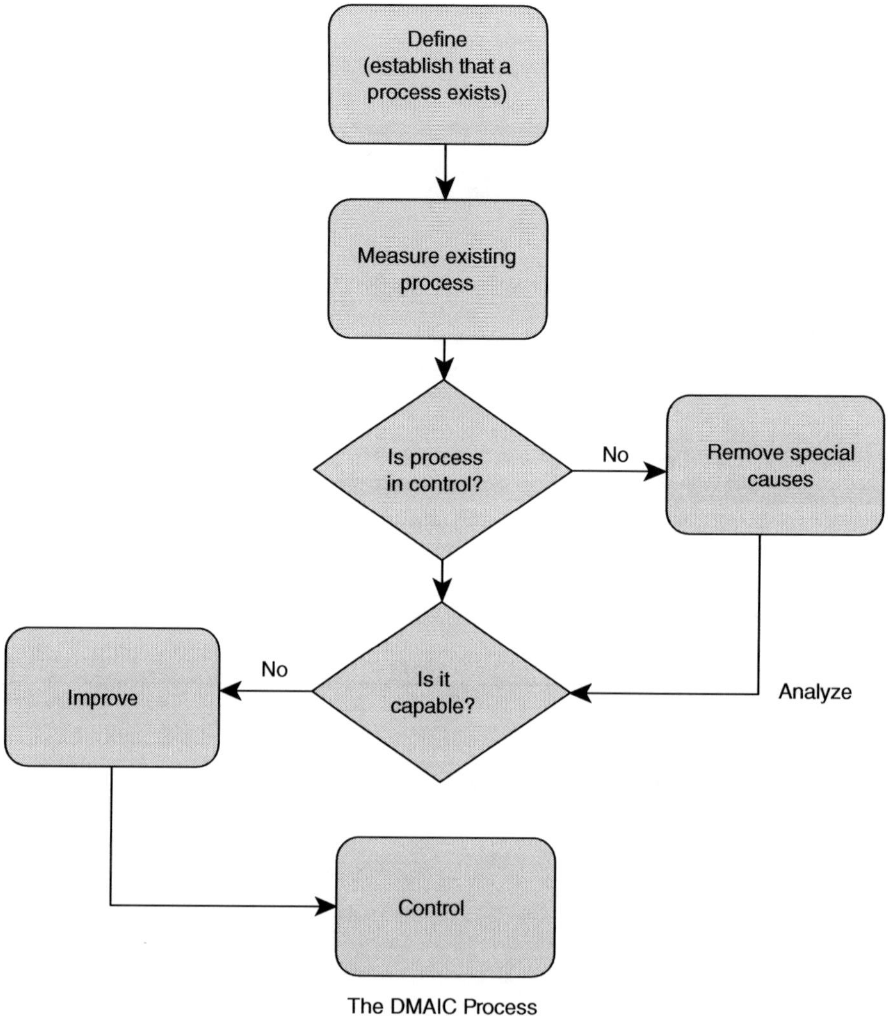

Figure 11.6 Flowchart of the DMAIC process

Measure: This phase involves developing operational definitions for each CTQ variable, ensuring the measurement system is valid, and establishing baseline capabilities. The Six Sigma approach asks the Black Belt project manager to quantify and benchmark the process using actual data. At a minimum, consider the mean or average performance and some estimate of the dispersion or variation (maybe even calculate the standard deviation). Trends and cycles can also be very revealing. Process capabilities can be calculated once there is performance data.

Analyze: The upstream variables (Xs) for each CTQ are identified with a flowchart. Once the project is understood, the baseline performance documented, and opportunities for improvement identified, an analysis of the process is appropriate. In this step, the Six Sigma approach applies statistical tools to validate root causes of problems. Any number of tools and tests can be used. The objective is to understand the process at a level sufficient to formulate options for improvement. The available options should be compared to determine the most promising alternatives. As with many activities, balance must be achieved. The

Six Sigma Black Belt is invaluable in striking the right balance by determining the extent to which each upstream variable influences each CTQ variable. This relationship may be expressed as:

$$CTQ = f(X_1, X_2, X_3, \dots X_k).$$

Where CTQ is the characteristic important to customers as clarified in the measure phase. X = ith variable that is hypothesized to influence the CTQ.

Improve: During the improve step of the Six Sigma approach, ideas and solutions are put to work. The Six Sigma Black Belt has discovered and validated all known root causes for the existing opportunity. The Six Sigma approach requires Black Belts to identify solutions. Few ideas or opportunities are so good that all are an instant success. As part of the Six Sigma approach, there must be checks to assure that the desired results are being achieved. Some designed experiments and trials may be required in order to find the best solution (i.e., the levels of the Xs that optimize the CTQ). When making trials and experiments, it is important that all project associates understand that these are trials and are part of the Six Sigma approach.

Control: Successful process revisions are formalized in conjunction with risk management and mistake proofing. Processes are documented and process owners are fully trained to minimize the possibility of variation returning to earlier levels. Process owners are taught the Kaizen approach that seeks to make everything incrementally better on a continuous basis. The sum of all these incremental improvements can be quite large. As part of the Six Sigma approach, performance tracking mechanisms and measurements are in place to assure that, at a minimum, the gains made in the project are not lost over a period of time. As part of the control step we encourage sharing with others in the organization. With this, the Six Sigma approach really starts to create phenomenal returns, ideas and projects in one part of the organization are translated in a very rapid fashion to implementation in another part of the organization.

DMADV steps

- Define the goals of the design activity. What is being designed? Why? Use QFD or analytic hierarchical process to assure that the goals are consistent with customer demands and enterprise strategy.
- Measure: Determine critical to stakeholder metrics. Translate customer requirements into project goals.
- Analyze the options available for meeting the goals. Determine the performance of similar best-in-class designs.
- Design the new product, service, or process. Use predictive models, simulation, prototypes, pilot runs to validate the design concept's effectiveness in meeting goals.
- Verify the design's effectiveness in the real world.

Six Sigma tools

As Six Sigma teams define and study business-related problems by the DMAIC or DMADV methodology, they use a variety of analytical tools and techniques. They include applications of:

- Mean, mode, median,
- Measurements of central tendency,
- Range, variance, and standard deviation,
- Graphical analysis,
- Histogram, run charts, Pareto charts,
- Correlation studies,
- Process mapping,
- The XY matrix,

- Measurement system analysis,
- Process capability tools,
- Multivariative study,
- Hypothesis testing,
- Failure mode effect analysis, and
- Design of experiments.

Roles in Six Sigma leadership

Six Sigma has several specific roles and responsibilities. The senior executive provides the impetus, direction, and alignment necessary for Six Sigma's ultimate success; must lead the executive committee in linking strategies to Six Sigma projects using a "dashboard". He/she has the all-important responsibility of leading

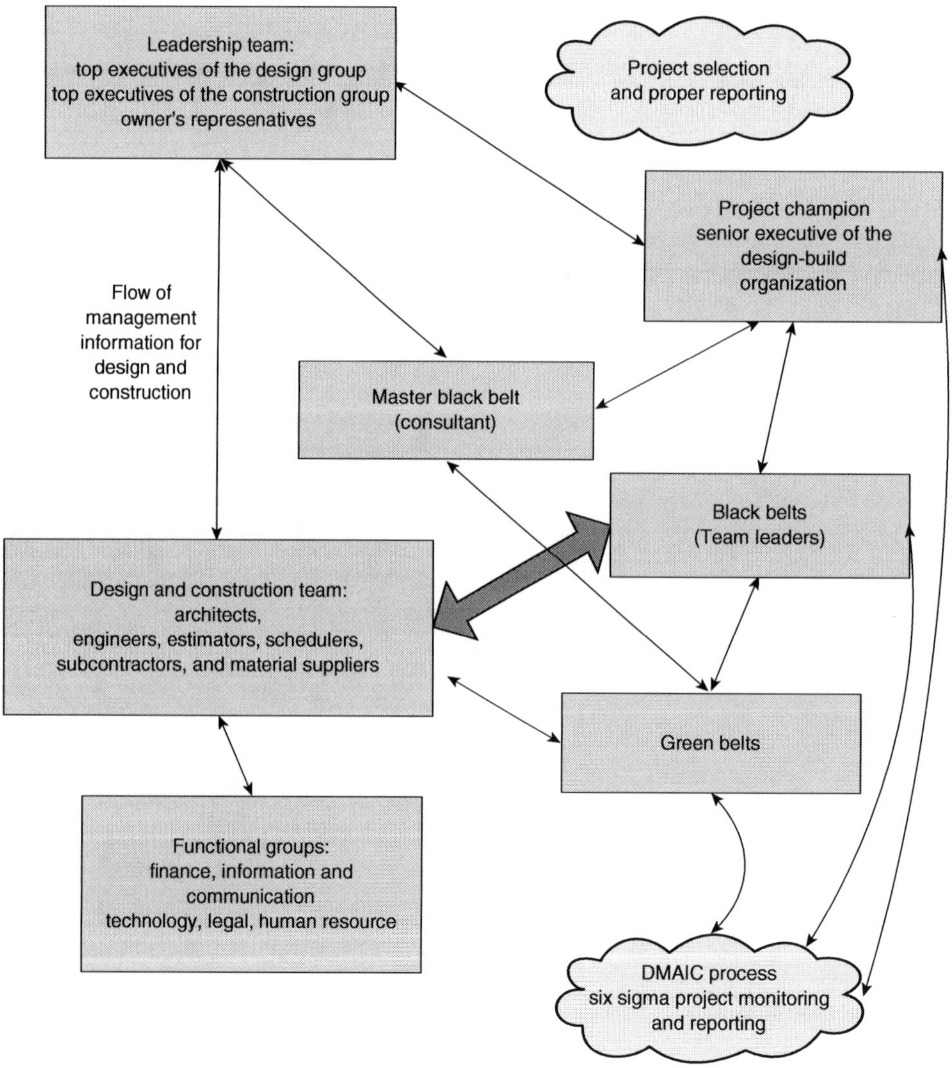

Figure 11.7 Six Sigma structure for a construction project

the overall initiative; must provide the resources necessary for success, and must approve the projects that are to be pursued by Black Belts and their teams. Very importantly, the senior executive must maintain an overview of the system to avoid sub-optimization and conduct project reviews.

The executive committee members must deploy Six Sigma throughout the organization, using dashboards; Prioritize and manage the Six Sigma portfolio; assign champions, black belts, and green belts to projects; and remove barriers and provide resources.

Champions are responsible for Six Sigma implementation across the organization in an integrated manner. The executive leadership draws them from upper management. Champions act as mentors to Black Belts, identify projects' impact on the organization, develop and negotiate project objectives and charters with the executive committee; remove political barriers or resource constraints, and keep the team focused on the project.

Master Black Belts, identified by champions, act as in-house coaches on Six Sigma. They devote 100% of their time to Six Sigma. They assist Champions and guide Black Belts and Green Belts. Apart from statistical tasks, their time is spent on ensuring consistent application of Six Sigma across various functions and departments.

Black Belts operate under Master Black Belts to apply Six Sigma methodology to specific projects. They devote 100% of their time to Six Sigma. They primarily focus on Six Sigma project execution, whereas Champions and Master Black Belts focus on identifying projects/functions for Six Sigma.

Green Belts are the employees who take up Six Sigma implementation along with their other job responsibilities. They operate under the guidance of Black Belts. Their assignments last for the duration of a Six Sigma project – typically 3–6 months, then return to their regular job responsibilities.

Process owners are the manager of a process. He or she has the responsibility for the process and has the authority to change it on his or her signature. The process owner must be accountable for the monitoring, managing, and output of his/her process; should empower the employees who work in the process to follow and improve best practices; and should accept, manage, and sustain the improved processes after the completion of the Six Sigma project.

Construction applications

The factors listed in Table11.2 are examples of variables in the construction environment that can be evaluated and improved in a Six Sigma program. For example, under the heading of lost time for accidents and delays, the Six Sigma team would collect data on the level of occurrence and the number of lost hours. Brainstorming could identify possible causes within the respective factors. Black belts can use statistical techniques to verify causality between upstream variables and these CTQ variables. Corrective measures would be implemented in operational procedures. Improved communication between upstream crews and successor crews could increase downstream awareness of potentially unsafe conditions. Procedural improvements would have upstream crews place safety barriers at critical areas and send emails to successor crews.

Deploying Six Sigma

There are a variety of strategies for implementing Six Sigma. The adoption of the Six Sigma methodology has to start with the leadership of a design or construction organization. As stated previously, the senior executive provides the impetus, direction, and alignment necessary for Six Sigma's ultimate success. The adoption of any major initiative such as Six Sigma, TQM, or lean construction must be integrated with the organization's strategic goals. These goals are typically documented in an organization's strategic plan; it serves as a road map to executives, managers, and employees for the actions they need to take to secure superior business performance.

Table 11.2 Summary of typical performance factors and measures

Performance Factors	*Performance Measures*
Construction quality	Customer satisfaction index Customer complaints Defects, failures, failed inspections Number of punch list items Warranty costs
Production issues	Lost time (accidents, delays)
	Rework costs (time, material)
	Material waste costs
	Percentage of projects complete (PPC)
	Labor productivity (PAR values)
Project delivery	Actual versus projected completion Actual cost versus projected cost
Project administration	Direct costs
	Rework labor hours
	Labor costs
	Change orders (number by source)
Human resources	Employee satisfaction index
	Training costs
	Number of employee suggestions
	Labor evaluation scores

Six Sigma adoption requires a long-term focus (although it produces results faster than TQM or ISO 9000) as it requires commitment to the cost of training and consultant support. As with other components of the strategic planning process, there should be goals, objectives, and expectations for a Six Sigma program. Responsibilities and time frames should be established and clearly communicated to all stakeholders. The following steps need to be taken in relation to the strategic plan:

1 Align executives and staff to Six Sigma objectives, deliverables, and time frames.
2 Mobilize by selecting project teams. Select projects based on importance and/or cost-benefit criteria.
3 Prepare and educate team members by conducting training (through consultant support).
4 Set up reporting systems to track team performance and ensure expectations are met.
5 Manage the Six Sigma process with an ongoing review at the senior executive level.

Some organizations have a corporate group that provides analytical skills – people are trained as black belts, green belts, and so on and are deployed to projects as needed. In some cases, people may be drawn from various business units, receive training, serve for a number of Six Sigma projects and then return to the respective business units. The Six Sigma culture can be gradually spread through an organization in this manner.

According to Gitlow (2006) there are five stages in the life of Six Sigma teams:

1 Forming: Team members get to know each other and establish ground rules.
2 Storming: Reflects efforts by the team to align goals and priorities in order to move ahead. Conflicts may arise and the direction of the team leader may be resisted. Team building exercises overcome these obstacles.
3 Norming occurs when team members display interdependence and develop cohesiveness. When common expectations are agreed for the team's activities.
4 Performing: The team works smoothly and directs combined efforts toward solving problems.
5 Adjourning: Teams disband voluntarily as their assignments have been completed.

The most successful teams direct their energies to the performing phase and lose very little time in the other four phases.

Case 1: Examples of Six Sigma application to construction

The Six Sigma process was applied successfully in the construction of the New Doha International Airport (NDIA) project in Qatar. The first phase of the project included the preparation of 22 square kilometers of land, of which approximately half was reclaimed from the sea. The scope included having 60 million cubic meters of sand and rockfill placed within 24 months for a contract value of E337 million. The hydraulically reclaimed fill was stabilized through a combination of compaction techniques, including high energy impact compaction (HEIC). The process of this novel compaction technique was optimized using the Six Sigma quality improvement scheme. It was first documented in Terra et Aqua Number 103, June 2006, by four engineers affiliated with the project (Avsar *et al.* 2006). Four companies partnered in the project. The major challenges of this project were an extremely short-schedule, requiring over 65 million cubic meters of fill to be placed in less than 24 months, with many important early milestone handover dates.

Why Six Sigma

The reclamation fill had many variables: quality, gradation, water content, and layer thickness; each condition requiring a varying number of passes by compaction rollers. In order to attain a uniform compaction in a non-uniform environment on an extremely congested site, a high level of QC was needed. FAA standards were incorporated in values to be met by a cone penetration test, in-situ dry density, and settlement at a specified design bearing pressure.

Overview of the process

Three techniques were used in the soil compaction process:

1 Hydraulic compaction, by the action of bulldozers driving over deposited material.
2 High energy impact compaction (HEIC) involves a non-circular module that is towed by a tractor and rises and falls repeatedly, striking the ground in the process. This method has a low frequency but high amplitude, generally leading to a greater penetration depth.
3 Conventional vibratory rollers: The impact rollers were driven in fixed patterns, changing from a clockwise direction to counterclockwise every 10 passes in order to achieve greater uniformity over the compaction area. The optimum speed of the impact rollers was determined to be 10–12 kph.

Implementing Six Sigma

The **DMAIC** process was used as a guideline to ensure that relevant data were collected, analyzed, and converted into information. The tools used in the framework of the Six Sigma implementation for the compaction process included: the compaction process map (or flowchart; see Figure 9.9) to identify potential causes, cause and effect diagram to generate a list of root causes, prioritization matrix of the most important root causes. The process map comprised four stages: initial placement with hydraulic compaction; watering, to provide optimal moisture content; rolling and compaction; and assessment, using geotechnical testing. Each stage involved field-based QC staff in direct contact with the operations foremen.

Root cause analysis

As a part of the Six Sigma deployment, a cause and effect diagram was prepared based on brainstorming by project staff from several operational functions. This was intended to identify possible root causes of variation in the compaction process in order to focus mitigating strategies on reducing it. Six categories were used to classify the ideas from the brainstorming exercise: manpower, methods, machines, measurements, materials, and geo-environment. The most important causes were identified as: manpower, optimal driving speed (for operator-driven equipment); machines, impact rate; and methods, roller track width, timing of added water. The in-situ dry density was the bottom line "effect" of the cause and effect diagram (Figure 11.8). Since, it was also a contractual requirement, it was the best variable to illustrate the influence of

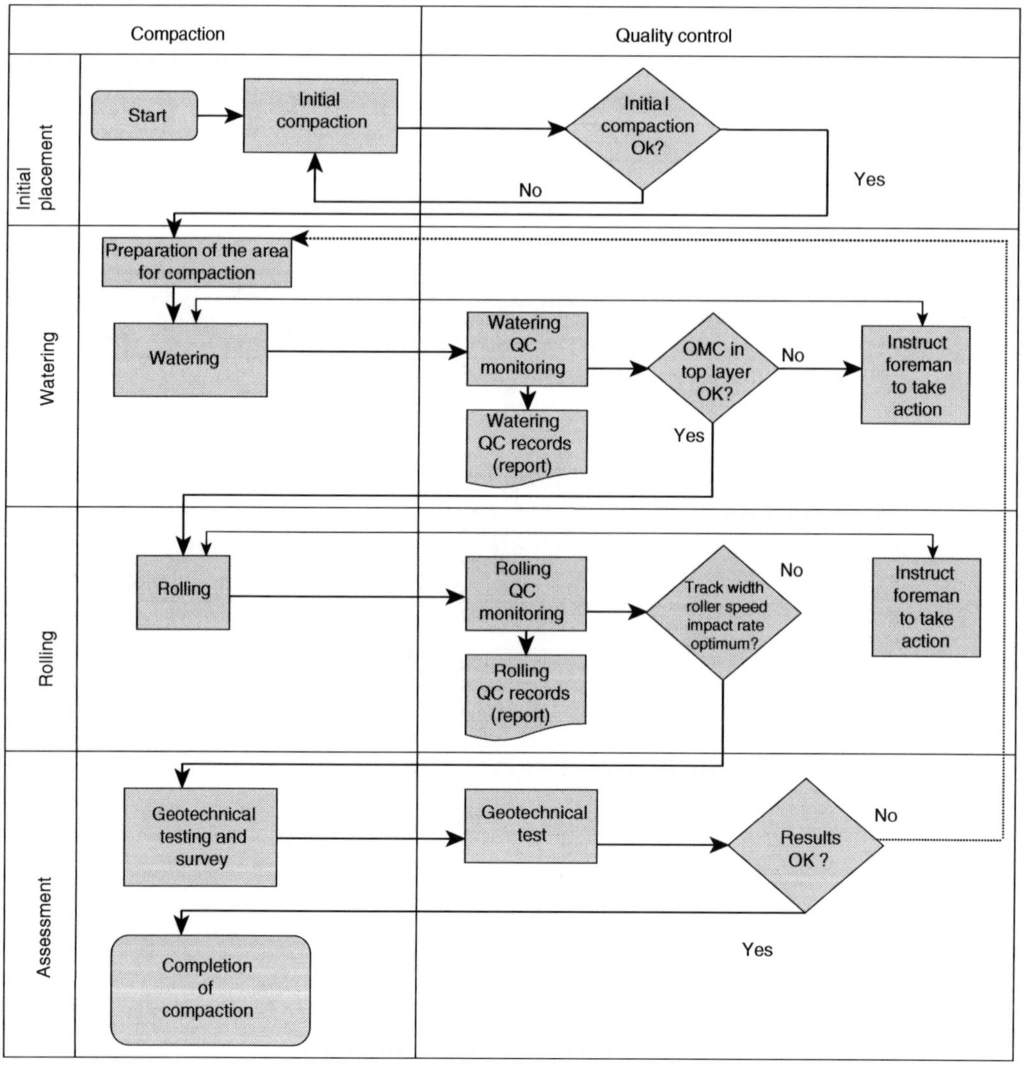

Figure 11.8 Map of the NDIA compaction process

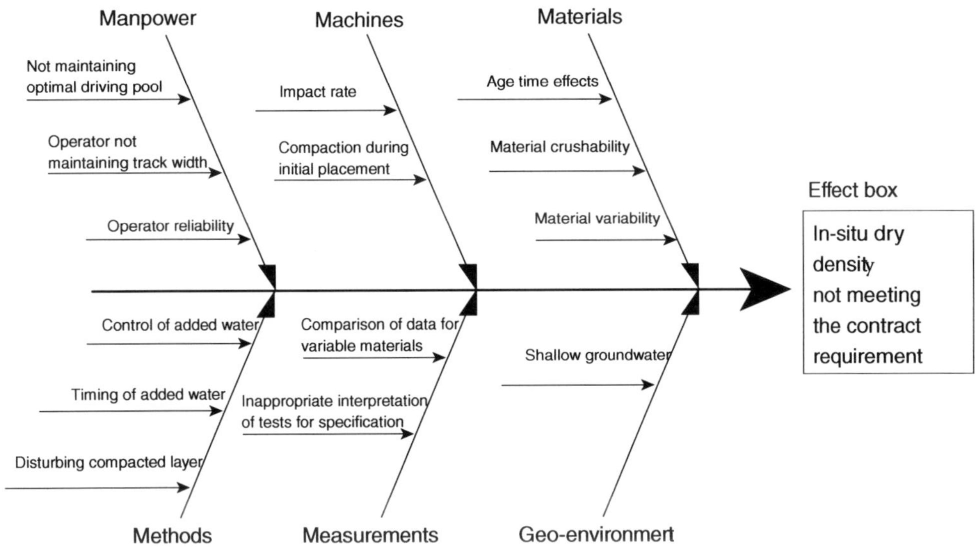

Figure 11.9 NDIA compaction cause and effect diagram

Six Sigma on the compaction process. It is noted that here, the in-situ dry density is expressed as a percentage of the maximum dry density (% MDD).

Prioritization matrix

The prioritized root causes were designated "upstream process indicators" (denoted by the symbol Xs). Optimum values and specification ranges were determined, as shown in Table 11.3. A procedure was established whereby quality controllers took ongoing readings and reported anomalies to the respective foremen. This facilitated prompt process adjustments, thus avoiding rework to achieve the required compaction levels. Two test areas were designated – an uncontrolled HEIC test area, and a Six-Sigma controlled area. In the controlled area, the variability of the sample results was significantly reduced and the process average increased, representing desirably higher values of dry density. As can be seen from Figure 11.10 the observations in the uncontrolled test area show more variability than the observations for Area B where the Six Sigma approaches were applied. The two areas are demarcated by a vertical dotted line. The mean compaction was higher for Area B – at 106.84% as compared with a value of 99% for the uncontrolled area. A direct result of the Six Sigma campaign on compaction was that the average in-situ dry density increased by 8% (from 99 to 107% MDD), and the variance decreased by 56% (from 83 to 36).

Table 11.3 Optimum Values of Process Indicators Xs

Upstream Process Indicators Xs	*Optimum Value*	*Specification Range*
Roller speed	11 kph	10–12 kph
Impact rate	2 impact/sec	1.8–2.2 impact/sec
Roller track width	2.6 m/track	2.4–2.8 m/track
Timing of water	20 min/watering session	15–25 min/watering session

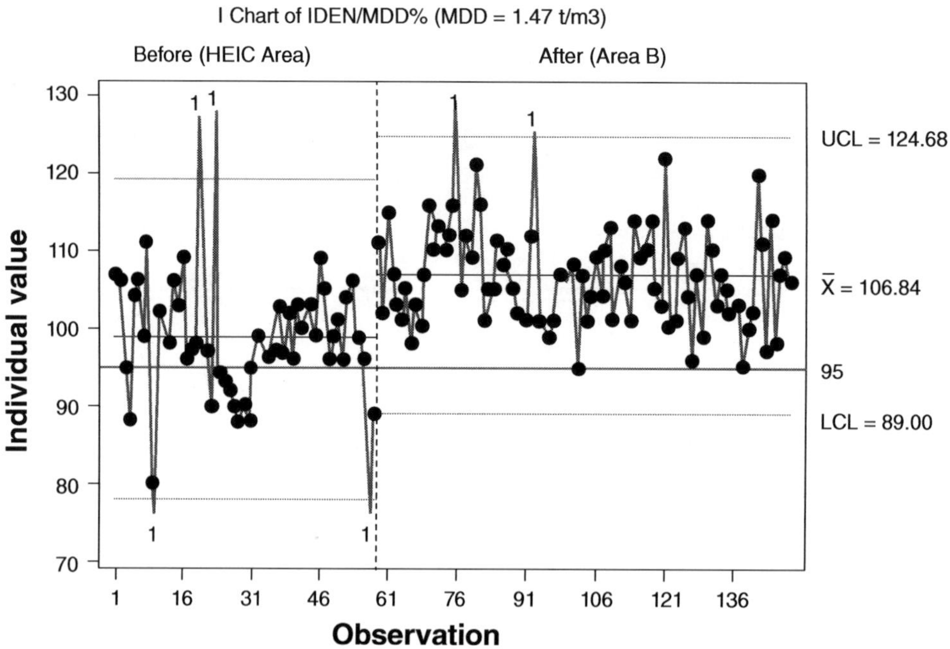

Figure 11.10 Values of dry density for the airfield substrate

Case 2: The DMAIC approach reduces costs for 'Construction XYZ' (Provided by Gilles Bélanger, MPM, PMP, MBB)

Introduction

Based in the Province of Quebec, Canada, *CONSTRUCTION XYZ* is an important player in the commercial, industrial and institutional field. They run most of their business in a French-speaking environment.

Market pressure in the Construction sector leads the organization to improve its internal processes. This story is about a CONSTRUCTION XYZ engineer using the DMAIC approach for improving a main process in the Construction industry: "Manage Change Notice".

Define

At the very beginning, a project charter was developed as shown in part below order to clarify the objectives of the study.

The other elements of a project charter are: Scope, Other deliverables, Budget, Schedule, Team and Communication plan.

Lead Time for a Notice Change (CN) in the construction industry is made of internal and external waiting time. The reality is that some time is required to process the Change Notice internally, but the owner representative must also approve the CN.

In every process improvement project, one of the first actions is to draw the SIPOC which establishes the perimeter of the project. It is often helpful to start first with the SIPOC. It may ease the task of building the Project Charter.

Table 11.4 Project description and objectives

SECTION	*DESCRIPTION*
BUSINESS CASE	CONSTRUCTION XYZ aims at reducing both Lead time and Process time of the Change Notice process which is one the company main process. **Year** **Quantity of Change Notice** 2016 1243 2017 1680 2018 Incomplete data
GOAL	Reduce CN management Lead Time to improve efficiency of construction projects
OBJECTIVE	Reduce the CN Lead Time by 40% for the period 2018–05–02 to 2019–01–01 compared to 2016–2017. The average CN Lead Time for 2016–2017 is 31 days, Process Time average for the same period is 1:38 h. The objective is to reduce the Lead Time to 19 days and the Process Time to 59 minutes.
Kpi'S	KPI's for this project are : **Lead Time** & **Process Time** for CN management, (Lead Time = Process Time + waiting time)

Table 11.5 Preparation of the SIPOC

SIPOC NAME:	*CREATE A CHANGE NOTICE (CN)*		*DATE:*	*31-Aug-18*
SUPPLIERS	*INPUTS*	*PROCESS*	*OUTPUTS*	*CUSTOMERS*
• Owner, • Subcontractors & Suppliers, • Supervisors, • Project Managers, • Professionals, • Government Agencies	• Change Description on CN Form, • Quotes from Subcontractors & Suppliers, • Invoices, • Subject Matter Expert Opinions, • Daily Report & Manpower Follow-Up, • Standards, Codes & Regulations	Analyze CN ↓ Assess resource requirements ↓ Ask quote to Subcontractors & Suppliers ↓ Feed database & prepare CN ↓ Revise CN ↓ Obtain CN approval from Project Manager ↓ Approve CN	• CN Approved or Refused • Agreement with Customer • Updated Payment Request • Contract Modifications • Budget Changes	• Owner, • Professionals, • Project Managers, • Subcontractors & Suppliers, • Administrative Assistant • Supervisor

Measure

Once the project limits were shared and approved, the next step was to understand the 'CREATE A CHANGE NOTICE (CN)' process in detail. The team used a large sheet of Kraft paper, several Post-it and shared their views about how that process steps were linked together. They searched for the process attributes to feed the PROCESS SHEET which is an enhanced SIPOC and is used to define the scope of a process and the process owner, process purpose and process scope, the customer requirements, related documents and process KPI's.

The team collected the time for each CN related task information to create the Pareto chart below:

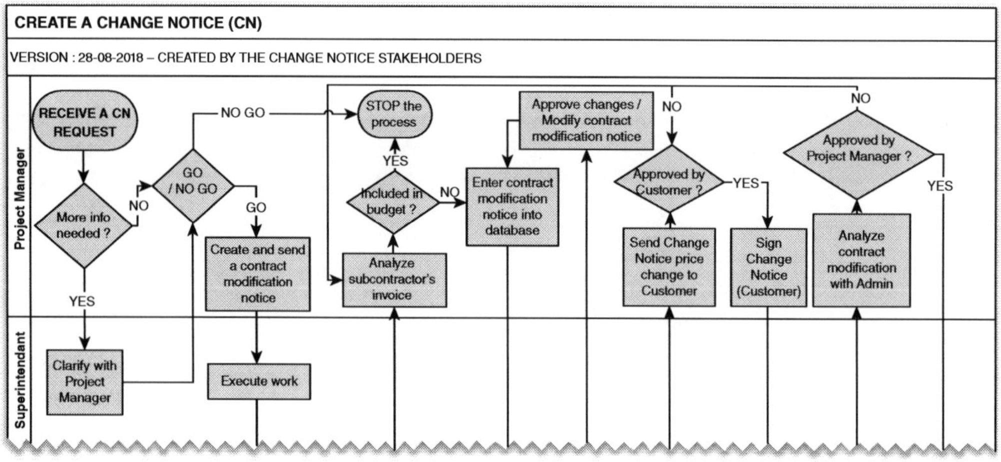

Figure 11.11 Flowchart for CREATE A CHANGE NOTICE (CN)

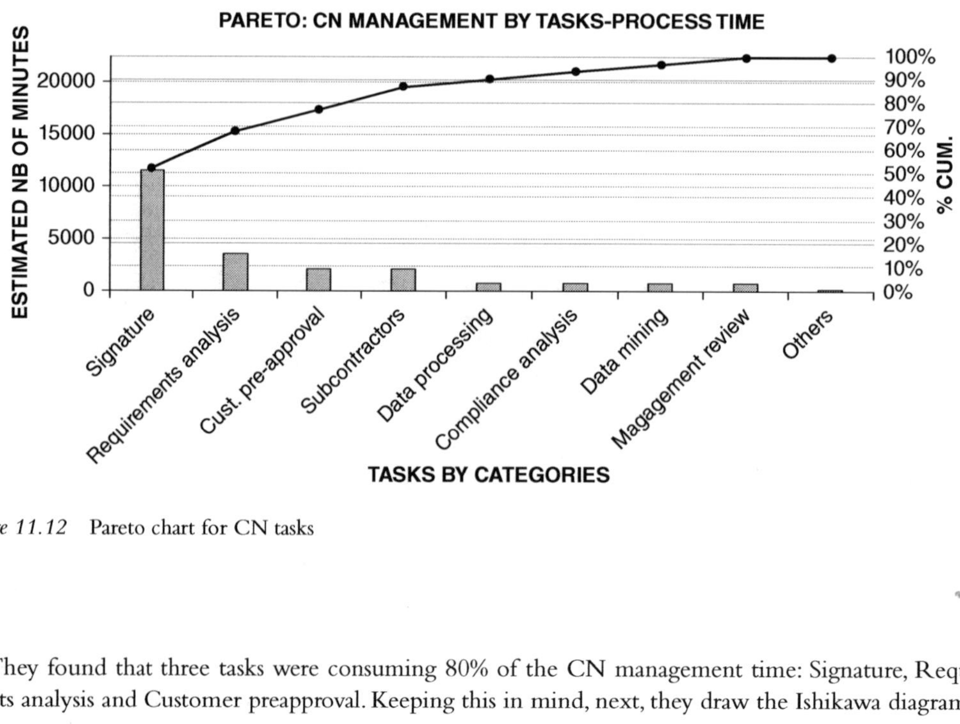

Figure 11.12 Pareto chart for CN tasks

They found that three tasks were consuming 80% of the CN management time: Signature, Requirements analysis and Customer preapproval. Keeping this in mind, next, they draw the Ishikawa diagram.

Analyze

Exploring the potential causes with the Ishikawa (AKA fishbone or Cause & Effect diagram):

Only 10 of the 13 potential causes were kept as real causes using a validation matrix.

The team used the 5 why's to find the root causes (partly shown below, see 'Improve' for root cause full list):

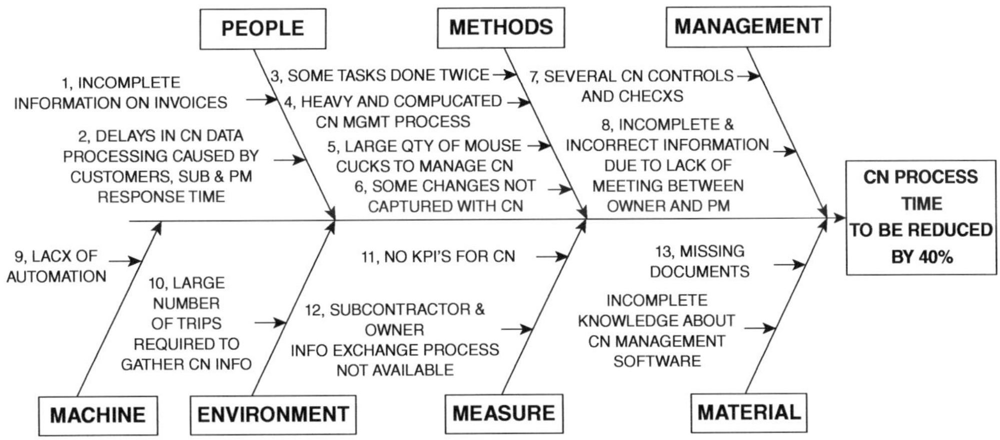

Figure 11.13 Ishikawa for 'CN process time to be reduced by 40%

Table 11.6 5 WHY root cause identification

5 WHY's: IDENTIFICATION OF ROOT CAUSES FOR CN TIME CONSUMPTION					
#	*REAL CAUSES*	*WHY #2*	*WHY #3*	*WHY #4*	*WHY #5*
1	Incomplete information on invoices	Cn applications and requirements provided by pm are not complete nor accurate	Little or no definition of CN requirements provided by project management leaders	**No requirements for cost breakdowns by tasks and specialties.**	
2	....	....	....	....	....

Six Sigma project leaders need to understand the importance of a direct link between each of the Why's as well as between the stated problem and its potential cause in the Ishikawa diagram. The following rules may help to generate root causes from which you can create efficient countermeasures:

1- You must have a direct link between each Why.
2- Avoid writing a why that is a disguised solution.
3- Stay away from any Why that is totally out of influence of your team.
4- Finally, you should be able to see a direct link by reading 'Why's' from the first to the last and return.

Improve

In this phase, the team using decision matrix, decided for the best countermeasures (or solutions) to address each of the root causes.

The effects of the Improvement Action Plan are shown below:

1 CN Lead time reduction was projected to be -25.9% which represents 65% of initial target
2 CN Process time reduction was projected to be 55.6% and that is 139% of the initial target

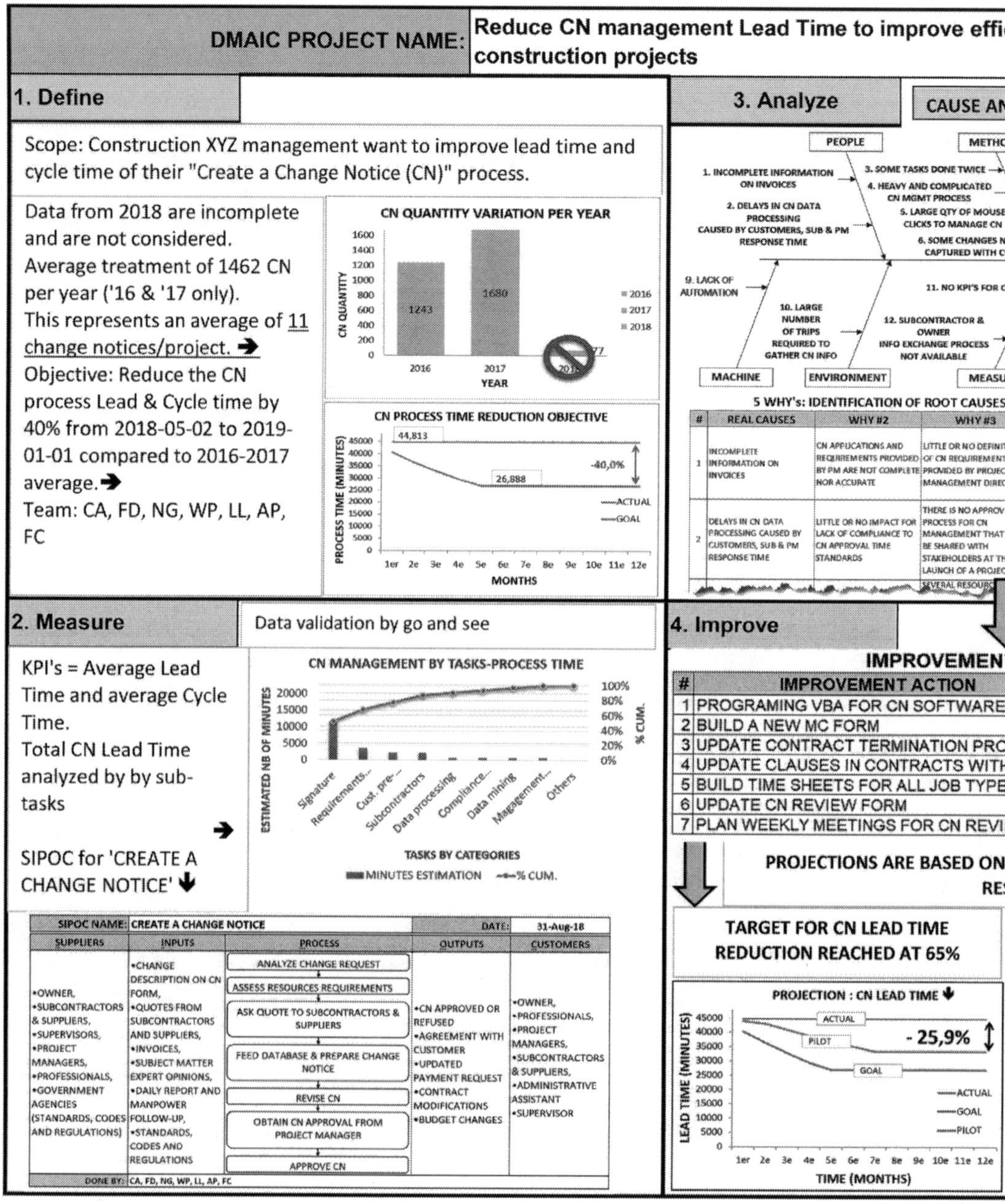

Figure 11.14 One page summary of the DMAIC process

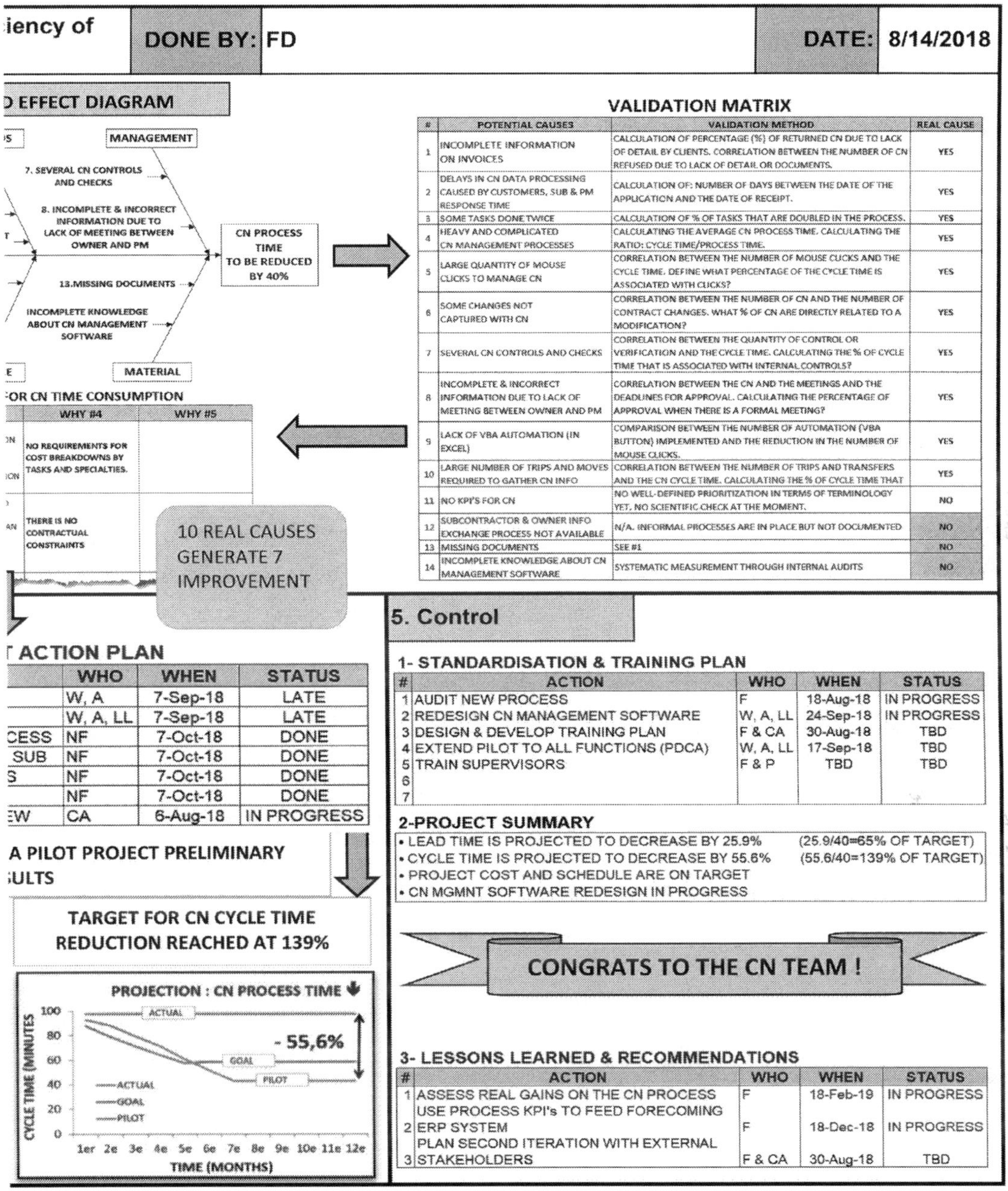

¡iency of

DONE BY: FD

DATE: 8/14/2018

D EFFECT DIAGRAM

S | MANAGEMENT

7. SEVERAL CN CONTROLS AND CHECKS

8. INCOMPLETE & INCORRECT INFORMATION DUE TO LACK OF MEETING BETWEEN OWNER AND PM

13.MISSING DOCUMENTS

INCOMPLETE KNOWLEDGE ABOUT CN MANAGEMENT SOFTWARE

E | MATERIAL

CN PROCESS TIME TO BE REDUCED BY 40%

VALIDATION MATRIX

#	POTENTIAL CAUSES	VALIDATION METHOD	REAL CAUSE
1	INCOMPLETE INFORMATION ON INVOICES	CALCULATION OF PERCENTAGE (%) OF RETURNED CN DUE TO LACK OF DETAIL BY CLIENTS. CORRELATION BETWEEN THE NUMBER OF CN REFUSED DUE TO LACK OF DETAIL OR DOCUMENTS.	YES
2	DELAYS IN CN DATA PROCESSING CAUSED BY CUSTOMERS, SUB & PM RESPONSE TIME	CALCULATION OF: NUMBER OF DAYS BETWEEN THE DATE OF THE APPLICATION AND THE DATE OF RECEIPT.	YES
3	SOME TASKS DONE TWICE	CALCULATION OF % OF TASKS THAT ARE DOUBLED IN THE PROCESS.	YES
4	HEAVY AND COMPLICATED CN MANAGEMENT PROCESSES	CALCULATING THE AVERAGE CN PROCESS TIME. CALCULATING THE RATIO: CYCLE TIME/PROCESS TIME.	YES
5	LARGE QUANTITY OF MOUSE CLICKS TO MANAGE CN	CORRELATION BETWEEN THE NUMBER OF MOUSE CLICKS AND THE CYCLE TIME. DEFINE WHAT PERCENTAGE OF THE CYCLE TIME IS ASSOCIATED WITH CLICKS?	YES
6	SOME CHANGES NOT CAPTURED WITH CN	CORRELATION BETWEEN THE NUMBER OF CN AND THE NUMBER OF CONTRACT CHANGES. WHAT % OF CN ARE DIRECTLY RELATED TO A MODIFICATION?	YES
7	SEVERAL CN CONTROLS AND CHECKS	CORRELATION BETWEEN THE QUANTITY OF CONTROL OR VERIFICATION AND THE CYCLE TIME. CALCULATING THE % OF CYCLE TIME THAT IS ASSOCIATED WITH INTERNAL CONTROLS?	YES
8	INCOMPLETE & INCORRECT INFORMATION DUE TO LACK OF MEETING BETWEEN OWNER AND PM	CORRELATION BETWEEN THE CN AND THE MEETINGS AND THE DEADLINES FOR APPROVAL. CALCULATING THE PERCENTAGE OF APPROVAL WHEN THERE IS A FORMAL MEETING?	YES
9	LACK OF VBA AUTOMATION (IN EXCEL)	COMPARISON BETWEEN THE NUMBER OF AUTOMATION (VBA BUTTON) IMPLEMENTED AND THE REDUCTION IN THE NUMBER OF MOUSE CLICKS.	YES
10	LARGE NUMBER OF TRIPS AND MOVES REQUIRED TO GATHER CN INFO	CORRELATION BETWEEN THE NUMBER OF TRIPS AND TRANSFERS AND THE CN CYCLE TIME. CALCULATING THE % OF CYCLE TIME THAT	YES
11	NO KPI'S FOR CN	NO WELL-DEFINED PRIORITIZATION IN TERMS OF TERMINOLOGY YET. NO SCIENTIFIC CHECK AT THE MOMENT.	NO
12	SUBCONTRACTOR & OWNER INFO EXCHANGE PROCESS NOT AVAILABLE	N/A. INFORMAL PROCESSES ARE IN PLACE BUT NOT DOCUMENTED	NO
13	MISSING DOCUMENTS	SEE #1	NO
14	INCOMPLETE KNOWLEDGE ABOUT CN MANAGEMENT SOFTWARE	SYSTEMATIC MEASUREMENT THROUGH INTERNAL AUDITS	NO

FOR CN TIME CONSUMPTION

	WHY #4	WHY #5
N ... ION	NO REQUIREMENTS FOR COST BREAKDOWNS BY TASKS AND SPECIALTIES.	
AN	THERE IS NO CONTRACTUAL CONSTRAINTS	

10 REAL CAUSES GENERATE 7 IMPROVEMENT

T ACTION PLAN

	WHO	WHEN	STATUS
	W, A	7-Sep-18	LATE
	W, A, LL	7-Sep-18	LATE
CESS	NF	7-Oct-18	DONE
SUB	NF	7-Oct-18	DONE
S	NF	7-Oct-18	DONE
	NF	7-Oct-18	DONE
EW	CA	6-Aug-18	IN PROGRESS

A PILOT PROJECT PRELIMINARY ULTS

TARGET FOR CN CYCLE TIME REDUCTION REACHED AT 139%

5. Control

1- STANDARDISATION & TRAINING PLAN

#	ACTION	WHO	WHEN	STATUS
1	AUDIT NEW PROCESS	F	18-Aug-18	IN PROGRESS
2	REDESIGN CN MANAGEMENT SOFTWARE	W, A, LL	24-Sep-18	IN PROGRESS
3	DESIGN & DEVELOP TRAINING PLAN	F & CA	30-Aug-18	TBD
4	EXTEND PILOT TO ALL FUNCTIONS (PDCA)	W, A, LL	17-Sep-18	TBD
5	TRAIN SUPERVISORS	F & P	TBD	TBD
6				
7				

2-PROJECT SUMMARY

- LEAD TIME IS PROJECTED TO DECREASE BY 25.9% (25.9/40=65% OF TARGET)
- CYCLE TIME IS PROJECTED TO DECREASE BY 55.6% (55.6/40=139% OF TARGET)
- PROJECT COST AND SCHEDULE ARE ON TARGET
- CN MGMNT SOFTWARE REDESIGN IN PROGRESS

CONGRATS TO THE CN TEAM !

3- LESSONS LEARNED & RECOMMENDATIONS

#	ACTION	WHO	WHEN	STATUS
1	ASSESS REAL GAINS ON THE CN PROCESS	F	18-Feb-19	IN PROGRESS
2	USE PROCESS KPI's TO FEED FORECOMING ERP SYSTEM	F	18-Dec-18	IN PROGRESS
3	PLAN SECOND ITERATION WITH EXTERNAL STAKEHOLDERS	F & CA	30-Aug-18	TBD

Figure 11.14 (Continued)

Table 11.7 Improvement Action Plan

IMPROVEMENT ACTION PLAN				
#	IMPROVEMENT ACTION	WHO	WHEN	STATUS
1	Programing VBA for CN software	W, A	7-Sep-18	Late
2	Build a new MC form	W, A, LL	7-Sep-18	Late
3	Update contract termination process	NF	7-Oct-18	Done
4	Update clauses in contracts with sub	NF	7-Oct-18	Done
5	Build time sheets for all job types	NF	7-Oct-18	Done
6	Update CN review form	NF	7-Oct-18	Done
7	Plan weekly meetings for CN review	CA	6-Aug-18	In progress

Control

The first goal of the control phase is to sustain the gains. The audit process is often used to do that. This is also the right time to develop and provide training plans pertaining to the changes introduced in the 'Improve' phase.

Project cost and schedule are on target. CN management software redesign is in progress. Team and management were very happy with the results.

Reflecting on the lessons learned is quite often a neglected area for many projects. In this case the team is looking to re-assess the gains 6 months later to make sure they are sustained. They plan for a second iteration with external stakeholders, which is a collaborative way to work with suppliers and subcontractors.

Table 11.8 Standardization/training plan

STANDARDIZATION & TRAINING PLAN				
#	ACTION	WHO	WHEN	STATUS
1	Audit new process	F	18-Aug-18	In progress
2	Redesign CN management software	W, A, LL	24-Sep-18	In progress
3	Design & develop training plan	F & CA	30-Aug-18	TBD
4	Extend pilot to all functions (PDCA)	W, A, LL	17-Sep-18	TBD
5	Train supervisors	F & P	TBD	TBD

Table 11.9 Lessons learned and recommendations

LESSONS LEARNED & RECOMMENDATIONS				
#	ACTION	WHO	WHEN	STATUS
1	Assess real gains on the CN process	F	18-Feb-19	In progress
2	Use process KPI's to feed forthcoming ERP system	F	18-Dec-18	In progress
3	Plan second iteration with external stakeholders	F & CA	30-Aug-18	TBD

Conclusion

The reader may ask if a rigorous methodology like DMAIC is applicable to organizations in the Construction industry. Well, this example proves that it is for a complex process like CN management.

Some of the success factors are

1- Select projects that are improving whole organization performance
2- Management to provide proactive support (provide resources and maintain the momentum)
3- Management to encourage cultural changes and provide training with continuous improvement in mind

DMAIC One-pager may be seen a kind of subset of A3 family one-page reports that are used to improve process performance. The one below is a communication tool that can be used in all phases of the DMAIC project. You may use it to conduct stand-up meetings to share the progress, constraints and roadblocks with all interested parties.

Conclusion

Both TQM and Six Sigma represent change in the way a design or construction organization carries out its mission. It is change for the better, but will come only to those who want to change and are willing to make the necessary commitment. Purists may take the position that TQM or its variants should not be attempted unless all the necessary ingredients are present, especially top management commitment. But such a textbook scenario is more the exception than the norm. It is better for an organization to make an incremental step toward improving quality than to maintain the status quo. Examples abound of quality initiatives that began at the middle management level, with small successes that were eventually sold upward and embraced by champions at the top management level (Forbes 1994). Organizations that have adopted Quality Management systems with some success will have acquired knowledge and tools that prepare them better than others to venture into lean construction.

Questions for discussion

1 Many construction projects are awarded based on the lowest bid. How is that practice viewed by the Deming philosophy?
2 Provide examples of three quality costs in construction projects.
3 What are the main barriers to TQM implementation? How can they be reduced, if not eliminated?
4 Discuss the similarities and differences between the objectives of TQM and the objectives of Lean Construction.
5 Does the prior existence of a TQM program in a construction organization inhibit the adoption of Lean Construction? Discuss.
6 Is Six Sigma easier to deploy in construction than TQM?
7 In what way does Six Sigma vary from the lean construction philosophy?
8 How can Six Sigma support lean construction applications?

References

Ahmed, S. M., and R. T. Aoieong. 1998. Analysis of quality management systems in the Hong Kong construction industry. *Proceedings of the 1st South African international conference on total quality management in construction,* November 22–25. Cape Town, South Africa, 37–49.

Avsar, S., M. Bakker, G. Bartholomeeusen, and J. Vanmechelen. 2006. *Six Sigma quality improvement of compaction at the new Doha International Airport Project/Terra et Aqua*, 14–22. Number 103. June.

Burati, Jr., J. L., and T. H. Oswald. 1992. Implementing TQM in engineering and construction. *Journal of Management in Engineering, ASCE* 9(4): 456–470.

Burati, Jr., J. L., and T. H. Oswald. 1993. Implementing TQM in engineering and construction. *Journal of Management in Engineering, ASCE* 9(4): 456–470.

Chase, G. W., and M. O. Federle. 1992. Implementation of TQM in building design and construction. *Journal of Management in Engineering, ASCE* 9(4): 329–339.

Crosby, P. 1979. *Quality is free: The art of making quality certain*. New York: Mentor.

Deming, W. E. 1986. *Out of the crisis*. Cambridge, MA: Massachusetts Institute of Technology.

Forbes, L. H. 1994. What do you do when your organization isn't ready for TQM? *National Productivity Review* 13(4), (Autumn): 467–478.

Gitlow, H. 2006. *Design for six sigma for green belts and champions*. Upper Saddle River, NJ: Prentice Hall.

Goetsch, D. L., and S. B. Davis. 2016. *Quality management for Organizational excellence: Introduction to total quality (8th Edition)*. Hoboken, NJ: Pearson Education Inc.

Harris, C. R. 1995. The evolution of quality management: An overview of the TQM literature. *Canadian Journal of Administrative Sciences* 12(2): 95–105.

Hayden, W. M. 1996. Connecting random acts of quality: Global system standard. *Journal of Management in Engineering, ASCE* 12(3): 34–44.

Juran, J. M. 1988. *Juran on planning for quality*. New York: Free Press.

Oberlender, G. D. 1993. *Project management for engineering and construction*. New York: McGraw-Hill.

Bibliography

Ahmed, S. M. 1993. *An integrated total quality management (TQM) model for the construction process*. PhD Dissertation. School of Civil & Environmental Engineering. Atlanta, GA: Georgia Institute of Technology.

Ahmed, S. M., L. P. Sang, and Z. M. Torbica. 2003. Use of quality function deployment in civil engineering capital project planning. *Journal of Construction Engineering & Management, ASCE* 129(4): 358–368.

Ahmed, S. M., P. Tang, S. Azhar, and I. Ahmad. 2002. An evaluation of safety measures in the Hong Kong construction industry based on total quality management principals. *CIB-W65/W55 10th international symposium construction innovation and global competitiveness*, September 6–10. Cincinnati, OH, 1214–1227.

Aoieong, R. T., S. L. Tang, and S. M. Ahmed. 2002. A process approach in measuring quality costs of construction projects: Model development. *Construction Management and Economics* 20(2): 179–192.

Arditi, D., and H. M. Gunaydin. 1998. Factors that affect process quality in the life cycle of building projects. *Journal of Construction Engineering and Management, ASCE* 124(3): 194–203.

Associated General Contractors of America. 1993. *Implementing TQM in a construction company*. Publication No. 1211. Alexandria, VA: The Associated General Contractors of America.

Besterfield, D. H. 1994. *Quality control*. Upper Saddle River, NJ: Prentice Hall.

Besterfield, D. H., C. Besterfield-Michna, G. Besterfield, and M. Besterfield-Sacre. 2003. *Total quality management*. Upper Saddle River, NJ: Prentice Hall.

Brown, S. 1992. *Total quality service*. Scarborough, ON: Prentice Hall Canada Inc.

Brown, S. 1995. *What customers value most*. New York: John Wiley and Sons.

Crosby, P. B. 1984. *Quality without tears*. New York: New American Library.

Deming, W. E. 1982. *Quality, productivity, and competitive position*. Cambridge, MA: Massachusetts Institute of Technology.

Deming, W. E. 1993. *The new economics*. Cambridge, MA: Massachusetts Institute of Technology.

Forbes, L. H. 1999. *An engineering management-based investigation of owner satisfaction, quality and performance variables in health care facilities construction*. PhD Dissertation. Coral Gables, FL: University of Miami.

Snee, R., and R. Hoerl. 2003. *Leading six sigma*. Financial Times. Upper Saddle River, NJ: Prentice Hall.

Tenner, A. R., and I. J. DeToro. 1992. *Total quality management*. Reading, MA: Addison-Wesley.

TQM Concept Map. 2003. Online at http://soeweb.syr.edu/faculty/takoszal/TQM.html

12 SUSTAINABLE CONSTRUCTION

Environmental sustainability and commissioning

This chapter covers two important topics: Part A addresses environmental sustainability in the construction environment, and Part B describes building commissioning that is used to support the design and construction of buildings.

Part A: Sustainability

Lean construction and green buildings

The First International Conference on Sustainable Construction, in Tampa, Florida, in 1994, was defined as "the creation and responsible management of a healthy built environmental base on resources efficient and ecological principles" (Kibert 1994). A sustainable condition for this planet as one in which there is stability for both social and physical systems, achieved through meeting the needs of the present without compromising the ability of future generations to meet their own needs.

According to the United States Energy Information Administration (EIA), US primary energy consumption increased by 4% in 2018 and superseded the previous record set in 2007 by 0.3%. The 2018 energy consumption increase was the largest since 2010 both in absolute and percentage terms.

The main driver was the consumption of fossil fuels such as petroleum, natural gas, and coal as it grew by 4% in 2018 and accounted for 80% of the total primary energy consumption. More specifically, natural gas consumption rose by 10% from 2017, driven by weather-related factors and by the power sector (higher demand for space heating and for air conditioning and for power generation) and offset a 4% decline in coal consumption. Coal-fired power plants continued to be replaced with gas-fired power plants. Initiatives such as Leadership in Energy and Environmental Design (LEED) established by the US Green Building Council (USGBC), Energy Star, and others promote green architecture and sustainable construction.

The worldwide consumption of fossil fuels is escalating rapidly, as is the presence of greenhouse gases. In 2008, the world consumption of energy was 15 terawatts (1 terawatt = 1,000 gigawatts). The world demand for oil was 31 billion barrels of oil per year. The growth in demand is forecasted to be 2% per year to 2025. This escalating situation suggests a need for prompt action by the construction industry to emphasize the design and construction of green buildings. Green buildings promote and support the concept of lean construction. While lean construction involves the minimization of waste, ensuring that the resources used in the building process enhance the value chain for a completed facility, green building design is based on design principles that promote environmentally beneficial long-term operation. Overall, a combination of these approaches significantly enhances sustainability. Not only is a lean building assembled using fewer resources (human, financial, energy, material, etc.), but it has a lower environmental impact in its on-going operation as well. Green practices may even reduce initial construction costs. The recycling of construction material reduces the amount of physical waste generated in a project, and the amount of pollution associated with the process is reduced as well. Sorting construction waste on a site is a recommended practice that qualifies for LEED credits. It can also reduce haulage costs for waste disposal as some of that material can be reused.

LEED requires clear documentation of design strategies and construction practices – as credits involve both stages of a project. This leads to better collaboration, including designers and constructors as well as owners. Sustainable design, therefore can dovetail very effectively with lean construction – as designers and contractors collaborate on constructability issues.

Benefitting from the "Triple bottom line": Ballard (2016) posits that sustainability is now commonly understood in terms of the triple bottom line: social, environmental, and economic. In his view lean has a major influence on all elements of sustainability; while its central focus is economic it has other significant impacts. The emanation of solid waste and greenhouse gases from construction sites has begun to receive attention, but more effort is needed to promote the social impacts of lean on the quality of working life and the creation of employment opportunities.

Importance of sustainable construction

In the United States, buildings and their landscapes are a major source of pollution accounting for a third of all energy consumed, more than a third of the carbon dioxide emissions, and almost half the sulfur dioxide emissions produced in the country. According to studies by the US Green Building Council (USGBC 2015), in the United States buildings account for:

- 72% of total US electricity consumption,
- 24 to 50% of energy use,
- 38% of total US carbon dioxide emissions,
- 548 million tons of construction and demolition waste (Source: www.epa.gov/smm/sustainable-management-construction-and-demolition-materials),
- 14% of potable water consumption, and
- 40% (3 billion tons annually) of raw materials used globally.

The barriers to sustainable projects

The primary barrier to the introduction of sustainable construction is the widely held belief that sustainable construction is more expensive and involves a higher risk. The higher costs are thought to be due to new technologies, specialized materials, and additional design. In many projects, cost projections emphasize initial costs, with less attention to life-cycle costs. Some facility owners are not motivated to generate lower operating costs for building occupants or tenants. The LEED process involves paying fees for certification as well as hiring commissioning services to verify that specific processes have been followed. A number of owners have adopted sustainability initiatives but have stopped short of seeking certification. In a 2013 survey of contractors (Holloway, S. and Parrish, K.), the obstacles to sustainable building were "higher initial cost" (90%) and "difficulty/cost of certification" (57%) as primary obstacles to sustainable building, even though 81% thought it was "the right thing to do". In the same survey, 88% of respondents reported the additional cost of LEED certification as less than 2% of overall project costs.

A wider understanding of sustainability and life-cycle costing, as well as a lowering of initial cost will greatly affect the acceptance of sustainable building (Koskela and Tommelein 2009). Nevertheless, recent

Table 12.1 Cost study summary adapted from Matthiesen and Morris (2004)

Certification level ⟹	*Platinum*	*Gold*	*Silver*
Average premium Costs – percentage of starting budget	8.57%	4.0%	1.87%

studies suggest that designers have begun to incorporate green practices in standard designs. Some contractors have begun to institute paperless jobsites and to practice waste diversion.

An earlier study by Matthiessen and Morris, (2004) looked at the influence of climate and location on LEED ratings and project costs. The regions selected were the mid-coastal (Santa Barbara and San Francisco), the California Central valley (Merced), the Gulf Coast (Houston), the Northeast Coast Boston), and the Rocky Mountains (Denver). As indicated in Table 12.1 above, the average percentage premiums were 1.87 for LEED Silver, 4.0 for LEED Gold, and 8.57 for LEED Platinum.

In a follow up study (Matthiessen and Morris 2007) the authors noted that construction costs had escalated, yet many buildings were able to obtain LEED certification without additional funding, especially if clear goals were established in the project inception phase.

The cost of registration is seen as a disincentive by some owners. In addition, most traditional projects only consider a short-term period; after that, all the benefits may not be received by the builder/developer/owner. Consequently, the long-term benefits from sustainable construction, including lower life-cycle costs, have been less than attractive to many developers or construction companies.

A wider understanding of sustainability and life-cycle costing, as well as a lowering of initial cost will greatly affect the acceptance of sustainable building (Koskela and Tommelein 2009). Nevertheless, recent studies suggest that designers have begun to incorporate green practices in standard designs. Some contractors have begun to institute paperless jobsites and to practice waste diversion.

The intersection between lean and green

The McGraw Hill Company's Smart Market Report conducted an extensive survey of the US construction industry in 2013. The report cited many lean benefits reported by survey respondents. While the benefits were not quantified in the survey, the percentages of respondents reporting medium and high levels of achievement were: Higher Quality Construction 84 percent, Greater Customer Satisfaction 80 percent, Greater Productivity 77 percent, and Improved Safety, 77 percent.

Lean construction projects have been demonstrated to lower both costs and schedules simultaneously, in some cases by 10% or more (CII Report 234-11). In lean projects, constructors and designers can increase profitability even at a lower cost to the owner because of fewer mistakes, fewer delays and less waste of human and other resources. Green construction has been shown to lower ongoing operational costs as well – electrical and water consumption are often reduced by 50% and 30 percent, respectively. Combined, these methodologies represent a powerful prescription for sustainability. Lean may enhance environmental benefits (Horman *et al.* 2004; Huovila and Koskela 1998). On the contrary, the lean emphasis on providing value may be at odds with sustainability initiatives if project owners do not declare the reduction of environmental impacts as a value (Cusumano 1994).

Green buildings promote and support the concept of Lean construction. While Lean construction involves the minimization of waste, ensuring that the resources used in the building process enhance the value chain for a completed facility, green building design is based on design principles that promote environmentally beneficial long-term operation. Overall, a combination of these approaches significantly enhances sustainability. Not only is a "lean" building assembled by using fewer resources – human, financial, energy, material, etc., but a "green" building has a lower environmental impact in its construction. This benefit continues in its ongoing operation as well. Green practices may even reduce initial construction costs. The recycling of construction material reduces the amount of physical waste generated in a project, and the amount of pollution associated with the process is reduced as well. Sorting construction waste on a site is a recommended practice that qualifies for LEED credits. It can also reduce haulage costs for waste disposal as some of that material can be reused.

The question of a premium for sustainable projects is still debated but has been measured in the range of 1 to 5% for LEED silver and Gold ratings as shown in Table 12.1 above, significantly less than typical lean construction savings. Clearly, lean and sustainability represent a smart business practice and social responsibility at the same time; they meet the "triple bottom line".

Sustainability standards

There are several standards that provide a framework for green building initiatives. They include:

- Energy Star,
- Green Globes,
- BREEAM, and
- LEED.

In terms of the degree of adoption, Energy Star is managed by the US Department of Energy and is applicable nationwide. Green Globes is growing nationwide, as is LEED, but the Florida Green Building Coalition applies in the state of Florida only. These systems vary in complexity and in the degree of verification required in order to have a building certified.

The complexity of Energy Star is moderate. A professional is required to verify the design intent, and follow this up 1 year after occupancy. Green Globes is also moderate in its requirements and includes the Energy Star criteria. The LEED system is complex – it uses a system of credits based on highly multifaceted and structured criteria. A formal application identifies the intent to have a project considered for certification and is preliminarily reviewed, but it is reviewed in detail at completion before the appropriate level of certification is granted (Table 12.1). The Florida Green Building Coalition has a similar two-stage review, but has a single certification level – " certified" or "not certified." Green Globes and LEED both have four levels of certification. Green Globes awards projects 1, 2, 3, or 4 globes based on a point-scoring system. LEED has four levels: certified, silver, gold, and platinum. Energy Star has a pass/fail system with a threshold of 75 out of a possible 100 points. All of these systems require a formal assessment by the rating agency for certification. The Energy Star system requires a third-party assessment by a professional engineer. Green Globe certification also requires a professional engineer's review.

The use of any of the foregoing standards will provide building designers with strategies for improving a building's environmental performance on several fronts, including energy efficiency, indoor air quality, potable water usage, and life-cycle cost. However, the LEED standards developed by the USGBC evidently accomplish this objective with a broader scope of influence and greater rigor. Furthermore, LEED certification involves the services of a LEED-accredited professional (LEED® AP) in various aspects of the design and/or construction. It also requires the formal process of commissioning to ensure that certain critical systems in a facility (such as HVAC) perform as required and meet the facility owner's project requirements (OPR). Consequently, LEED is proposed as the most appropriate standard for enhancing the lean performance of buildings on a lifecycle basis.

Building Research Establishment Environmental Assessment Methodology (BREEAM)

BREEAM is the world first sustainability assessment method for buildings. It was created by the Building Research Establishment (BRE) Group in 1989 in the United Kingdom. There are more than 2 million

BREEAM registered buildings in 81 countries. BREEAM is an international scheme that provides independent third-party certification of the assessment of the sustainability performance of individual buildings, communities and infrastructure projects. BREEAM is also used in a country-specific format, for example, Germany, Sweden, Austria and the Netherlands.

Assessment and certification can take place at a number of stages in the built environment life cycle, from design and construction through to operation and refurbishment. The main output from a certified BREEAM assessment is the rating. A certified rating reflects the performance achieved by a project and its stakeholders, as measured against the standard and its benchmarks. The rating enables comparability between projects and provides reassurance to customers and users, in turn underpinning the quality and value of the asset. The BREEAM ratings range from Acceptable (In-Use scheme only) to Pass, Good, Very Good, Excellent to Outstanding and it is reflected in a series of stars on the BREEAM certificate.

BREEAM is based on the following applications:

1 Communities (Master planning),
2 Infrastructure (Civil Engineering and Public Realm),
3 New Construction (Homes and Commercial Buildings),
4 In-Use (Commercial Buildings),and
5 Refurbishment and Fit-out (Homes and Commercial Buildings).

BREEAM measures sustainable value in a series of categories, ranging from energy to ecology. Each of these categories addresses the most influential factors, including low impact design and carbon emissions reduction; design durability and resilience; adaption to climate change; and ecological value and biodiversity protection.

The BREEAM Categories are as follows:

1 Energy,
2 Health and Wellbeing,
3 Innovation,
4 Land Use,
5 Materials,
6 Management,
7 Pollution,
8 Transport,
9 Waste, and
10 Water.

Each category is subdivided into a range of assessment issues, each with its own aim, target and benchmarks. When a target or benchmark is reached, as determined by the BREEAM assessor, the development or asset score points, called credits. The category score is then calculated according to the number of credits achieved and its category weighting. Once the development has been fully assessed, the final performance rating is determined by the sum of the weighted category scores.

In the case of BREEAM, third-party certification involves the checking – by impartial experts – of the assessment of a building or project by a qualified and licensed BREEAM Assessor to ensure that it meets the quality and performance standards of the scheme. At the heart of this process are certification bodies – organizations with government approval (through national accreditation bodies) to certificate products, systems and services.

www.breeam.com/discover/how-breeam-certification-works/

Sustainability and the US Green Building Council

The US Green Building Council (USGBC) is a nonprofit membership organization whose vision is "a sustainable built environment within a generation." Its membership includes corporations, builders, universities, government agencies, and other non-profit organizations. Its mission is: "The USGBC is the nation's foremost coalition of leaders from across the building industry working to promote buildings that are environmentally responsible, profitable and healthy places to live and work." Since USGBC's founding in 1993, the council has developed a comprehensive family of LEED® green building rating systems, and an expansive educational offering. There is also the industry's popular Greenbuild International Conference and Expo (www.greenbuildexpo.org), and a network of local chapters, affiliates, and organizing groups.

Table 12.2 Summary of Selected Green Building Rating Systems

Green Building Rating System Comparison	*Building Initiatives Green Globe Rating*	*Building Research Establishment Environmental Assessment Methodology (BREEAM)*	*Leadership in Energy and Environmental Design (LEED)*	*Energy Star*
Online assessment	Yes	No	Yes	Yes
Third-party assessment (separate from rating agency)	Required only for Green Globes certification	Yes	No	Yes (by a professional engineer)
Fundamental commissioning	Yes	Yes	Yes, required	No
Enhanced commissioning	Yes	Yes	Yes	No
Rating agency required?	Third-party verification for globe award	Third party assessment required	Yes	Yes (Energy Star)
Complexity	On-line system may improve efficiency and reduce certification cost. Includes Energy Star criteria	Licensed BREEAM Assessor assesses projects according to the BREEAM Standards. Assessor then submits their assessment to the Certification body for a verification decision	Complex weighted average calculation needed	Moderate. Professional verify design intent, follow-up 1 year after occupancy
Degree of adoption	Growing nationwide	Worldwide	Worldwide	Nationwide – Dept. of Energy
Certification levels	Four Globe levels (1000 points): 1. globe = 35–54% 2. = 55–69% 3. = 70–84% 4. = 85% and over	Acceptable (In-Use scheme only) Pass Good Very Good Excellent Outstanding	Four levels: Certified = 40–49 Silver = 50–59 Gold = 60–79 Platinum = 80+	Pass/fail 75/100 points or more

The USGBC has led a national consensus to produce a new generation of buildings that deliver high performance through measurable systems. Council members collaborate to develop LEED products and resources. For more information, visit www.usgbc.org.

LEED (Leadership in Energy and Environmental Design)

Members of the USGBC from all segments of the building industry developed the LEED system starting in 1995 to provide a national standard and continue to contribute to its on-going evolution and growth. Its development included representatives from all sectors of the building industry with funding from the US Department of Energy, and unique in having had its development open to both industry and public scrutiny.

According to the USGBC, the LEED green building rating system is a voluntary, consensus-based national standard for developing high-performance, sustainable buildings. The LEED certification program is a feature-oriented rating system that awards buildings points for satisfying specified green building criteria. The six major environmental categories of review include: Sustainable Sites, Water Efficiency, Energy and Atmosphere, Materials and Resources, Indoor Environmental Quality, and Innovation and Design.

LEED provides a framework for assessing building performance and meeting goals for sustainability. It is based on proven scientific standards and promotes the use of state-of-the-art techniques for sustainable site development, energy efficiency, water conservation, materials selection, and indoor environmental quality. LEED promotes the proliferation of green building knowledge and expertise by recognizing achievements, and through a program of project certification, professional accreditation. The program also provides training and technical resources. LEED:

- Establishes a common standard of measurement to define "Green" building,
- Promotes integrated, whole-building design practices,
- Raises consumer awareness of green building benefits,
- Encourages competition among industry members,
- Recognizes environmental leadership in the construction industry, and
- Transforms the construction industry/market.

LEED can be applied to all building types including new construction commercial interiors, core and shell developments, existing buildings, homes, neighborhood developments, schools, healthcare and retail facilities. There are several subsets of LEED standards. The latest version of LEED is v4.1 and the most updated LEED Rating systems are as follows

- LEED BD+C for Building Design and Construction,
- LEED ID+C for Interior Design and Construction,
- LEED O+M for Building Operations and Maintenance,
- LEED Residential, and
- LEED Cities and Communities.

Facilities are certified to a variety of levels based on a credit system that reflects the degree of compliance with environmentally based standards. LEED standards are continually being upgraded. The levels prevailing in 2019 were: Certified (the basic level), Silver, Gold, and Platinum levels of LEED green building certification and are awarded based on the total number of credits points earned within each LEED category. The distribution of credits was changed with each update.

LEED performance areas are highly detailed and can only be presented in outline form in this publication. Readers are encouraged to contact the USGBC for the most up-to-date documentation of the LEED standards A brief overview of the eight categories in the LEED rating system is as follows:

1 Location and Transportation: This category is a new category introduced to all the rating systems of LEED v4. The LT category is the most recent addition to the suite of categories that have been the core of the LEED rating systems. This category places more emphasis and attention on reducing one of the main contributors to global warming: transportation. The LT category includes strategies and ideas to reduce the costs, pollution, and depletion of resources related to the daily transportation of people and to and from a building. The location of a building is as important as how it is built. Through its location, a building can meet the needs of the local community, support active street life, and promote healthy lifestyles. Building location also plays an important role in reducing greenhouse gas emissions. If people can take public transportation, ride bicycles, or walk to the building, the project helps reduce the carbon emissions associated with commuting. A project that is connected to the community by pedestrian paths and bicycle lanes encourages people to walk or bike instead of drive, not only helping to reduce air pollution, but also promoting physical activity.

2 Sustainable sites (SS): By selecting an appropriate project location, designers can reduce urban sprawl, promote the use of mass transportation, and reduce the dependence on personal transportation. Urban infill areas are examples of this practice, as they may already have the needed infrastructure. Site sustainability can also be enhanced by retaining existing trees and landscaping. Plants that have minimal water and pesticide requirements are highly desirable. Recycling of paving material materials and mulches promotes the concept of sustainability.

3 Water efficiency: Water usage can be reduced by 30% or more in commercial buildings. In a typical 100,000 square foot office building with 650 occupants, low-consumption fixtures equipped with sensors can save over one million gallons of water per year, assuming that each occupant uses 20 gallons per day. The strategies used include water efficient landscaping, irrigation efficiency, use of recovered rainwater and recycled wastewater, and non-potable water treated by a public agency.

4 Energy and Atmosphere (EA): Improvements in the energy efficiency of buildings reduce operating costs and enhance comfort. Greater energy efficiency also reduces pollution from power generation equipment. Passive design strategies such as building shape and orientation, passive solar design and the use of natural lighting have measurable benefits. Natural lighting, in particular, has a positive impact on the well-being and productivity of building users. High-efficiency lighting systems provide the maximum amount of lighting per kilowatt; task lighting is more cost effective than area lighting. Advanced lighting controls such as motion sensors and dimmable systems are also highly cost effective. The heating or cooling system for the facility should be properly sized and as energy efficient as possible to maintain adequate temperature and humidity conditions. The careful use of building orientation with reduced glass exposure on the east and west building faces reduces the solar gain of facilities. High R-value insulation and reflective roof and wall colors reduce the cooling load of facilities. Alternative energy sources such as solar cells and fuel cells should be considered.

5 Materials and Resources (MR): Recycled materials reduce the impact of waste in landfills, and also reduce the impact of natural resource consumption to produce new construction materials. Reuse and recycling of materials from demolition are incentivized by the LEED credit structure. The approved strategies include: storage and collection of recyclables such as metal, glass, plastic, paper, and so on; use of salvaged materials from other locations; sourcing construction materials within 500 miles to minimize transportation-related pollution; using wood certified to come from well-managed forests; in one project 3,966 cubic yards of waste material resulted from the initial demolition (over 99% of this waste

was reused or recycled); and 9.23 tons of obsolete window glass and 3.25 tons of concrete waste dust were converted into over 8,000 concrete blocks.

6 Indoor Environmental Quality (EQ): According to the Environmental Protection Agency (EPA) the levels of pollutants indoors are often many times greater than in outdoor air. Buildings that provide for increased levels of filtered outdoor air have lower levels of pollutants and appear to provide a better environment for users. Fewer cases of illness are documented. Hospitals that are designed with lower levels of pollutants and that have been built based on green principles have observed lower levels of site-borne illnesses. The average length of hospital stays (ALOS) has been shown to be shorter than with non-green facilities.

7 Regional Priority (RP): The RP category first appeared in LEED in 2009 to incentivize project teams to earn credits that meet specific needs for specific regional climates. Each location has their own particular climate, population density, and regulation. Because of this, some environmental issues are more critical than others and affect each project differently. One point is awarded for each Regional Priority credit achieved. No more than 4 credits identified as Regional Priority credits may be earned. USGBC has prioritized credits for projects located in the United States, Puerto Rico, the US Virgin Islands, and Guam. All other international projects should check the database for eligible Regional Priority credits.

8 Innovation (IN): The Innovation category has been part of every LEED rating system since the very first version. It is typically the last category in most of the rating systems on the project checklist. However, the Regional Priority (RP) category is a most recent addition to LEED.

The Innovation strategies to meet innovation credit requirements rely on the Project Surroundings and Public Outreach knowledge domain concerning the environmental impacts of the build environment and the values of sustainable design. The IN Credit – Innovation one to five points can be achieved through one of three options. These options are innovation, pilot credit and exemplary performance.

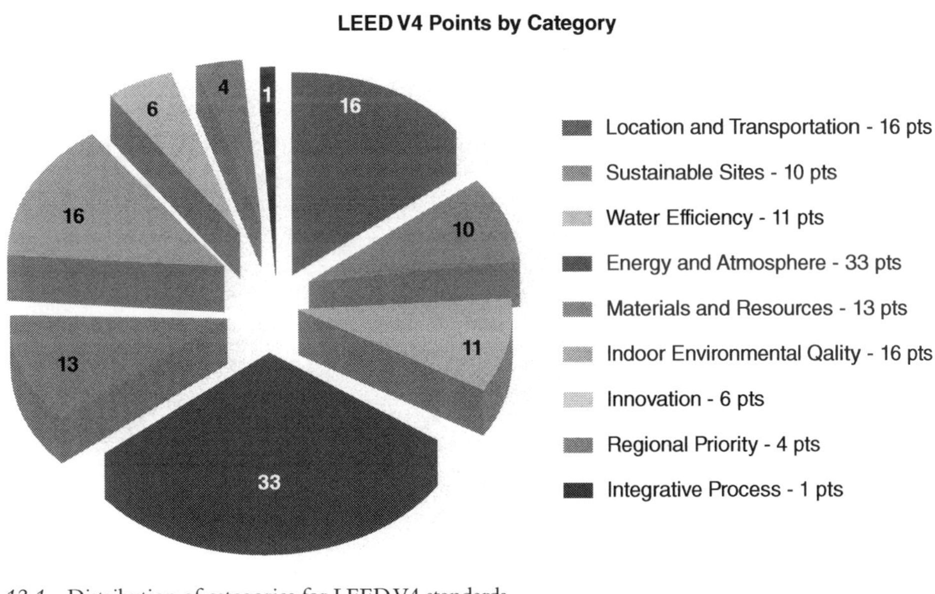

Figure 12.1 Distribution of categories for LEED V4 standards

Source: From US Green Building Council, *New construction and major renovation reference guide*. Reproduced with permission from the USGBC

Examples of operational benefits

Energy- and water-efficient buildings have been able to reduce their operating costs significantly. Use can be cut to less than half than that of a traditional building by employing aggressive and well-integrated green design concepts.

- Employees in buildings with healthy interiors have less absenteeism and tend to stay in their jobs. The Internationale Nederlanden (ING) Bank headquarters in Amsterdam uses only 10% of the energy of its predecessor and has cut worker absenteeism by 15%. The combined savings equal \$3.4 million per year.
- More than 17 million Americans suffer from asthma, and 4.8 million of them are children. 10 million school days are missed by children each year because of asthma, which is exacerbated by poor Indoor Air Quality (IAQ).
- A healthy indoor environment can reduce the likelihood of lawsuits and insurance claims. In *Bloomquist v. Wapello* (500 N.W.2d 1, Iowa 1993), plaintiffs successfully sued employers and builders for creating an unsafe work environment due to inadequate ventilation and pesticide applications.
- Insurance companies are using climate change protection activities as a means to manage risk and maintain profitability.

The USGBC reports that energy efficiency measures reduced the operating expenses of the Denver Dry Goods building by approximately \$75,000 per year. At a grocery store in Spokane, Washington, 48 tons of waste was recycled during construction. It would otherwise have incurred high-disposal costs. Students in day-lit schools in North Carolina have consistently scored higher than students in conventionally lit classrooms.

Several studies point to the energy savings derived by sustainable construction based on LEED principles and the ENERGY Star system. Two studies, one by the New Buildings Institute (NBI) and one by CoStar Group, have verified that third-party certified buildings outperform their conventional counterparts across a wide variety of metrics, including energy savings, occupancy rates, sale price, and rental rates.

In the NBI study, [1] the results indicate that new buildings certified under the USGBC's LEED certification system are, on average, performing 25–30% better than non-LEED certified buildings in terms of energy use. The study also demonstrates that there is a correlation between increasing levels of LEED certification and increased energy savings. Gold and Platinum LEED certified buildings have average energy savings approaching 50%. The NBI Study confirms that newly constructed LEED certified buildings use significantly less energy than their conventional counterparts and that they perform better overall. The report also underscores that monitoring a building's on-going operations and maintenance, as required in LEED for existing buildings: operations and maintenance and Energy Star, are equally important. LEED and Energy Star provide building owners and operators with a valuable structure to maintain high performance and deliver savings over time.

Energy savings under the EPA's Energy Star program are equally impressive: buildings that have earned the Energy Star label use an average of almost 40% less energy than average buildings, and emit 35% less carbon. But beyond the obvious implications of reduced energy use and reduced carbon emissions, the results from both studies strengthen the business case for green buildings as financially sound investments.

According to the CoStar study,[2] LEED buildings command rent premiums of \$11.24 per square foot (2008 dollars) over their non-LEED peers and have 3.8% higher occupancy. Rental rates in Energy Star buildings represent a \$2.38 per square foot premium over comparable non-Energy Star buildings and have 3.6% higher occupancy. And, in a trend that could signal greater attention from institutional investors, Energy Star buildings are selling for an average of \$61 per square foot more than their peers, while LEED buildings command a remarkable \$171 more per square foot. The group analyzed more than 1300

Table 12.3 Financial benefits of green buildings: cost per square foot based on 20-year net present value (NPV) calculation

	Category	*20-yr NPV (U.S. Dollars)*
1	Energy value	5.79
2	Emissions value	1.18
3	Water value	0.51
4	Construction waste (1 year)	0.03
5	Commissioning operations and maintenance value	8.47
6	Productivity and health value certified and silver	36.89
7	Productivity and health value gold and platinum	55.33
8	Less green cost premium	(4.00)
9	Total 20-year NPV (certified and silver)	48.87
10	Total 20-year NPV (gold and platinum)	67.31

LEED Certified and Energy Star buildings representing about 351 million square feet in CoStar's commercial property database of roughly 44 billion square feet and assessed those buildings against non-green properties with similar size, location, class, tenancy and year-built characteristics to generate the results. The USGBC cites studies of green buildings in which workers had productivity gains of up to 16%, including better work quality and lower absenteeism.

Energy Star is a prerequisite in LEED for existing buildings and owners and tenants place a premium on green buildings. LEED also provides financial and environmental benefits through intelligent site selection, water conversation, improved indoor air quality, waste reduction, and environmentally friendly material selections.

Healthy indoor environments can increase employee productivity according to an increasing number of case studies. Since workers are by far the largest expense for most companies (for offices, salaries are 72 times higher than energy costs, and they account for 92% of the life-cycle cost of a building), this has a tremendous effect on overall costs. (See *Green Development* by the Rocky Mountain Institute for more information.)

Project cost savings

Green Building projects that are well integrated and are comprehensive in scope can result in lower or neutral project development costs (Table 12.3). Rehabilitating an existing building can lower infrastructure and materials costs. Integrated design can use the payback from some strategies to pay for others. Energy-efficient building envelopes can reduce equipment needs – downsizing some equipment, such as chillers, or eliminating equipment, such as perimeter heating. Using pervious paving and other runoff prevention strategies can reduce the size and cost of storm water management structures.

Examples of green construction projects

The following case studies illustrate successful applications of LEED principles in construction projects.

Case Study #1

The Michigan State University's Shaw Hall Project

(with special recognition of Mary E. Kremposky of CAM Magazine, January 2014, Dr. Tariq Abdelhamid, and Architect Amr Abdel-Azim, both of Michigan State University.)

Figure 12.2 Partial view of the Shaw Hall dining area at work inception

Michigan State University (MSU) is a public research university in East Lansing, Michigan, United States. MSU was founded in 1855 and served as a model for land-grant universities in the USA. MSU is recognized internationally as a top research university and a leader in international engagement.

Since the early 2000s MSU began updating its dining hall facilities to meet the growing needs of the student body. After other dining halls had been upgraded as traditional projects, they set about renovating The Vista at Shaw Hall, using the IPD methodology. They also sought to have the benefits of certification by the US Green Building Council (USGBC). The completed project exemplifies a successful Lean and LEED GOLD project, driven by a knowledgeable owner. It was the first IPD project carried out at a public university in the entire USA. This renovation project was complicated, due to the fact that the original facility was built in the late 1940s and contained asbestos insulation on ceiling surfaces in the basement – as was the standard practice at that time. Any disruption of the asbestos would require abatement that would be over the reach of the budget.

The project team:

Architect of record: Neumann/Smith Architecture, Southfield
Design architect: Mesher, Shing, McNutt, of Seattle, Washington
Construction manager: Clark Construction Company, Lansing, Michigan
Mechanical/electrical engineer: Peter Basso Associates, Inc., Troy, Michigan
Civil engineer: Nowak & Fraus Engineers, Pontiac, Michigan
Structural engineer: Desai/Nasr consulting engineers, West Bloomfield
Asbestos abatement – Quality Environmental Service, Inc.

MSU solicited a combined proposal from the architect and contractor that became an IPD contract. Neumann/Smith and Clark also joined forces with Peter Basso Associates. The Architect of record, contractor and owner then worked under one contract, sharing both risks and rewards throughout this collaborative venture.

MSU's management team consisted of:

Amr Abdel-Azim, Architect III, MSU Physical Plant Engineering and Architectural Services

Jack Mumma, Construction Contract Administrator.

Tariq Sami Abdelhamid, PhD, MSU's Integrated Program Organization Advisor, Division of Residential and Hospitality Services (RHS).

Dr. Abdelhamid served as an internal coach, guiding the project team through the Lean IPD process. He provided eight days of workshops over a four-month period, reportedly giving the project team "homework" in the form of reading assignments to optimize their understanding and implementation. of lean principles.

During the design phase, the architect, engineers and trade contractors collaborated with MSU's chef to make critical decisions about the layout of the facility. In keeping with the collaborative approach of IPD, Neumann/Smith Architecture, was joined at the design table not only by Clark Construction Company, but by the chef, three trade contractors and two engineering firms.

MSU solicited a combined proposal from the architect and contractor as part of the IPD contract. Neumann/Smith and Clark also joined forces with Peter Basso Associates, mechanical and electrical engineers, to secure the project. The architect of record, contractor and owner then worked under one contract, sharing both risks and rewards throughout the project. Mesher, Shing, McNutt served as the design architect based in Seattle, Washington.

MSU selected Lean/IPD with an integrated form of Agreement (IFOA), with contingencies for hidden existing conditions, and construction risk. Building information modeling (BIM) was used for clash detection, and notably the contractors were active participants in the design phase.

Scope of work

Shaw Hall/s dining facility was originally built in 1949 but had only had minor cosmetic changes. In order to match other dining areas that had been previously upgraded, the scope of work included a new kitchen and dishwashing area, new student lounges, and new East and West lobbies.

To raise the dining experience to a level comparable to MSU's other renovated facilities, the $12 million project called for, a new loading dock and recycling center addition, new MEP systems and a dramatic transformation of the entire dining area. To avoid operational disruption, MSU established a temporary dining facility on the lower level during construction. The ceiling was raised by removing the old ceiling and fluorescent lamps to provide more volume; the ceiling now has an exposed style with a variety of soffits. The project team placed great emphasis on avoiding any interference with the asbestos insulation in the building, deliberately avoiding any need for mechanical, electrical, or plumbing lines that would need to be supported and could potentially create a problem.

Design features

The project was executed with sustainable construction practices and incorporated a variety of sustainable building features. It was LEED Registered with the certification goal of Silver. Sustainable design measures included: Energy-efficient light fixtures and controls;

Figure 12.3 Partial view of interior after completion

A high-efficiency HVAC system

Automatic ventilation control during periods of low occupancy;

Water-saving plumbing fixtures; daylight controls;

variable air volume for kitchen hoods; variable volume pumps;

sub-metering of utilities for improved accounting; high-quality, environmentally friendly seating made of recycled materials; composting of food waste (pre and post); low-flow plumbing fixtures to achieve 20% water savings; and low (volatility) VOC paints and coatings were used. Architecturally, what was once a large open area was converted into three restaurants – the Breadbox, the Garden Wok and Mainstreet – offering two different food venues. Each also offers its own unique design flavor, along with clusters of varied seating types and even different flooring surfaces. Back-of-house kitchen areas were converted into the Breadbox, a breakfast and lunch venue.

MSU's Dr. Tariq Abdelhamid, a proponent of both lean and sustainability, assisted contract administration staff to guide the application of the lean process. The Vista at Shaw Hall was a renovation project completed in May 2013 at a cost of $12 million and was the first Lean/IPD project at a US public university.

The process

The IPD contract called for a significant departure from the status quo in MSU's project delivery. At a price in the range of 12 million it was considered manageable f or learning purposes without too much risk. It also had a basis for comparison with other renovation projects at MSU. The architect of record Neumann Smith was happy to continue a successful working relationship with the builder – Clark Construction. While they lacked IPD experience, they were enthused about working with MSU.

The team used the lean practice of co-location to work at the site with easy access to each other for rapid decision-making. They used a multi-party agreement with two small contingencies in the project budget; MSU reserved 2% for hidden existing conditions, and 2% construction risk contingency. The contingencies were considered to be "planned waste" and were collaboratively developed by anticipating all foreseeable conditions. They were subdivided by 65% for the construction management goals of time, cost, and quality. The 35% was planned for release based on the team's success in applying the Five Big Ideas that have enabled the success of many lean projects (see the Glossary for a definition of the Five Big Ideas). The project team was evaluated on a monthly basis to determine how well the ideas were implemented, and it clearly experienced a high level of success.

Dr. Abdelhamid guided the project team effectively, utilizing Target Value Design (or delivery?) to continually adjust their targets for dollar cost, time or quality. The team also used pull planning in the Last Planner® System, working backwards from set milestones such as the project due date in order to predetermine the handover of activities from one discipline to the next. The trade contractors were very collaborative in helping to design the project.

Project outcomes

The outcomes were outstanding – the seating capacity of the facility was almost doubled from 455 to 720 persons. The project was completed 3 weeks early, at a price 15% lower than comparable projects; the contract administration cost was 28% less. In addition, 84% of the construction waste was diverted

Figure 12.4 A partial view of the completed facility

from landfills. 11% was recycled – flooring, concrete, and metals. In keeping with LEED standards, 21% of construction materials were obtained within a distance of 500 miles. According to the owner's representative there were only 12 requests for information (RFIs), as compared with 144 that would normally be expected in a project of this size. Architecturally, a new window curtain wall provides a picturesque view of a nearby river; it was not initially in the scope of work but it became possible due to savings resulting from the TVD process. Abdelhamid credits lean design and construction with "achieving green features at reasonable and competitive first costs to the owner."

The year-and-a-half invested in the collaborative design process translated into a less costly construction administration phase. "From an architectural standpoint, we spent a lot more time upfront in design, but we spent a lot less time in construction administration," said Neumann/Smith Principal Stanley E. Cole, AIA, LEED AP BD+C. "Because everything was already figured out and the tradespeople were working together closely, we were 28% under what we typically spend on this type of project for construction administration. That money ended up going back into the project."

The Vista at Shaw Hall facility certainly offered exciting new ways of approaching a project, including lean techniques, IPD, and even LEED strategies. The project attained LEED Gold certification. This project reportedly gained an appreciative user audience – the stakeholders are confident in their knowledge of the lean strategies, while creating a delightful dining experience at the same time. All participants have sharpened the "tools of their trade," learned new strategies for the delivery of quality projects, and at the end of the day, have created a truly wonderful dining experience for MSU students.

Case Study # 2

Okanagan College Trades Center, Kelowna, BC, Canada

This addition to an existing facility was built in British Columbia, Canada between 2012 and 2016. It is contributed by Murray Guy formerly of Integrated Designs, who served as a commissioning agent in assisting the project to meet several environmental initiatives including NetZero. The prime contractor was PCL Construction, and the project was required to meet LEED Gold and LEED Platinum standards as noted in the Conditions of Satisfaction below. Gary McEwan of PCL contacted Integrated Designs for assistance, and was introduced to a Lean Commissioning Cx (LeanCx) approach. As Murray describes it, LeanCx is a proactive and disciplined commissioning system that utilizes lean systems to accomplish an early completion. Murray refers to this action as "**Getting to Done is Done**."

Conditions of Satisfaction – highlights:

- A new addition approximately 5,400 square meters: (classrooms, shops, labs, administration space, student space, food services);
- Renovation of approximately 10,000 square meters: (trades shops & classrooms, shops storage yard);
- Budget: $33M cash flow over four years;
- A LEED Platinum Addition, a LEED Gold renovation, Living Building Challenge Petal Recognition;

Reduced Operations and Maintenance costs

- Minimal disruption to an active campus;

Lean Project Organizational Structure, Procurement & Commercial Terms:

Lean IPD was introduced to the project at the Schematic Design phase. The project organizational structure used Construction Management.

The Project Team was procured through a Request for Quotation (RFQ) and a request for proposal (RFP) for design, design assist (fee for service) for major trades partners with construction at a fixed price.

Lean project delivery methods & management (operating system)

An early Collaborative Team was established, with the owner, Designers, CM, Major Trades & Commissioning Agent. Lean Project Delivery facilitation was provided. Boot Camp training was held through several project phases. Target Value Design was used with the design team, and the Last Planner® System (LPS) was also implemented.

Accommodations were made on the site for a Big Room and a common digital documents room. A virtual Big Room was also provided. Choosing by Advantages (CBA) was used by the project team for better decision making. Study Action Teams were used to promote discussion and learning during the course of the project.

- Prefabrication elements were incorporated in the supply chain.
- 5 S techniques were used by the project team.
- Lean Commissioning was implemented, this included the "Done-is-Done" practices.

Green – Sustainability standards achieved

Leadership: 2nd Living Building Challenge Project
Education: Sustainable Construction Management Program
Innovation: Early verification of Net 0 energy through Lean commissioning
Economics: Net 0 Energy at no added cost
Community: Utilization of waste heat from City district sewage treatment plant

Estimates of savings with lean Cx

For the Okanagan College Project, an analysis was completed to determine if the additional cost of $50,000 or $1.00 per square foot could be justified.

- Design commissioning savings were estimated at $150k (1% lower contingency)
- Labor & overhead savings were $144k for completing 6 months early
- Energy savings were projected at $125k (10% energy saving for 5 years)
- Increase in productivity for the mechanical and electrical work would realize labor savings of $45k

Based on this assessment the project could realize $464,000 in savings for an investment of $50,000 representing an almost 10 times return on investment. LeanCx was adopted for this project, was successful and was presented at two National Lean Construction Institute conferences.

Case Study #3

The Alberici St. Louis Office Building

St. Louis, Missouri
Architects: Mackey, Mitchell Associates

The headquarters of the Alberici Corporation was completed in 2005 and received the highest possible rating from USGBC (i.e., a Platinum rating). This accomplishment was due to a highly efficient design that reduced energy usage by 60% relative to a standard design. The building featured high-efficiency HVAC, a better thermal envelope, increased daylighting and lower-lighting power densities. The building also produced renewable energy through a 65 KW wind turbine, equivalent to 18% of the building's regulated energy cost. Green strategies:

- Material and Resources: 20% by cost manufactured within 500 miles
- Over 50% of the wood specified was certified in accordance with the Forest Stewardship Council
- Over 10% of materials were salvaged or refurbished
- Construction/Demolition waste: 93% (6000 tons) diverted from landfills, pre-existing building used as structural fill
- Indoor environment: 75% of spaces naturally daylit, operable windows provide cross-ventilation. 90% of occupants have a direct outdoor view
- Underfloor air distribution
- Hydro powered, sensor-activated faucets
- Natural linoleum flooring, Recycled rubber flooring

Note: In addition to being granted a LEED Platinum rating, this facility also received four Green Globes of a possible four.

Case Study # 4

Recertification of Adobe's San Francisco Office

601 Townsend
San Francisco
LEED Platinum

Adobe's San Francisco office is an adaptive reuse project that retrofitted a building completed in 1908 to operate at a high level of performance. In 2008 the project team earned a LEED platinum certification for Existing Buildings: Operations and Maintenance. The team recently recertified the project at the platinum level under LEED v2009. Cutting energy use in a plant with 900 workers, 1,800 personal computers, and continuously operating data servers was a challenge. To meet energy goals, the project team used a web-based monitoring system, developed partly with Adobe Software.

The software permits energy and water use to be tracked-and adjusted-in detail, and reports total CO2 emissions. To date, 41 energy conservation measures and related strategies have been implemented, resulting in an ENERGY STAR score of 100, the maximum possible score. Electricity usage has been effectively reduced by 63% over a seven-year period.

Water use at 601 Townsend has been reduced to 62%. Faucet and shower head aerators reduce maximum water flow to just two quarts per minute. Restrooms feature a mix of high-efficiency flush toilets, dual flush toilets, and waterless urinals.

The diversion of solid waste from landfill through composting and recycling has risen from 23% to 98%. Every desk now has a second smaller "side saddle" wastebasket so that compostable and recyclable items stay separated. Adobe standards mandate that all products from copy paper to carpet must contain high recycled content. Even take-put food service products (paper, plates, napkins, cups, etc.) are compostable.

Commissioning and LEED

Commissioning is a critical component of the lean construction processes. The CxA is a third-party professional that reports to the owner directly and acts as an extra pair of eyes and ears on the owner's behalf. Commissioning is a planned, systematic, quality control process that involves the owner, users, occupants, O&M staff, design professionals, and contractors. It is most effective when begun at the project's inception. It is essential for a project to be LEED-certified.

LEED recognizes two classes of commissioning: fundamental commissioning and enhanced commissioning. Fundamental commissioning involves having the CxA verify that a building's energy-related systems are installed, calibrated, and perform according to the OPR, BOD, and construction documents. Enhanced commissioning involves a similar task but is begun early during the design process and involves additional activities that continue after building completion. The impact of commissioning on LEED is most visible in the category of Energy and Atmosphere (EA):

EA Prerequisite 1: Fundamental Commissioning of the Building Energy Systems
EA Credit 3: Enhanced Commissioning

Goals of the commissioning (Cx) process are to:

- Improve the quality and functionality of the building and its key systems;
- Ensure consideration of facility operation and maintenance throughout design and construction;
- Enhance the quality and coordination of design and construction documents;
- Facilitate transition from construction to occupancy and use;
- Achieve higher-than-normal energy efficiency with a payback period of 7–8 years or less; and
- Support sustainability goals and LEED certification on a project-specific basis.

The Commissioning Authority serves as an objective advocate for the owner and is responsible for:

a. Directing the commissioning team and process in the completion of the commissioning requirements,
b. Coordinating, overseeing and/or performing the commissioning testing, and
c. Reviewing the results of the system performance verification.

The Commissioning Authority shall report results, findings and recommendations directly to the owner.

Full commissioning services are to be provided in five (5) phases:

Predesign, Design, Construction, Acceptance, and Post-acceptance

Where projects involve LEED certification, the CxA shall provide both fundamental and enhanced commissioning for selected construction projects in order to meet the requirements of LEED in two places: EA Prerequisite 1, and EA Credit 3, as described in the USGBC's handbook.

Cx for Designated Building Systems – New Construction

The building systems included in the scope of the commissioning responsibility will include all or some of the following systems:

- HVAC systems
- Energy management systems (EMS)
- Lighting control systems
- Electrical power distribution systems
- Fire protection
- Fire alarm
- Smoke control
- Security alarms
- Telephone and intercommunications systems

Part B: Commissioning

The National Conference on Building Commissioning has officially defined "Total Building Commissioning" as: a "Systematic process of assuring by verification and documentation, from the design phase to a minimum of one year after construction, that all facility systems perform interactively in accordance with the design documentation and intent, and in accordance with the owner's operational needs, including preparation of operation personnel."

In effect, building commissioning is an excellent quality assurance mechanism. For new construction projects it is a cradle-to-grave systematic process. It is optimally applied to all phases of a construction project – programing/predesign, design, construction/installation, acceptance, occupancy, and post-occupancy. It focuses on verifying and documenting that the facility and all of its systems and assemblies are planned, designed, installed, tested, operated, and maintained to meet the OPR. *The documentation of the owner's goals and needs; that is, project requirements (OPR) is of paramount importance. It clarifies and quantifies the needs, then records the extent to which they are met or exceeded.*

It must be noted that the commissioning agent or authority is not necessarily a single individual. As many disciplines are involved in a project, a commissioning firm may have representatives who are experts in different disciplines. They provide CxA service in their respective areas of expertise.

Categories of commissioning

1 Total Building Commissioning:Verifies attainment of owners' goals and needs with regard to all building features/systems, from predesign/planning through acceptance/occupancy and post-occupancy.
2 Commissioning for equipment start-up and acceptance: The scope usually includes major MEP systems: HVAC, control systems, and electrical systems, fire suppression/alarm systems), final inspections, tests, and basic operating performance.
3 Commissioning for certification (LEED, Energy Star, Green Globes, etc.): The scope includes design reviews and construction inspections to certify performance goal attainment.
4 Commissioning for functional performance testing: Testing of selected building systems including the building envelope to verify that the respective performance expectations have been met.

For existing buildings, there are two main approaches: (1) Where a building/facility was provided with building commissioning at the time of construction, recommissioning ensures that the established performance levels are sustained. (2) Where building systems were never formally commissioned, retro-commissioning describes the application of commissioning procedures to existing building systems.

Commissioning and lean construction

Lean construction requires a collaboration of all design and construction stakeholders. There is evidence that the early involvement of contractors in the design process, for example, leads to more constructible designs. By the same token, commissioning agents improve both the design and construction processes by enabling designs that use resources cost effectively, and that benefit from the quality assurance provided by the commissioning process. This approach minimizes waste, not only in the design of facilities, but also in their subsequent operation.

In renovation and addition projects it is important to implement recommissioning procedures. Today's and future construction must be based on energy-efficient standards; while an addition might be based on up-to-date energy standards, it would be counterproductive not to bring the preexisting construction up to similar performance levels.

The engagement of commissioning services needs to be dovetailed with the requirements for sustainability classifications. In LEED standards, additional points are obtained when enhanced commissioning is implemented. Enhanced commissioning has the Cx involved to a greater extent in the predesign/planning phase as well as in the acceptance and post-acceptance phases.

An important benefit of the commissioning process is the quality assurance provided by ensuring that installation work is faithful to the plans and specifications. Problems and conflicts are detected early when the CxA asks installers to verify installation details at a time when corrective work can be done cost-effectively if it is needed. In the case of HVAC systems, for example, this can be very beneficial; changes to ductwork configuration may avoid conflicts, but airflow quantities may be very difficult to compensate for once ceiling spaces are filled with various plumbing, sprinkler systems, and electrical raceways. The Cx review process ensures that user/client satisfaction is secured with the final product, in keeping with lean principles.

Building performance can be validated by using the commissioning process for quality assurance beyond the requirements of LEED standards. For example, CxA review of the building envelope is not required by LEED but obligates the contractor to demonstrate that there are no leaks or early points of failure.

Lean commissioning (LeanCx)

Murray Guy, a Canadian Commissioning Authority (CxA) is a proponent of lean commissioning. He has integrated the steps involved in providing commissioning services with the various project activities by using the Last Planner® System as a framework. He offers the guidelines listed below:

The LeanCx approach proposed here (Collaborative Science Research Building, Client: University of Saskatchewan, Author: Integrated Designs Inc., Murray Guy, P. Eng, LEED AP ID+C, Date: January 12, 2017) is based on the industry standard ASHRAE Guideline 0 – The Commissioning Process. The main difference is that there is more upfront planning and coordination required that helps to reduce time and improve efficiency leading to a much more effective process. LeanCx (Guy, M., 2017) requires a highly collaborative and disciplined work effort, beginning in the design stage. It involves identifying potential issues early on and a commitment from all team members to resolving them in a timely and systematic manner. The goal is to produce a fully coordinated design that meets the owner's Project Requirements and causes minimal changes or delays in the field during construction.

During construction the Last Planner® System (LPS) is used to collaboratively plan, coordinate and execute the delivery of commissioning work. As Murray points out "For our lean commissioning system to work we need the whole team to buy in to using the Last Planner® System. This will help ensure that functional performance testing and integrated system testing are completed in a systematic manner prior to substantial completion as opposed to projects that have greatly delayed schedules."

Pull planning or reverse phase scheduling is described in Chapter 4 in the context of a typical lean project. The goal of pull planning with the commissioning process is to make the flow of work more reliable, and efficient to eliminate the waste associated with waiting, wrong pieces of equipment, system interfaces not working, sensors not being calibrated, sequences not being programmed properly, graphics not being completed. Waste is also caused by communication errors, systems not being ready for testing, and not having enough people to complete the work on schedule. The process includes collaborative **pull-based planning** to develop a master plan, **looking ahead** to ensure that commissioning tasks are constraint free, **collaborating** and agreeing on what tasks are next, **monitoring production** to keep activities on track and **measuring progress** as a percentage of promises kept and tasks completed.

The LeanCx approach expands the "black box" (representing commissioning) that you normally see on a project to be a proactive process that holds the team accountable for a much higher level of performance. It requires a different way of thinking and full commitment.

The process calls for:

- Early formation of the commissioning team,
- Adoption of the Last Planner® System,
- Collaborative project planning & delivery, and
- Proactive planning and management of work commitments.

Tracking progress

An important aspect of delivering a lean project is to establish a system of accountability for work commitments. The team shall use a simple scorecard to monitor and track team progress in completing the commissioning work, dealing with project issues and tracking the percentage of promised work completed:

- # Check Sheets,
- # Functional Tests,
- # Quality Issues,
- # Performance Issues,
- % Promised Work Completed, and
- Estimated time of completion.

To achieve LEAN production, the team needs to deliver on commitments. The Percentage of Planned Work Completed (PPC) is a key metric for determining the success of a lean project. One cause of late delivery at this stage is team leaders overcommitting. It is in everyone's best interest to prevent this from happening. There are two rules for collaborative production planning:

1 If you promise to do it, get it done.
2 If it cannot be done, don't promise to do it.

By breaking the work plan into manageable pieces and holding regular, short, energetic, action-oriented work sessions we can ensure that the work is being delivered in a timely systematic fashion that is best for the team and the project. On a regular basis the team gathers to review promises completed and establish the network of promises needed to move the project forward. The following questions are asked of each team member.

- What commitments did you complete?
- What commitments will you complete?
- What concerns or constraints do you have?
- Are you on track to meet the overall plan?
- What percent of the project is complete?

Our success as a team will depend upon how well we plan, coordinate and are accountable to each other for delivering the project to a collaboratively developed commissioning plan.

Table 12.4 Commissioning Milestones for LeanCx

Tasks	*Commissioning Milestones for LeanCx*
1	Owner's Project Requirements Document with Updates
2	Incorporate Design Commissioning into the Process
3	Track Design Issues
4	Provide Commissioning Specifications
5	Provide Shop Drawing Reviews
6	Provide and Manage Cx System Database (Check Sheets & Functional Tests)
7	Static Checks (Verify Installations meet Requirements)
8	Hold Regular Cx Coordination Meetings
9	Equipment Start-Ups
10	Functional Performance Testing
11	Integrated System Test
12	Systems Performance (7-day Trends)
13	O&M Manuals
14	Training complete
15	Seasonal Testing
16	Final Commissioning Report
17	Measurement and Verification

Traditional commissioning practices

Rationale for commissioning

Commissioning reduces a building owner's total cost of ownership. Energy costs are a major component of a building's operating costs. It is likely that these costs will increase in the future as utilities respond to market uncertainties. Even though a facility may have been designed to be energy efficient, contractors focus on getting equipment installed in place to meet project deadlines. It is easy for errors to occur that prevent equipment from performing as specified. Commissioning is a formal requirement for facilities to qualify for LEED certification. It provides assurance that a facility's systems are functioning as designed, and that LEED or other certification is justified. With regard to scope, commissioning may be applied to a complete facility or to individual building systems such as HVAC.

Other benefits of commissioning

- It clarifies owners' project requirements (OPR). These up-front requirements are an important foundation for project design. The CxA's input on the consequences of these requirements helps to improve the overall probability of desired system performance.
- Project designs have fewer errors/imperfections.
- Project cost is reduced, as change orders are avoided.
- Delays, waste, and rework are minimized.
- Punch lists are reduced.
- Decision-making and construction are speeded up.
- The operation and maintenance of facilities are improved, especially with regard to electrical, mechanical, and plumbing. Operating costs (including energy) are reduced because of optimum performance.
- There is trouble-free turn over – systems are fully functional.
- There are fewer warranty calls to contractors.
- Equipment life is extended, as it works within normal limits.
- Air quality is improved and occupants experience fewer illnesses.

COMMISSIONING VERSUS TESTING, ADJUSTING, AND BALANCING

Testing, adjusting, and balancing (TAB) are routinely performed on newly installed HVAC systems, but this action does not replace the need for the formal commissioning of these systems. TAB specialists check water and air flows under normal working conditions but are not required to test these systems under the extreme conditions that may occasionally occur. Their scope of work does not include validating that the OPR have been met; rather, they help to adjust systems such as the HVAC so that they function satisfactorily.

Full building commissioning includes a much wider variety of building systems than does the test and balance process; TAB focuses on HVAC and electrical, and mechanical systems. On the other hand, full building commissioning includes the building envelope – roofing, doors, windows, communications systems, and life-safety systems. The qualifications of the CxA exceed the requirements of TAB specialists. In a typical new construction project, the CxA is brought on board as early as the predesign stage. A competent CxA is expected to interact with the owner and designers to define project requirements for the owner that also promote the best environmental practices.

COMMISSIONING COST/BENEFIT ANALYSIS

Estimates of commissioning cost vary among different sources; however, there is evidence that these costs are recovered within a relatively short period of time from the savings in operating costs. Guidelines published by the Commissioning Group (ACG) of the Associated Air Balance Council (AABC) identify the cost of commissioning HVAC systems as between 2 and 5% of HVAC construction cost. Other estimates place costs between 2–3% of total mechanical cost for commissioning HVAC and control systems and 1 to 2% of total electrical cost for commissioning electrical systems. Overall, commissioning HVAC and electrical systems in typical buildings costs 1.5–2.5% of total construction cost (Table 12.6).

The University of Wisconsin's (UW) Commissioning training program points out that the commissioning process should clearly document that its total cost through 1 year of occupancy has been recovered by the end of design or early construction. UW indicates that if this is not the case, then the process is not being properly implemented. For each $80,000 paid for commissioning, returns have been as high as $500,000. UW points to many design-based savings in addition to a smoother construction process such as

- Needs identification early in the process,
- Proper system/component selection,
- Reduction of requests for information (RFIs), and
- Problems are discovered earlier when easier to correct.

Overall, the commissioning process benefits all construction stakeholders: owner, occupants, designers, contractors, subcontractors, testing agents, manufacturers/vendors, bonding companies/insurers, and LEED professionals.

SUMMARY OF THE COMMISSIONING PROCESS

The following factors ensure a successful commissioning process:

- Commitment by the owner/client to the process, and ongoing support. Many contractors do not attach great importance to the commissioning process and may not respond to communications in a timely manner.
- Early clarification of the CxA's role and requirements in the contract documents as well as in preconstruction meetings.
- An experienced project manager for commissioning that represents the authority.
- A clearly defined commissioning plan that all the parties understand.
- A cooperative attitude by all the parties to the project.

Commissioning Agent Responsibilities (Figure 12.5, Table 12.5):

1 **Predesign phase:** Assist owner and project team to develop OPR and BOD

   a. The OPR should detail the functional requirements of a project and expectations with regard to the systems that will be commissioned.
   b. The BOD provides a narrative describing the design of the systems to be commissioned, including design assumptions. Review scope of equipment and systems.

   Prepare a predesign commissioning outline. It shall contain:

   – A description of the commissioning process
   – Identification of the commissioning team's structure, with a summary of the roles and responsibilities of each team member
   – Identification of the systems to be commissioned
   – Preliminary commissioning time requirements for each project phase
   – Commissioning cost estimates

2 **Design phase**

   - Perform, at a minimum, one commissioning design review of the OPR, BOD
   - Review design to ensure it accommodates commissioning
   - Write commissioning specifications defining contractor responsibilities
   - Write the commissioning plan

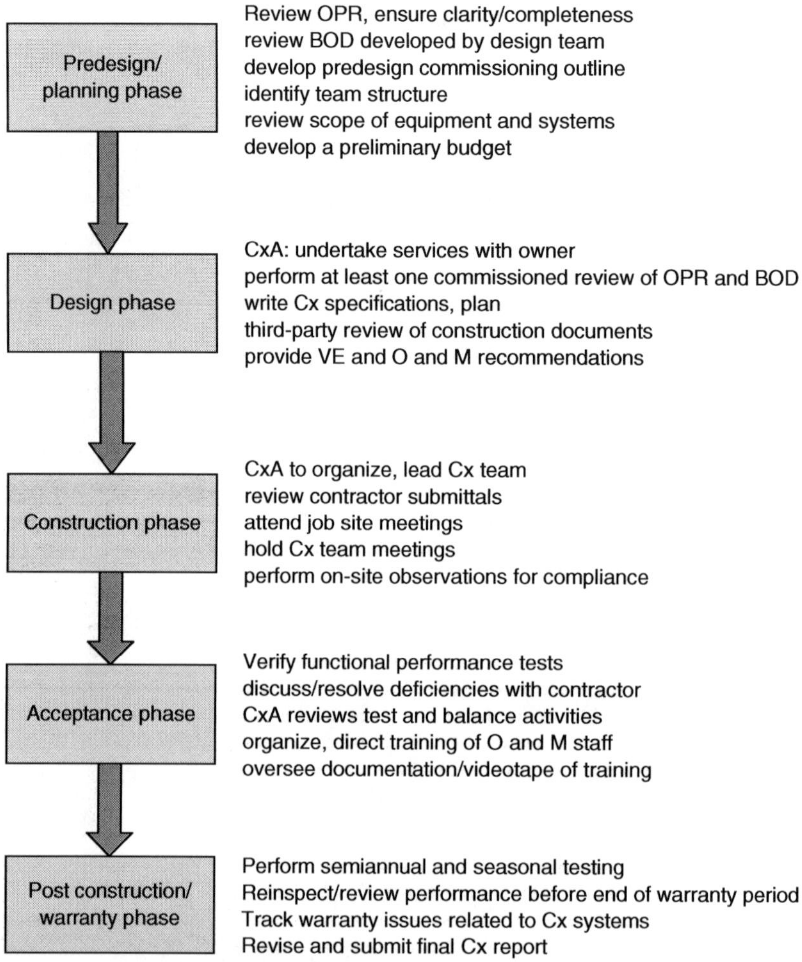

Figure 12.5 Process flow for commissioning

Table 12.5 Comparison of Fundamental and Enhanced Commissioning

	Tasks	*Fundamental Cx*	*Fundamental and Enhanced Cx*
1	Develop basis of design	Design team	Design team: CxA may assist
2	Incorporate commissioning requirements in construction documents	Project team or CxA	Project team or CxA
3	Conduct Cx design review before midconstruction docs	N/A	CxA
4	Develop and implement a Cx plan	Project team or CxA	Project team or CxA
5	Review Cx-related contractor submittals	N/A	CxA
6	Verify installation and performance	CxA	CxA
7	Develop a systems manual for Cx items	N/A	CxA
8	Verify training requirements met	N/A	Project team or CxA
9	Complete a summary Cx report	CxA	CxA
10	Review building operation within 10 months after completion	N/A	CxA

- Establish the project schedule for commissioning
- Perform plan reviews during the design phases of the project, identifying items that cause problems in obtaining the final balance conditions required by the design. Upon completion of the review, a written report of recommendations shall be submitted.
- Conduct a third-party review of the construction documents prior to bidding

3 **Construction phase**

- Organize and lead the commissioning team
- Review contractor submittals applicable to the systems being commissioned for compliance with the OPR and BOD. This review shall be concurrent with A/E reviews and submitted to the design team and the owner.
- Update the commissioning plan to reflect equipment and controls data from the submittals.
- Perform on-site observations during construction for compliance with the specifications and prevailing industry standards. Discuss deficiencies with the appropriate contractor's representative and report findings to the owner's representative.
- Maintain an issues log, tracking all issues that arise, to ensure they are resolved.
- Carry out and document system verification checks
- Carry out and document system start-ups
- Witness equipment and system start-ups as deemed necessary; ensure documentation is complete.
- Review test and balance reports

4 **Acceptance**

- Review test and balance (TAB) activities as specified for the project
  - Carry out functional performance tests (FPT) on all designated systems
  - Collaborate with the contractor to retest repaired items
  - Organize and direct the training of the O&M staff for effective, on-going O&M of all systems
  - Videotape O&M staff training sessions
  - Provide full documentation

5 **Post-acceptance**

- Perform and document any required "off season" functional performance tests
- Review building operations within 10 months after substantial completion with O&M staff and occupants. Include a plan for resolution of outstanding commissioning – related issues
- Revise and submit final commissioning report
- Follow-up with post-warranty validation and/or recommissioning.

Table 12.6 Estimated commissioning costs

	Commissioning Scope	*Estimated Cost Range*
1	Whole building: controls, electrical, mechanical; predesign through warranty	0.5–3% of total construction cost
2	HVAC and automated controls systems only	1.5–2.5% of the mechanical contract cost
3	Electrical system only	1–1.5% of the electrical contract cost
4	Energy-efficiency measures	\$0.23–0.28/square feet (\$2.48 – \$3.01/ square meter

To be fully effective, commissioning should begin with the predesign stage of the project and continue beyond the warranty phase. As stated previously, the commissioning authority can help the owner/client to define the OPR to take advantage of technological requirements as well as cost-effective construction methods. This involvement enables designers to adopt a holistic design that incorporates energy efficiency and sustainability, minimizing environmental pollution, and maximizing life-cycle cost effectiveness.

The building commissioning scope of work should be interwoven into the project's goals for performance, quality, safety, and innovation. The CxA should be the objective "eyes and ears" of the owner/client throughout the project to ensure that construction delivery decisions do not diminish the functional performance of the equipment specified for the project. At the substantial completion stage, the CxA oversees functional performance tests to ensure that the facilities' systems work as promised. The CxA ensures that the O&M staff are properly trained by the contractor in all the procedures and equipment needed for reliable operation. In the post-warranty stage the CxA may recommission a facility to ensure that systems continue to work cost effectively.

Commissioning requirements in construction documents

(Reproduced with permission from the USGBC *New Construction and Major Renovation* :

- Commissioning team involvement,
- Contractors' responsibilities,
- Submittals and submittal review procedures,
- O&M documentation/manuals,
- Meetings,
- Construction verification procedures,
- Start-up plan development and implementation,
- Functional performance testing,
- Acceptance and closeout,
- Training, and
- Warranty review site visit.

Sample of commissioning contract requirements

A. General Scope of Services:

1 Serve as the Commissioning Authority for Building Commissioning (CxA) for a specified variety of major building systems, including HVAC systems, in construction projects.

2 Be responsible for

- directing the commissioning team and process in the completion of the commissioning requirements
- coordinating, overseeing, and/or performing the commissioning testing, and
- reviewing the results of the system performance verification. The Commissioning Authority (CxA) shall report results, findings, and recommendations directly to the owner.

3 Full commissioning services are to be provided in five (5) phases: predesign, design, construction, acceptance, and post acceptance. One or more of these phases may be excluded on a project-specific basis.

4 Where projects involve LEED certification, the CxA shall provide both fundamental and enhanced commissioning in keeping with the requirements of the USGBC.

5 The building systems included in the scope of the commissioning responsibility will include all or some of the following systems where applicable:

- HVAC systems
- EMS
- Lighting control systems
- Electrical power distribution systems

- Fire protection
- Fire alarm
- Smoke control
- Security alarms
- Telephone and intercommunications systems
- Cable TV and CCTV systems
- Elevator controls
- Building envelope

Retro-commissioning

Retro-commissioning seeks to ensure the functionality of equipment and systems and also to optimize how they operate together in order to reduce energy waste and improve building operation and comfort. Thus, the goal of ensuring comfort and productivity of the building occupants accompanies the goal of cost savings.

Retro-commissioning occurs after construction, as an independent process, and its focus is usually on energy-using equipment such as mechanical equipment, lighting, and related controls. It may or may not emphasize bringing the building back to its original intended design.

Retro-commissioning begins with the planning phase, which consists of identifying project objectives, targeting systems for improvement, and defining tasks and responsibilities. A plan for conducting the work should then be prepared. An investigation phase follows, in which on-site assessment and testing are conducted. This phase allows deficiencies to be found and the scope of work to be refined. Once the scope is finalized, the improvements are then carried out in an implementation phase, and their success is validated. Finally, the completed improvements are "handed-off" to the owner along with information and knowledge gained during the process to help ensure long-term performance for the owner.

Retro-commissioning (EBCx) steps: The CxA will carry out the following duties:

1 Planning phase

   Develop retro-commissioning objectives for approval by the Board
   - Review available documentation and obtain historical utility data,
   - Develop the retro-commissioning plan, and
   - Prepare commissioning cost estimates for the specific project including the (a) cost of services to be provided, and (b) payback from energy and other savings.

2 There is no design phase.

3 Investigation phase:

   - Perform site assessment,
   - Obtain or develop missing documentation,
   - Develop and execute functional test plans,
   - Analyze results,
   - Develop master list of deficiencies and improvements,
   - Recommend cost-effective improvements for implementation, and
   - Identify the scope of major repairs/installation to be performed by others.

4 Implementation phase:

   - Implement minor repairs and improvements /or conduct field visits, inspections, and oversight of tests (by others) to review work performed by others,

- Review Pre-Functional and Functional Testing and the interoperability of systems and components (Performance Testing) to confirm compliance with the retro-commissioning plan,
- Retest and re-monitor for results,
- Fine tune improvements if needed, and
- Revise estimated energy savings calculations.

5 Project handoff and integration phase:

- Prepare and submit final report,
- Compile or update an indexed systems manual, which contains important building documentation such as construction record documents, specifications, submittals, narrative description of the sequence of operations, training materials, ongoing optimization guidance,
- Perform deferred tests (if needed),
- Develop and provide an O&M plan for the facility, including guidelines for implementing a new preventive maintenance plan,
- Develop a comprehensive training plan for O&M staff. Train staff on system operation and maintenance. Prepare a video recording of this training for future reference, and
- Develop a recommissioning plan/schedule (3-year schedule), a plan for operational sustainability and ongoing commissioning.

6 Persistence phase:

- Implementation of systems and tools to support both the persistence of benefits and continuous performance improvement over time,
- Monitoring and tracking of energy use, and
- Trend key system parameters to detect problems early and assess system performance.

Building systems to be retro-commissioned (EBCx) existing construction

The building systems included in the scope of the recommissioning responsibility will include all or some of the following systems:

- HVAC systems,
- EMS,
- Electrical power distribution,
- Lighting control systems,
- Emergency power generators and automatic transfer switching,
- Uninterruptible power supply systems,
- Fire protection,
- Fire alarm,
- Smoke control,
- Plumbing systems, water distribution, sanitary,
- Telephone and intercommunications systems,
- Cable TV and CCTV systems, and
- Elevator controls.

Qualifications

The commissioning authority (CxA) requires a broad and deep set of skills to be effective. Technical expertise is needed in building design, operations, and diagnostics. This is especially important with the ongoing emergence of sophisticated electronic control systems for communications, security, alarms, energy management, building automation systems, and so on. In particular, today's CxA should be LEED accredited in order to meet the qualifications for LEED-certified projects.

Cutting-edge approaches involve the use of computer modeling to visualize the expected performance of a system, then verifying that performance in the field. Management and communications expertise is also needed to enable the CxA to work effectively with others and maintain cooperation and collaboration. Expertise is also essential in documenting procedures as well as providing training to facility O&M staff and users. The following factors should be considered in selecting a commissioning provider:

1 **Related experience:** At a minimum, the company's qualifications and experience should include the following:

   A. Preferably at least 10 years' experience with the types of building, HVAC, and control systems included in the subject project.
   B. Experience in commissioning at least two projects of similar size and of similar equipment to the current project. This experience should include writing functional performance test plans and assembling a complete commissioning plan.
   C. Experience in the testing, design, specification, or installation of commercial building mechanical and control systems and other systems being commissioned.
   D. Experience working with project teams, project management and conducting scoping meetings; good team building skills, strong communication skills, especially documentation.
   E. Experience in planning and delivering O&M training.
   F. Direct responsibility project management of at least two (2) commercial construction or installation projects with mechanical costs greater than or equal to current project costs.
   G. Experience in design installation and/or troubleshooting of direct digital controls and energy management systems (EMS) if applicable.
   H. Demonstrated familiarity with metering and monitoring procedures. Knowledge and familiarity with air and water testing and balancing.

2 **Professional/technical staff:** Typical requirements, depending on project size would include the CxA's organization having a staff of no fewer than five (5) field technicians. Personnel performing the work should preferably have a mechanical engineering degree from an accredited college or university or a state-issued mechanical contractor's license and/or certified by either the Building Commissioning Certification Board (BCCB), AABC, or the National Environmental Balancing Bureau (NEBB).

Certification organizations

It is important to ensure that a commissioning agent/authority has taken an appropriate course of study, and has the experience necessary to provide the quality of expertise needed to justify the cost of commissioning services. Listed below are four organizations that offer commissioning certification and certify commissioning providers. Each organization had its own set of requirements and a different title for the providers it certified. The organizations and certifications are listed below. The organizations may be consulted on their requirements for certifications.

1 Building Commissioning Association (BCxA) Certification offered through the Building Commissioning Certification Board (BCCB), Certified Commissioning Professional (CCP), Associate Commissioning Professional (ACP), Certified Commissioning Firm (CCF).

2 National Environmental Balancing Bureau (NEBB), NEEB Firm Certification, NEBB Certified Professional (CP), NEBB Certified Technician (CT).
3 The AABC Commissioning Group (ACG), Certified Commissioning Authority (CxA), Certified Commissioning Technician (CxT).
4 (Provides a noncredit certificate only: Qualified Commissioning Process Provider Certificate) Association of Energy Engineers (AEE), Certified Building Commissioning Professional (CBCP).

Building Commissioning Certification Board (BCCB)

The BCCB confers the following designations:

1 The Certified Commissioning Professional (CCP)

The BCCB recognizes education combined with experience in CCP qualification criteria. Completion of higher education is accepted in lieu of experience and advanced levels of experience are accepted in place of higher education. Applicants are required to meet a minimum of education and experience as outlined in the table below.

Table 12.7 CCP qualification criteria

Education	*Experience*
Licensed Architect or Professional Engineer	3 years commissioning experience
Bachelor's Degree in the building sciences or equivalent military training/service	5 years commissioning experience
Associate/Technical/Vocational 2-year degree, non-building related Bachelor of Science Degree, or equivalent military training/service	8 years' building industry experiences** with at least 5 of those years spent working in building commissioning
High School Diploma, or GED	10 years' building industry experience** with at least 5 of those years spent working in building commissioning.

2 Associate Commissioning Professional (ACP)

Basic Requirements

Recognizing a combination of education and experience, where completion of higher education requirements is accepted in lieu of experience and higher levels of experience are accepted in lieu of greater education, applicants must meet a minimum of education and experience as outlined in the table below:

Table 12.8 ACP Qualification Criteria

Education	*Experience*
Licensed Architect or Professional Engineer	1-year building industry-experience
Bachelor's Degree in the building sciences* or equivalent military training/service	2 years building industry-experience
Associate/Technical/Vocational 2-year degree, non-building related Bachelor of Science Degree, or equivalent military training/service	3 years building industry experience or 2 full years in providing building commissioning
High School Diploma, or GED	7 years building industry experience with at least 5 of those years spent working in building commissioning

3 Certified Commissioning Firm (CCF)

Candidate firms are required to have one active Certified Commissioning Professional (CCP) who has been actively involved in the practice of commissioning in the last 24 months, able to show three completed projects with client references. Firms must demonstrate how they adhere to the BCxA Essential Attributes and offers continuing education to all their employees.

To be eligible your firm must:

Have been providing commissioning services for a minimum of 24 months
Employ at least one Certified Commissioning Professional (CCP)
Provide three completed commissioning projects with client references

National Environmental Balancing Bureau (NEBB)

1 Firm Certification

The NEBB certification covers HVAC systems commissioning, plumbing systems, and fire protection systems (as of 2007). To be eligible for certification, a firm must have operated continuously as a legal entity, actively engaged in the field for which it is seeking NEBB Certification, regularly providing services in one (1) or more of the disciplines in which NEBB provides certifications, for a minimum period of twelve (12) months prior to applying for certification. Firms that meet the specified criteria can apply for NEBB certification. A firm must designate a Certified Professional employed within their firm to supervise all NEBB related work. The designated individual will be designated as the NEBB Certified Professional. Firms must also determine in which NEBB discipline(s) they are interested in becoming certified.

2 Individual Certification

NEBB Certified Professional (CP) and NEBB Certified Technician (CT)
To obtain these certifications, a combination of education and work experience is required. Education requirements can range from High School Diploma, Associate Degree to bachelor's degree in Engineering from an accredited college or equivalent military Training. Work experience can range between 2–8 years, depending on the education level. To learn more about specific requirements for each certification, the reader is encouraged to visit the NEBB website.

The AABC Commission Group (ACG)

1 Certified Commissioning Authority (CxA)

The AABC confers the Certified Commissioning Authority (CxA) designation to professionals and their employers. The CxA program serves the industry by recognizing independent building commissioning professionals who demonstrate the technical, management, and communications skills required for competent performance. ACG's CxA certification is available to industry professionals who lead, plan, coordinate and manage commissioning teams to implement commissioning processes in new and existing buildings. CxA certificants:

- Manage commissioning projects
- Prepare commissioning documentation
- Conduct commissioning activities
- Coordinate operator training activities
- Complete warranty phase activities

- Conduct existing building commissioning
- Conduct on-going commissioning
- Are independent and work on behalf of building owners

EDUCATION AND EXPERIENCE REQUIREMENTS

Minimum education and experience requirements, as outlined in the table below, must be met by all applicants:

Table 12.9 CxA Qualification Criteria

Education	*Experience*
Licensed Architect or Professional Engineer	3 years commissioning experience
Bachelor's Degree in building sciences or equivalent military training/experience	5 years commissioning experience
Associate/Technical/Vocational 2-year degree, non-building sciences bachelor's degree, or equivalent military training/experience	8 years' building industry experience**, with 5 of those years in commissioning experience
High school diploma or GED	10 years' building industry experience, with 5 of those years in commissioning experience

2 Association of Energy Engineers (AEE)

The Association of Energy Engineers (AEE) confers the Certified Building Commissioning Professional (CBCP) designation. To be eligible for this certification, applicants must meet one of the following eligibility criteria:

Table 12.10 AEE Qualification Criteria

Eligibility		*Experience*
4-year related* degree OR Professional Engineer (PE) OR Registered Architect (RA)	AND	3+ years related experience
4-year unrelated degree	AND	5+ years related experience
2-year associate degree	AND	5+ years related experience
None	AND	10+ years related experience
Current status of Certified Energy Manager (CEM®)		

Questions for discussion

1 Describe the benefits of sustainable buildings
2 What is the purpose of having a LEED-Accredited Professional (LEED® AP) in the design team for a lean project?
3 Some members of the construction industry believe that implementing both LEED and lean in a project may inflate costs. Discuss.
4 Why should an owner pay for commissioning, when fees are already being paid to architects and engineers?
5 What specific services would you expect from the CxA at each phase of design and construction?
6 What are the minimum qualifications you would expect the CxA to have?

7 If an owner is not seeking LEED certification or another environmental certification, is the cost of having a commissioning authority (CxA) justified?
8 How does enhanced commissioning differ from fundamental commissioning?
9 In what way is sustainability linked with lean construction?
10 In what way does commissioning support lean construction?

Notes

1 The NBI study was funded by USGBC with support from the US Environmental Protection Agency and can be accessed at: www.usgbc.org/DisplayPage.aspx?CMSPageID=77#usgbc_publications.
2 For more information on the CoStar study: www.costar.com/News/Article.aspx?id=D968F1E0DCF73712B03A099E0E99C679.

References

Ballard, G. 2016. Lean construction, in D. J. Powell and T. Netland. The Routledge Companion to Lean Management. Abingdon: Routledge.
Bloomquist v. Wapello County. 1993. Iowa Supreme Court. https://law.justia.com/cases/iowa/supreme-court/1993/90-1371-0.htm
BREEAM. *How BREEAM certification works*. Online at www.breeam.com/discover/how-breeam-certification-works/. Accessed on April 4, 2019.
Cusumano, M. A. 1994. The Limits of 'Lean'. *Sloan Management Review* 35(4): 27–32.
Guy, M. 2017. Lean Commissioning (Lean Cx) Plan. (Personal Correspondence, June 22, 2019)
Horman, M.J., D. Riley, M. H. Pulaski, and C. Leyenberger. 2004. "Lean and Green: Integrating Sustainability and Lean Construction." Toronto, Canada: CIB World Building Congress.
Huovila, P., and Koskela, L. 1998. Contribution to the principles of lean construction to meet the challenges of Sustainable development. *Proceedings of the 6th Annual Conference of The International Group for Lean Construction (IGLC-6).*
Kamposky, M. 2014. A Recipe for Lean/IPD at MSU's dining hall. *CAM Magazine*, January 2014 desainasr.com/downloads/MSU_ShawHall_Magazine.pdf
Koskela, L., and I. Tommelein. 2009. "The Economic Theory of Production Conceals Opportunities for Sustainability Improvement". *Seventeenth Annual Conference of The International Group for Lean Construction (IGLC-17).*
LEED Core Concepts Guide. 2014. *An introduction to LEED and green building*, 3rd ed. Washington, DC: USGBC.
Matthiessen, L and Morris, P. (2004) "Cost of Green". *Davis Langdon Report* http://www.usgbc.org/Docs/Resources/Cost_of_Green_Full.pdf
U.S. Green Building Council (USGBC). 2007. *New construction and major renovation reference guide, version 2.2*, 3rd ed. Washington, DC: USGBC.

Bibliography

Heschong. 1999. *Skylighting and retail sales: An investigation into the relationship between daylighting and human performance.* Gold River, CA: Heschong Mahone Group. Online at www.hm-g.com/Daylighting.
Kibert, C. J. 1994. *Proceedings of the 1st international conference on sustainable construction.* Center for Construction, FL: University of Florida.
Kibert, C. J. 2016. *Sustainable construction: Green building design and delivery*, 3rd ed. Hoboken, NJ: John Wiley & Sons.
Kremposky, M. E. 2104. "A Recipe for Lean IPD at MSU's New Dining Hall". *CAM Magazine*, January, pp. 68–73.
Landman, M. 1999. *Breaking through the barriers to sustainable building: Insights from building professionals on government initiatives to promote environmentally sound practices.* Medford, MA: Department of Urban and Environmental Policy, Tufts University.
Odom, J. D., and G. DuBose. 1999. *Commissioning buildings in hot humid climates.* Meridian, CO: CH2MHILL.
Roodman, D. M., and N. Lessen. 1995. *World watch paper 124: A building revolution.* Washington, DC: Worldwatch Institute.

13
SAFETY MANAGEMENT

Introduction

The construction industry has a reputation for being unsafe, due to its historically high rate of occupational accidents (Bureau of Labor Statistics 2016–2017). In the year 2017 alone, construction accounted for 9.5% of all deaths and 3.1% of all disabling injuries/illnesses in private industry in the United States. But no industry has a good record, some are just worse than others. It is often claimed that the nature of construction work makes accidents inevitable, but that is an unacceptable position to take.

The relationship between safety management and lean management has been explored and researched. Results show that safety risks can be mitigated by integrating lean principles and tools in the production planning.

Developed by Toyota in the 1950s, lean in its essence is "a way to design production systems to minimize waste of materials, time, and effort in order to generate the maximum possible amount of value." – Koskela *et al.* (2002, p. 221). Lean therefore aims to produce the best quality at the lowest cost with the shortest lead time while concentrating on best safety practices and highest morale among workers. These lean principles and techniques have been adapted in the construction industry as a new philosophy of production to improve its performance and reduce the considerable amount of waste that tends to exist within the industry. "Lean principle of not releasing defective or incomplete work into the process also guides us to take care of the safety issues as a real part of production and not just do things as a camouflage." (Maki and Koskenvesa 2012). Therefore, safety in construction can improve using the lean concepts and principles.

In the context of lean construction, injuries and fatalities have consequences that work in opposition to a lean philosophy. While lean seeks to reduce or eliminate waste and to deliver more value to the customer, accidents have many negative consequences including lost work hours. Minor injuries may cause a temporary distraction to treat the affected individuals. Not only does an injured worker reduce the capacity of the work force, but other work crews generally lose additional time in the treatment of that individual. Serious injuries and fatalities inevitably cause major work stoppages and dampen workers' morale as well as their productivity.

Safety management is a method of controlling safety policies, procedures, and practices within a company. It is currently being implemented by many construction companies to limit their liabilities and costs, thereby making them more competitive in the construction marketplace. Safety management is a dynamic process operating in a constant state of change. Therefore, the process must be constantly monitored and adjusted to achieve the desired goals.

Most of the accidents that occur on a job site could be prevented if contractors treated safety as a priority. Safety programs can help to prevent accidents with injuries and deaths at the job site, and these programs provide an attractive return on investment through higher-production rates, lower workers' compensation premiums, and so on. They also improve a company's reputation and provide motivation for the Labor force. The decision to implement and run a good safety program rests solely in the hands of a company's top management. They are the only ones responsible for designing and implementing procedures to reduce accident rates.

The total cost of occupational fatal and non-fatal accidents in 2004 was $49.7 billion representing 10.72% of the total turnover of the construction industry. The total cost of occupational injuries/illnesses and fatality incidents can threaten the survival of a construction company in a highly competitive environment. For example, a company operating at a 4% profit margin would have to increase contract prices by $400,000 to pay for a $16,000 injury, such as the amputation of a finger.

Safety improvement begins with a systems-thinking approach, using a methodology that is similar to the approach of total quality management (TQM). In fact, existing safety management systems can be significantly improved with the adoption of core TQM principles and procedures. Weinstein (1997), recommends several applications of TQM principles that are helpful to occupational or work safety.

Construction site safety is an area of concern for employers of construction workers, this concern for safety is extended to the employees of subcontractors as well. The concern for safety has intensified in the past two decades, primarily due to the escalating costs of workers' compensation insurance (WCI), the rise in the number of liability suits, the intensification of safety regulations, and the mandate by owners to address construction worker injuries. As a result, there has been a relatively steady decline in the incidence of fatalities and disabling injuries. An issue of particular concern is the attitude that construction work is inherently unsafe, and a project's design has no effect on safety. This attitude must be dispelled.

How safe is construction?

According to the US Department of Labor, Bureau of Labor Statistics (BLS) during 1995 through 2017 there were 5,247,711 occupational injuries/illnesses among construction workers, representing approximately 9.03% of the total injuries/illnesses involving days away from work in the private sector as shown in Table 13.1.

Dividing the construction industry into three major divisions as per the North American Industry Classification System (NAICS Code), Table 13.2 represents the statistics for non-fatal incidents and Table 13.3 shows the statistics for fatal incidents recorded for General Building Contractors, Heavy Construction Contractors except Building Contractors, and Specialty Trade Contractors. The data indicate that specialty contractors accounted for the highest average incident rates; that is, 67.11% of all injuries/illnesses and 59.33% of all fatalities.

The BLS has reported that the average number of days away from work for construction workers caused by non-fatal occupational injuries was approximately 4.5 in 2004. The average number of injuries and illnesses per 100 full-time workers can be calculated using Equation 13.1 and is termed the *Incidence Rate*. In construction from 1995–2004, the average incidence rate is found to be 4.1. This indicates that about 1 of every 25 construction workers had an injury/illness during this period. For the entire private sector, the average incidence rate was approximately 3.0 for the same period, supporting the earlier finding that the accident rate in construction is higher than in other industries.

$$IR = \frac{N}{EH} \times 200{,}000, \tag{13.1}$$

where IR = incidence rate, N = number of injuries and illnesses, and EH = total hours worked by all employees during the calendar year. The base for 100 equivalent fulltime workers (working 40 hours per week, 50 weeks per year) is 200,000.

Table 13.1 Fatal and non-fatal occupational accidents

Year (1)	Non-Fatal Cases			Fatalities		
	All Private Industries	Construction	Percentage (%)	All Private Industries	Construction	Percentage (%)
	(2)	(3)	(4)	(5)	(6)	(7)
1995	2,040,929	190,591	9.34	5,497	1,055	19.19
1996	1,880,525	182,334	9.70	5,597	1,047	18.71
1997	1,883,380	189,839	10.08	5,616	1,107	19.71
1998	1,730,534	178,341	10.31	5,457	1,174	21.51
1999	1,702,470	193,765	11.38	5,488	1,191	21.70
2000	1,664,000	194,400	11.68	5,347	1,155	21.60
2001	2,031,098	222,811	10.97	5,281	1,226	23.22
2002	1,952,591	204,436	10.47	4,978	1,125	22.60
2003	1,936,639	197,731	10.21	5,043	1,131	22.43
2004	1,836,485	199,975	10.89	5,177	1,224	23.64
2005	3,971,700	404,600	10.19	5,734	1,192	20.79
2006	3,857,400	402,500	10.55	5,840	1,239	21.22
2007	3,796,400	371,700	9.79	5,657	1,204	21.28
2008	3,508,700	314,200	8.95	5,214	975	18.70
2009	3,111,500	244,200	7.85	4,551	834	18.33
2010	2,906,100	195,900	6.74	4,690	774	16.50
2011	2,878,000	190,100	6.61	4,693	738	15.73
2012	2,871,400	184,500	6.43	4,628	806	17.42
2013	2,855,000	197,888	6.93	4,585	828	18.10
2014	2,809,800	196,300	6.99	4,821	899	18.65
2015	2,765,300	199,600	7.22	4,836	937	19.38
2016	2,719,800	197,700	7.27	5,190	991	19.10
2017	2,685,100	194,300	7.24	5,147	971	18.87
Total	47,780,977	5,247,711	–	119,067	23,823	–
Average	2,077,433.76	228,161.35	9.03	5,176.83	1,035.78	19.93

Source: Bureau of Labor Statistics, *Fatal and non-fatal Occupational Statistics from 1995–2017* Online at www.bls.gov. Accessed on May 1, 2019

Equation 13.1 can also be used to estimate the incidence rate of fatalities with N representing the number of fatalities. The analysis shows that during a 10-year period from 1995 to 2004 approximately one of every 3900 construction workers died on the job. It is often claimed that the nature of construction work makes accidents inevitable. However, many companies have improved their safety plans, including developing "Zero Accidents Programs," and a great improvement can be seen in their current safety records (Bureau of Labor Statistics 2006)

Cost of occupational injury/illness accidents

The costs associated with fatal and non-fatal occupational accidents can be classified as economic costs and non-economic costs (Dorman 2000). Non-economic costs are associated with the deep human emotions that arise when life is unnecessarily shortened or impaired. It is hard to quantify these costs in numbers.

Many studies have been conducted to estimate the real cost of accidents for the construction industry. In 1979 the Department of Civil Engineering at Stanford University, under contract with the Business

Table 13.2 Non-fatal occupational accidents involving days away from work in private construction classified by the NAICS Code

Year (1)	*General Building Contractors*		*Heavy Construction Contractors Except Building*		*Specialty Trade Contractors*	
	Total (2)	*Percentage (%) (3)*	*Total (4)*	*Percentage (%) (5)*	*Total (6)*	*Percentage (%) (7)*
1995	40,315	21.15	27,857	14.62	122,421	64.23
1996	37,161	20.38	24,778	13.59	120,395	66.03
1997	36,947	19.46	26,132	13.77	126,760	66.77
1998	38,551	21.62	25,638	14.38	114,152	64.01
1999	38,968	20.11	23,379	12.07	131,418	67.82
2000	36,992	19.03	24,244	12.47	133,164	68.50
2001	35,873	16.10	32,308	14.50	154,630	69.40
2002	32,845	16.07	30,735	15.03	140,856	68.90
2003	31,125	15.74	28,764	14.55	137,842	69.71
2004	30,405	15.20	30,613	15.31	138,957	69.49
2005	79,700	19.70	48,600	12.01	276,300	68.29
2006	81,500	20.25	48,800	12.12	272,100	67.60
2007	76,500	20.58	46,200	12,43	248,900	66.96
2008	62,900	20.02	39,100	12.44	212,300	67.57
2009	48,500	19.86	31,500	12.90	164,200	67.24
2010	38,700	19.75	29,000	14.80	128,200	65.44
2011	38,400	20.20	27,400	14.41	124,200	65.33
2012	38,100	20.65	25,700	13.93	120,800	65.47
2013	37,900	19.15	26,400	13.34	133,400	67.41
2014	40,400	20.58	26,200	13.35	129,800	66.12
2015	40,000	20.04	25,300	12.68	134,300	67.28
2016	37,500	18.97	25,900	13.10	134,400	67.98
2017	42,100	21.67	24,000	12.35	128,200	65.98
Total	1,021,382	–	698,548	–	3,527,695	–
Average	44,407.91	19.40	30,371.65	12.94	153,378.04	67.11

Source: Bureau of Labor Statistics, *Fatal and non-fatal Occupational Statistics from 1995–2017*. Online at http://www.bls.gov. Accessed on May 1, 2019

Table 13.3 Fatal Occupational accidents in private construction classified by the NAICS Code

Year (1)	*General Building Contractors*		*Heavy Construction Contractors Except Building*		*Specialty Trade Contractors*	
	Total (2)	*Percentage (%) (3)*	*Total (4)*	*Percentage (%) (5)*	*Total (6)*	*Percentage (%) (7)*
1995	176	16.68	246	23.32	618	58.58
1996	184	17.57	248	23.69	606	57.88
1997	194	17.52	252	22.76	648	58.54
1998	213	18.14	272	23.17	680	57.92
1999	183	15.37	280	23.51	710	59.61
2000	188	16.28	282	24.42	685	59.31
2001	199	16.23	302	24.63	725	59.14
2002	164	14.58	288	25.60	673	59.82
2003	171	14.50	290	25.64	670	59.24
2004	190	13.97	307	25.08	727	59.40

(*Continued*)

Table 13.3 (Continued)

Year (1)	*General Building Contractors*		*Heavy Construction Contractors Except Building*		*Specialty Trade Contractors*	
	Total (2)	*Percentage (%) (3)*	*Total (4)*	*Percentage (%) (5)*	*Total (6)*	*Percentage (%) (7)*
2005	227	19.04	244	20.47	677	56.80
2006	219	17.68	224	18.08	724	58.43
2007	249	20.68	219	18.19	690	57.31
2008	206	21.13	190	19.49	567	58.15
2009	154	18.47	169	20.26	487	58.39
2010	159	20.54	147	18.99	447	57.75
2011	129	17.48	156	21.14	423	57.32
2012	142	17.62	174	21.59	473	58.68
2013	154	18.60	165	19.93	494	59.66
2014	180	20.02	140	15.57	564	62.74
2015	175	18.68	148	15.80	595	63.50
2016	182	18.37	159	16.04	631	63.67
2017	196	20.19	152	15.65	610	62.82
Total	4,234	–	5,054	–	14,124	–
Average	184.09	17.80	219.74	21	614.10	59.33

Source: Bureau of Labor Statistics, *Fatal and non-fatal Occupational Statistics from 1995–2017*. Online at http://www.bls.gov. Accessed on May 1, 2019

Roundtable (BR), found that accidents cost the construction industry $8.6 billion or about 6.5% of the $137 billion spent on industrial, utility, and commercial construction (Fullman 1984). The researchers estimated that adequate safety programs could reduce annual accident costs by approximately $2.75 billion. The probable cost of such programs would be about $86 billion, making them cost effective by a ratio of 3.2–1. Everett and Frank (1996) estimated that the cost of occupational accidents to the construction industry in 1994 was $31.94 billion, representing 11.9% of the total turnover of the construction industry in that year.

Direct costs

Workers' Compensation Insurance (WCI) premiums represent the greatest direct cost of accidents paid by any industry (Fullman 1984). The premiums can vary for each state and also among insurance companies. A contractor who is concerned about the high cost of such insurance premiums must recognize that these costs are directly related to each company's safety records. Insurance companies are usually willing to give substantial discounts to contractors who maintain good safety records. These discounts are often higher than the cost of a good safety program (Wilson and Koehn 2000).

Other direct costs associated with an occupational accident in a construction company include: an increment in liability insurance premiums, replacement or repair of lost or damaged equipment, and loss of key personnel time (laborer, foreman, superintendent, project manager, and others). Worker's compensation premiums come from the experience rate plans, based on the following formula:

$$\text{Standard premium} = \text{base} \times \text{payroll} \times \text{EMR}. \tag{13.2}$$

The base or manual rates are the average cost of accidents plus administrative costs and profit for insurers per $100 of straight time wages paid for each of 600 work classifications. The payroll units can be calculated by dividing an employer's straight time direct Labor costs by $100. The experience modification ratio (EMR) of a company is based on its own accident records and is calculated by rating bureaus or advisory organizations. Depending on the frequency and severity of the injuries suffered by the contractor's employees during a given year, the EMR can vary dramatically. Most states accept the EMR ratings provided by the National Council on Compensation Insurance (NCCI) but a few states have their own rating bureaus. The NCCI is a private corporation, created and funded by member insurance companies. California, Delaware, Hawaii, Indiana, Massachusetts, Michigan, Minnesota, New Jersey, New York, North Carolina, Pennsylvania, Texas, and Wisconsin have their own government-run rating bureaus that are separate from the NCCI.

Direct costs are defined as the benefits paid to and on behalf of injured employees by WCI (Everett and Frank 1996). According to worker's compensation data tabulated for Engineering News Record (ENR) by broker Marsh USA Inc., New York City, the national average premium for the three key crafts – carpenters, masons, and structural ironworkers per \$100 payroll in the year 2004 was \$23.74 (Bureau of Labor Statistics 2006).

In a report obtained by Everett and Frank in 1996 from the NCCI, the mean EMR for the 37 states was 96.3%. On an average, 25% of the total project costs were attributable to Labor.

Therefore:

$$\text{WCI}_{(\text{construction industry 2004})} = \$463.6 \times 23.74\% \times 25\% \times 96.3\% = \$26.5 \text{ billion.} \qquad (13.3)$$

Where \$463.6 billion is the construction volume of work in the year 2004 (Bureau of Labor Statistics 2006), 25% is the percentage of construction volume that is direct Labor, 23.74% is the average manual rate for the three key trades, and 96.3% is the mean EMR. However, the insurance benefits received by injured workers represent only about 65% of the insurance costs (Fullman 1984). Everett and Frank (1996) evaluated the BR report that established this percentage and they found that this value still remains the same (65/35).

$$\text{direct cost}_{(\text{WCI insurable})} = \$26.5 \times 65\% = \$17.22 \text{ billion.} \qquad (13.4)$$

Indirect costs

Indirect costs of accidents in the construction industry are normally expressed as a function of the direct costs (Hinze *et al.* 2000). Everett and Frank (1996) estimated that the indirect costs are roughly double the direct costs; a research study at the Stanford University indicated that this ratio is 1.6 (Fullman 1984), and the Hinze and Applegate (1991) found this ratio to be approximately 4.2.

The indirect costs are all costs resulting from the injuries that are not covered through insurance coverage. Some indirect costs normally borne by a construction company are:

1 Lost productivity:

  a. Job shutdown at the time of injury
  b. The injured worker's reduced capacity upon return to work
  c. Coworkers at the time of the injury: watching and helping the injured
  d. Coworkers who are shorthanded following the injury
  e. Coworkers who must train a replacement worker
  f. Supervisor/management time hiring or retraining a temporary or permanent replacement worker
  g. Management time lost in investigating and reporting the incident (to government, insurance, and news media representatives)

2 Fines
3 Extra wage costs

4 Damaged equipment and the costs of repairing or replacement
5 Clean up
6 Lawsuits
7 Damage to the company's image and reduced competitiveness
8 Reduced workers' morale
9 Transportation costs
10 Cost to reschedule the work
11 Liquidated damages for rescheduling the work

On the other hand, lean has the goal to deliver successful projects by maximizing value to the project owners. Adapting the lean principles in construction can help reduce the overall cost of the project by reducing the number of accidents, wasted time and lost work days.

Safety and lean construction

Several of the lean production concepts and techniques have been successfully implemented in the construction industry from which effective lean construction tools, such as the Last Planner® System, have been developed. The improvement of construction safety is therefore an important component of lean construction practices. Safety has a significant impact on construction-related waste. Labor hours lost to illness or job-related injuries do not add value to the construction process. In fact, there is the added penalty of increased insurance costs to builders that have an imperfect safety record, even if they have a record of otherwise successful performance. Although significant improvements have been made in the United States since 1996, construction work still results in the injury or death of over 8% of its workers. Lean, however, focuses on reducing waste and poor safety can be considered a source of waste.

Conventional construction approaches result in self-interest by the parties that leads to optimization at the task level as individual subcontractors seek the highest level of productivity. This action tends to reduce the predictability with which work is released downstream. In turn, this reduces trust, complicates coordination, and increases project durations. It also reduces the safety of workers on site. Lean construction, on the other hand, improves the predictability of work flows and the effectiveness of work planning, and reduces workers' exposure to hazards on site.

Research has shown that 45% of all accidents are associated with planning-related failures. In Danish research, projects that are managed with The Last Planner® System have been shown to experience approximately 50% of the accident rate and up to 70% fewer lost work days due to illness, when compared with traditional projects. Additionally, implementing the Last Planner® System as a lean practice can enhance safety practices such as management commitment, pre-project planning, and pre-task planning.

Lean Principles emphasize the elimination of waste and consistent with this principle, it is best to eliminate the safety hazard if possible because doing so will remove the risk associated with the hazard. Accidents and injuries can be considered a type of waste because the accidents/injuries consume extra cost and time and can have significantly negative social and jobsite culture impacts. Eliminating this waste is consistent not only with the principles of safety management but also with those of lean production. Furthermore, accidents introduce variability and disruption into the work process. Safety management emphasizes eliminating accidents, thereby reducing variability in the work. Lean construction also aims to minimize variability to stabilize the work flow. Therefore, the principles of lean construction and safety management appear to overlap.

Best practices have improved national accident statistics (Howell *et al.* 2002), in 2000 OSHA reported an injury rate of 4.3 per 1,000 employees versus an industry-wide rate of 8.2. However, these rates have improved as better practices have been implemented. OSHA reports that worker injuries and illnesses are down – from 10.9 incidents per 100 workers in 1972 to 2.8 per 100 in 2017. Private industry employers reported nearly 45,800 fewer non-fatal injury and illness cases in 2017 compared to a year earlier, according to estimates from the Survey of Occupational Injuries and Illnesses (SOII). Lean construction advocates recommend that further improvements need to be made, beyond the prevailing best practices. The first run studies approach in lean construction is based on simulation and experimentation. It can measure the impact of better planning of work assignments to ensure they are capable of meeting safety, quality, cost, and time standards before they are due to be implemented.

A Norwegian case study shows that lean construction can improve safety and overall project performance (Andersen *et al.* 2012). The study was conducted on the St. Olav's Integrated Hospital in Norway, which examined the second phase of the project. "The overall results of the empirical study substantiate positive results in several attributes and dimensions; keeping the building time in phase 2 on par with phase 1 despite increase in complexity, 3.4% cost reduction per m^2, improved build quality with 55.1% reduction in warranty costs per m^2, better HSE performance with a 56.5% reduction in H value, better cooperation, more content, etc."

There are several strategies for improving safety in construction projects. They are discussed next.

Organizational cultural change

Safety has a paramount importance in the construction industry. To address safety by utilizing the lean principles, companies need to change the prevailing company's culture with respect to production management. However, implementing lean practices and tools solely cannot accomplish a company's cultural change without addressing and communicating how to think in a lean manner. The implementation of lean practices necessitates the understanding of those principles and management support for them.

Nesensohn *et al.* (2012) argue that integration of lean principles requires not only long-term thinking and vision but also guidance on how to achieve it. The researchers propose that this guidance should include deciding what practices to benchmark, which also allows for assessing the extent to which the industry is advancing towards a lean system (Etges *et al.* 2012).

Protection through design

Construction site safety has been an area of concern for employers of construction workers. This concern is extended to the employees of subcontractors as well, and it has intensified in the past two decades due to several factors:

- The escalating costs of workers' insurance,
- The rising number of liability suits,
- Intensification of safety regulations, and
- The decision by owners to address construction workers' injuries.

This has led to a steady decline in the incidence rate of fatalities and disabling injuries. The attitude that construction work is inherently unsafe and that a project's design has no effect on safety, is of concern. This attitude must be dispelled.

Physical work conditions do not have to harbor numerous safety hazards for an unsafe condition to exist. Additional hazards may exist due to no fault of the contractor; in some cases, the contractor is bound by a design that places specific hazards on the construction site. If designers fully understood their effect on construction safety, that would help in making job sites safer for construction work. They are often unaware of the extent to which their design decisions may impact construction safety. Construction worker safety is often overlooked until the start of the construction phase. As owner concerns about construction worker safety increase, it is anticipated that designers and contractors will be more heavily involved in ensuring safety in the future. While the involvement of contractors in safety is well established, little information exists on how designers can be involved.

Project delivery method

The construction industry has evolved over the years and so have its project delivery methods. Understanding the connection between lean design/construction and safety management is especially of interest given the industry's current movement towards more integrated project delivery methods. Such methods are designed to integrate design and construction knowledge and tasks with the goal of creating higher quality facilities in a more efficient manner. The potential impacts that the design can have on construction, and especially on construction worker safety, are well known (for example, Gambatese *et al.* 2008). Additionally, delivery methods that promote the collaboration of designers and constructors prior to the start of construction and include contractual arrangements that enable the integration of construction knowledge into the design have been shown to result in improved safety performance

Worker behavior

Both safety management and lean management address worker's behavior. Research shows that human errors play a major role in the occurrence and reoccurrence of accidents in the construction industry. Human error has the potential of reducing effectiveness of work performed and compromises safety on the job site. The focus on completing a task in a timely and efficient manner should not overplay the risks and hazards associated with completing this task.

There is no doubt that the worker is an active and important contributor to the safety system. Therefore, workers are encouraged to be proactively involved in any safety efforts as part of their overall job performance. The recommended safety measures and practices aim to enhance the situational awareness of workers. Situational awareness describes the process of detecting hazards, comprehending the identified hazards, and predicting the risk associated with the hazards. Enhancing situational awareness at the work face, along with promoting risk-averse decision-making, is a foundational component of contemporary safety management. Lean principles, likewise, promote enhanced situational awareness through standardization and transparency of work operations and conditions.

Safety and building information modeling (BIM)

Building information modeling is considered an approach that is helping the construction industry to reduce document errors and omissions, reduce rework and facilitate real time construction tracking and reporting. BIM helps identify potential safety hazards and thus corresponding prevention methods can be applied to help prevent accidents from occurring. This can happen during construction planning and in certain cases, during design. Research shows that developing a safety rule-checking platform for fall hazard detection and prevention in building information models can help reduce accidents by detecting the location of potential fall hazards and thus provide installation guidelines of corresponding fall protection equipment that solve the identified hazards virtually in a BIM.

Role of design professionals in construction safety

The most significant role of a design professional has traditionally been to design a building or structure such that it conforms to accepted engineering practices, local building codes, and is safe for the public. The safety of construction workers is generally considered to be in the hands of contractors. However, design professionals can improve construction safety by making safer choices in the design and planning stages of a project. It would result in fewer on-site decisions by the contractor that may create hazards for the work force.

Research presented by Behm (2005) suggests that designers can have a strong influence on construction safety. In 1985, the International Labor Office recommended that designers consider the safety of workers that were involved in erecting buildings. In 1991, the European Foundation for the Improvement of Living and Working Conditions concluded that about 60% of fatal accidents in construction were the result of decisions made before the beginning of site work. In 1994, a study of the United Kingdom's construction industry found a causal link between design decisions and safe construction.

Figure 12.1 depicts the ability to influence construction safety at various stages of a project. The ideal time to influence construction safety is during the concept and design stages. The ability to influence safety diminishes as the schedule moves from concept toward start-up. Unfortunately, safety usually is not addressed until construction begins.

The clearest example of design professionals' influence on safety is in the design of a parapet wall. The International Building Code paragraph 704.11.1 requires that a parapet wall be at least 30 inches high. OSHA 1926 Subpart M requires a 42-inch guardrail or other fall protection when working at elevated heights. This means that if the parapet wall were designed to be between 30 inches and 42-inches high, a temporary guardrail at a height of 42 inches or other fall protection would have to be used during construction and future roof maintenance. A decision would have to be made at the site concerning fall protection. This leaves open the possibility of an injury if fall protection is inadequate, workers are not trained, or if fall protection is not used at all. However, if the designer specifies a 42-inch high parapet wall, not only does the design comply with the building code (safe for the public), but the risk of a fall injury during the lifetime of the structure is eliminated because fall protection would not be required.

Designing for Safety can be termed as a formal process that adopts hazard analysis at the beginning of the design stage. Identifying hazards is the first step of the procedure. Engineering measures are then applied to eliminate these hazards or reduce their risk. The design measures start with trying to eliminate the hazards by engineering design. If the hazards still prevail, then safety devices are incorporated. If these methods do not eliminate the risk then warnings, instruction, and training are used as a last resort. This process has been applied to the design of products, equipment, machines, facilities, buildings, and job tasks.

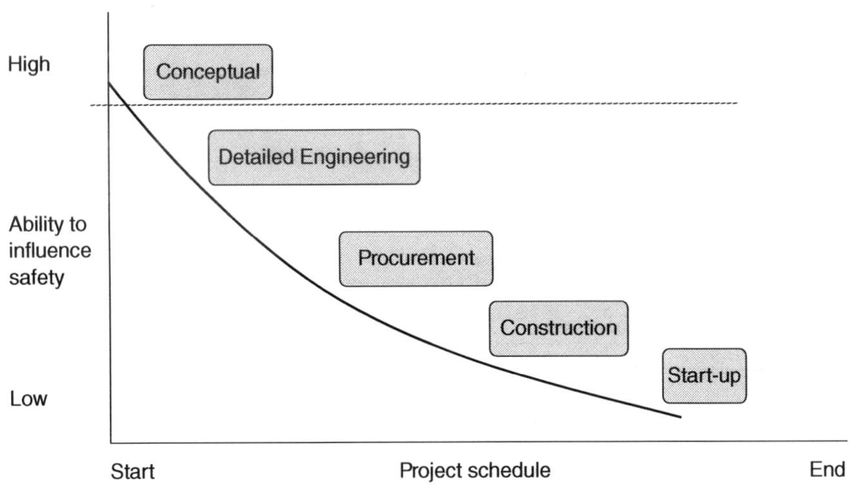

Figure 13.1 Time/safety influence curve

Source: Adapted from Behm, M., *Safety Science*, 43(8), 589–611, 2005. With permission from Elsevier

Designing for construction worker safety (DfCS) is the extension of the DFS process to construction projects. This process applies to the design of a permanent building, facility, or structure. It does not address methods to make construction safer, but helps in making a project safer to build. The use of a fall protection system, for example, is not part of the DfCS process. It would come into play in influencing design decisions that could eliminate or significantly reduce the need for fall protection systems during construction and maintenance. An ability to identify potential hazards associated with construction and maintenance workers in the design stage of a project is required. It is the skill of the design professional to eliminate the hazard by incorporating the appropriate design features.

The involvement of design professionals, specifically engineers, is not totally new to construction safety. Many of OSHA's construction regulations require an "engineer" or "engineering controls." The design expertise would be extended to include the safety aspects of permanent structures, including maintenance, rather than designing temporary structures and systems for construction. The most important feature of this process is the input of site safety knowledge into design decisions. Progress reviews would ensure that safety is considered throughout the design process. The key feature of this process is the input of site safety knowledge into design decisions (Figure 13.2).

Enhancing design to improve site safety

In order to improve the design process, suggestions can be taken from targeted resources such as safety design manuals and checklists, ideas generated by the researchers, and interviews with industry personnel. Manuals and checklists typically address hazards during the startup, operation, and maintenance stages of a project. The on-site workforce and others who visit the job site generally provide the most design suggestions. They are involved in the day-to-day construction activities and are frequently exposed to job site hazards. Any effort that designers put to addressing construction worker safety is voluntary, since OSHA does not place any great responsibility for safety on them.

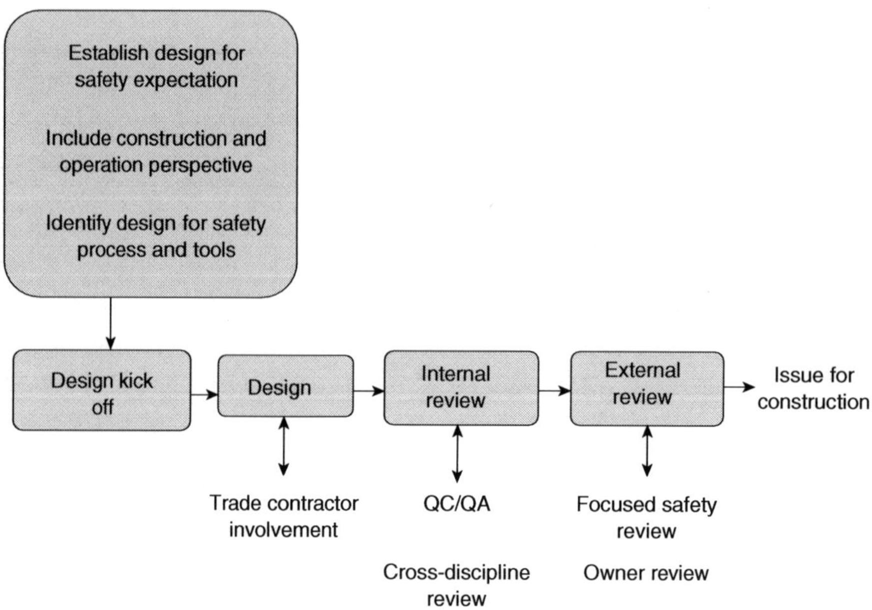

Figure 13.2 A typical design for safety process flowchart

The great diversity of the construction industry has led to the accumulation of design suggestions that reflect all types of design disciplines. The suggestions that deal with both the architectural and construction management disciplines constitute most of the design suggestions. Hence, architects may have the most to gain from the database of design suggestions. The suggestions can be sorted according to the project components, construction site hazards, and project systems. The following list documents possible design suggestions their purposes.

1 Suggestion: The design of prefabrication systems should be based on units that can be built on the ground and erected in the required place.

Purpose: It would reduce the worker's exposure to falls and being struck by any falling objects.

2 Suggestion: Design underground utilities such that they can be placed using trenchless technologies.

Purpose: This would ensure the elimination of the safety hazards associated with trenching.

3 Suggestion: There should be adequate clearance allowed between structures and power lines.

Purpose: Overhead power lines are hazardous when operating cranes.

4 Suggestion: Permanent anchorage points should be designed in the residential roofs.

Purpose: Provide fall protection anchorage for roofing contractors during future maintenance.

5 Suggestion: Cable type lifeline system are required to be designed for tower structures.

Purpose: These will allow workers to hook onto the structure and move up and down during any future maintenance.

6 Suggestion: The window sills should be designed 42 inches above the floor.

Purpose: This will eliminate need for fall protection during construction and future maintenance.

7 Suggestion: Design permanent guardrails around skylights.

Purpose: Prevent workers from falling through skylights.

8 Suggestion: Design components to facilitate prefabrication in the shop or on the ground so that they may be erected in place as complete assemblies.

Purpose: Reduce worker exposure to falls from elevation and the risk of workers being struck by falling objects.

9 Suggestion: Design steel columns with holes in the web at 0.53 and 1.07 m above the floor level to provide support locations for guardrails and lifelines.

Purpose: By eliminating the need to connect special guardrail or lifeline connections, such fabrication details will facilitate worker safety immediately upon erection of the columns.

10 Suggestion: Design beam-to-column double connections to have continual support for the beams during the connection process by adding a beam seat, extra bolt hole, or other redundant connection point.

Purpose: Continual support for beams during erection will eliminate falls due to unexpected vibrations, misalignment, and unexpected construction loads.

11 Suggestion: Minimize the number of offsets in a building plan, and make the offsets a consistent size and as large as possible.

Purpose: Prevent fall hazards by simplifying the work area for construction workers.

12 Suggestion: Design underground utilities to be placed using trenchless technologies.

Purpose: Eliminate the safety hazards associated with trenching, especially around roads and pedestrian traffic surfaces.

13 Suggestion: Design roadway edges and shoulders to support the weight of construction equipment.

Purpose: Prevent heavy construction equipment from crushing the edge of the roadway and overturning.

14 Suggestion: Position mechanical, piping, and electrical controls away from passageways and work areas, but still within reach for easy operation.

Purpose: Controls that protrude into passageways and work areas, or are hard to operate, hidden, or inaccessible, create safety hazards for construction and maintenance workers.

15 Suggestion: Allow adequate clearance between the structure and overhead power lines. Bury, disconnect, or reroute existing power lines around the project before construction begins.

Purpose: Overhead power lines that are in service during construction are hazardous when operating cranes and other tall equipment.

16 Suggestion: Route piping lines that carry liquids below electrical cable trays. Purpose: Prevent the chance of electrical shock due to leaking pipes.

17 Suggestion: Do not allow schedules with sustained overtime.

Purpose: Workers will not be alert if overtime is maintained over a sustained period.

Behavior-based safety

This term is used for programs focused on changing the behavior of workers in order to prevent occupational injuries and illnesses. Such programs make workers accountable for their injuries and illnesses, and rely on the belief that most workplace safety and health problems are the result of voluntary acts. Behavior-based programs target specific worker behaviors, enlist hourly employees and management in monitoring these behaviors on the shop floor, and use a checklist to document workers actions. Related to behavior-based safety are safety incentive programs that reward individual employees, entire departments, and/or workplaces for lower injury and illness rates.

There has been a rise in adapting behavior-based safety as a way of side-stepping the safety and health risks associated with increased line speeds, work duties, mandatory overtime, and other forms of work restructuring. In the year 1931, H.W. Heinrich, an insurance company executive, reviewed accident reports completed by plant supervisors. Because supervisors tended to blame workers for injuries and illnesses, their conclusions backed Heinrich's belief that most industrial accidents are caused by unsafe acts. He later developed a model for explaining the factors contributing to workplace accidents (1959). Often, union members and representatives look to energize existing safety programs and are eager for any attention from management about safety. Their desire to improve cannot be held at fault. They may be unaware of the negative consequences associated with behavior-based programs. Behavior-based safety systems may vary from one organization to the next, depending on its form and complexity. At the basic level they have the following elements in common.

- Identifying behaviors that impact safety,
- Defining these behaviors precisely,
- Development and implementing mechanisms for measuring those behaviors in order to determine their current status and set reasonable goals,

- Providing appropriate feedback, and
- Reinforcing progress.

Due to the failure of behavior-based safety to address the root causes of injuries and illnesses, it is important for union members and representatives to demand resources for safety programs targeting workplace hazards. Behavior-based safety is a bargaining issue just like other management proposals. To substantially improve safety, one must also improve the larger organization.

Worker attitudes toward safety

It is commonly thought that the best way to control and prevent accidents is to have close control over workers. This approach might work with children but it will not work with any workforce. Behavioral scientists have suggested that employees are more committed if they have a voice in setting work goals and controlling safety. Working people always want to contribute to the accomplishment of a worthwhile effort. They seek purposes and principles that lift, inspire, empower, and encourage them to do their best. In order to meet these needs, more companies are moving toward employee empowerment and self-directed work units in the decision-making process. Along with improving safety performance it will also provide positive momentum to the overall work and quality. Companies must acknowledge that employees can perform the job provided they receive appropriate training.

There is a need for a new approach that blends safety attitudes and behavior patterns with production and profit realities. With a change in priorities this approach must stress safety as a human value rather than as a priority. Even if this approach is easy to understand it takes time to apply. The paradigm needs to be shifted along with a change in culture. Culture dictates how employees act and how they are treated. It also demonstrates the values of the company and its employees, and determines whether a job will be performed safely. Culture influences business functions like safety, construction, quality, and scheduling.

There are many possible ways to improve this culture. Top management's philosophy flows down through an organization and impacts all functions. Therefore an organization's philosophy is a projection of the CEO's personal values, beliefs, and ideals as understood and applied by middle management. The deployment of these attributes to middle management and throughout the organization requires careful strategic planning; it is often not done with the rigor it deserves.

Worker attitudes can best be shaped through a culture that involves everyone, at all levels, in developing safety practices and procedures for the organization. Each person within the organization must be visible in the safety process and empowered to do what is needed, and that promotes the sense of self-worth and value. In such an environment safety achieves a value status, hence these values must be based on a mutually accepted set of principles that effectively govern all the actions.

The strengthening of employee safety values requires a supportive approach to promote positive attitudes and behavior modification. It has gained much more importance in today's workplace, as more people are working without direct supervision. The following methods should help to develop a safety-oriented environment.

- Safety awareness: Promote the importance of critically observing safety practices, and understanding how to identify and heed early warnings of safety.
- Changing risk behavior: Increase awareness of attitudes, beliefs, and values to promote behavioral change.
- Thinking process: Learn to manage automatic responses, and how to manage attitudes for the appropriate behavior.
- Individual responsibilities: Increase the level of personal responsibility for safety. This process is bound to intensify the concept of being accountable for one's actions.

- Commitment and values: The employees must be reminded consistently what is at stake for one's self, family, and colleagues and how to make the workplace safe for themselves and others.
- Leadership: The most important aspect of leadership is the acceptance of responsibility, which is absolutely true in the field of safety management.

Contractor's role and contribution to safety

Contractors have been the focus of many studies involving the construction industry, construction management firms, and design/build firms. But the reality is that specialty contractors, often working as subcontractors, perform most construction. The chances of a construction project involving an injury to a construction worker or to a member of the public are continually increasing, Construction industry professionals have to assume that any resulting litigation is likely to involve a jury trial, leading to an uncertain outcome. Moreover, in cases of this type there is always an expert that is willing to advance the opinion that the defendant was negligent. The key to improving this state of affairs is to draft each construction contract in a manner that shifts the responsibility for site safety and the safety of the public to the organization or individual that is best able to control the associated risk. In typical projects, the contractor organization is the most appropriate entity.

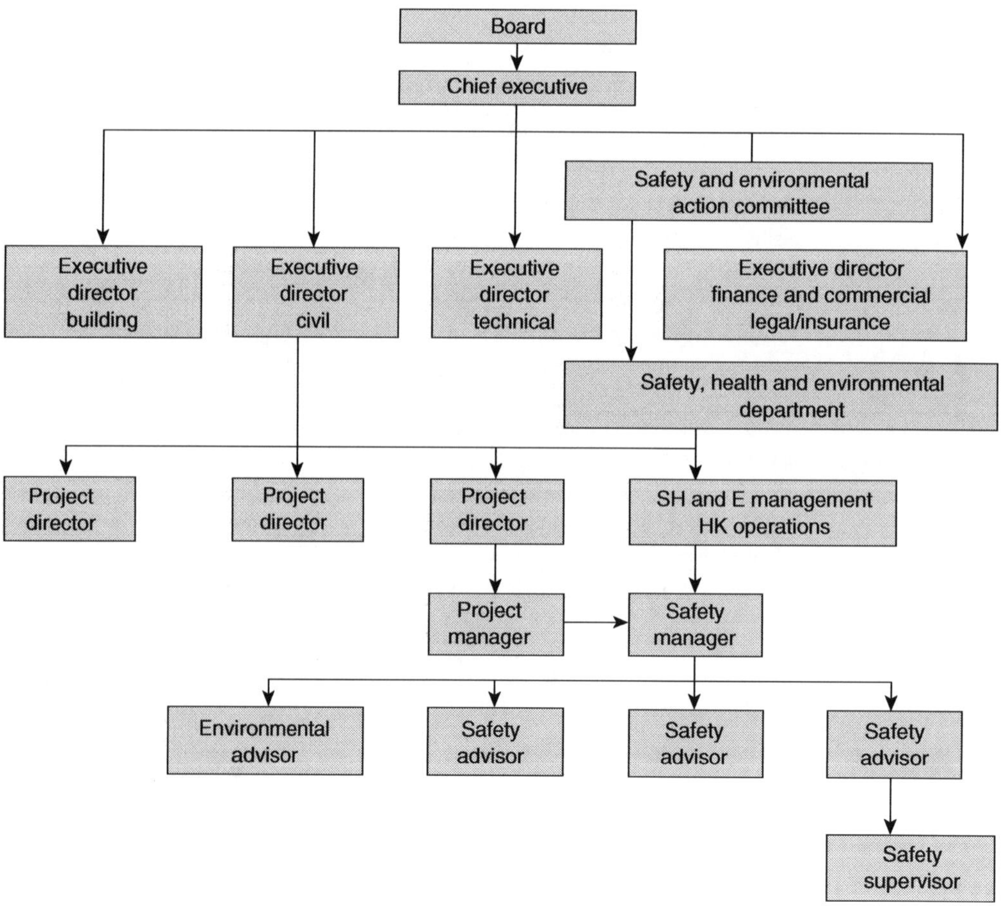

Figure 13.3 Typical safety organization structure of a contractor

Several contractor associations provide a variety of safety-related services and information; these resources have been effective in reducing the incidence of worker injuries. In projects where trade associations have been involved, safety performance has been better than in cases where this support was unavailable. Specialty contractors and subcontractors such as mechanical contractors may benefit especially from this approach, given their comparatively smaller size and structure.

The AGC and ABC are the largest of these associations, with nationwide offices; depending on the respective geographic location, their capabilities include a broad variety of services including non-safety issues such as skills testing and technical training. They are also equipped to screen workers and conduct drug testing, as a reliable third party. The use of these services increases the likelihood that costly staff turnover can be minimized, as the screening and testing process identifies those people that are best suited for working with the contractor organization

Research findings suggest that safety incentives are not a guarantee of good safety performance, and should be employed with caution. As construction activity increases, the injury rate tends to increase as well. Formal safety programs should be introduced in keeping with construction activity, so that site safety can be managed before it becomes a problem.

Safety through regulatory practices

A failure to manage site safety effectively results in worker injuries and leads to financial losses, human conflicts, and civil penalties. The government sets regulations to provide a safe working environment for workers. There are two kinds of safety regulations: prescribed safety regulations, where the government states the minimal working conditions that must be provided in a construction site; and self-regulatory, where the government states general duties for employers, thus allowing them to determine the best way of achieving the objectives of the legislation in their organization. The knowledge of a safety engineer is of great importance in both situations. In a prescribed system, government regulation cannot address all of the possible safety hazards that might occur at construction sites. In a self-regulatory system, on the other hand, an organization must develop its own approach to providing adequate working conditions and therefore the experience of safety engineers is of equally great importance.

To capture knowledge, it is important to know the typologies of knowledge. Two such typologies have been defined. Explicit knowledge is defined as knowledge that is precisely and formally articulated and is often codified in databases of corporate procedures and best practices, whereas tacit knowledge is understood and applied unconsciously. Tacit knowledge is based on practical experience, while explicit knowledge is theoretically based.

Explicit knowledge

Explicit knowledge of construction site safety exists in accident records, safety regulations, and safety guidelines. Accident records contain information on actual accidents reported at construction sites. These are useful for risk assessment by categorizing the safety hazards in terms of frequency-consequence. An organization can use this knowledge to prioritize its responses to these respective hazards, and better manage its resources to address them.

In Figure 13.4, hazards are categorized as Type I must be prioritized while Type IV hazards can be assigned a lower priority. Type I hazards are those that may have major consequences and occur frequently in building construction projects, such as objects falling through an unprotected opening in a floor slab. Type II hazards may have major consequences but do not occur as frequently as Type I hazards. A worker's fall from the upper floor of a building would be such an example. Type III hazards may have moderate consequences and seldom occur, such as a worker being struck by a tower crane. The categorization of hazards may vary in different organizations since the frequency and consequences of accidents depend on several factors.

		Frequency		
		Often	Moderate	Seldom
Consequences	Major	I	I	II
	Moderate	I	II	III
	Minor	II	III	IV

Figure 13.4 Frequency-consequence level for categorizing safety hazards

Safety regulations can also represent a type of explicit knowledge in site safety. Examples of these are the Occupational Safety and Health Acts from the United States and Construction Site Safety Regulations (CSSR) of Hong Kong. These regulations affirm the minimum required conditions that must be followed in a construction project; however these conditions might not be enough to provide a safe working environment. This is especially true for contractors working in countries encouraging self-regulation through the implementation of a safety management system. The scheme provides general duties for employers, allowing them to achieve the objectives of the legislation in an approach best suited to their organizational culture.

Tacit knowledge

Tacit knowledge of construction site safety is of paramount importance for organizations in the self-regulatory system, since the knowledge of safety engineers and managers influences the level of on-site safety. Hence the knowledge of these professionals must be harnessed; it relates to the senses, instinctive bodily responses, individual perception, experience, rules of thumb, and intuition. Safety hazard recognition is an important actualization of tacit knowledge in construction site safety. Safety hazard recognition is considered a tacit knowledge because it relies on the safety engineer's experience. If management does not recognize the hazards that may occur on a site, then management cannot provide relevant training or procedures to handle uncertain conditions.

Knowledge management is usually manifested as a business system that is enabled by an array of technologies. It requires both the explicit and tacit knowledge of construction site safety personnel to derive the following benefits:

1 Developing effective safety programs that recognize actual safety hazards. After capturing the available knowledge, an organization can ensure that safety engineers and operational units develop a common definition of the respective hazards. This common definition enables practical safety programs that directly address the hazards that are likely to be met.
2 Establishing an effective training program that improves workers' skill related to the actual safety hazards identified. Once the knowledge is captured by the safety management team, they can provide an effective training program for workers that is tailored to the hazards that are likely to be encountered.

Problems with capturing knowledge

Explicit knowledge is easier to capture than tacit knowledge. Explicit knowledge can be captured from existing theories and axioms written in books, regulations, company records, and guidelines. However, tacit knowledge is obtained from the experience of an individual (i.e., safety engineer or manager), therefore it is difficult to capture since the knowledge is stored in **a** unique way in each individual's mind. Two conventional mechanisms are available to capture the knowledge of safety engineers or managers. First, it can be explored by conducting discussions. Second, it can also be observed at an actual site and provide better safety information related to the actual construction components and the processes that may have safety hazards.

This approach has the advantage of reflection, reminding safety engineers or managers of an earlier experience they have had. But this mechanism has the following limitations:

1 It may be unsafe for workers since they are exposed to construction operations.
2 It may obstruct workers in carrying out their tasks.
3 It can only be undertaken during the construction stage.

Management responsibility

The Occupational Safety and Health Administration (OSHA) specifically requires that employers be responsible for providing workers a place of employment free from recognized hazards. This may be interpreted as a safe place in which to work. The type of work in which the company is engaged influences a construction firm's safety practices. A contracting firm may employ hundreds of workers as employees and subcontractors. Consequently, the general contractor may have the responsibility of managing the safety of its own employees as well as the employees of the various subcontractors utilized for the project. This often places the general contractor in an awkward position, since the firm may not be competent in managing the safety practices of all subcontractors. In most cases the general contractor defers the responsibility of safety to the individual subcontractors and may never take part in ensuring that they are actually utilizing all the measures necessary for a safe working environment.

Subcontractors face similar problems to those of general contractors as they may employ large groups of workers. They are challenged to implement a safety program that will satisfy both the requirements of the general contractor and the standards outlined by OSHA. However, a subcontractor's emphasis on safety is often influenced by the size of its organization. Although some smaller firms may have excellent safety programs and records, many find it a difficult process because of the economies of scale. Safety training is often left to an on-the-job learning exercise or taught by the employees' union or trade organization.

According to J. Hinze at the University of Florida, the owner of the project and the project designers can also have an impact on the safety practices encountered on the job site. Moreover, engineers have a professional and moral obligation to take safety, health, and welfare under consideration. As described above, today's trend in the construction industry is to enhance safety by design.

However, it is important to note that the owner is best positioned to make safety a priority in a project by including safety management in the design and performance criteria. In fact, there may be a decrease in cost due to reduced insurance rates, which in turn will lower the contractors' expenses. As this method of design gains acceptance, it is almost certain to improve overall construction safety along with the safety records of construction companies.

Safety practices

The safety practices that are encountered on construction sites are as diverse as the sites themselves. All construction sites have their own distinctive aspects of safety. The larger the construction projects the better they tend to be organized from a safety standpoint. These types of projects are often high profile. Companies involved in such visible construction projects have reputations to sustain as well as safety records to maintain and hence are generally better prepared to manage the safety aspects of a project. Having full-time safety personnel relieves the pressure on the on-site construction project team. They can keep team members informed of possible safety problems and ensure that subcontractors are implementing their safety programs adequately, freeing the project personnel to focus their efforts on the project itself.

Small to medium projects tend to hold the greatest potential for improvement in safety management. Projects of this type usually involve smaller firms that may not have an adequate safety program or the personnel to oversee safety criteria. Implementation of their safety management programs is usually left up to the foreman or the project superintendent. It is on these types of projects that good safety management practices, implemented properly, can have a very positive impact on overall project safety and improve the company's profits by keeping injuries and claims to a minimum.

Case study: Safety implementation at the workspace

An effective method of safety management was used on a small-to medium-sized project that experienced many of the problems characteristic of smaller projects. While the project was not difficult to construct, it had a tight schedule, which resulted in the project team directing the majority of their efforts in attempting to meet construction deadlines. As a result, the safety management on the project did not obtain the attention that it should have received. At peak times, this firm had approximately 20 employees working on the project. The company required daily "tool box" safety talks each morning for their employees. These talks lasted about 10 minutes and included a description of the work to be accomplished that day. In addition, the foreman or job superintendent discussed appropriate safety information for the task(s) with the team.

This approach worked well at the beginning of the project, since the small number of subcontractor employees on the job generally attended the contractors' morning safety sessions. However, as construction progressed it became increasingly difficult to maintain control of the on-site safety practices. The company had a full-time safety department, but, as it was a branch office of a Canadian construction firm, the site was often not visited by the safety officer. Therefore, little safety support was given to the superintendent. The difficulty encountered was that subcontractors on the project did not, at times, place the proper emphasis on safety. In fact, they sometimes overlooked safety violations in order to maintain the project schedule.

As the total number of workers on the site increased to approximately 50, and subcontractors began holding their own safety talks, it became clear that changes needed to be made. In addition to the tool box meetings, a weekly safety conference of foremen was also held by the general contractor. Appropriate safety information and project scheduling were reviewed in those meetings. In addition, possible conflicts between various trades were discussed in order to decrease potential problems and increase the safety level at the job site. It was first suggested at this weekly safety/project meeting to assign different trades the task of conducting safety inspections. Therefore, instead of the general contractor being solely responsible, the enforcement of safety was to be shared. The weekly safety inspections would be conducted on a rotating basis by the various on-site subcontractors.

This method also educated employees on the safety concerns of the various trades and how they could affect the project. The overall results achieved by this method of safety management were excellent. In particular, the safety inspections were generally conducted the day before or the morning of the weekly meeting with the foremen. With this approach, any needed corrective action could be quickly taken. In addition, the foremen and superintendent were exposed to the concerns of various trades that increased their knowledge of a variety of different safety practices. It should be noted that the subcontractors did not request payment for the time spent making the inspections, and they were not concerned with possible liability problems.

As with any new concept, there was some difficulty in implementing the process. Older workers, in particular, perceived that they were being criticized for their actions and initially rejected the plan. As the project progressed, the various trades actually looked forward to their assigned week to conduct the safety inspections and, surprisingly, most of the subcontractors welcomed the feedback on their safety practices. Another positive benefit of this approach was that it required the involvement of the workers, not just the foremen and superintendents. This meant that the apprentices' comments and observations were just as important as those of the journeyman or the foreman on the project.

Ways to improve workplace safety

There are several factors that can help to improve safety in the workplace. These factors, on their own or combined, provide a comprehensive method for developing an improved safety program.

Stretching exercises

Reese and Eidson (2006) explained that scientific research has not yet determined whether exercise programs are an effective prevention technique. However, many companies that have implemented exercise programs report that warm-up programs have helped them reduce back injuries and other cumulative trauma disorders. Stretching exercises can help to keep involved muscle groups from tightening up; workers may be unaware of gradual changes that eventually lead to chronic medical problems. Stretching exercises are most effective if they are done on a daily basis, as such routine activity help workers to strengthen major muscle groups, especially the ones that are important for maintaining a healthy back.

Daily safety huddles

Safety huddles are short safety meetings covering a specific safety topic or type of hazard. They have been used in the industry for many years. A strong safety culture is created by keeping employees focused on safe work practices and behaviors. A daily safety huddle program is the tool that helps to keep the workforce talking and thinking safety. A huddle can be held with a group of workers to review:

1. Hazards of a job that is about to be started;
2. Causes and corrective action for a recent accident or "near miss";
3. Job hazards that should be emphasized along with safety rules that will prevent accidents; and
4. General problems concerned with both on-the-job and off-the-job safety.

The ideal size for a safety huddle is 6–10 workers. Larger groups can be used; however, the workers will be less likely to enter into any discussions. A huddle normally lasts approximately 10 minutes if workers are not seated, if the topic is well chosen, and if the discussion does not get off on a tangent. In preparing for a huddle, if it appears that the interest will last more than 10 minutes, consider a series of three or four huddles or hold the meeting in a room where the workers can sit comfortably and where noise and other distractions are absent. Normally a supervisor leads a huddle, simply because it is his or her responsibility to know the employees, the nature of the job to be undertaken, and its associated hazards. There is no reason that a worker cannot conduct a huddle provided he or she (1) takes the assignment seriously, (2) is given sufficient time to prepare, (3) is not too nervous about the idea, and (4) is given encouragement and help, if it is needed. A small huddle is generally better for a person who is not used to speaking to a group. There is no one best way to conduct a huddle. However, the following suggestions should give a good chance of success.

1. Make a clear announcement of the time, place, and reason for the huddle. Start on time.
2. Explain why the huddle is being held.
3. Keep the huddle from going off on a tangent. If a worker hits on a thought that has merit, promise to have a huddle on that topic at a later date. Keep the promise!
4. Allow time for discussion and questions. If you don't know the correct answer, don't guess at it. Call your safety representative who probably has the answer.
5. Appeal to the workers' desire for approval. Point out things that are being done right as well as problem areas. Also, you can appeal to the workers' competitive instinct. Build a good "sales pitch" and show some enthusiasm!

6 Use a real accident case to emphasize a point. The more recent and the more close (geographically) to your location, the more effective the example will be.

lean construction changes the way we work and communicate also in the meetings. Lean principles, methods and tools help to create collaboration between subcontractors, foremen and superintendents to plan the overall project schedule and delivery (Howell and Macomber 2006). Leaders should encourage conversation to get commitments made and enhance co-ordination and co-operation (Sull and Spinosa 2007). Lean tools promote communication, standardized work and the reduction of waste, all of which help make our working environment and conduct on the sites safer.

Safety officer/liaison

Duties vary, but generally an OSHA safety officer oversees the planning, creation, and implementation of the occupational health and safety program of a company or governmental organization. Duties include administrative, supervisory, and hands-on tasks.

Program design

The officer designs a safety program consistent with OSHA requirements that sets up processes to reduce accidents and work-related injuries. The program should assign safety responsibilities to various employees. The officer designs procedures and forms for internal reporting of compliance with safety procedures as well as program violations, curative steps, accidents, and injuries. She/he also maintains Materials Safety Data Sheets (MSDS) and other OSHA forms.

The officer analyzes workforce accidents and injuries to determine where process changed, communication or training needs improvement, and develops and manages reports for company management. He/she also recommends and implements program changes where necessary. He/she ensures OSHA requirements are followed and prepares the company for audits. The safety officer acts as a liaison between the company and OSHA.

The OSHA safety officer supervises staff safety officers and ensures coverage when the company operates. An OSHA safety officer may be responsible for developing safety procedures for new company products, services, and manufacturers.

Incentive programs

Peer and 360 reviews measure employee strengths and development areas. Peer reviews offer candid feedback to employees from those that work with them the most. A 360 review is the most comprehensive type of review and includes peer reviews, self-ratings, and higher management reviews (skipping the employees' direct manager). Many companies conduct peer and 360 reviews on employees at the senior management levels.

Reward programs such as annual or off-cycle promotions, salary increases, and bonuses that are tied to employee performance are used by most firms. In addition, there are certificate, anniversary, and achievement award programs. Depending on the structure of the company's human resources corporate initiatives, reward programs may be used weekly, monthly, quarter, biannually, or annually to recognize outstanding employee contributions.

Skip level lunches are a tool employers use to encourage, recognize, and motivate employee performance. During skip level lunches, senior managers meet with top employees within their department or division. Employees are encouraged to talk freely with senior managers, ask questions, network and, of course, enjoy a free meal.

Another form of an employee motivation incentive program is the employee networking program. Employee networking programs bring in prominent guest speakers to kick off events such as a women's conference or a department manager's off-site meeting. Some companies create employee networks that focus on culture, history, and current achievements, physical, and other employee differences (Black employee network, Hispanic employee network, employees with disabilities network). The networks are generally open to all employees at the firm.

Effects

Employee motivation incentive programs increase employee awareness about how performance criteria are measured at the company. The programs also motivate an employee to take responsibility for the success of his/her career.

Summary

There has been a growth of interest in construction worker safety. Rather than being just a concern of constructors that is enforced by OSHA regulations, job site safety has also attracted the attention of owners. These owners have recognized that the cost of construction accidents and injuries eventually increases construction costs, and they have begun to place safety on a higher level of priority. However, this attitude has not yet been adopted by the design community, and designers typically avoid any responsibility for the safety of construction workers to minimize their liability exposure. Existing codes and design standards reflect this attitude, and worker safety rests on the constructors' shoulders. Moreover, designers are not traditionally educated about DfCS. No reference standards exist to bridge the gap between existing design standards and construction worker safety.

Design professionals can enhance project safety, mitigate common safety hazards, and reduce the number of worker injuries by using the DfCS approach. It offers the benefit of reducing the need for redesign work, and can help to avoid the types of accidents that could lead to the need for liability claims on all parties, including the designers themselves. Design suggestions have been accumulated that provide designers practical examples of how to design for construction worker safety. And since designers traditionally have not been educated on site hazards and do not continually observe them, they have not been able to provide as much input into the list of design suggestions.

A gap exists between constructors and designers in the knowledge of and commitment to job site safety. As a result, information on designing for construction site safety must ultimately be drawn from construction personnel and transferred to designers. Construction workers are unique facility users and their safety warrants the attention of designers. Addressing safety in the construction phase not only requires soliciting and publishing design suggestions, but also requires designers to change their traditional mindset. Thus, owners must provide the initial impetus – by requesting or requiring by contract terms – that designers consider construction worker safety in their designs. Lean design and construction facilitates this paradigm shift through a closer collaboration between designers and contractors, and through a relational form of contract that apportions risk fairly. Adapting the lean principles and techniques can lower costs by eliminating waste and improve the overall safety in construction.

Questions for discussion

1. Why is it important to consider the safety aspect in the construction industry?
2. What role do design professionals play in the workers safety? Elaborate.
3. What connection is there between safety and lean construction?
4. What part do contractors and owners play in the safety of the construction project?

5 How do workers' attitudes to safety affect the overall safety of the construction?
6 Describe behavior-based safety and its role in worker safety.
7 Discuss the pros and cons of attempting to build a project using lean methods without a specific safety component?

Appendix: OSHA checklists (selected items)

Note: Due to space limitations, an abbreviated checklist is provided below purely for illustrative purposes. It was abstracted from a more comprehensive checklist developed by the Occupational Safety and Health Administration (OSHA 1992). Readers are advised to consult with safety professionals to gain access to *current* checklists that are appropriate for field application.

Employer posting

- Is the required OSHA workplace poster displayed in a prominent location where all employees are likely to see it?
- Are emergency telephone numbers posted where they can be readily found in case of emergency?
- Where employees may be exposed to any toxic substances or harmful physical agents, has appropriate information concerning employee access to medical and exposure records and "Material Safety Data Sheets" been posted or otherwise made readily available to affected employees?
- Are signs concerning "Exiting from buildings," room capacities, floor loading, biohazards exposures to X ray, microwave, or other harmful radiation or substances posted where appropriate?

Recordkeeping

- Are all occupational injuries or illnesses, except minor injuries requiring only first aid, being recorded as required on the OSHA 200 log?
- Are employee medical records and records of employee exposure to hazardous substance or harmful physical agents up-to-date and in compliance with current OSHA standards?
- Are employee training records kept and accessible for review by employees, when required by OSHA standards?

Safety and health program

- Do you have an active safety and health program in operation that deals with general safety and health program elements as well as the management of hazards specific to your worksite?
- Do you have a safety committee or group made up of management and labor representatives that meet regularly and report in writing on its activities?
- Are you keeping your employees advised of the successful effort and accomplishments you and/or your safety committee have made in assuring they will have a workplace that is safe and healthful?

Medical services and first Aid

- Is there a hospital, clinic, or infirmary for medical care in proximity of your workplace?
- If medical and first-aid facilities are not in proximity of your workplace, is at least one employee on each shift currently qualified to render first aid?
- Are medical personnel readily available for advice and consultation on matters of employees' health?

- Are emergency phone numbers posted?
- Are first-aid kits easily accessible to each work area, with necessary supplies available, periodically inspected, and replenished as needed?

Fire protection

- If you have a fire alarm system, is it certified as required?
- Are fire doors and shutters in good operating condition?
- Is the maintenance of automatic sprinkler systems assigned to responsible persons or to a sprinkler contractor?
- Are fire extinguishers recharged regularly and noted on the inspection tag?
- Are employees periodically instructed in the use of extinguishers and fire protection procedures?

Personal protective equipment and clothing

- Are approved safety glasses required to be worn at all times in areas where there is a risk or eye injuries such as punctures, abrasions, contusions, or burns?
- Are protective gloves, aprons, shields, or other means provided and required where employees could be cut or where there is reasonably anticipated exposure to corrosive liquids, chemicals, blood, or other potentially infectious materials. See 29 CFR 1910.1030(b) for the definition of "other potentially infectious materials."
- Are hard hats provided and worn where danger of falling objects exist?
- Is appropriate foot protection required where there is the risk of foot injuries from hot, corrosive, poisonous substances, falling objects, crushing, or penetrating actions?
- Do you have eye wash facilities and a quick drench shower within the work area where employees are exposed to injurious corrosive materials?

General work environment

- Are all worksites clean, sanitary, and orderly?
- Are work surfaces kept dry or appropriate means taken to assure the surfaces are slip-resistant?
- Is combustible dust cleaned up with a vacuum system to prevent the dust going into suspension?
- Are all toilets and washing facilities clean and sanitary?
- Are all work areas adequately illuminated?

Walkways

- Are aisles and passageways kept clear?
- Are wet surfaces covered with non-slip materials?
- Are holes in the floor, sidewalk, or other walking surface repaired properly, covered, or otherwise made safe?
- Is there safe clearance for walking in aisles where motorized or mechanical handling equipment is operating?

Floor and wall openings

- Are floor openings guarded by a cover, a guardrail, or equivalent on all sides (except an entrance to stairways or ladders)?

- Are skylight screens of such construction and mounting that they will withstand a load of at least 200 pounds?
- Are grates or similar type covers over floor openings such as floor drains of such design that foot traffic or rolling equipment will not be affected by the grate spacing?
- Are manhole covers, trench and similar covers, plus their supports designed to carry a truck rear axle load of at least 20,000 pounds when located in roadways and subject to vehicle traffic?

Stairs and stairways

- Are standard stair rails or handrails on all stairways having four or more risers?
- Do stairs have landing platforms not less than 30 inches in the direction of travel and extend 22 inches in width at every 12 feet or less of vertical rise?
- Are stairway handrails located between 30 and 34 inches above the leading edge of stair treads?
- Are stairway handrails capable of withstanding a load of 200 pounds, applied within 2 inches of the top edge, in any downward or outward direction?
- Is material on elevated surfaces piled, stacked, or racked in a manner to prevent it from tipping, falling, collapsing, rolling, or spreading?

Exiting or egress

- Are exit signs provided with the word "EXIT" in lettering at least 5-inches high and the stroke of the lettering at least 1/2-inch wide?
- Are at least two means of egress provided from elevated platforms, pits, or rooms where the absence of a second exit would increase the risk of injury from hot, poisonous, corrosive, suffocating, flammable, or explosive substances?
- Are there sufficient exits to permit prompt escape in case of emergency?
- Are special precautions taken to protect employees during construction and repair operations?
- Where ramps are used as part of required exiting from a building, is the ramp slope limited to 1 foot vertical and 12 feet horizontal?

Exit doors

- Are doors that are required to serve as exits designed and constructed so that the way of exit travel is obvious and direct?
- Are exit doors openable from the direction of exit travel without the use of a key or any special knowledge or effort when the building is occupied?
- Where exit doors open directly onto any street, alley, or other area where vehicles may be operated, are adequate barriers and warnings provided to prevent employees stepping into the path of traffic?

Portable ladders

- Are all ladders maintained in good condition, joints between steps and side rails tight, all hardware and fittings securely attached and moveable parts operating freely without binding or undue play?
- Are non-slip safety feet provided on each ladder?
- Are ladder rungs and steps free of grease and oil?

References

Andersen, Bjørn, Alemu Moges Belay, and Eva Amdahl Seim. 2012. Lean construction practices and its effects: A case study at St Olav's integrated hospital, Norway. *Lean Construction Journal*, 122–149.

Antillon, E. I. 2010. *A research synthesis on the interface between lean construction and safety management.* Civil Engineering Graduate Theses & Dissertations. 31. Online at https://scholar.colorado.edu/cven_gradetds/31

Behm, M. 2005. Linking construction fatalities to the design for construction safety concept. *Safety Science* 43(8): 589–611.

Bureau of Labor Statistics. 2006. *Fatal and non-fatal occupational statistics from 1992–2004.* Online at http://www.bls.gov. Accessed on March 1, 2006.

Bureau of Labor Statistics. 2019. *Fatal and non-fatal occupational statistics from 1995–2017.* Online at www.bls.gov. Accessed on May 1, 2019.

Dorman, P. 2000. *The economics of safety, health, and well-being at work: An overview. In Focus program safe work, international labor organization (ILO).* Geneva/Olympia, WA: The Evergreen State College.

Etges, B., T. A. Saurin, and I. R. Bulhões. 2012. Identifying lean construction categories of practices in IGLC. *Proceedings of the 20th annual conference of the international group for lean construction, international group for lean construction (IGLC).* Oslo, Norway.

Everett, J. G., and P. Frank, Jr. 1996. Cost of accidents and injuries to the construction industry. *Journal of Construction Engineering and Management, ASCE* 122(2): 158–164.

Fullman, J. B. 1984. *Construction safety, security, and loss prevention.* New York: John Wiley and Sons, Inc.

Gambatese, J., Toole, T.M, and Behm, M. 2008. Prevention through design practice and research: a US construction industry perspective. *Proceedings of the CIB W99 International Conference.* Hinze J, Bohner S, Lew J, eds., Gainesville: University of Florida, Mar. 9–11: 93–102.

Heinrich, H. W. 1959. *Industrial accident prevention*, 4th ed. New York: McGraw-Hill.

Hinze, J., and L. Applegate. 1991. Costs of construction injuries. *Journal of Construction Engineering and Management, ASCE* 117(3): 537–550.

Hinze, J., R. Coble, and T. Haupt. 2000. *Construction safety and health management.* Upper Saddle River, NJ: Prentice-Hall Inc.

Howell, G. A., G. Ballard, T. S. Abdelhamid, and P. Mitropoulos. 2002. Working near the edge: A new approach to construction safety. *Proceedings of the 10th annual conference of the international group for lean construction (IGLC-10).* Gramado, Brazil.

Howell, G. A., and H. Macomber. 2006. What should project management be based on? *Proceedings of the 14th annual meeting of the international group for lean construction.* Santiago, Chile.

Koskela, L., G. Howell, G. Ballard, and I. Tommelein. 2002. The foundations of lean construction. In *Design and construction – Building in value*, eds. R. Best, and G. De Valence. Oxford: Elsevier Science Ltd, 211.

Mäki, T., and A. Koskenvesa. 2012. An examination of safety meetings on construction sites. *Proceedings for the 20th annual conference of the international group for lean construction (IGLC).* July 17–22, San Diego, CA.

Nesensohn, C., S. T. Demir, and D. J. Bryde. 2012. Developing a 'true north' best practice lean company with navigational compass. *Proceedings of the 20th annual conference of the international group for lean construction (IGLC).* Oslo, Norway.

OSHA Checklists. 1992. Adopted from OSHA Publication #2209 – *OSHA handbook for small businesses* (revised). Washington, DC: OSHA.

Sull, D. N., and C. Spinosa. 2007. Promise-based management – The essence of execution. *Harvard Business Review*, April. pp. 78–86.

Weinstein, M. B. 1997. *Total quality safety management and auditing.* Boca Raton, FL: Lewis Publishers.

Wilson, J., and E. Koehn. 2000. Safety management: Problems encountered and recommended solutions. *Journal of Construction Engineering and Management, ASCE* 126(1): 77–79.

Bibliography

Gambatase, J., C. Pestana, and H. W. Lee. 2017. Alignment between lean principles and practices and worker safety behavior. *Journal of Construction Engineering and Management* 143(1).

Gambatese, J., J. Hinze, and C. Haas. 1997. Tool to design for construction worker safety. *Journal of Architectural Engineering, ASCE* 3(1).

Kelley, R. 1996. Worker psychology and safety attitudes. *Professional Safety, Journal of the American Society of Safety Engineers* (July): 14–17.

Krizan, W. 2000. Insurance: Party is over for cheap workers' compensation coverage. *Engineer News Record ENR* 245(13): 44–45.

OSHA Work zone safety Quick Card. 2007. 3284–05R-07. *Working safely in trenches safety tips QuickCard™* US Department of Labor. Online at www.osha.gov.

Reese, C. D., and J. V. Eidson. 2006. *Handbook of OSHA construction safety and health*, 2nd ed. Oxford: Taylor & Francis Group, 545.

Zhang, S., K. Sulankivi, M. Kiviniemi, I. Romo, C. M. Eastman, and J. Teizer. 2015. BIM-based fall hazard identification and prevention in construction safety planning. *Safety Science* 72: 31–45.

Zhang, S., J. Teizer, J. Lee, C. Eastman, and M. Venugopal. 2012. Building information modeling (BIM) and safety: Automatic safety checking of construction models and schedules. *Automation in Construction* 29, January 2013: 183–195. Elsevier.

14
GENERAL LEAN TOOLS

Several lean tools are placed in this chapter to have them in one convenient location. Several of these tools and techniques are illustrated with examples to reinforce understanding. They are:

1 The PDCA cycle
2 Pareto analysis
3 The Ishikawa diagram
4 The 5 Why diagram
5 Swim lane diagrams
6 Plus Delta analysis
7 The 5 Step plan (5S)
8 The Kaizen methodology
9 Choosing by Advantages
10 Value Stream mapping
11 The A3 report
12 First Run studies
13 Visual management

1. The PDCA cycle

The PDCA cycle is an improvement cycle based on the scientific method of proposing a change in a process, implementing the change, then measuring the results and taking appropriate action. The four stages are respectively. Plan – identify the problem and develop a plan for improvement, Do – implement the plan on a trial basis, Study or Check – assess whether the plan is working, then Act – Institutionalize the improvement, then continue the cycle.

Continual improvement seeks to eliminate waste in all forms, improve quality of products and services and responsiveness to customers while reducing costs.

Note – problem solving is not necessarily improvement; it is maintenance. For example, if a process has quality levels of 95% and they are elevated to 99%, that represents improvement. If on the other hand quality levels fall to 50% and a team effort restores them o 95% then that would be maintenance.

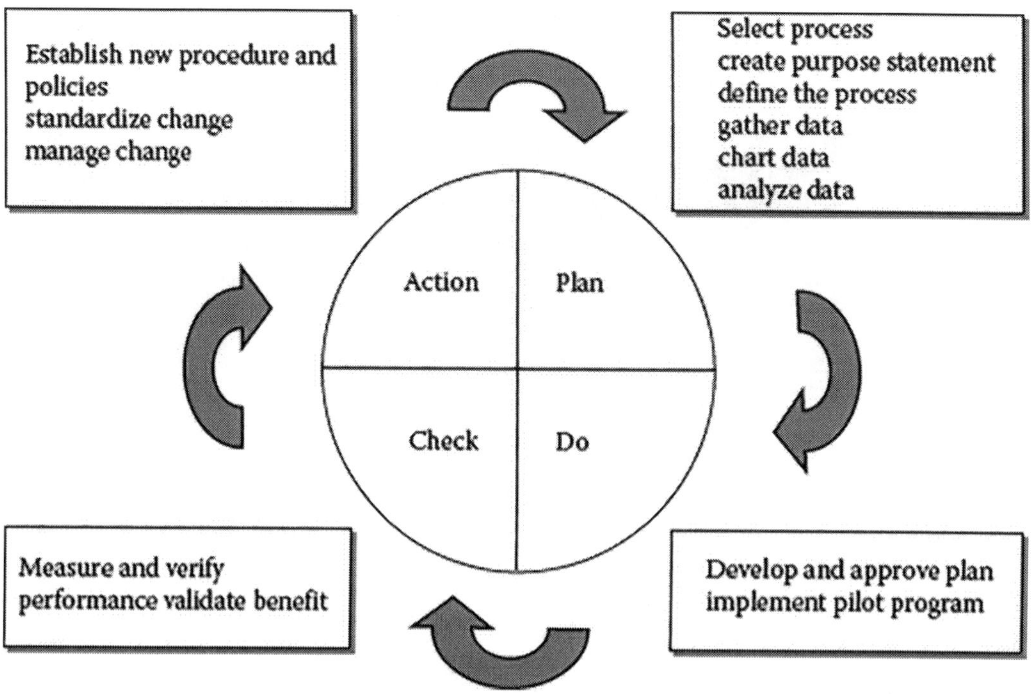

Figure 14.1 Plan-Do-Check-Act Cycle

2. The Pareto chart

The Pareto chart is named for an Italian economist, Vilfredo Pareto (1848–1923), who studied the distribution of wealth in Europe. He proposed an economic theory based on the observation that a few people were very wealthy, while large numbers of people had relatively little means. Essentially, 20% of the people in Italy held 80% of the wealth, hence the "80–20" rule. Dr. Joseph Juran applied this concept to the field of quality, distinguishing between the "vital few" and the "useful many." The Pareto chart helps users to identify those 20% attributes that represent 80% of the benefit. Faced with a variety of problems, a user can find the 20% that, when addressed, provide 80% of the benefit.

The example shown below represents a window installation operation. The data provided, list the number of problems relating to window installation and their unit cost. For example, when a window is installed in a manner that violates the code, it costs $1,100. Four occurrences are priced at $4,400. In performing Pareto analysis, one must decide whether to use the number of occurrences or the costs of these occurrences. In this case, the analysis was based on total costs.

NOTE FOR PROBLEM SOLUTION. When the data for these problems are provided they may be in random order. Before an analysis can be performed, the values of interest must first be listed in descending order of magnitude. It is best to use a table or matrix to record the values.

The procedure for drawing the chart is as follows:

1 Set up a matrix like the one below, with the respective headings as shown.
2 List the values of the attributes in descending order of magnitude (i.e., the largest cost first and the smallest last). The category "Not finished on time" costs $19,625 By contrast, the category "crew very rude" costs $800 and is listed last.
3 The cost column is totaled to $44,824.10.

4 For each category, divide its cost by this total to obtain a percentage.
5 Use the right-hand column to create a cumulative total.
6 Draw the chart with two "y" axes. The left-hand vertical scale is used for the cost of the transactions. The right-hand vertical axis is used for the cumulative value.
7 Draw the cumulative curve, using the right-hand scale and terminating at a value of 100%.

The completed chart (Figure 14.2a) shows that the first four categories represent 80% of the cost of complaints.

Item	*Category*	*Quantity*	*Cost per Occurrence*	*Total Cost*	*%*	*Cumulative %*
c	Not finished on time	25	785.00	19,625.00	44%	44%
g	Wrong window installed	2	3500.00	7000.00	16%	59%
f	Job not done properly	4	1250.00	5000.00	11%	71%
a	Code infraction	4	1100.00	4400.00	10%	80%
h	Damage during window installation	1	3500.00	3500.00	8%	88%
e	Not starting on time	15	150.00	2250.00	5%	93%
b	Leave job site dirty	18	124.95	2249.10	5%	98%
d	Crew very rude	8	100.00	800.00	2%	100%
		77		44,824.10	100%	

Figure 14.2a Analysis of Customer Complaints

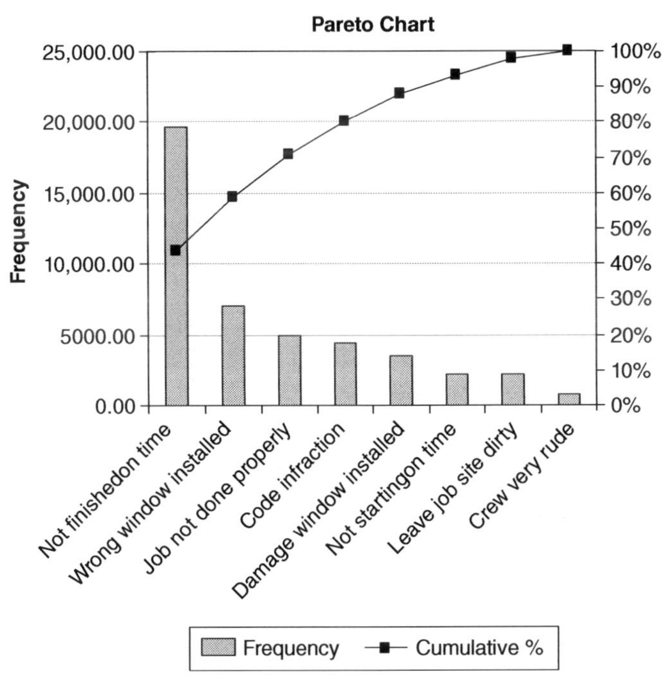

Figure 14.2b Pareto chart of customer complaints

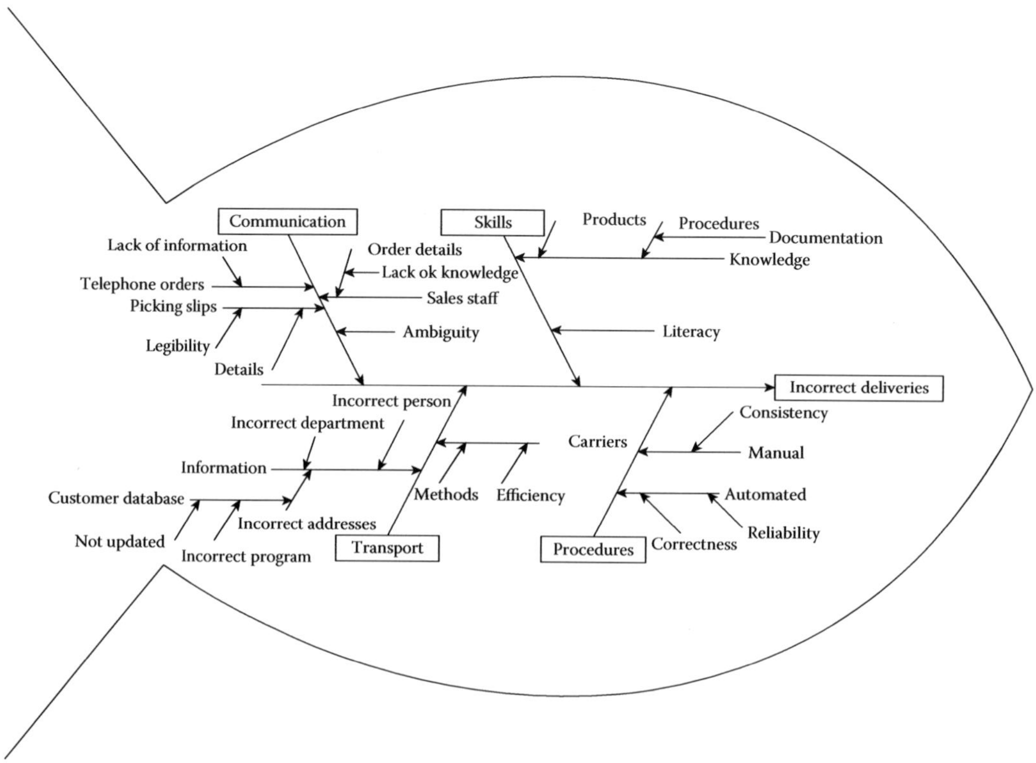

Figure 14.3 The Fishbone Diagram

3. The Ishikawa diagram

The Ishikawa (cause-and-effect), or fishbone diagram

The Fishbone Diagram (Figure 14.3) or cause and effect diagram was developed by Dr. Kaoru Ishikawa. The diagram also bears his name. It serves as a tool to find the sources of quality problems, or in fact, almost any type of problem; it focuses attention on causes, not just symptoms. The diagram looks like the skeleton of a fish, with the problem represented by the head of the fish. The ribs represent possible causes, and the smaller bones are sub-causes. The process works best with a group of people who are familiar with the problem, or who may be involved with it. A facilitator works with the group, helping them to work backward from the observed effect caused by the problem, asking "why?" in a manner like five-Why Analysis. The diagram serves as an important communications tool as users examine target processes in detail in order to find a solution to the problem that is associated with it.

A generic form of the chart may be constructed with four bones radiating from the central spine. They are typically labeled machines, materials, methods, and people, as many problems can be traced to the influence of one or more of those factors. Guidelines for problem solving with the diagram are:

1. State the problem clearly in the head of the fish.
2. Draw the backbone and the ribs. If the team can think of possible causes, other than machines, materials, and so on they can label the ribs accordingly.
3. Working with the facilitator, work backward from the head of the fish, and along each rib to identify sub-causes that branch from each rib. Asking "why" five times often helps to accomplish that.

4 Have the team agree on the most significant possible causes.
5 Identify strategies and set goals to address the core causes.

In the example above, a construction project is affected by many incorrect deliveries. The project team has a fair idea of the sources of the problem, and has selected the ribs or branches communications, skills, transport, and procedures. Brainstorming by the team has identified few sub-causes on the skills rib such as knowledge, procedures, and documentation. On the other hand, many sub-causes have been identified on the communication and transportation ribs. Communications seem to be affected by such issues as lack of information, ambiguity, and lack of knowledge in placing material orders. Remedies could include more training for staff and verification of any orders placed.

4. 5 Whys – root cause analysis

Comments on 5 why diagrams

The five-Why analysis procedure is a simple, yet effective procedure for finding the root cause of a problem. It is a part of the Toyota problem-solving process and focuses on involved processes rather than on people. The steps in the process are:

- Ask why a process deviates from expectations.
- Ask why the answer is as stated.
- Repeat the questions: asking why until a root cause is found. (If needed, this may extend beyond five questions.)
- Other quality tools such as a Pareto analysis and Fishbone diagrams may be used instead.

Example

1 Why wasn't the drywall installed in room 302 last week as scheduled?

Answer: The conduit work for electrical and communications was not complete.

2 Why wasn't the conduit work completed?

Answer: The electrical subcontractor did not have the material to complete the work.

3 Why wasn't the material available?

Answer: The materials were ordered late.

4 Why was the material ordered late?

Answer: There was a discrepancy between the plans and specs regarding the configuration. There was an open request for information (RFI).

5 Why did this RFI remain open?

Answer: The communications specialist did not spot the discrepancy until the middle of the week.

The root cause of this problem should prompt the electrical foreman and general foreman to improve the constraints analysis process. Shop drawings should be prepared on a timely basis and approved through the project manager and designer to avoid last minute surprises. Reliable, sound commitments can only be made when all constraints have been removed.

Comments

According to one construction manager (and student) his 3-year-old child uses the technique all the time and gets favorable results constantly. By repeatedly asking "why" the analyzer will have a great chance of peeling away trivial layers and eventually uncovering the root cause of the problem. The beauty of this tool is its simplicity – it is easy and does not require formal training like other quality tools.

It is important to write down the answer for previous questions – it helps in developing the next "why" questions. As pointed out there is no magic number for how many times one may ask – they should continue until a root cause is determined.

Additionally, the Root Cause Analysis (RCA) takes everything a little bit deeper with some additional steps:

RCA Process Steps:

1 Define the problem (describe the event)
2 Gather data and evidence
3 Five-Whys
4 Classify causes into causal factors
5 Reveal multiple root causes
6 Identify all harmful factors (might be causes)
7 Identify corrective actions & solutions`
8 Implement them – PDCA (Plan, Do, Act, Check)

5. Swim lane diagrams

Lean practitioners are aided by tools such as *value-stream maps* and *swim-lane diagrams* to document and analyze *current states* of processes and to propose desired *future states* that eliminate constraints and wasteful practices. For example, a swim-lane diagram documenting the request for information (RFI) process required by traditional, design-bid-build contracts reveals unnecessary causes for delay (Figure 14.4). In this

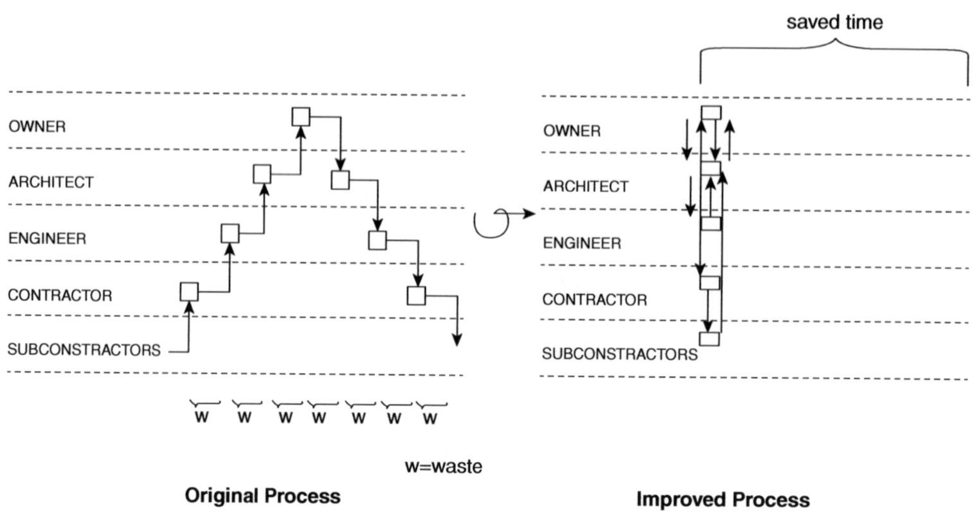

Figure 14.4 Swim lane diagrams comparing the RFI communication process during traditional design-bid-build projects (left) versus a typical lean project delivery Big Room meeting (right)

Source: Adapted from Rybkowski (2012)

Note: The horizontal axis represents time.

circumstance, a subcontractor working on site, finding an error, cannot query an owner directly, but must communicate in writing along the document chain required by a legal contract. By contrast, if critical stakeholders instead meet collectively in a *Big Room* meeting, communication is liberated, and issues can be resolved much more quickly. Lean tools such as swim-lane diagrams help lean project managers visualize and quantify before-and-after impacts of their interventions. In the RFI example just discussed, project managers can collect data from time stamps on documents to benchmark their system's current state – and then map a recommended and improved future state. In this way the swim lane diagram can help a manager quantify actual savings.

6. Plus-delta analysis

Plus-Delta analysis is an important component of the learning process. What produced value? Items in this category are listed in the "Plus" column. What might produce more value? This points to items in the delta column (i.e., those that could be improved).

Plus-Delta analysis can be used for a wide range of topics, including construction meetings related to The Last Planner® System, such as the Weekly Work Plan meeting. Plus-Delta enables continuous improvement in the way the meetings are conducted by having meeting participants ask and answer the foregoing questions. One common use of the Plus Delta is in Big Room meetings where major project issues are discussed. Near the end of the meeting, the moderator of facilitator will ask attendees "(a) what went well, which events would you consider a plus. Following that, attendees are asked to name (b) the issues that could have been improved".

The responses are recorded, either on a flip chart, or preferably on a Smart Board. At the subsequent meeting these responses are displayed as a reminder for continuous improvement.

Jim Shug, a consultant, cites the extensive use of Plus/Deltas in the military after tactical/training/combat operations as after-action reviews (AAR) and lessons learned meetings. He expresses the opinion that feedback is a gift and great leaders want to know what others think and how to improve. Their intentions may be great, but the perceptions of their actions are reality to those around them.

Shug views Plus/Deltas as a significant step toward developing a learning organization. He recommends that there should be not only be a discussion, but some concrete facts as well to show "reality". "In military training events, they could play back radio transmissions from the battle and snapshots of vehicle/squad positions. This really becomes a wakeup call to leaders on the ground and takes shape in a similar way for projects."

Regarding the construction process, Plus-Delta may be applied to performance measures such as those listed in Chapter 2. Ed Anderson, a consultant, emphasizes the BIG FIVE: cost, schedule, quality (scope/spec), safety, and reliability. He states that there can be many subsets of these, but lean advocates simplicity (KISS) first, then when and where needed, to add more detail. Anderson cautions that while measurement is important, proactive lean behavior involves having systems in place that concentrate on eliminating non-conformances, as opposed to measuring and tracking everything.

Some possible measures are listed here for convenience.

Table 14.1 Plus-Delta highlights

Plus	*Delta*
Solution oriented team worked well	Conflicting service layouts caused re-work
Speed of construction	Supply chain coordination – Late equipment
Trade to trade communication – best ever	Unclear instructions on changes caused re-work
Good selection of systems – quick install	Just in time design was challenging
Scheduling and Last Planner Coordination	Some trades just doing what's best for them

Source: This chart represents a project review by Kinetic Construction Ltd.

7. The 5 Step plan – 5S

The Five-Step plan is a Japanese approach to improving operations. It helps to create an organized environment in a facility that promotes the application of lean thinking. It involves a sequential process based on the five Ss.

1 *Seiri:* (Straighten up) Segregate and Discard. This involves getting rid of unnecessary items in order to be more organized. This step is applied to such areas as tools, work in process, products, documents, and so on.
2 *Seiton:* (Put things in order) Neatness is obtained by proper rearrangement of the work area and identification of proper locations. The object of the exercise is not just to maintain a neat appearance. Workers can reliably find what they need to do a job without wasting time by looking. They should place tools and equipment in a manner that will improve the flow of work. Everything should be kept in their designated locations – equipment should be returned to its proper place at the end of each job.
3 *Seiso:* (Clean up) Cleaning and daily inspection to avoid the confusion of an untidy work area. Unnecessary items should be disposed of properly in order not to create clutter that inhibits the effective execution of work tasks. After each job, tools and equipments should be cleaned and restored to their proper locations. The root causes of waste and uncleanliness are investigated to ensure that they do not reemerge.
4 *Seiketsu:* (Personal neatness) This refers to standardizing locations for tools, files, equipment, materials, and so on. Color coding and labeling can help to standardize locations. Employees' creativity is actively involved in the development of standardized systems.
5 *Shitsuke:* (Discipline) It provides motivation for employees to sustain the other four Ss. Management can use recognition programs to provide motivation.

8. The Kaizen process

Kaizen is the Japanese concept of continual incremental improvement. Literally, Kai means change, and zen means good; hence, Kaizen stands for ongoing changes for beneficial reasons on a never-ending basis. It involves all members of an organization actively participating in making improvements on an ongoing basis. Masaki Imai (1986) of Japan credits Kaizen as the "single most important concept in Japanese management, and Japan's competitive success." Kaizen is rooted in Japanese philosophy – they see improvement as being closely related with the concept of change. According to Imai in the Japanese way of thinking "change is a basic condition of life and should be incremental in order to be healthy, whereas sudden change is seen as unnatural."

The Kaizen approach encourages employees/stakeholders to develop and implement ongoing improvements to the systems that they are most involved with, thereby improving their job performance. It seeks to standardize processes and eliminate or reduce waste. It starts with recognizing a problem, and subsequently a need for improvement. It has been said that complacency is the archenemy of Kaizen.

One important consideration for the adaptation of Kaizen to construction-related activities is that improvements are expected to be followed by a period for new methods/processes to be standardized. As described, the Kaizen methodology is appropriate for a specialty contractor or supplier. On a construction site the time frames may need to be reduced from 1 week to 2 or 3 days.

Roles in the Kaizen process

Senior executives/managers: They must first embrace Kaizen as an organizational strategy and communicate this policy to all stakeholders. As with other performance improvement initiatives, top management's commitment, leadership, and active support are critical to the success of Kaizen.

Middle managers: They deploy the policies established by senior executives. They ensure that staff members have the appropriate training and preparation. They establish implementation milestones for supervisory staff.

Supervisors: They work closely with line workers to implement Kaizen at the functional level. They are actively involved in collecting Kaizen suggestions for improvement from employees and coach the respective teams to promote success.

Employees: They are expected to actively pursue self-development through education and training in order to support Kaizen team activities.

Kaizen practitioners study a process firsthand by visiting the involved work area (*gemba*) and observing the activities involved. This approach clarifies the difference between value-added and non-value-added process steps. Kaizen teams reduce non-value-added steps through a prescribed approach: to combine, simplify, or eliminate to reduce waste.

There are two primary approaches to Kaizen improvement activities: Flow Kaizen and Process Kaizen. Flow Kaizen emphasizes value stream improvement while process Kaizen focuses on the elimination of waste. While Kaizen involves ongoing, long-term dedication to improvement of the organization and its people, the process periodically conducts short-term activities as "Kaizen events" or "Kaizen breakthrough" methodology. These events are a cross-functional, team-based process for rapid improvement, and are often a week long in duration. They involve a bias for action that harnesses creativity to obtain results. The areas selected for Kaizen events have direct impact on organizational performance

Employees' must contribute and focus 100% to the Kaizen philosophy for the needed organizational transformation. Kaizen events are selected for maximum visibility and impact. They have several advantages:

- Benefits are obtained very rapidly.
- Quick implementation sends a positive message to all stakeholders.
- They provide hands-on training for everyone.
- They allow management to gauge workers' resistance to change.

Typical Kaizen structure

Preliminary activities:

Select area/work unit for Kaizen

Based on business requirements
Based on potential for improvement

Establish project objectives

Document baseline data
Set metrics for improvement – numerical goals

Appoint Kaizen team members

Team leader and subteam leader

Publicize the event – meet with the involved work unit team

Explain roles and expectations
Engage hearts and minds for the event

Day 1:

Teams meet for conceptual training and problem identification. Training topics include:

- Lean theory and lean production systems
- Kaizen breakthrough methodology
- Team dynamics – multipurpose
- The lean transformation

Day 2:

Evaluate/analyze current processes:

- Document existing processes
- Identify value-added versus non-value-added activities
- Measure cycle times
- Discuss improvement options
- Formulate process improvement – rearrangement of layouts
- Initiate process improvement

Day 3:

- Continue process improvement
- Apply lean tools
- Focus on delivering value to the customer
- Recheck cycle times

Day 4:

- Refine process/system improvements
- Establish new work/process standards
- Document new standard operating procedures
- Recalibrate new cycle times

Day 5:

- Present recommendations and results to management
- Celebrate Kaizen event accomplishments

Kaizen events are short-term in duration but have proven to be very effective. They focus everyone's effort for a few days, but the benefits in improved performance far outweigh the cost of lost production.

9. Choosing by advantages

Introduction

CBA is a collaborative, sound decision-making method developed by Jim Suhr around 1981while working for the US Forest Service. CBA provides a powerful tool to structure meaningful conversations and make collaborative decisions. It is a system composed of principles, methods, and models. New project delivery forms like IPD and Lean Practices require decisions to be made collaboratively. Furthermore, more complex projects require more specialists and teams to use cross-disciplinary analysis in making decisions.

The example below, provided by Lucie Parrot, P. Eng. M.Eng., CVS-life, FSAVE, shows a tabular example with two alternatives.

How CBA works

The Choosing By Advantages methodology allows a few scenarios to be compared against criteria that are difficult to measure on a comparison scale, as in a typical multi-criterion grid. This is a three steps method:

1 We compare the scenarios one by one,
2 The facts about each attribute are explained in each scenario,
3 We identify which of the scenarios is more advantageous.

In a second phase, a score can be awarded to each advantage. This phase was not carried out in this example.

Example: The following table illustrates an evaluation of two different heating systems in the renovation of an office building:

A. Heating and air conditioning is done entirely by new fan coils unit on the walls, under each window, in the space currently occupied by the existing Fan coil unit to be replaced or a little larger when a larger Fan Coil unit is needed to bear the full burden
B. Heating is mainly done by radiant heating on the ceiling, and new Fan Coils are installed for air conditioning and will be able to heat in times of extreme cold to compensate for the lack of radiant heating capacity in some cases.

Attributes	**Fan coils only (larger if necessary)**	**Ceiling Radiant heating (9 to 20 ft high) with fan coils to meet peak demand**
Comfort level felt by occupants	Hot air blows throughout the heating period	Radiation first and hot air blow if necessary
		More comfortable
Stability	Hot air heating varies a lot in temperature	Radiant heating keeps the temperature more constant
		More comfortable
Maintenance	At a person's height	At or above a person's height
	Easier	
Installation	At a person's height	At or above a person's height
	Easier	
Noise	Fan noise all the time	Less noise, Fan runs only to meet peak demand
		More pleasant
Risks	Presence of water on the ceiling	Presence of water on the ceiling
	No advantage	No advantage
Architectural integration	In the current volumes on the walls or a little more	In current volumes and on the ceiling
		A little easier to integrate
Cost	1 system to install	2 systems to install
	Cheaper	More expensive

Figure 14.5 Steps to implement the CBA tabular method. Example by Lucie Parrot, P. Eng. M.Eng., CVS-life, FSA

As in both cases, the air conditioning would be done by the Fan Coil units, the question is mainly about radiant heating: is it advantageous to install it if anyway, it is necessary to install the Fan Coil units for air conditioning in the summer?

The two systems are therefore compared on each of the attributes of a heating system, namely the comfort level felt by the occupants, the consistency in the temperature of the premises, the maintenance of the systems, the installation of the systems, the noise generated by each alternative, the risks associated by each, the architectural integration into the existing building and finally, the cost.

Such a benefit comparison can be achieved on a significant number of attributes. The benefits are highlighted in green, based on the information associated with each attribute. The advantages of radiant heating are at the occupant level. Since the renovation project is a result of complaints from occupants, it is normal for the importance to be given to the attributes of the heating system that benefit occupants rather than installers or maintenance.

Radiant heating has four (4) advantages rather than two (2) for Fan Coil units. The cost may not be considered as both options are on budget.

10. Value Stream mapping (VSM) at "*GROUPE XYZ*"

By: Gilles Bélanger, Engineer, MPM, PMP, MBB
President, Gilles Bélanger & Associates Inc.

Introduction

Based in the Province of Quebec, Canada, *GROUPE XYZ* is a major player in the commercial and industrial construction field. They run their business in a French-speaking environment.

The company started their Lean journey in 2015. I am thankful for the privilege I have been granted to train a selected group of talented people in this organization. Several training sessions were held with stakeholders representing the following functions: Project Managers, Quality Manager, Health & Safety agent, Finance, Legal and Upper Management.

Value Stream mapping was first introduced to the groups with the drawing below:
With the following explanations:

> VSM Purpose: Visualize wastes in the information and materials flow. The goal is to create countermeasures to improve the performance of a cross-functional important process that has customers and suppliers.

Figure 14.6 As-is Value Stream mapping

The three components: **information flow**, **workflow** (raw materials and equipment required plus data) are drawn according to the **time** (represented by the clock) that elapses to create the product for the customer. The whole thing creates the as-is VSM.

This is a team activity; the participants at the bottom of the lighthouse share their visions. Note the different bubbles shape for each member of the team which reflect their individual and sometimes quite different perceptions of their processes. The tide symbolizes the continuous changes in business.

The light generated by these inputs guides the teams to improve the performance of the mapped processes, it also stands for the Future State VSM.

In the upper right corner, one guesses 6 stars in the shape of a pan, it's a reference to the Polar Star that guides the navigation at sea. VSM provides a True North map for improving major processes.

The introduction was followed by the conventional description of the icons and architecture associated with VSM's.

Creating the as is (current state) VSM

In 2016, the company aimed at improving their "Quote" process after realizing their rate of success was not high enough to maintain their sales level for the next couple of years.

A dedicated group first created an A3 report titled "Improve our ability to achieve quality estimates for future projects". They defined several countermeasures that includes a VSM as illustrated below:

Table 14.2 Countermeasures and action plan

COUNTERMEASURES	*WHO*	*WHEN*
Define quality criteria for a quote	ML	October 30, 2015
Define criteria for judging the relevance of a project to be estimated by *GROUPE XYZ*	PA	January 30, 2016
Create the as is (current state) and the future state VSM for "Create a quote"	PA	April 28, 2016

I worked with them to create a VSM charter to make sure stakeholders were properly identified as well as their expected roles in drawing the as-is and future state VSM and to define the expected gains.

We worked with a team of five stakeholders to identify the main processes for one day and half. The group decided to focus on the constraints that each process was facing. It was also decided not to draw the timeline, but to work instead with the summarized Lead Time (LT) and Process time (PT)

The As-iss-is VSM generated by the team efforts is shown in the figure below:

Definitions

The red triangles represent the inventory found before each process (in this case: number of documents to be treated).

The curved arrow stands for 'Pulled' information.

The black and white arrow is for 'Pushed' information.

CONSTRAINTS= In this effort, the team wanted to capture the constraints and obstacles that each process was facing
LT = Lead Time
PT = Process Time
&C&A = Percentage of transactions that are Complete and Accurate

Here is legible information for each process in the as is (current state) VSM from left to right:

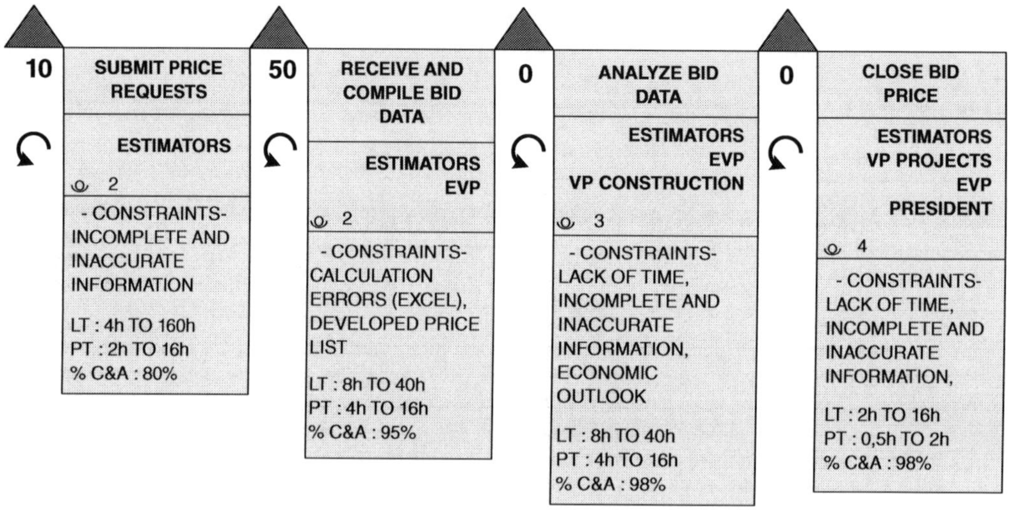

Figure 14.7A As-is Value Stream mapping detailed process

Followed by:

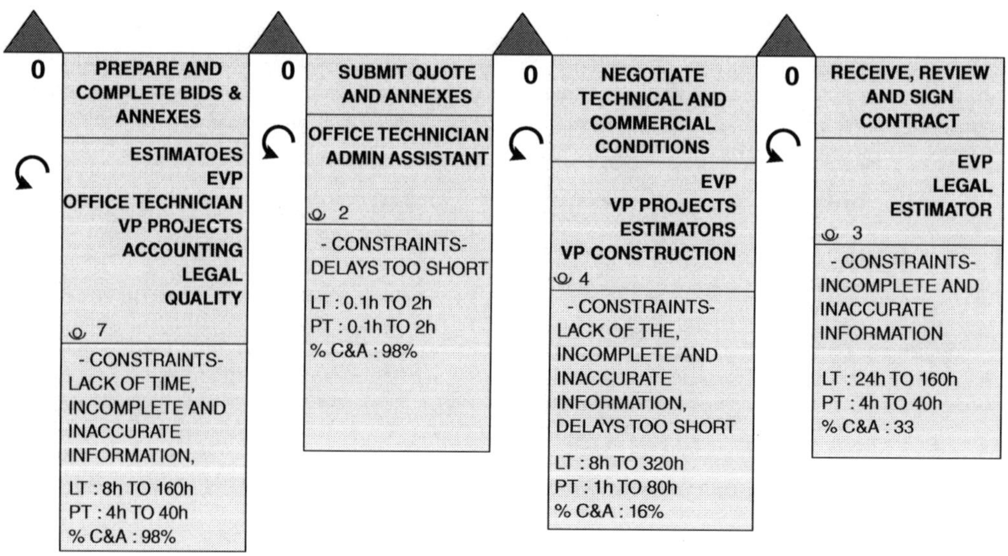

Figure 14.7B As-is Value Stream mapping detailed process

And finally:

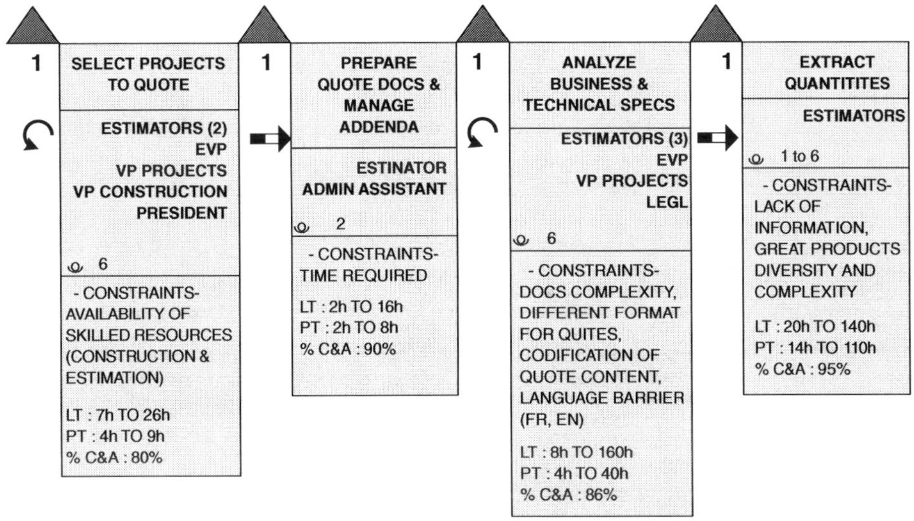

Figure 14.7C As-is Value Stream mapping detailed process

A couple of weeks later we met again to draw the future state for the "Create a quote" process. Drawing the Future State VSM. It took some time to initiate the flow of ideas to create the VSM future state. We had a situation where nobody wanted to provide new ideas, maybe because it could have been understood as a critic to the present situation. After a couple of passionate and rich discussions, I just followed my intuition and I started to propose different alternatives that nobody really liked until, finally, one person stood up and said, we are going nowhere with this effort, here is what we should do!

I said alright, you tell us what we should do, I'll sit down, and you drive the meeting from now on. During that rich exchange, the team realized that one weak process that was partly addressed in an inconsistent manner was the transfer of a contract to the Construction team. Many good ideas were put on the table, we captured them all (see yellow bursts) and started to draw the "Create a quote" Future State VS. That led to the addition of an important process in any construction company: 'Transfer Contract Knowledge To The Construction Team'.

To reflect the addition of this process, he Value Stream name was changed to "GET A CONTRACT'.

The forecasted metrics were calculated as shown in the table below:

FUTURE STATE VALUE STREAM PERFORMANCE METRICS

VALUE STREAM NAME	GET A CONTRACT	PREPARED BY:	PA, VC
DATE:	25-Nov-15		

METRICS	Initial values	Projected future valuse	Projected % of improvement
Total Lead Time	669.5 h	354 h	**-47.0%**
Total Process Time	211.3 h	162.5 h	**-26.5%**
Activity Ratio	31.50%	46.00%	**46.0%**
Rolled C&A %	2%	3.3%	**165.0%**

Figure 14.8 Future state Value Stream performance metrics

We then created an action plan to execute each of the improvements (see bursts) identified in the future state, a small part of them are shown on the figure below:

VALUE STREAM ACTION PLAN

VSM NAME:	GET A CONTRACT		CEATION DATE:	29-Oct-15
← APPROVALS →			SCHEDULED REVIEW DATES	
EXECUTIVE SPONSORS	VALUE STREAM CHAMPION	VSM FACILITATOR	13-Nov-15	
NAME: JP	NAME: PA	Nom: Gilles Bélanger	Dec 16, 2015 with PA	
SIGNATURE:	SIGNATURE:	SIGNATURE:	16-Jan-16	
DATE:	DATE:	DATE:		

FUTURE VSM BURST #	MEASURABLE TARGET	COUNTERMEASURE DESCRIPTION	REALIZATION METHOD*	OWNER	IMPLEMENTATION SCHEDULE (2015-2016) 10	11	12	1	2	3	4	5	6	7	8	9	STATUS
1	Number of criteria defined and approved	Define criteria for judging the relevance of a project to be estimated by *GROUPE XYZ*	Proj	PA													60%
2	Update Checklist	Refresh the 'CHECKLIST'	JDI	CG													50%

*** REALIZATION METHOD: Proj (Project), KE(Kaizen), JDI(Just do it)**

Figure 14.9 Value Stream Action Plan

See Future Start VSM (showing countermeasures in bursts) for 'GET A CONTRACT' on the figure below.

After several months the Future State VSM becomes the updated as-is VSM.

Conclusion

The team found the VSM process to be both very demanding as well as rewarding. One should keep in mind that VSM is not a topic for Kaizen simply because a Kaizen is an effort to put in place some known solutions in a very intensive manner. VSM on the contrary is an exploration of unknown territories. Quite often it's also better to add SME's in the Future state VSM team: the people who know the process well and have a clear vision about where they need to go.

The last process, "Transfer Contract Knowledge to The Construction Team," proved to be more difficult to develop and to implement than originally forecasted. After several months, a formal structure was developed and tried. The new process follows the spirit of the Last Planner® System, which promotes an early involvement of suppliers and subcontractors. The process owners are having high expectations on contract compliance to schedule, budget, specifications and profit levels.

Most importantly, the whole exercise has improved their quote rate of success by more than 400% between 2017 and 2018. This in part due to a much better definition of their core expertise which is aligned with corresponding projects.

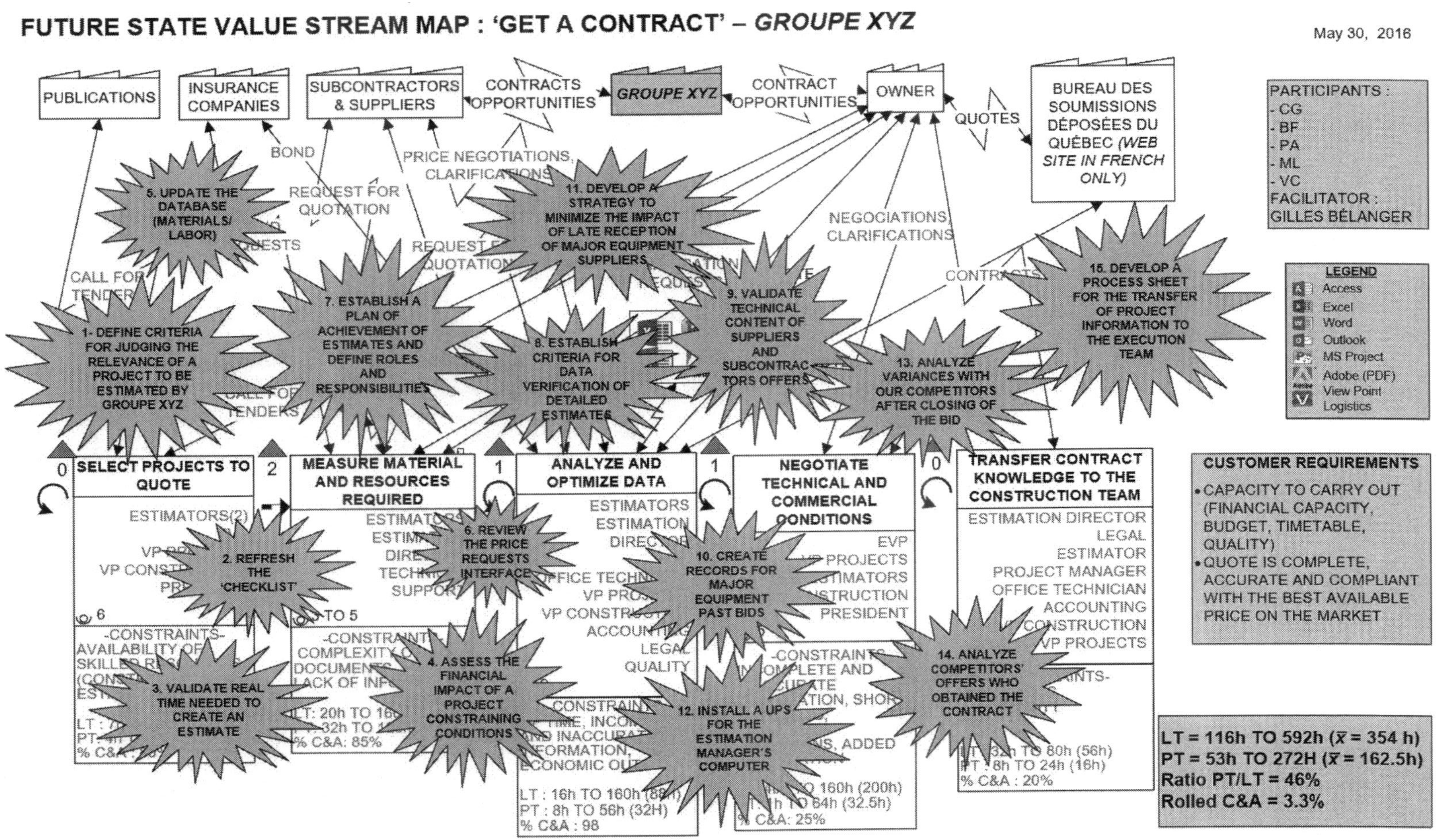

Figure 14.10 Value Stream map

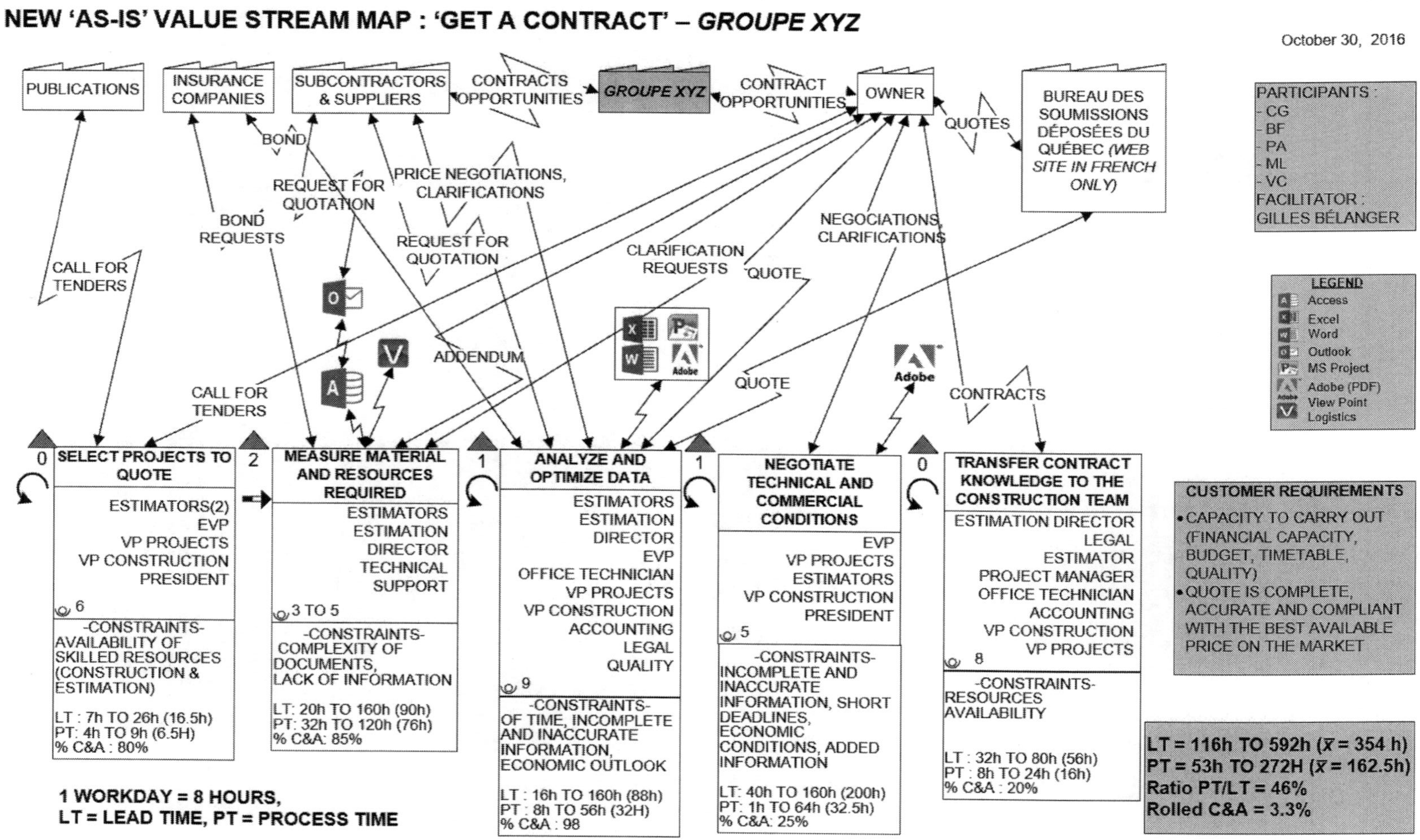

Figure 14.11 Updated Value Stream map

11. The A3 report

The A3 report is a system for implementing PDCA management, which is simple yet disciplined and rigorous (Sobek and Smalley 2008). The report used by Toyota fits on a single sheet of A3 paper, measuring approximately 11 × 17 and documents the results of the PDCA cycle in summary form. It has a template that guides users in the steps needed to determine the root causes of an organization's problems and to address them systematically. The A3 report concisely describes problems and the available information to understand their ramifications quickly, focusing on improving processes and solving the problems (Figure 14.12).

The development of the A3 report is thought to have been influenced by Taichi Ohno's dislike of reports that were longer than one page. It must be understood that the A3 report is not by itself a complete system for problem solving. In Toyota's case their success in process improvement comes from their underlying philosophy about collaboration and teamwork. The A3 report serves as an information and facilitation tool to support a very practical problem-solving process. A3 thinking as practiced by Toyota is based on seven elements (Sobek and Smalley 2008).

Logical thinking process: Logical thought processes are employed to recognize the most important problems, using the 80/20 rule, and recognizing the ramifications of the possible courses of remedial action.

Objectivity: Participants are encouraged to express their viewpoints; no single viewpoint drives the process, and consequently there is little opportunity for biases and misconceptions. A3 thinking considers multiple perspectives to promote objectivity and find the solution that benefits all stakeholders.

Results and process: Both results and the process used to obtain them are equally important. The involvement of people in the process leads to personal development and to more lasting outcomes. The process also provides a mechanism for future refinement, as opposed to a spontaneous idea.

Synthesis, distillation, and visualization: The A3 format is concise and compresses large quantities of information in a graphical format; the visualization it provides aids the synthesis and distillation process.

Alignment: A3 reports enable a practically oriented consensus – three-dimensional communication flows between the problem-solving group and others. Horizontal communication considers others at the same level that may be affected by proposed changes; vertical communication examines broader organizational issues. These approaches are inclusive and lead to better alignment between people system wide

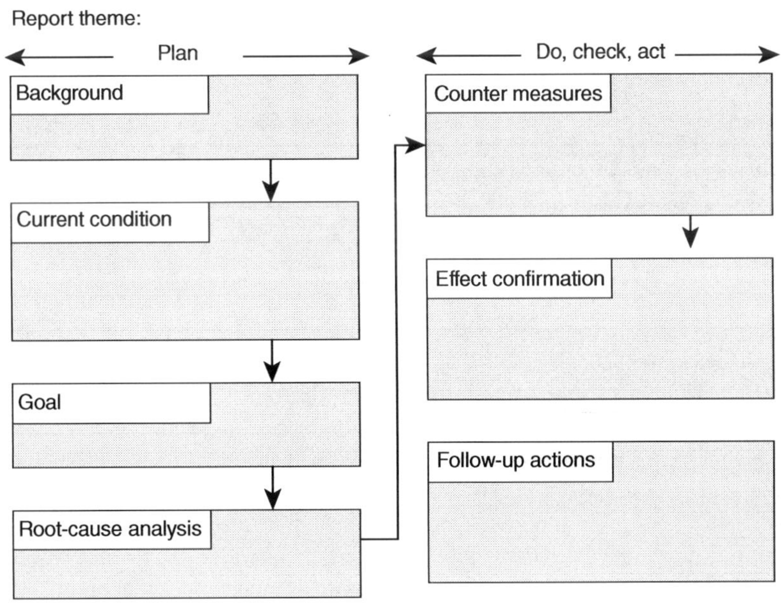

Figure 14.12 Typical flow of the problem-solving A3 report

Coherence within and consistency across: A3 reports have a logical flow that focuses on root causes of identified problems and provides a coherent approach. This increases the effectiveness of problem resolution.

Systems viewpoint: People are guided (and mentored) to develop a deep understanding of: (1) the purpose of the course of action, (2) how the course of action furthers the organization's goals, needs, and priorities, (3) how it fits into the larger picture and impacts other organizational units.

Overall, the A3 report is not considered a documentation format, but really is an important adjunct to the informed use of the PDCA cycle.

The three most frequently used examples of reports are: (1) the problem-solving report, (2) the proposal report, and (3) the status report. The problem-solving A3 sets the stage for analyzing and improving operations. It consists of seven sections; that is, rectangular boxes that are used for recording information. The sections are:

1 Background
2 Current condition and problem statement
3 Goal statement
4 Root-cause analysis
5 Countermeasures
6 Check/confirmation of effect
7 Follow-up actions

Sections I through IV are usually stacked vertically on the left side of the page; Sections V through VII are similarly stacked on the right side.

Recommendations for report content are as follows:

Background: Provide a clear theme that is relevant to the objectives of the project or project team.

Current condition and problem statement: Describe the problem clearly and visually. Distinguish between facts, observations, and opinions.

Goal statement: Describe the goal or target clearly and state what is to be accomplished. State how the goal will be measured or evaluated. State what you expect will improve, by how much, and when.

Root-cause analysis: Show that the analysis was thorough and that it examined the appropriate issues deeply. Use the cause and effect diagram, based on all relevant factors. These may include people, machines/equipment, materials, measurement, and the environment. Use five-Whys/one-How analysis.

Countermeasures: Show that the countermeasures relate specifically to the root causes identified. Show clear steps for these actions. Explain how the effects will be verified, and how problem reoccurrence will be prevented. Provide an implementation plan – who does what and when.

Confirmation of effect: Explain how effectiveness of the countermeasure will be measured. Show that the PDCA "check" item aligns with the previous goal statement. Provide a "plan B" if performance is not improved and show what may have been missed.

Follow-up actions: Describe what is left to be accomplished Show how problem reoccurrence will be avoided. Explain how the improvement will be standardized and communicated. Show how other organizational units will be advised.

The Proposal A3 report begins with a thematic title and presents a plan for consideration by decision makers. The Status A3 report presents a logical snapshot of a project or problem-solving effort and how it is progressing. The report describes the situation with a theme and explains accomplishments as well as any work that remains to be done. A recommended format is as follows:

Background Issues: Provide a clear theme that is relevant to the objectives of the project or project team.

Current condition and problem statement: Describe the problem clearly, visually. Distinguish between facts, observations, and opinions. Quantify the situation as much as possible. Show the progress made and the action steps taken to date.

Results: Indicate the results obtained to date. Confirm that improvement has occurred and quantify it as much as possible. Show that the metrics used are appropriate. State whether any negative side effects have occurred. If results fall short of expectations, explain possible causes.

Unresolved issues/follow-up action items: Describe any problems that remain. Explain actions needed to meet planned progress. Explain how gains may be sustained.

Decisions. See A3 report below

The flexibility of the A3 report makes it a remarkable tool based on PDCA (Plan-Do-Check-Act, learn by doing). It can be used to:

1 Define a problem
2 Get consensus on countermeasures
3 Follow-up an action plan
4 Obtain approval for an investment project (and the signatures of stakeholders). The A3 format needs to be modified to reflect expense justification requirements and the signature section must be rigorously managed. All signatures must be visible before spending money.
5 An A3 report can be used in many situations where multiple stakeholders are involved, and a decision must be made with their support. Important information is concisely compressed on one page, which becomes easy to share and manage.

The following case by Gilles Bélanger, MPM, PMP, MBB, illustrates the use of the A3 report to solve a problem at a Canadian company referred to as GROUPE XYZ. Gilles is an engineer, who provides consulting services as the President of Gilles Bélanger & Associates Inc.

Background

Based in the Province of Quebec, Canada, *'GROUPE XYZ'* is a major player in the commercial and industrial construction field. They run their business in a French-speaking environment.

The company started their Lean journey in 2015. The A3 report was first introduced as a tool to capture discussions and decisions about a problem. Late in 2016, they observed a downward trend in their estimating quote rate of success which triggered a problem-solving project using an A3 report.

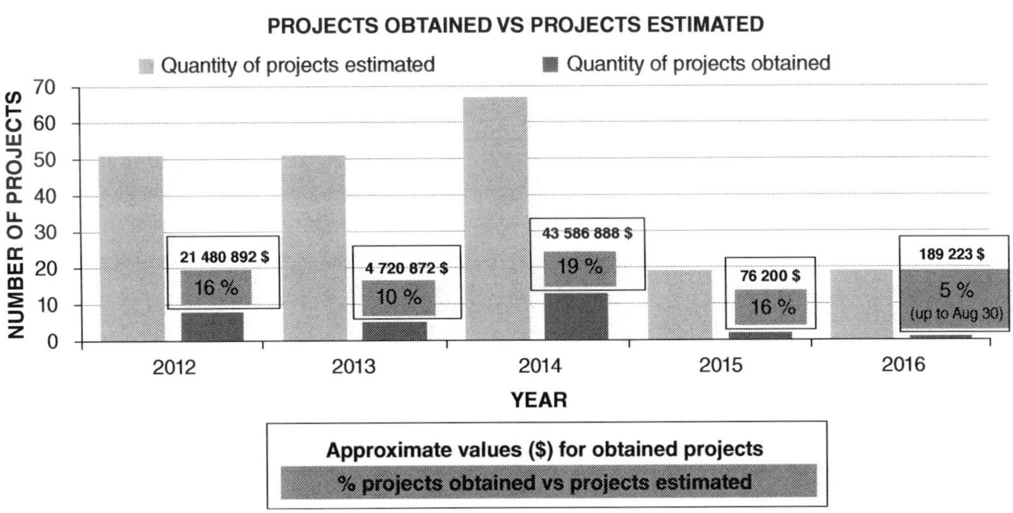

Figure 14.13 Projects obtained vs. projects estimated

They first did a graphical analysis of their quote success rate. For the first 8 months of 2016, it went down to 5% as shown on the graph below. In any commercial and industrial construction company, the cost of preparing quotes is a major expense that requires well-paid experts with advanced knowledge in many areas.

A team was formed to create an A3 report titled: Reverse the downward quote rate of success at '*GROUPE XYZ*' (since 2014). Their goal was to use the A3 format to capture the current situation, the desired situation and the plan of action that will fill the gap.

Goal

The team established the following goal; they felt it would be a breakthrough:

"Increase to 20% our success rate for projects obtained compared to those estimated."

Current conditions

Then they ran a data search program to understand where they were when they lost a bid. See the chart below:

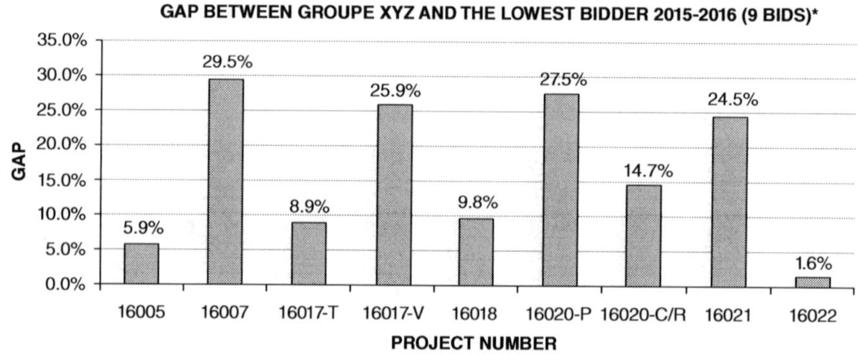

Figure 14.14 Gap between bidders

* When the gap between the lowest bidder and the 2nd lowest bidder was more than 5%, the second bidder's price was used to calculate the difference.

Once they saw how much difference they had with the lowest bidder, they felt the need to see how they were spreading their $ values over a project bid. Seven types of expenses were identified and using that frame they did a thorough analysis of all bids produced in 2015 and the first half of 2016. See pie chart below:

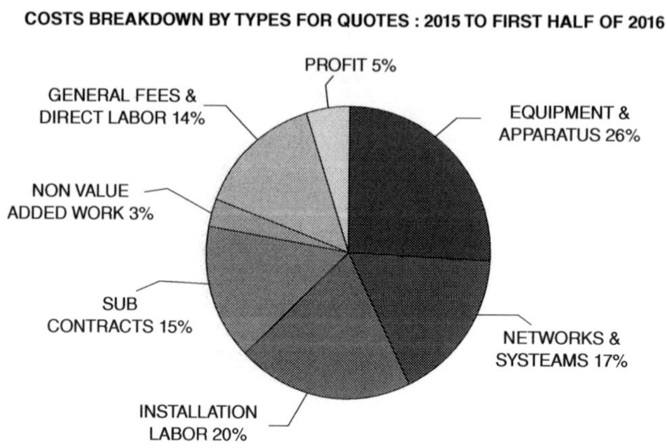

Figure 14.15 Cost Breakdown by types for quotes

Root cause analysis

Using the knowledge gained from previous data collection analysis, the team created a cause-and-effect diagram (Ishikawa) and used the 5 Why's document (please see the DMAIC section for illustration of these tools) to find the root causes below:

1. Gaps in estimating software proficiency (how to meet installation needs for labor and material: electrical, plumbing, HVAC).
2. Insufficient knowledge of estimating software data management methods
3. Incomplete and inaccurate estimates are produced due to short lead time
4. The knowledge from experienced construction leaders in the company are not shared with estimators

Countermeasures

After several brainstorming sessions with the stakeholders, they agreed on the list below:

Table 14.3 Countermeasures and benefits

CAUSE #	*COUNTERMEASURES*	*BENEFITS*
1 & 2	Develop an ongoing training program for our estimation resources.	Cycle time and Lead time ↓, ↑ % for C&A information
3	Creates KPI's for estimates data validation.	↑ Data accuracy make gaps visible
4	Review estimation processes by adding a data review step (by experienced dedicated resources).	↑ Knowledge transfer and data C&A (complete & accurate)

The team realized the benefits to the organization in defining its core business and the absolute need to align its offerings with its core expertise.

This effort mobilized the whole management team and proved to be a major contribution to the gains achieved by this problem-solving activity.

Developing the countermeasures led to the action plan in the section below.

Action plan

The action plan was rolled out at the end 2016 and the first half of 2017.

One of their actions was to review the swim lane flowchart for "Create and Estimate". Please refer to following flowcharts before (2016) and after (2018) below:

DESCRIPTION	SCHEDULE (2016-2017)												RESPONSIBLE	STATUS
	9	10	11	12	1	2	3	4	5	6	7	8		
Train estimators in construction installation methods	■	■	■	■	■	■							Estimating Director	100%
One site visit/mth for the next 6 mths (construction site and prefab factories) by estimators	■	■	■	■	■	■							Estimating Director	100%
Train estimators: how data from quantity extractions are processed by the estimating software	■	■	■	■	■	■							Estimating Director	100%
Mapping current and future processes	■	■											PA	100%
Create meaningful KPI's to validate bid data	■	■	■	■	■	■	■	■					PA	100%

Figure 14.16 Action plan

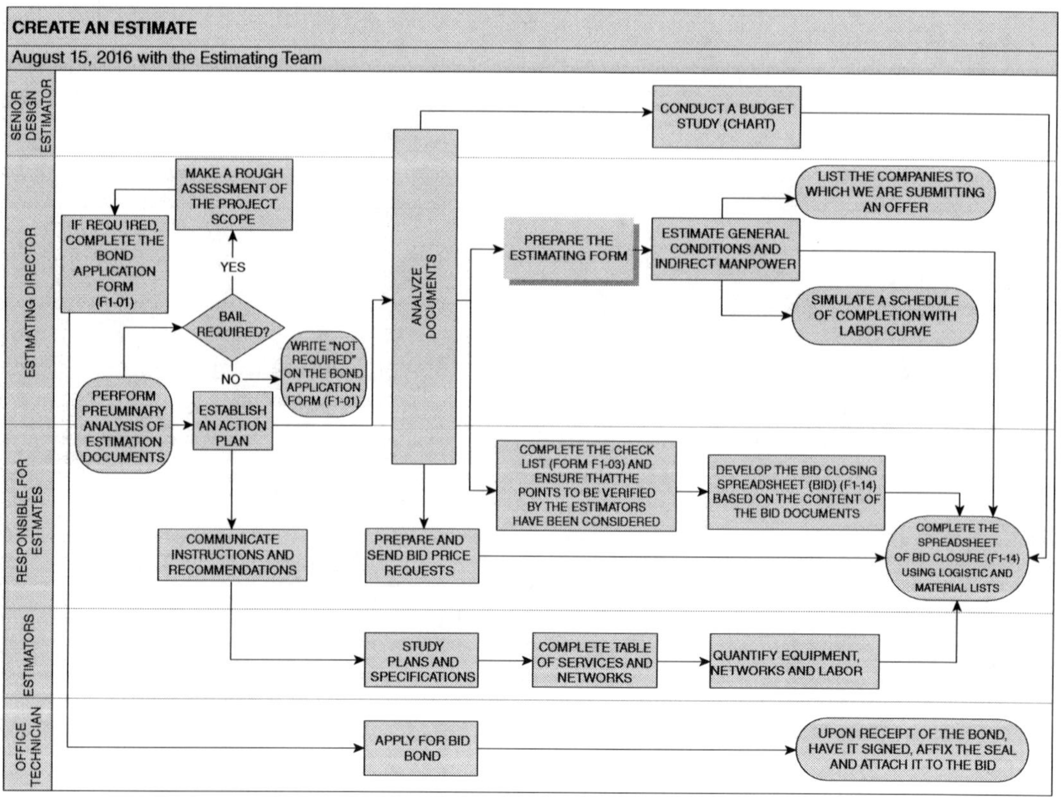

Figure 14.17 Create and estimate before 2016

The updated (2018) flowchart reflects the Team focus on data completeness and accuracy:

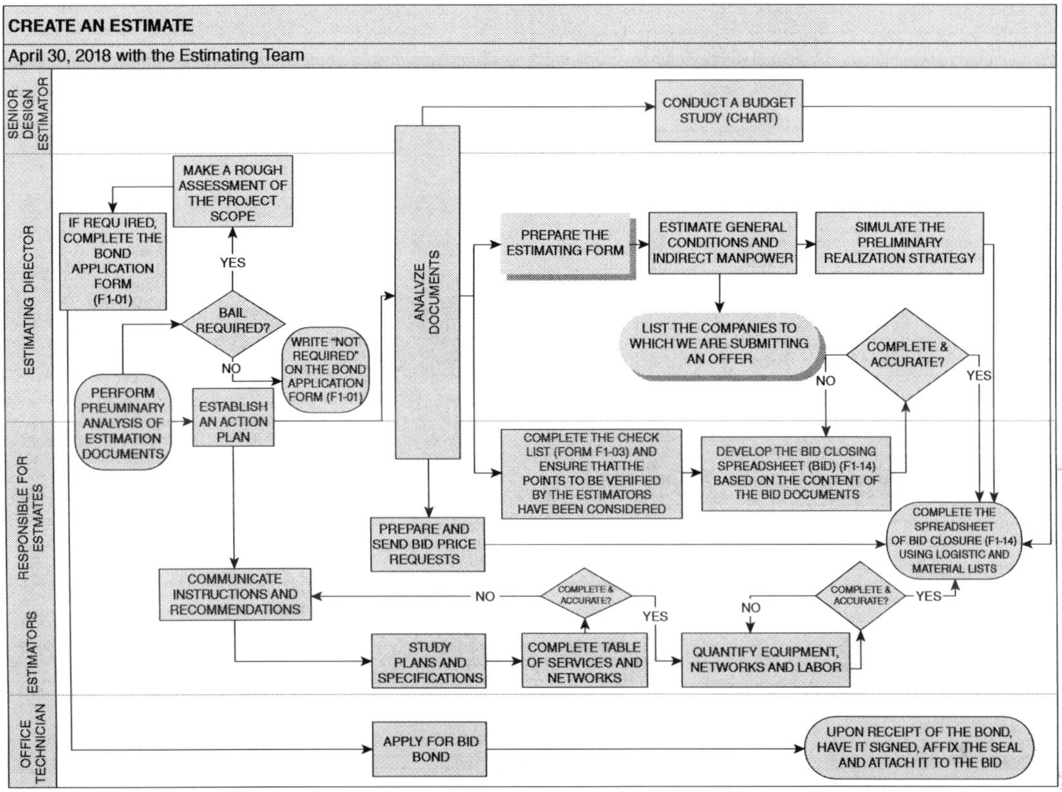

Figure 14.18 Create and estimate after 2016

(Sub-processes that are shaded have their own flowcharts).

Follow-up actions

Results were measured, of course. See below the reduction over time of the gap between *'GROUPE XYZ'* bids and their competitors:

This resulted in a significant improvement in the bids rate of success.

The team realized that even though the 20% success rate for the quotes original goal was not achieved, the result was interesting. For the first 9 months of 2018, 47% of the dollar value of the quotes was converted into a contract. As a result, the dollar value of the offers obtained is now a key performance indicator.

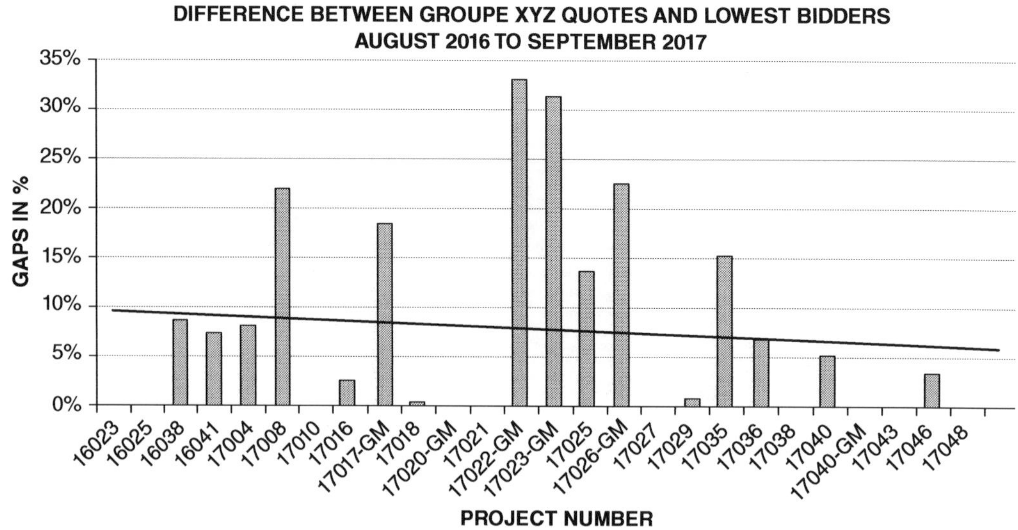

Figure 14.19 Difference between group quotes and lowest bidders

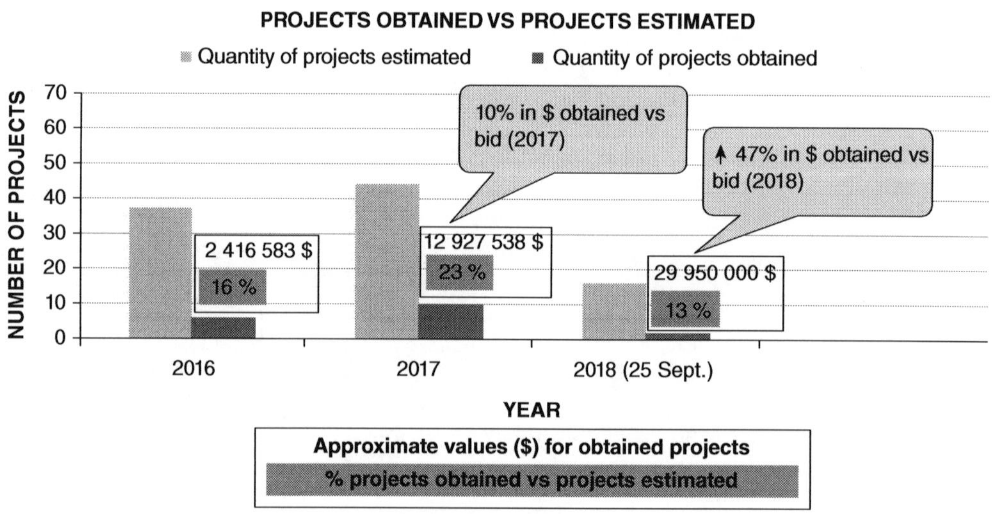

Figure 14.20 Projects obtained vs projects estimated

Conclusion

Team members learned a lot during this process improvement project. The A3 report proved to be a valuable tool for capturing meeting discussions.

A3	OPPORTUNITY: REVERSE THE DOWNWARD QUOTE RATE OF SUCCESS AT GROUPE XYZ (SINCE 2014)	START: 2016-08-26 END: 2017-04-28
	DONE BY (OWNER): PA	APPROVED BY: JPG

BACKGROUND

Since 2014, GROUPE XYZ bid results show a decline in our competitiveness.

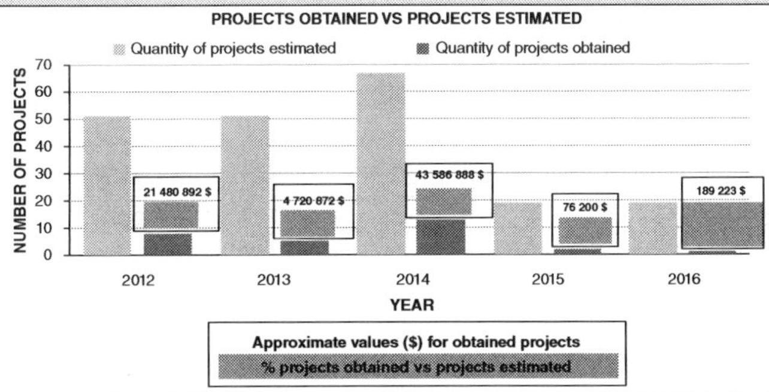

CURRENT CONDITIONS

* When the gap between the lowest bidder and the 2nd lowest bidder was more than 5%, the second bidder's price was used to calculate the difference.

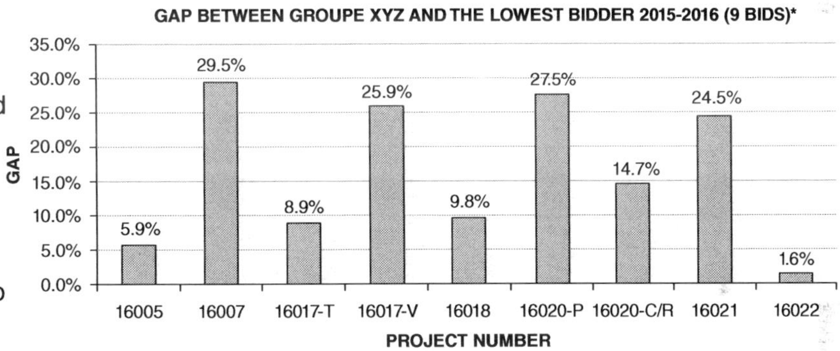

Cost breakdown analysis: understand how money is spread o estimates.

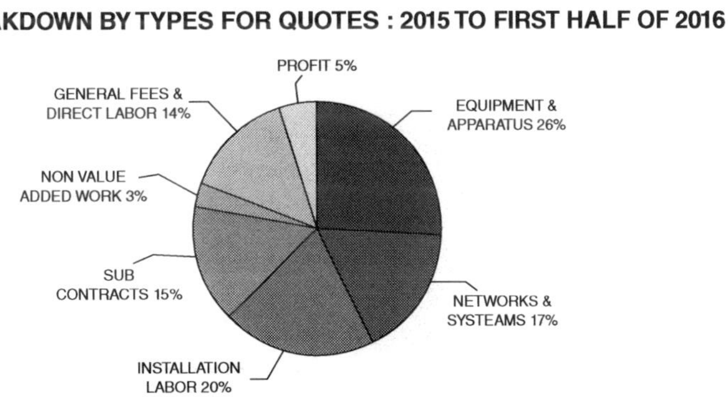

GOAL

Increase to 20% our success rate for projects obtained compared to those estimated.

Figure 14.21 A3 report

STAKEHOLDERS NAMES: VC, YB, CG, JPG	**VERSION & DATE:** 03 (2017-04-28)

ROOT CAUSE ANALYSIS

The following root causes are the results of a Cause and Effect diagram followed by a 5 why:

1. Gaps in estimating software proficiency (how to pull installation needs for labor and mat erial: electrical, plumbing, HVAC).
2. Insufficient knowledge of estimating software data management methods
3. Incomplete and inaccurate estimates are produced due to short lead time
4. Knowledge from exper ienced construction leaders is not shared with estimators

COUNTERMEASURES

CAUSE#	COUNTERMEASURES	BENEFITS
1 & 2	Develop an ongoing t raining program for our estimation resources.	Cycle time and Lead time ↓, ↑ % for C&A information
3	Creates KP I's for estimates data validation.	↑ Data accuracy, make gaps visible
4	Review estimation processes by adding a data review step (by experienced dedicated resources).	↑ Knowledge transfer and data C&A (complete & accurate)

ACTION PLAN — DATE OF UPDATE : 2017-04-28

DESCRIPTION	SCHEDULE (2016-2017)												RESPONSIBLE	STATUS
	9	10	11	12	1	2	3	4	5	6	7	8		
Train estimators in construction installation methods	■	■	■	■	■	■							Estimating Director	100%
One site visit/mth for the next 6 mths (construction site and prefab factories) by estimators.	■	■	■	■	■	■							Estimating Director	100%
Train estimators: how data from quantity extractions are processed by the estimating software	■	■	■	■	■	■							Estimating Director	100%
Mapping current and future processes.	■	■											PA	100%
Create meaningful KPI's to validate bid data.	■	■	■	■	■	■	■	■					PA	100%

FOLLOW-UP ACTIONS

Follow-up by measuring the success rate of the projects obtained vs those estimated, as well as by measuring the differences between GROUPE XYZ estimates and those of the lowest bidders (see the first 2 charts on the left to be updated).

Figure 14.21 (Continued)

12. First run studies

This involves trying new construction ideas on a pilot basis and applying the PDCA methodology. To determine its success and identify the best means, methods, and sequencing for a specific activity. These studies are carried out a few weeks in advance of their actual use in a project, with enough time to obtain needed materials.

A study documented by Salem *et al.* (2006), reviewed the installation of bumper walls and construction joint installation. Selection occurred in the "PLAN" phase. The test installation was observed carefully in the "DO" phase and videotaped in the "CHECK" phase. The finished work was examined carefully by the project manager, foreman, and crew to identify possible improvements. In the "ACT" phase various recommendations were tested, resulting in a 38% reduction in the cost of bumper walls, and 73% reduction in the cost of construction joints.

DPR Construction – a world-renowned construction organization has taken to concept to another level with First Run videos.

First Run Studies	*Trying New Construction Ideas on a Pilot Basis*
Value stream mapping: (VSM)	The documentation of specific actions to create finished products materials to meet customer demand.
Kaizen:	Continuous improvement philosophy
5S:	It is a system for workplace organization—it promotes lean thinking and application.
Visual management, visual controls	The placement in plain view of all indicators and activities allowing all to understand the status of the system at a glance.
Kanban:	Pull system central to the Just-In-Time (JIT) philosophy.
Poka-yoke:	Error proofing
Preventive and predictive maintenance	Keeping all equipment functional and maintained so it can operate reliably.

Figure 14.22 First-Run studies process

13. Visual management

In lean manufacturing, resources that do not contribute to the desired production are considered to be waste and are excluded from the system (Table 14.4). Visual controls provide transparency; as construction has mobile workstations, the placement in plain view of all indicators and activities allows everyone to understand the status of the system at a glance. The work flow and the respective action plans can be seen on the job site (Moser and dos Santos 2003). Commitment charts are typically used to secure everyone's agreement to actions that support lean endeavors. They have been used to identify best practices for job site safety. Having a document signed by all responsible parties in plain sight provides high visualization; this improves safety compliance. Safety signs that are designed/selected through a team effort also seem to obtain higher compliance.

Project milestones are typically maintained in a trailer under the watchful eyes of project staff. A visualization approach involves posting milestones in the work area for everyone to see. It leads to higher levels of involvement among crews. Similarly, PPC charts in plain view make everyone accountable. One nationally known contractor maintains an electronic display board in the main site trailer, near an entrance way

Table 14.4 Mechanisms for increased visualization
Examples of mechanisms for increased visualization

Requirements for Increased Visualization	*Criteria/Change*
Commitment charts	Visualization
Safety Signs	Team effort
Mobile signs	Knowledge
Project milestones	Communication
PPC charts	Relation with other tools

Source: Adapted from Salem, O., Solomon, J., Genaidy, A., and Minkarah, I., *Journal of Management in Engineering*, ASCE, October, 2006. With permission.

that everyone uses. It is approximately 4 × 3 and displays an automated countdown of the number of days remaining on the project. It is very effective in keeping people aware of time frames and the overall commitments of the project.

Exercises to test your understanding

1 A contractor organization is concerned about the quality of its housing construction and does an analysis of the sources of failure to take corrective action. The failures are categorized by the affected components. The cost of each failure is listed below.

a. Develop a Pareto chart by hand based on this information.
b. Show what building elements that should be targeted for corrective action, based on the 80/20 rule. Show your calculations, including cumulative percentages. You may use the reverse side of each page.

Element	*Number of Failures*	*Cost Per Failure ($)*
(A) Electrical	4	90
(B) Woodwork	2	47
(C) Plumbing	8	78
(D) Roof	1	83
(E) HVAC	2	55
(F) Doors	3	11
(G) Damaged walls	12	8

2 Using an on-site materials storage trailer as an example, draw a cause and effect diagram to represent a problem that is reducing productivity. (Trades staff have to queue up to get materials before starting a task and lose a lot of time in the process.) Sketch the structure of the diagram. Start with an observed effect "Long waiting time." Show possible causes on four bones on the chart labeled Materials, Workers, Methods, and Environment. To save time, show only two causes on each bone.

3 Pareto Chart: Draw a chart using the following information. Construct a cumulative curve to show the categories that account for 80% of the occurrences.

Item	Number of Occurrences	$ Cost Per Occurrence
A. Roof	8	30
B. Doors	7	9
C. Windows	10	90
D. Ceiling	5	16
E. Carpet	20	5
F. Paint	30	7

References

Fauchier, D. 2015. "Root Cause Analysis." Lean Construction, The Realignment Group of California, LLC, 13th Oct. *www.leanconstruction.org/media/docs/2015-congress/presentations/T05%20Fauchier.pdf.*

Imai, M. 1986. *Kaizen*. New York: McGraw Hill.

Moser, L., and A. dos Santos. 2003. Exploring the role of visual controls on mobile cell manufacturing: a case study on drywall technology. *Proceedings of the Annual Conference (IGLC-11)*. International Group for Lean Construction, Blacksburg, VA. http://strobos.cce.vt.edu/IGLC11

Rother, M., and J. Shook. 1998. *Learning to see: Value stream mapping to add value and eliminate muda*, V. 1.1. Brookline, MA: The Lean Enterprise Institute.

Rybkowski, Z. K. (2012). *Lean Construction Training Modules*, Department of Construction Science, Texas A&M University Construction Industry Advisory Council (DVD).

Salem, O., J. Solomon, A. Genaidy, and I. Minkarah. 2006. Lean construction: From theory to implementation. *Journal of Management in Engineering, ASCE*, 168–175.

Sobek, II, D., and A. Smalley. 2008. *Understanding A3 thinking: A critical component of Toyota's PDCA management system*. Boca Raton, FL: CRC Press.

Bibliography

Sarkar, D. 2006. *5S for service organizations and offices: A lean look at improvements*. Milwaukee, WI: ASQ Quality Press.

Suhr, J. 1999. *The choosing by advantages decisionmaking system*. Westport, CN: Quorum.

15

LEARNING FROM PROJECTS AND ENHANCING

Lean project delivery and IPD

Introduction

This final chapter draws on the knowledge provided in the preceding chapters in the context of learning from projects and using those lessons to improve future performance, especially with lean-based projects.

a. Retrospectives are structured, reflective events that enable a project team to have a common understanding of their project experience. It represents the "Check" element of the PDCA cycle and prepares the team for new action for continuous improvement.
b. For both lean and traditional projects, Post Occupancy Evaluation (POE) is discussed as a means of learning those lessons that are only available after the completion of a project or facility, i.e. those that focus on how well it meets the everyday needs of its users. "Post mortem" surveys are presented as a subset of POE. An improvement is recommended to adapt POE to lean-based projects.
c. Enhanceme nts to the lean project delivery system and IPD processes are suggested based on knowledge gained from projects.

Overview

Both retrospectives and Post Occupancy Evaluation (POE) provide excellent feedback on the conduct of projects. This information is very helpful to project teams and other stakeholders in providing a learning experience that promotes improvements in the conduct of projects in the next cycle of operations.

There is one major difference between POE and Retrospectives, and it is the timing of these events. POE by its very definition is conducted after a facility has been turned over to the owner at the completion of a project. It provides highly valuable learning lessons on large scale issues that can influence a subsequent project, but unfortunately corrective action at that point would be inordinately expensive or simply impossible. Retrospectives, on the other hand may be conducted at any time during a project and allow the benefit of lessons learned to be used while the project is still in progress.

Retrospectives

As recommended by the Lean Construction Institute, there are three types of formats recommended for retrospectives

1 A quick format – at the end of each meeting; these are usually held as plus deltas
2 Regularly occurring activities, such as every three weeks, or at the end of work cycles
3 Event-based activities, such as at the completion of major milestones. A major breakdown would be such an event.

The purpose of retrospectives in reviewing a project is to determine:

What went well.
What was different from expectations.
What did **not** go well.

By using this information, a team can determine:

What to keep doing
What to stop doing
What to start doing

As pointed out by the firm Lean Project Consulting, a retrospective allows reflection on a team's experience in a structured, open forum to improve effectiveness in the future. People learn from the situation, learn from each other, decide on both actions to take and the expectations for those actions. This action "tightly couples learning with action" in keeping with the "Five Big Ideas" presented in Chapter 2 and others.

William Seed, a lean construction leader and a Director of the Lean Construction Institute describes retrospectives in the book "Don't Conform – Transform" as a means of checking the alignment around the larger purpose of a program or the Conditions of Satisfaction (CoS). He thinks that retrospectives should lean more toward intentional contemplation than to utilitarian objectives. Notably, he points out that "silence produces more innovation than noise ever will".

In terms of organizing a retrospective, facilitation is a good place to start. It is highly beneficial to utilize a facilitator to guide the process. The process involves extensive discussion, and the sharing of ideas. It is important for all voices to be heard; some individuals may be highly vocal, but may not necessarily have the best ideas. An effective facilitator maintains a neutral position and creates a positive, blame-free environment where ideas can be expressed without fear of criticism.

Participants should be advised to review all available data (before a retrospective) in connection with the area being studied; this ensures that meaningful, substantive discussions can be held, leading to recommendations that are specific and actionable.

Action steps should be listed at the end of each meeting. It is best to have a consensus on the actions to be taken, and the responsible individual(s) for each item. This information should be clearly documented, and should be part of the record for the project. It should be made available as quickly as possible after each meeting. Smart Boards can be very effective in keeping track of this information, allowing it to be readily retrieved at the beginning of the next meeting.

Time requirements

A Plus/Delta review at the end of a typical meeting may only require 10 minutes. With a project milestone, one to one-and – a half hours should be adequate.

The location:

> A Big Room would be a good environment for conducting a retrospective since it would have wall-mounted project and schedule information. The LCI recommends a One hour Retrospective (documented by Kathleen A. Iberle) for a group of about 25 people. Index cards are provided as well as well as "Sharpie" fine-point markers. A large table is divided into two sections labeled "good" and "needs improvement". The team is provided with index cards – about 200, 3 inches by 5 inches in size; in two different colors, and twenty cards 4 inches by 6 inches, of another color. A facilitator manages the process.

An agenda is essential for ensuring that all needed steps are identified, and also carried out. With many issues to cover, there is the potential for things to be missed, or not covered adequately. It is important for attendees to identify problems, and not try to solve them at the retrospective.

A draft agenda

a) 5 minutes – an overview of the rules, followed by having attendees write things that went well on 3 by 5 cards of one color, and less successful events on the other 3 by 5 cards.
b) 10 minutes – participants brainstorm on the good and bad outcomes, writing silently on the rear of each card, then placing them on a table or white board.
c) 5 minutes – the cards are reviewed, and duplicates are grouped together.
d) 20 minutes – The facilitator helps the group to find solutions for specific "needs improvement" issues.
e) 5 minutes – the team votes on the issues to pursue promptly.
f) 5 minutes – people are assigned to follow up on specific issues.

The process should ensure that people are given the time and resources to resolve the issues that have been identified. Without that, they are unlikely to succeed.

It is advisable to have ongoing retrospectives so they become a regular event, with positive results.

Learning from lean construction projects

Companies should first understand themselves before rushing out to hire a consultant to make a lean transformation. Companies that understand and that have practiced self-transformation can benefit from the fresh ideas brought by a consultant. Those that have not tried to change from within may delude themselves by expecting consultants to bring about change extrinsically, when it needs to occur intrinsically. A lean consultant would benefit a company most by helping it to learn how to change, but a consultant is "always an outsider" and the people in the company must take the initiative to bring about change. Ballard recommends that companies start with the LPS as it identifies improvement opportunities while freeing up resources to address them. The Five Big Ideas for Lean Project Delivery serve as a so-called True North and can be applied to virtually any large, complex project.

The importance of Conditions of Satisfaction

By definition, the CoS are a set of criteria or expectations that specify how the success of a project will be determined. It cannot be over-stressed how important they are. As a project progresses many events may occur and people can easily lose sight of what different parties are expected to do, and what they expect others to provide to them. In extreme cases the lack of documented CoS has led to failures and litigation. As described by Chris Dierks in Chapter 7. The Conditions of Satisfaction outline the success factors and the expected behaviors that are essential for the implementation of a successful lean project. This example relates to contractors primarily. The owner's CoS would emphasize the owner's priorities, but overall, the project team should mutually agree on the CoS. By signing a written document, team members acknowledge their commitment to collaborating in a spirit of trust for the best interests of the team and the project.

They promise to commit to the principles and practices of lean project delivery as outlined in a reference guide – in this example "LPD 101 Lean Project Fundamental Guide". They commit to working continually in a spirit of trust, transparency and collaboration to eliminate waste and improve the performance of the team.

Examples of project success as viewed by the team are:

- Everyone is profitable
- The project is delivered to the mutually agreed Target Cost
- The project is delivered to meet a mutually agreed schedule
- All team leaders will have excellent knowledge of their work and their scope
- RFIs will be only those where information is clearly missing in the documents or site conditions require resolution
- RFIs will be issued and answered in a timely manner to avoid becoming project constraints
- Change notices resulting from legitimate RFIs will be processed in a timely manner.

- The overall objective will be to minimize RFI's and Change Notices
- There will be exceptional teamwork, and minimal rework due to poor quality work
- All schedules will be developed and executed using the Last Planner® System

Overall, the superior performance of lean projects is in part due to such clearly understood, and agreed, CoS.

Recommendations from lean projects

The following recommendations are based on lean project cases and discussions with various construction professionals.

Procedural recommendations

Ed Anderson, of Lean Implementation Services of Florida has several recommendations on procedural issues in his publication *Lean Planning and Production Control (LPPC): Tips, Tricks, & Lessons-Learned*, Rev. Feb. 2018, V24.

Ed stresses the importance of the Big Room – he views it as an investment. For a construction site it should be at least a double wide unit and it should be equipped with dry-erase boards, communications systems and restroom and other amenities. Breakout rooms facilitate small meetings to resolve issues without slowing down the main meeting. Stakeholders should attend every meeting, without exception. Even if stakeholders are traveling they should be in contact with the big room and be able to participate.

Lean stakeholders need to see reliability as a 2 way street. If a CM or owner wants specialty subcontractors to be reliable, then he or she must be reliable as well. On the issue of reliability, Ed speaks of PPC values in terms of language that was used on the Heathrow Terminal 5 project years ago, they used the acronym "CRI" for Commitment Reliability Index

Superintendents must buy in to the lean process and the Last Planner® System – unless they buy in the plans may not be viable.

Ed sees the communication from sticky notes as true information. To be meaningful and actionable, commitments must be in writing, and the sticky notes accomplish that requirement.

Ed observes a major hurdle that has to be overcome with Superintendents, and that is the question of "pushing for reliability". Traditionally, Superintendents have "pushed for speed" in order to meet critical milestones, and have often done this for many years for generous salaries. It requires great effort and persistence to get them to change their perceptions, even after exposure to simulations, and prioritize commitments and reliable promising instead.

Every task and constraint must have an explicit owner. These responsibilities must be assigned in writing. "If everyone is responsible, then no-one is responsible".

Ed proposes an innovation to replace the Sticky Note process. It is the NxtNote system, a web-based technology that generates printed sticky notes. While handwritten notes can be effective, their legibility depends on each person's handwriting. Transposition to a data base can be very time consuming. The NxtNote equivalent can receive data from computers and handheld devices, and can be scanned at 1,000 error-free words per minute. The data can be transferred across global boundaries, enabling collaboration even at great distances.

If safety problems persist after the Last Planner® System has been implemented, Ed views their occurrence as an indication that the LPS is not being implemented correctly. To be effective, supervisors must develop Job Safety Analyses (JSAs) ahead of time, and train crews on implementing them. He stresses that supervisors should look for and point out positive safety behaviors.

Training experiences

The feedback from successful lean programs and practitioners confirms that workshops are most effective for getting projects started on the right footing. For example, DPR Construction is recognized domestically and in other countries as a foremost practitioner of lean construction. Chris Dierks is a Project Executive with DPR, and East Coast Lean & IPD Leader. Chris is a strong believer in workshops at the onset of a project. Projects start with teaching and learning and Chris plans these activities to best match the dynamics of a project by varying the duration and agenda for each workshop.

Workshops typically includes various games and simulations, such as the "Airplane Game" which demonstrates the advantages of continuous flow over the batch processing of work assignments. Such a workshop explores a number of key questions and covers several important topics relating to lean concepts. While educational content is presented, it is a prelude to several other questions pertaining to how the project team will function. Examples of such questions are:

How can we develop a lean mindset? What general learning do people need?

What are the different processes involved? It is important to give everyone a base knowledge of lean to prepare them for working as a lean organization.

What will provide the highest value to the entire team? How do we organize as a team?

Chris likes to link this activity with having team members meet each other, and also get to know about each other as it provides transparency and an understanding of expectations.

Chris considers it very helpful to the direction of the team when an owner or an architect can say to them "Here is what the program looks like, Here is where we want to go" because it provides clarity to everyone.

Other project team members should also explain the decision-making process in their own organization so that everyone understands how other parties operate and what limitations they have to work with.

Chris finds it beneficial to lean projects to use the kickoff training to clarify the Conditions of Satisfaction for the project. The CoS are important and may vary by the needs of different stakeholder groups. At the end of the workshop Chris endeavors to have a draft of the CoS as a point of reference for the team. It provides transparency and clarity. It alerts project managers to how well communication is working, and whether course corrections are needed – while they are still possible.

A goal setting exercise with the project team also focuses stakeholders on the team's priorities. This approach has worked with many of DPR Construction's lean projects.

As the team begins work on the project, further training may be needed depending on its size and complexity. Surveys of team health are conducted to gauge how well the team remains integrated and focused on its objectives.

JE Dunn Construction is very committed to lean construction principles and practices through the advocacy of Rebecca Snelling, VP and National Lean Director. Even though most of the company's projects are conventional in nature, Rebecca treats them as if they were lean projects. Her team typically helps a project team with lean facilitation at the inception of each project. They present simulations that illustrate important lean principles. On a typical project, every member of the project team, including craft persons is provided with a one-day training class that focuses on lean thinking and on the specific tools that a project will use. Separately from that, all employees (project and non-project) go through a one-day lean course that is focused on lean principles and tools that they can apply to their own work.

Usually, a Lean Specialist demonstrates pull planning to a superintendent or project manager (PM) on an actual project in three ways: first, by facilitating a pull plan in front of the superintendent or PM; second, by co-facilitating a pull plan with the same superintendent or PM; and third, by observing the superintendent or PM administer a pull plan on his or her own. The results of this initiative have been outstanding – traditional projects have been delivered early, and on budget or less.

JE Dunn's experiences demonstrate that lean practice does not have to depend on having an ideal lean project with an integrated form of agreement. If an organization is committed and sets an example, others will follow.

With regard to owner organizations, Sutter Healthcare has determined that the training of project teams as well as in-house staff is highly beneficial to their construction program. One day per month is reserved for staff training, although it is primarily process-driven. Digby Christion, Director of Integrated Lean Project Delivery, has coordinated in-house resources for IPD best practices but utilizes LCI meetings and conferences for training on the Last Planner® System and Target Value Design. Sutter also often hires consultants to get new project teams started.

The majority of lean training is hands-on at the project level. In particular, Last Planner training and education are done at the start of projects, including the design and construction partners. In selecting trade partners, preference is often given to companies that have had exposure to lean practices such as the LPS.

Christian's training is held once for each Project Manager, and a sequel to that program is being developed. He thinks the more integrated your team is the more likely you are to achieve your goals no matter how ambitious those goals are, and the more likely you are to achieve them with a team that is energized and happy.

Project lessons

Sutter HealthCare has managed one of the largest facilities construction programs in the USA, starting in the early 2000's. Since 2006, Sutter has completed 23 IFOA contracted projects totaling over $4.1 billion. They were delivered on time, on budget, and without scope compromises with only one exception where a five year project was a few months late and a few percent over budget. Currently, As of 2019, Sutter has another $500 million in IFOA projects in progress and does not intend to move away from the contract model and delivery processes. Consequently, they have had the opportunity to observe the workings of the Integrated Form of Agreement (IFOA), and give it credit it for their success.

Some examples of experiences with the IFOA that have influenced lean practices:

1) Limited grounds for change orders, only allowing them for: owner changes, unanticipated changes in the law, unanticipated decisions by authorities having jurisdiction, and unexpected discovered conditions. Errors and omissions in the work of the design team are not allowed grounds for a change order. Reportedly, this has transformed the quality of the work provided during the design phase of IFOA projects.
2) Prefabrication can greatly improve project efficiency – most of the walls (about 90%) on the 3rd and 4th floors were pre-assembled offsite. Whereas the original schedule for on-site framing was 40 days per floor, the actual onsite framing duration was 4 days per floor, leaving the floor ready for final inspection and drywall.
3) Because of the accuracy of the information and the use of offsite pre-assembly, on-site productivity increased, quality improved, and overall onsite material waste was reduced. The framing contractor's productivity doubled compared to the on-site construction baseline; there was less than 1% material waste compared to 15–20% when stick building the same types of wall. Decoupling the framing and the overhead construction resulted in the more efficient installation of the overhead systems and less trade crowding on the floor. It also eliminated the onsite in-wall construction phase.
4) Steel detailing for interior spaces was greatly improved through the use of 3D visualization. The other disciplines were enabled to perform a comprehensive parallel review of their work against the steel fabrication model. Other process steps would eliminate clashes with MEP systems, steel plates, and large slab openings; coordination with skin, and coordination with the edges of slab, for example. Eventually, the steel design model was completely replaced with the 3D steel detailing model
5) The results were very positive. The entire structure was erected with fewer than 130 RFIs relating to steel design – far fewer than comparable projects. The IPD team realized $1.5 million in savings related to the steel due to labor savings and reduced rework. Similar to structural steel, this simple change in the process resulted in significant gains. It saved the team approximately 30-days of non-value added time per slab pour which was about the time required to complete the coordination of the hangers ahead of each pour. The MEP team then streamlined the conversion process from 3D models to field installation.
6) IPD allows teams to identify waste across the supply chain, while they are working on the project, and to be able to take corrective action. These projects set the stage well for the projects that followed later.
7) With regard to information technology, when BIM and lean are implemented there are significant benefits as compared to conventional approaches. This allowed most of the fears to be discussed openly. The lessons learned from a workshop explored the design for fabrication potential and efficiencies of model-centric processes that minimize the use of 2D information. At the end of the workshop the entire team chose to implement the same model-centric design to fabrication workflows as well as the rigorous planning approach used with the hospital project, without increasing the overall project cost. These types of concerns inhibit some IPD teams from fully exploring innovation opportunities.

JE Dunn Construction learned important lessons by facilitating lean design as CM for an Iowa State University (ISU) project. Rebecca Snelling, VP and Lean Director coordinated with the architects and the engineering team for the project. The last Planner System for Design was used in the design phase of the project. It was worthy of note that the design team had not participated in the pull planning of design activities before the inception of this project. The lessons learned by JE Dunn included the following: It is absolutely necessary to have the right people in the room who are equipped to make decisions.

There should be no formality that inhibits speaking up – titles are not important.

Having people meet face to face rapidly increases their accountability. Having a lean specialist lead each planning session is all important – pull planning in design is different from pull planning in construction, and the specialist needs to be skilled in the design environment. The meetings on this project clarified an understanding of the overall project goals. It also underscored the importance of accountability and commitments.

Impact of lean on construction projects

JE Dunn's project experience reinforces the fact that much can be accomplished with the lean methodology, even in situations where there is no formal requirement for a lean contract, such as an integrated form of agreement. They credit Lean practices implemented by the project teams for the significant schedule savings, along with savings returned to each owner.

Pull planning sessions were held 2–4 weeks before planned activities. Sessions involved trade partners' personnel that were actually performing the work.

Together and with JE Dunn, they created a plan not only effective for themselves, but also for all trade partners involved in the specific scope.

Other major impacts that directly benefited the project include:

Expedited Schedule Skyloft saved a month on the schedule, allowing the owner to stock the building and hire the facility manager earlier. Aspen West saved two months on schedule with similar opportunities.

Budget Savings Both projects were completed under budget, resulting in a net savings to the owner.

Both project teams aggressively managed the change order process. As a result, change orders were limited to unforeseen issues at Aspen West and zero change orders for Skyloft.

Yet another project, the HT Residence Halls Renovation project succeeded because of JE Dunn's commitment to Lean principles. Continuous communication, precise coordination between the trades, and respect and accountability among all the stakeholders were all key factors that allowed the team to meet and exceed project requirements. Despite unforeseen conditions and a lack of reliable as-built documentation, JE Dunn was able to complete the project to the owner's satisfaction. They received an Excellence in Construction award for the work at HT, the highest project specific honor bestowed by the Associated Builders and Contractors.

The team has seen the Last Planner system in action, can verify its positive impact to a project and is excited to tell the story so others can experience the same benefits.

Specific actions to support the lean concept

JE Dunn provided a lean specialist for training, coaching, and support. The Regional President of the company had established an expectation that all project teams would use the Last Planner® System. The lean specialist was effective in providing the needed training. The specialist role is sometimes rotated through different staff members in order to maintain continuity, and could be a good model for other companies to consider.

Kinetic Construction Limited – lean successes and lessons learned

Kinetic Construction Ltd. is a mid-sized General Contractor – Construction Manager operating out of Victoria, British Columbia (BC) since 1984. Kinetic has three branches on the west coast of British Columbia. Kinetic adopted the lean methodology in recent years; lean is not as well known in Canada as it is in the US. Kinetic has done an outstanding job of implementing lean. Their CEO embraced the concept and supported it with staff who are strong believers in lean concepts, and who were unrelenting in having lean practices become the way of doing business. As Trevor Coyle, their Lean Integration Manager explains in Chapter 10, they had a rocky start in implementing lean. They had consultant support initially and tried the recommended methods for implementation, with an earlier project but without success. Kinetic optimistically viewed the earlier lean project as an excellent learning experience. It was started with a training workshop and a team health survey was conducted shortly after. The results indicated that the bootcamp message had an impact on people's feelings of happiness and their belief that this was in-fact a good collaboration effort. The event itself was an excellent team building exercise. (It is sometimes necessary to let people fail first in order for them to accept new ideas such as lean construction.)

Kinetic embarked on the Brookes School Project, armed with the lessons learned from the first project.

The Brookes School project was a great success Site clearing started in July 2017 and critical construction was completed on schedule in August 2018, just in time for the new school year. The project followed Lean principles from design through construction completion.

Kinetic Construction was able to deliver a project that should have taken 17 months but in just 13 months instead, on time and on budget. The project also gained the attention of Award magazine who published the story of this highly ambitious and collaborative achievement. Tom Plumb, president, and CEO of Kinetic Construction was rightfully proud of this achievement – he pointed out that the lean approach allowed the $25M school to start providing exceptional educational opportunities for 300 local and international students without delaying the start of term-1.

Manager's reflection: As Trevor Coyle notes: When we introduced the LPS as a tool we didn't fully understand the magnitude of effort required up front and the difficulties we would be faced with to sustain the results. After all, we had a project to build. Because we had not seen the potential of the system, we were reluctant to "enforce" mandatory attendance when trades were extremely busy. The feeling that the busiest people on the project should not attend the planning sessions when they are behind is a natural one. People are inclined to feel that if they can just get around the next corner (figuratively speaking), they'll be caught up and the issues will subside. However, the reality is always the same in the end. A common saying we hear these days is; If you fail to plan – You plan to fail.

Although we were coached on the importance of complete collaboration – users naturally interpret a powerful tool such as LPS, as a solution to solving all of our problems. In retrospect, we believe that we over-sold the system without fully understanding the cultural foundation and consistent early coaching required to succeed. We learned that participation is always required by every major or critical scope to ensure the value of the effort is returned to the last planners and the project.

Learning from green initiatives

With regard to "green" approaches to construction, the renovation project at Michigan State University clearly indicated that such projects can be built with the IPD method and deliver the "triple bottom line" that Glenn Ballard has referenced. Not only was the benefit of environmental sustainability obtained, but initial costs were lowered and the schedule reduced by the lean methodology. The completed project exemplified a successful Lean and LEED GOLD project, driven by a knowledgeable owner. It was the first IPD project carried out at a public university in the entire United States.

Although owners drive the design and construction process, the OAEC community (Owners, Architects, Engineers, and Contractors) should give serious thought to the lessons taught by this project in light of all the concern about the state of the environment. If this project was attainable in a public sector setting then it could easily be applied in private sector projects. Not only are initial costs reduced with lean construction, but operating savings continue to accrue through reduced operating costs.

Lean commissioning

Commissioning is an important process that verifies that systems are installed, tested, and documented to show that they conform to the owner's requirements and the designer's intent. It is closely aligned with green projects as it is a requirement for certification (such as LEED certification). Murray Guy formerly of Integrated Designs in Canada has had great success in combining Commissioning with the Last Planner system. Murray determined that commissioning activities need to be put on the lookahead planning board, and weekly plans, otherwise unplanned activities become constraints during the project. Old fashioned sticky notes can meet that need, even if that information is entered in a program later. This process has been used in delivering highly sustainable buildings such as netzero projects, complying with all required environmental standards and on time and within budget by integration with the Last Planner® System. This approach indicates that both lean and sustainability initiatives are attainable in a project, and do not have to be traded off against each other. The lean commissioning approach elevates what could be a passive activity to a proactive process that holds the project team accountable for a much higher level of performance. It was used successfully on the Okanagan College Trades Center, Kelowna, BC, Canada. This addition to an existing facility was completed around 2016. Murray Guy formerly of Integrated Designs, served as a commissioning agent in assisting the project to meet several environmental initiatives including NetZero.

As described in Chapter 12 it requires a different way of thinking and full commitment. The success of Lean Commissioning (Lean Cx) is enabled by: Early formation of the Commissioning Team, development of a milestone schedule for commissioning activities, and integration with the milestone schedule for the overall project.

Overview of post-occupancy evaluation (POE) for continuous improvement in construction

The process of post-occupancy evaluation (POE) provides a structured and systematic method for learning lessons from past construction projects. Preiser *et al.* (1988) define POE as the process of evaluating buildings in a systematic and rigorous manner after they have been built and occupied for some time. Although one may intuitively suspect that a facility succeeds (or fails) in serving its users well, one cannot fully appreciate and measure to what extent this occurs without conducting a POE. Construction professionals have come to realize that those who forget the "sins" of the past are doomed to repeat them again and again. POE can provide redemption from these mistakes if it is used correctly.

The quality of constructed facilities is important to its users. People experience more satisfaction within their environment if it is kept in an esthetically pleasing manner. Preiser *et al.* (1988) noted that "spatial attributes, the sequence, location, relationships, size, and detail of a facility's spaces have been shown to affect occupant behavior." POE provides a means for improving the quality of future facilities by learning from past projects.

A POE involves the measurement of the functioning of a facility as compared with its purpose as defined in a formal program, and by the objectives of the architect/designer. As these objectives should mirror the agreed-upon wants and needs of the client/owner, POE requires that a systematic methodology be used to compare these expectations with the effects of the facility on its users. The results of the POE can identify the extent to which the design intent has been met; this feedback can also help to identify "best practices" that can be used to improve future designs. The quality advocate, Peter Senge, in his book *The Learning Organization*, advocates that organizations follow a cyclical process in order to learn from their actions and use that knowledge to improve their performance.

In order to apply POE effectively, it is good for design and construction professionals to start with introspection, to look inward and ask why POE is being considered, and what one hopes to derive from it. As with so many other endeavors, if one does not plan where one is going, one is likely to arrive at an unexpected or unwanted destination. The experiences of world-class organizations indicate that analysis can provide invaluable information, enabling one to manage analytically, not just intuitively. In the design and construction arena, this analytical approach leads stakeholders to examine and improve the processes involved in both construction design and delivery.

Scope of evaluations

The scope of POE studies varies with the evaluation objectives. The evaluation may focus on:

- A specific part of a facility
- An entire facility
- A building program
- Programming/design
- Construction procurement and delivery

Categories of POE: historical, comparative, longitudinal, and quasi-experimental

Ex-ante evaluation involves an analysis of facility performance "before the fact." It may be informal, as conducted by some architects in visioning how a facility may be used. It is a standard part of the design cycle – imaging, drawing, and testing. It often does not include users' inputs but it should, as the users can project themselves into the use of a facility with more realism than others can. On the other hand, simulation methods evaluate the utility of a facility in terms of travel distances, and so on, as well as perceptual criteria.

There are several categories of POE. They include:

- Historical: Studying the facility in retrospect to determine if actions taken during the design/construction process have been effective.
- Comparative: Contrasting two situations such as two similar facilities after one has been specifically changed.
- Longitudinal: Taking baseline measurements before changes are made. Changes are then initiated and differences attributed to them.
- Quasi-experimental: Using statistical approaches to compare experimental and control situations.
- Post-mortems: A re-visitation of the design and construction processes themselves provides critical process-related lessons.

Many different models have been developed for the process of POE. A frequently used one is the environmental design model that is based on the performance concept. It involves comparing specific performance criteria that have been established for a given type of facility, such as a hospital or a school, with the actual performance that is observed or measured by the building's occupants and evaluators. The performance measures generally include: the spatial aspect, the physical interaction between the individual and the facility; the social aspect, the social significance of the environment; and the service aspect, the environmental conditions necessary for safety and health.

Role of programming in POE

Programming is a critical reference point for POE studies.

- Programming is a statement of a project's problems or constraints as visualized before design work is begun.
- Programming is analysis and design is synthesis.

A well-designed program treats users' requirements in terms of form, function, and economy. It also provides a methodology to evaluate the success of the design procedure. The subsequent POE compares project outcomes with the program and its influence on design.

Using an educational facility by way of an example, a POE of facilities may comprise four headings: user satisfaction, economy, function, and performance. User satisfaction will address the question: How satisfied are the users with the facility? Are there distinct likes and dislikes with regard to a particular building feature? Economy is concerned with whether or not the facility is economical to operate and maintain. It considers the efficient use of space, time, and resources, ease of cleaning, maintenance, and repair; the use of appropriate materials and technology for the educational environment; the relative quality of materials and construction. Function addresses whether or not the facility functions as intended; if it helps or hinders the essential activities such as learning and physical activity. The factors considered are: flow, orientation, accessibility, and supervision of circulation of students; comfort, safety, and convenience; the actual use of space for programmed and non-programmed instructional activities. Performance addresses the application of creativity and innovation in design and construction: the performance of electrical, mechanical, structural, and other building systems; the quality of temperature and humidity controls, light and sound controls; color, texture, space, and perceived aesthetics; the social image and character of the facility; user satisfaction, and the adaptability of the facility for changes in use.

Planning for the POE

A facilitator must be designated to coordinate the POE; this should be a seasoned professional with a background in engineering, architecture, or construction. A background in quantitative analysis would also be helpful. It is highly recommended that this facilitator should not be a member or affiliate of the design team or the construction team, in order for the POE process to be unbiased and objective.

As the facilities to be studied are likely to belong to institutions or large owner organizations such as retail chains, it should be feasible for them to use a member of their technical staff to serve as a POE facilitator. In projects based on lean-based delivery methods, an independent facilitator such as a lean coach could manage the POE process.

A POE survey instrument is developed to address the factors of user satisfaction, economy, function, and performance in the project/facility being reviewed. It uses a five-point Likert scale ranging from "very dissatisfied" to "very satisfied," with a midpoint indicating "neither satisfied nor dissatisfied." One version of the survey is developed for facility users (who are familiar with a facility) and another for design, construction, operations, and maintenance personnel. This latter group may represent a design/technical team that can address a very broad range of issues related not only to the owner's intent, but also to long-term performance of a facility.

In the case of an institution such as a college or university, design/technical team members are drawn from several departments such as: maintenance, construction, safety, energy management, communications, information technology, campus police, cafeteria services, and transportation, to name a few. In the case of a hospital, the POE survey should include many users – administrators, doctors, nurses, and most importantly, patients. Patients are seldom asked for their opinion and that runs counter to the lean philosophy. Other sources may be used for data gathering: the analysis of drawings, building observations through photography and video, and the analysis of operating cost. Maintenance records are one valuable source of operating cost and energy consumption data.

Based on the degree of need, a POE may be tailored to include such factors as general satisfaction, aesthetics, layout, air quality, privacy, durability of construction, maintainability of equipment, energy, and efficiency.

A candidate facility is selected for review; ideally, it should have been occupied for between one and two years, long enough not to be distracted by short-term final adjustments by the contractor, but not so long that shortcomings become accepted as part of "living with" the facility.

POE procedures

- Invitations are sent to the team members for two distinctly different evaluation meetings:
  a. a facility user meeting, and
  b. a design/technical team meeting.
  The second group is especially critical as it is the best source of remedial action information. It is usually more convenient to have separate meetings on different occasions as the meeting agendas are somewhat different; the survey instruments, in particular, are different. The design/technical team meeting requires having team members acquainted with a facility they may be seeing for the first time, so more time has to be spent in providing guidance on the layout and features to be examined in detail.
- The facilitator leads the evaluation meetings. The design/technical team is instructed on the approach needed to make building observations and complete the survey documents.
- A general discussion is held on the background and conduct of the project. The owner's project staff and the architect/engineer (A/E) of record as well as other decision makers provide background on the inception of the project and its subsequent conduct.
- The POE participants walk through the entire facility in small groups and individually complete the surveys.
- Following the site visit the data from the design/technical team's surveys are collated and analyzed.
- The scored and narrative responses for both users' and design/technical surveys are combined with historical building performance information. Maintenance records are checked for operating costs, breakdowns, and malfunctions.
- A comprehensive POE report is prepared. The survey findings are reconciled with current standards and specifications to develop meaningful recommendations.
- The report is distributed to all involved parties.

Once a POE is completed and documented, it is critical that additional steps be taken to apply the lessons learned to the design and construction of future facilities. These steps represent the stages in the continuous improvement cycle shown in Figure 15.1, and are required for systematic improvements to occur.

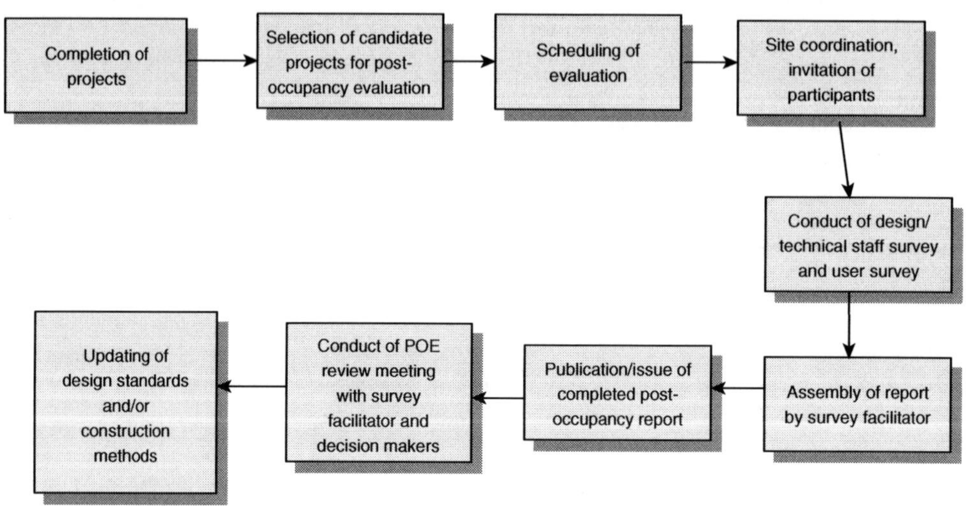

Figure 15.1 Flowchart of the post-occupancy process

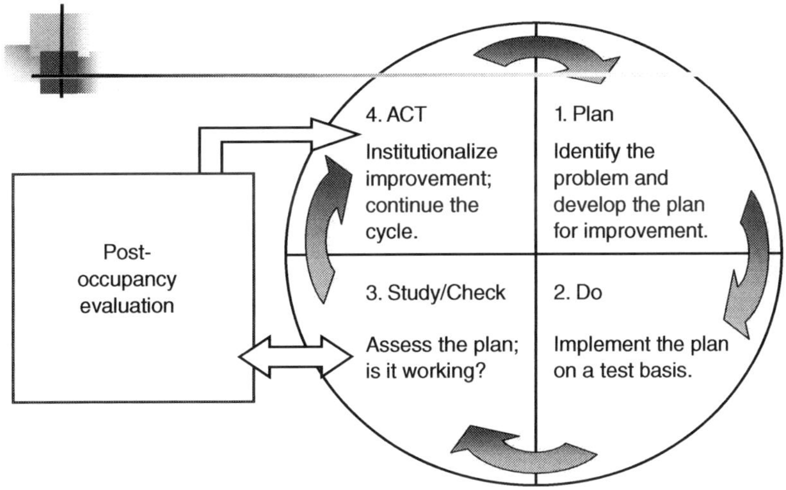

Figure 15.2 The Deming Cycle (PDSA Cycle) and Post-occupancy Evaluation

Source: Reprinted from Forbes, L., *Proceedings of the 2002 IIE Annual Conference 2002*, Orlando, FL, May 19–22, 2002. With permission.

A meeting is conducted with selected decision makers, in which the survey findings and recommendations are presented.

Approved changes are subsequently made in the appropriate reference documents, so that the specifications and construction procedures for future projects can have enforceable requirements to utilize identified best practices. These documents may include the master specifications, design criteria, and construction procedures manuals that many large organizations maintain for reference purposes.

As indicated in the foregoing figure, the post-occupancy process serves as the "Study/Check" phase of the PDSA cycle (Figure 15.2). The recommended changes are implemented in the "Act" phase. In subsequent iterations of designing and building a particular building or building feature, the use of the PDSA cycle enables the owners and designers to evaluate the impact of POE-based improvements.

Quality score calculations

The quality score is not mandatory in most cases. In special cases where such a level of detail may be needed, the following equation provides for the determination of a "Quality Score" for each surveyed facility or project. The questionnaire responses are combined to obtain mean ratings (R) for each question. A weight (W) is assigned to reflect the importance of a particular objective. For example, under the heading of the objective "Function," respondents could be asked to rate the adequacy of a specific office space. The same question, when asked of different spaces may have different weights applied. The composite Quality Score would be based on the factors in the equation below: R = rating and W_{jk} = relationship of question j to objective k.

$$\text{Quality Score}\left(QS_k\right) = \frac{\sum_{j=1}^{n} R_j \times W_{jk}}{\sum_{j=1}^{n} W_{jk}}.$$

Quality professionals can further enhance the POE process through life cycle analysis to assess building/ facility performance on a wider scale that includes not only user satisfaction with everyday utilization, but also issues of operating cost, durability, and reliability. User satisfaction questionnaires provide valuable information on the perceptions of the client (construction owner) with respect to the finished product and, very importantly, the nature of the associated processes. A typical questionnaire would address contractor performance measures relating to such issues as timeliness, responsiveness, communication, empathy, cost, and construction quality. The performance of designers could be evaluated in a similar fashion.

Overall, the delivery of design and construction services can be significantly enhanced through the use of "lessons learned" and "best practices." The information collected through the POE application should also address how to design processes to improve design quality, cycle time, transfer of learning from past projects, and provide effective performance measurement systems for understanding, aligning, and improving performance at all levels. The quasi-experimental type of POE should be investigated for application where two or more facilities may be compared with each other.

Other POE approaches

Several other POE approaches can be used within organizations, especially designers and contractors, to measure performance. The Construction Industry Institute (CII), headquartered in Austin, Texas, has produced important construction research findings that can be used very effectively in the POE process. Their research has looked into the performance of project teams that are used on certain types of construction projects where team membership changes during the transition from conception to completion. They have identified that project leadership behaviors have a great influence on project cost performance, and that the adoption of best practices influence schedule performance.

A POE could examine the extent to which the respective success factors had been incorporated in a project. Leadership behaviors, for example include: focus on objectives, team values and environment, empowerment and decision making, and external/internal relations. Best practices include: planning and project controls, scope management and project alignment, and quality.

Sample questions

1 Focus on project objectives. Please respond to the following items, using a scale of 1 = not at all, to 7 = to a great extent.

   To what extent has your team leader

   a. Communicated team goals and objectives?
   b. Focused the team members?
   c. Aligned team member goals with the project goals and objectives?

2 Empowerment and decision making: Please respond to the following items, using a scale of 1 = not at all, to 7 = to a great extent

   To what extent has your team leader

   a. Allowed the team to make its own decisions?
   b. Supported the team's decisions?
   c. Promoted a spirit of empowerment among team members?

Client satisfaction with design and construction services

Many owners in both the public and private sectors have begun to use criteria other than low prices for selecting design and construction providers. In considering candidates for new business, these owners are increasingly referring to the nature of their interaction with designers and contractors on earlier projects. This past performance is rated and included in an evaluation of candidates for new work.

A performance evaluation questionnaire is an excellent adjunct to a POE. It addresses what is perhaps the most important part of the design and construction process; that is, the perceptions of the client (construction owner) with respect to the finished product and, very importantly, the nature of the associated processes. It goes without saying that the owner's level of satisfaction with the outcome of a project is an important measure of success. A typical questionnaire would address performance measures relating to such issues as timeliness, responsiveness, communication, empathy, cost, and quality. A typical survey could address these issues as follows.

Owners

Use the following scale to reflect how the following factors influence your satisfaction with a contractor organization. 1 = no importance to 5 = extreme importance.

1 Timeliness: Does the contractor (for miscellaneous jobs):

   a. Provide a fair estimate of cost and schedule?
   b. Give small jobs high priority?
   c. Complete jobs quickly, once they start?
   d. Respond promptly when asked for work status?

2 Communications: Does the contractor:

   a. Update me on jobs and their status?
   b. Explain each job before starting it?
   c. Notify me of and explain project delays?
   d. Follow up to ensure satisfactory completion?

3 Quality: Does the contractor:

   a. Emphasize the facility's performance characteristics?
   b. Strive to build in durability in the facility?
   c. Provide high quality workmanship, beyond code requirements?
   d. Treat quality as an essential element of owner satisfaction?

Other factors may be added to the foregoing lists as needed. In the case of completed projects, it is important to ask how well the Conditions of Satisfaction were met.

Best practices to derive the best results from POE

A number of best practices will help construction stakeholders to derive the best results from the use of POE. Utilize the services of an experienced professional such as a lean facilitator to interface with the involved parties to ascertain the organization's information needs and develop an appropriate survey instrument (questionnaire). The POE questionnaire should be validated to ensure that it asks the right questions and will be interpreted in a consistent manner in repeated use.

- Develop an organization-wide commitment to continuous improvement. A commitment to the lean construction philosophy would be ideal for harnessing lessons learned to deliver greater value and eliminate waste. Incorporate the Plan-Do-Study-Act (PDSA) cycle in the area of facilities construction activities. Top management must take the lead and encourage organization-wide participation in efforts to continuously learn and improve performance. Empower employees at all levels to recommend/implement improvements.

- Begin the journey toward adopting a new business model for facilities construction/management based on the principles of the Malcolm Baldrige National Quality Award (MBQNA) Criteria. Implement a training program that may start with conducting staff orientation sessions to learn about Baldrige principles.
- As a part of adopting the new business model, consider administering a self-assessment survey to staff involved with facilities design or construction. Voluntary and confidential staff responses to a battery of well-designed questions can provide senior management with the current state of the organization based on several criteria. These may range from employee morale to the effectiveness of the processes used to conduct business, as well as the extent to which some best practices are used.
- Devote the resources necessary to administer or participate in the POE activity. Develop the infrastructure necessary for a successful program; institute quality assurance reviews in the field and require that contractors address them, even if code compliance is otherwise achieved.
- Tailor the POE Program to the organization's needs; this will depend on whether your organization is an owner, architect, engineer, or contractor/builder. One size does not fit all.
- Develop quantitative performance measures and monitor them over time. Integrate the POE with the PDSA cycle. The findings from POE become inputs to the study phase of the cycle.
- Include the "voice of the customer" (the end user) in future projects as much as possible. Incorporate POE findings/recommendations in specifications, standards, and procedures promptly.
- Be willing to change.

Enhancing the Lean Project Delivery System™ and IPD

Collectively, many lessons have been learned from a number of lean projects that have been executed in recent years. At the same time, the structure of lean methodologies has been evolving. For example, BIM technology has become more accessible, more capable, and more affordable. Sustainable construction has become a standard requirement as there is increasing interest in reducing the carbon footprint and in reducing operating costs. Above all it has become clear that having a multiskilled team working in close collaboration from project definition, design, and eventual construction is the best way to have a facility that not only meets the owner's requirements, but that exceeds them in terms of both initial costs and operating costs. With the passage of time, new approaches must be considered for governing the design and construction processes. The effectiveness of lean project delivery methods is greatest when lean methodology is incorporated at the very beginning of the design process. The owner's project requirements (OPR) need to be defined at the project definition phase as they are an integral part of defining the purpose of a project, and later determining how well the owner's needs have been met. Sustainable construction involves designing and building a facility based on a system such as Leadership for Energy and Environmental Design (LEED). LEED considers many factors such as the location of a facility relative to the community in order to reduce transportation costs. It promotes the reuse of material from building demolition, and the recycling of various categories of building materials from off-site sources. LEED minimizes the use of potable water and seeks to maximize energy efficiency while improving indoor air quality for the benefit of building occupants.

Design based on these factors involves major long-term decisions and cost trade-offs on the part of project owners; initiatives such as LEED are feasible at project inception, and much less so at a later stage. The OPR should also be clarified relative to the configuration of a facility; the space requirements should be analyzed relative to the use of a facility for lean operations. Recent projects have been able to accommodate 90% of proposed operations in 70% of the space that was initially budgeted.

Commissioning should be expanded beyond its customary role in facility acceptance. As described in Chapter 12, commissioning is an important adjunct to sustainability initiatives; LEED standards require the support of a commissioning agent to ensure that both the design and execution comply with documented requirements.

Building information modeling (BIM) has been shown to enhance projects in several ways, especially when integrated with a project as early as the conceptual design phase. A Stanford study revealed improvements in a cross section of projects:

- Up to 40% elimination of unbudgeted change
- Cost estimation accuracy within 3%
- Up to 80% reduction in the time needed for a cost estimate
- Contract price savings of up to 10% through clash detection
- Up to 7% reduction in project duration of each project team

Therefore the utilization of BIM in predesign team activities is highly cost effective.

Lean construction has been shown to work best when the design has been planned to facilitate it. In an integrated team, construction representatives collaborate with designers to promote constructability. Because of the newness of lean construction, many projects have been designed in the traditional mode and lean has been adopted only during construction when the project has fallen behind schedule.

Depending on the level of maturity of each project team, a number of specialist functions should be considered to support the design and construction professionals with lean deployment. They include:

a) A lean facilitator: This individual provides orientation, training, measurement, and ongoing support with lean initiatives throughout the project.
b) A sustainability/LEED professional assists designers in determining the level of certification that is feasible for a given project.
c) An ICT/BIM manager: This professional uses information technology including BIM to support the entire design and construction process, further facilitating lean supply and prefabrication in keeping with project requirements.
d) The commissioning agent (CxA) plays a vital role in project design, ensuring that required systems are specified appropriately. The CxA provides quality assurance by verifying that functional performance tests confirm that design conditions are met by completed construction. The CxA also confirms sustainability requirements have been met, as required by LEED and other related standards.

Importance of champions

As mentioned throughout this book, lean requires transformation, both at the personal level and the organizational level.

> *We cannot become what we need to be by remaining what we are.*
>
> Max de Pree

This transformation has to take place in the hearts and minds of all stakeholders – owners, designers, contractors, subcontractors, and suppliers to name a few. This challenge is especially great in the design and construction environment, where people are creatures of habit, and often resist change. It falls on the lot of special individuals within stakeholder organizations to lead others toward a new way of doing things. To be effective, champions must be highly placed in their respective organizations, so they can get the ear of leaders and also have control over resources that support the transformation. The champion should be a member of the executive committee, or at least a trusted direct report of a member of the executive committee. He or she should have enough influence to remove obstacles or provide resources without having to go higher in the organization. Clearly, the champion has to have enduring commitment to the lean concept, and a combination of the passion and motivational skills needed to engage others – at all levels – to embark on an unfamiliar course of action.

The importance of leadership in the lean transformation

Throughout this book it has been repeatedly stressed how important leaders are in the adoption of the lean methodology. A fitting example is offered by the case of Kinetic Construction Limited of Canada. After embarking on their lean journey for less than three years they achieved outstanding outcomes with the Brookes School project which proved to be a great testament to their commitment to lean and their leadership support. Tom Plumb, Kinetic's CEO, made a presentation at the annual conference of the Lean Construction Institute of Canada in Vancouver in April 2018. An excerpt from Tom's heartfelt address is provided below:

Leaning your organization – the CEO's perspective

By: Tom Plumb, CEO of Kinetic Construction Ltd. in Victoria, British Columbia, Canada.

So lean sounds awesome, like its something your company and the industry really needs. So how do you actually do this? All I can tell you is what made sense to me and what we are doing.

Before you decide what you want to do about lean or how you are going to do it, you really need to figure out why. I first discovered the lean concept in 2016 at a Lean Construction Institute conference in Calgary. The atmosphere and energy of the conference were geared to shedding old ways and improving or evolving, and just felt right. On the flight home, I made a commitment for Kinetic to become a lean organization, and identified the individual in our organization that I could task with this mission.

Over the next year, I studied what the organization really needed to look like. I saw that there were tools you could use, but I concluded that it was the mindset that was the real mission. This was going to be good for the people that you value in your organization. I saw that if we could build a better engine by building on our culture and turning people from doers to purpose-focused problem solvers, everything we needed to achieve as a business would come, but money couldn't be the primary focus.

Be brutally honest with yourself about the state of your company and your industry. The industry has suffered bankruptcies, litigation, and many examples of inefficiency. I developed a "healthy dissatisfaction". The sensei Shingo states "those who are satisfied can never improve" Understand your culture and what it means to you and your organization. If our industry and the way it functions were an employee, would you continue to employ him or her?

Commit to becoming a student again – even in midlife. Admit that you don't know everything and be a sponge. Read and study like you've never done before. Understand the power of humility. A humble spirit is a fundamental of the lean mindset. Will lean give you a good culture? – not by itself. If your group has a fixed mindset reliant on the past and focused on immediate profits and not on your people and continuous improvement, I personally think you will waste your time. But if you have a good culture, the lean mindset can be like rocket fuel. Lean is a long game, It's a marathon, not a sprint. Knowing this is important – because only when your people know why they are doing something can they truly take on problem solving and continuous improvement.

So here are some steps we followed:

1 Start with your leadership team. They must be the first believers. If they don't believe it, how do you expect the rest of the company to really follow?.

2 Assign someone as your lean implementation manager to the primary learning and training role.
3 Develop the mindset to use lean tools effectively – i think there are two primary aspects to lean – i). True lean, "the Toyota Way" (the way you learn to think differently) we went to a consultant for this. ii). Lean tools and applications for construction. – LCI has resources on these for its members.
4 Develop a roll-out plan for your organization – we called it a lean road map
5 Abolish the word change – use the word improve
6 Plan and manage the improvement – there will be no other shift in your organization as big as this
7 **Invest in training. Don't dabble, invest** – time, money, energy – plan for this carefully.
8 Embrace technology – it's lean
9 Do something – try something, don't be afraid to fail at it. Learn from it, improve from it. Do it better next time.
10 Seek out resources – others have been doing lean longer and successfully – find them and, learn from them.

I'd like to leave you with these thoughts.

I believe that business success does not primarily come from a desperate mission for profit. True success comes from having a long term vision, a focus on people and their development, creating a culture that is based on your values, and on purpose-focused people. In short, a better engine. I believe that everything you need financially for your organization to succeed will come from having a better engine. Everything that I read about being a better CEO and company, dovetailed into everything I read about lean. To me it was a perfect fit.

Lead from the top, work from the bottom – become a support system for your people who will be making it happen. Remember, people don't work for you, your job is to care of them.

Evangelize – don't be afraid to suggest something completely different. You may be surprised at the feedback. If there is one thing that binds us together in this industry, it's misery. You will find we all have a common experience and enemy.

And finally, and most importantly, don't give up. Don't fool yourself, this is hard, and will take a lot of effort. As my grandmother used to say " anything of great value only comes through great effort". So go ahead, *lean* into it.

[Tom Plumb, President and CEO of Kinetic Construction Ltd was appointed in 2016, taking over for the company's founder. As described in Chapter 10, Kinetic is a mid-sized General Contractor – Construction Manager operating out of Victoria BC since 1984. As the new leader of Kinetic, Tom felt obliged to seek out progressive solutions that he could apply within the organization but also that would have an impact on the local industry as well. He established Kinetic's lean construction program in 2016 under the direction of Trevor Coyle, who holds the CM-Lean certificate of the Associated General Contractors of America, (AGC). Award Magazine praised Kinetic's highly successful lean project with the Brookes School facility.]

Sustaining lean construction

James Womack, co-author of *The Machine That Changed the World* (1990), is one of the great influences in today's lean movement. He notes a tendency for organizations to "backslide" to the old ways of working after making significant improvements with lean) He notes that this is often the result of failing to connect the improvement efforts to the way the organization is managed. The most frequent cause is that of middle management resistance to change. Womack thinks that the root cause of that problem has two factors:

a. Confusion in the organization about priorities
b. The failure to make someone responsible for the performance of each so-called value stream.

As middle managers do their jobs they may not be equipped, or have the time to monitor gaps in performance between current levels and the level improvement efforts could make. In the lean environment this is called value stream management. Womack feels that fixing the root causes of poor performance requires the participation of everyone, *and a change in their behavior*. In answer to the question "What can we do about our sustainability problem?" Womack proposes an approach that would not require significant organizational changes; instead a manager in another area could be assigned to audit the performance of improvement initiatives as reflected in the horizontal flow of value along the value stream. That individual could serve as a liaison between responsible managers and top management to gain agreement on who must do what by when to achieve a sustainable leap in performance that will benefit the customer and the organization. He suggests that to prevent regression, someone needs to periodically clarify priorities for each value stream and identify the performance gap between what the customer needs and what the value stream is providing. The person taking responsibility then needs to engage everyone touching the value stream to carefully capture the current condition (the "current state") of the value stream that is causing the gap. The next step is to envision a better value stream and determine who will need to do what, by when, to bring it into being and move to a higher level of performance. Finally, the value stream leader needs to determine what will constitute evidence that the performance gap has been closed and collect the data to demonstrate this. Womack points out that this exercise is, of course, nothing but Dr. Deming's Plan-Do-Check-Act cycle conducted repetitively by the responsible person, ideally employing A3 analysis.

This approach implies that the organization has a degree of maturity in its lean journey, when a constituency of lean converts (and advocates) exists. This situation often occurs in the manufacturing environment, but very infrequently in the construction environment. Construction stakeholder organizations (i.e., designers, contractors, subcontractors, and owners) need to think beyond the needs of the immediate project and work toward a future state in the value stream that represents higher levels of performance, quality, safety, and customer satisfaction. By working with a champion within each stakeholder organization, a lean facilitator may be an effective catalyst. He or she can help stakeholders in acquiring the necessary knowledge and in adopting the beliefs and behaviors that are needed for "leanness."

Womack issues a caution that would serve the lean construction community well; Other organizations are different from Toyota and each other, hence experiments with value stream management methods are needed. If the sustainability problem is not addressed, he points out, the current surge of interest in lean – driven by the success of Toyota – may become just another episode in the long history of unsustainable management improvement campaigns.

The way forward

In the construction environment, the top managers of stakeholder organizations (i.e., owners, designers, construction managers/contractors and suppliers) should embrace the practice of continuous learning and improvement that is central to the lean construction philosophy. Above all, as Tom Plumb recommends, they should accept the fact that change takes time, and they should make the investment of the human and financial resources needed to make it occur.

> *He who would learn to fly must first learn to walk and run and climb and dance; one cannot fly into flying.*
>
> Friedrich Wilhelm Nietzsche

As studies have shown, various forms of waste in the construction supply chain account for as much as 30% or more of the construction dollar. There are many opportunities for improvement that will not only reduce construction costs and deliver value for owners, but also improve profitability for designers and constructors. Owners, in particular, need to recognize the superiority of lean methodologies such as lean project delivery and IPD in securing the foregoing objectives. They should seek to have their projects delivered through these methodologies, and follow the example set by large owners such as the Sutter Health Care System in California that garnered significant reductions in cost and schedule, with high levels of owner/user satisfaction. Forms of contracts have been developed to facilitate lean projects, such as the IFOA or various Consensus Docs versions. In closing, the authors take the position that owners are in the strongest position to move the industry toward an adoption of lean project delivery.

Questions for discussion

1 What are the benefits of retrospectives?
2 Who should participate in the process, and why?
3 What can an owner learn from a POE?
4 What can a designer learn from a POE?
5 How can information from retrospectives benefit a POE?
6 How do post-occupancy studies relate to lean construction?
7 Why does interest in new initiatives wane after initial implementation?
8 How can lean construction initiatives be sustained
    a. by designers?
    b. by contractors?
    c. by owners?

Appendix: reduced examples of "lessons learned" recommendations from post-occupancy evaluation of a new educational facility

Design factors

- 1,460 student stations, 1,310 design capacity
- Gross square feet: 160,113 on a 17 acre single story campus plan.
- The design avoided dead-end halls and staircases to improve security.
- Open corridors reduced conditioned space by 15%.
- A thermal energy storage system (TES), ice production used to reduce energy cost.

Design/construction recommendations

In future facilities, investigate benefit of smaller footprint/2-story design. Confirm inclusion of review comments from specialists, maintenance, quality control staff, plant operations.

- Fast track review comments to avoid time delays and incorporate end users' comments.
- Ensure test/balance compliance before acceptance.
- Encourage designers, design builders to review problems in existing facilities and improve current designs.
- Involve all postconstruction stake holders in the design/review process.

Classroom recommendations

- Provide adequate electrical outlets to anticipate expansion/overcrowding.
- Provide adequate (or complete) separation between classrooms and the administration area to minimize interruptions to the instructional process.
- Conduct HVAC test and balance with computers in working mode. In computer labs the heat load can affect comfort conditions.

Restroom recommendations

- Position staff restrooms to minimize travel distance.
- Improve the privacy of locker rooms.
- Consider using electrically operated hand dryers to reduce paper towel use.

Auditorium recommendations

- Consider designing auditorium lighting with adequate levels to promote dual use as classroom space.
- Design HVAC air distribution system to avoid high velocities/noise levels.
- Provide adequate restroom facilities in the auditorium for stand-alone use.

Security recommendations

- Utilize card access systems to control access to critical areas.
- Install swing gates with secret latches in administration office areas.
- Place fire alarm pull stations near doorways, avoiding swinging doors.
- Promote designs that facilitate visibility and supervision of entrances.

Handicapped accessibility recommendations

- Design main counters for accessibility from both sides (36" h × 36" w).
- Ensure that accessible restrooms are labeled at 60" above finished floor level.
- Ensure ADA accessibility for all sinks and water coolers.
- Install doors with a latch side clearance of 18" to 24".

HVAC recommendations

- Allow for over-enrollment of 30% or more when sizing systems to ensure that ice depletion does not occur, resulting in punitive electrical demand charges.
- Ice storage systems for air conditioning.
- Design chiller controls to alternate the starting sequence for longer life.
- Size air return openings/ducts to minimize noise levels.

References

Anderson, E. 2018. *Lean planning and production control (LPPC): Tips, tricks, & lessons-learned*, revised. February, V24. Lean Implementation Services of Florida.

Forbes, L. 2002. Continuous learning through quality-based post occupancy evaluation. *Proceedings of the 2002 IIE annual conference*, May 19–22. Orlando, FL.

Preiser, W. F. E., H. Rabinowitz, Z. Rabinowitz, and E. T. White. 1998. *Post occupancy evaluation*. New York: van Nostrand Reinhold.

Seed, W. R. 2018. *Don't conform, transform! A guide to better project outcomes*. Lean Construction Institute.

Womack, J., D. Jones, and D. Roos. 1990. *The machine that changed the world*. New York: Harper Collins.

Bibliography

Ballard, G. 2008. The lean project delivery system™: An update. *Lean Construction Journal* 1–19.

Ballard, G. 1999. *The challenge to change*. From selected readings. Lean Construction Institute.

Ballard, G. 2000. The lean project delivery system™. *LCI White Paper-8*, September 23. (Revised 1).

Forbes, L. 2005. Industrial engineering applications in construction. In *Handbook of industrial engineering*, ed. A. Badiru. Boca Raton, NY: Taylor & Francis, 31–37.

Garsden, B. 1995. Postconstruction evaluation. *Journal of Construction Engineering and Management* 121(1): 37–42.

Hill, K., K. Copeland, and C. Pikel. 2016. *Target value delivery: Practitioner guidebook to implementation*. Lean Construction Institute.

Lichtig, W. A. 2005. Sutter health: Developing a contracting model to support lean project delivery. *Lean Construction Journal* 2(1, April): 105–112.

Matthews, O., and G. Howell. 2005. Integrated project delivery an example of relational contracting. *Lean Construction Journal* 2(1): 46–61.

Senge, P. M. 1990. *The fifth discipline: The art and practice of the learning organization*. New York: Doubleday.

16

ADVANCED LAST PLANNER® SYSTEM (LPS) METRICS FOR CONSTRUCTION PLANNING AND CONTROL

*Farook R. Hamzeh, Salam Khalife, Ghali El Samad, Lynn Rizk, Hisham Abou-Ibrahim, Malak Al Hattab, and Samir Emdanat**

Introduction and research background

With the advancement of technology and computer-based tools, the methods employed in planning and control of construction projects are becoming more potent. The main driver behind these methods, however, is the human knowledge through which the interpretation of the data and information at hand is of utmost importance in steering projects towards their successful delivery. Throughout this technological evolvement, researchers have introduced several metrics to guide planners and control teams in their assessment of project performance and control as the project progresses. These metrics are framed within the Last Planner® System (LPS) of production planning and control which has been increasingly gaining attention in recent years.

The LPS system is founded on lean principles to improve planning reliability and mitigate the detrimental effects of uncertainty and variability in construction projects (Ballard 2000). Some of the major lean principles include: increasing value to customers, identifying value and removing non-value adding steps, organizing the work to have smooth flow based on pull from the customer, and implementing continuous improvement on projects. In this regard, the LPS advocates reliable planning on construction projects based on the pull technique and involving downstream players in upstream decisions.

Several metrics have been developed to complement the LPS system in monitoring and improving the planning efforts on projects as explained in the following sub-sections. However, construction projects still encounter failures as a result of planning deficiencies. This raises concerns regarding the capability of the current planning metrics to capture the full reality and dynamics on projects. This research investigates the available metrics used and highlights the need for more advanced metrics that describe, monitor, and control the various dynamic aspects of construction planning. Consequently, the authors propose novel metrics, based on Hamzeh, El Samad *et al.* 2019; Hamzeh, Al Hattab *et al.* 2019; Abou Ibrahim *et al.* 2019, and a new framework that combines the suggested metrics along with the previously existing LPS metrics. This framework forms a global platform inclusive of capacity planning and other evaluation measures at both the project and team levels. The presented chapter contributes to the construction planning knowledge through offering new methods for improved planning practices and assessment tools.

Construction planning and project performance

Project management encompasses various actions undertaken during the course of construction projects to maximize the chances of success. Among these actions, three major activities constitute the top of the list: the communication behaviors, the control mechanisms, and the planning efforts (Chan *et al.* 2004). Planning on construction projects is the act of determining what needs to be done, how, when, and by whom, while taking into consideration the necessary resources and their accompanied costs (Laufer and Cohenca 1990). Planning is thus performed by making early decisions, mainly regarding the work tasks, their interactions, the specified time schedule, and the required resources. These decisions eventually become direct assignments or guidelines for site management to base their operational decisions on (Laufer and Tucker 1987). Accordingly, proper planning can reduce the number of inherent project risks that may otherwise have remained unforeseen and not accounted for (Aziz and Hafez 2013). While planning sets the course to achieve project objectives, the control process ensures that the project remains in alignment with its scope and achieves its targets. Project control thus intends to minimize the gap between project planning and project execution. As such, control mechanisms involve monitoring and evaluating the project to enable corrective actions to take place for achieving the intended project objectives (Kerzner 1998; Ballard and Tommelein 2016).

Conventional planning practices recognize the relationships between tasks using the Critical Path Method (CPM). However, the main problem with these practices is the deficiency in acknowledging the workflow between tasks, whether it is the flow of trades, space, or resources (Howell *et al.* 2010). Even though CPM succeeded in the development of high-level plans, it failed to be a practical communication tool, a daily operational tool for production planning and control, and a tool for root cause analysis of problems at the task level (Birrell 1987).

Similarly, conventional control systems, such as the commonly used Earned Value Management (EVM), have demonstrated several limitations, one of which is the inability to focus on work flow (Cândido *et al.* 2014). EVM calculates schedule and cost variances, then forecasts the new expected completion date and the cost at completion. Although this is an important exercise that needs to be performed throughout the course of construction, this approach alone does not provide information about the reliable release of work downstream (Kim and Ballard 2000). Additionally, EVM can produce wrong percentages of work completed and incorrect estimates for completion forecasts (Demachkieh and Abdul-Malak 2018). Therefore, new methods shall be considered to accommodate for those deficiencies.

The Last Planner® System

Managing production flow is an essential process of construction planning and control on a project. Consequently, developing reliable and achievable plans is also a critical aspect of proper planning. However, the emergence of problems is inevitable on construction projects and is largely due to unforeseen circumstances and inherent variability that are not accounted for earlier. Therefore, planning needs to include the proactive analysis of potential problems that might arise (Junnonen and Seppanen 2004). The Last Planner® System emerged with the aim to enhance planning reliability and effectiveness. This leading system is based on lean principles and promotes regular monitoring of planning reliability during project implementation. It calls for the involvement and commitment of relevant stakeholders; therefore, plans would evolve collaboratively and hold a promising alignment of short and long-term plans (Ballard 2000). This alignment signifies that predictions are met, and as a result, work can flow as desired while having minimal waste produced.

In this context, Ballard (2000) indicated that each project stage requires a different level of planning effort and control. Early project stages undergo long-term perspective planning followed by short-term perspective planning as the project progresses forward and new information becomes available. As such, major project milestones are set during the early stages and are then broken down into phases. Look ahead plans are generally based on a 6-week timeframe; these plans are then broken into Weekly Work Plans (WWP) (Ballard 2000). Therefore, planning is performed in greater detail the closer it gets to the start of an activity (Hamzeh and Langerud 2011). In addition, the LPS involves the following planning practices: (1) developing the work plan with those who are going to execute the work, (2) identifying and removing work constraints ahead of time to make work ready and increase reliability of work plans, (3) making reliable promises and driving work execution based on coordination and active negotiation with trade partners and project participants, and (4) learning from planning failures by detecting root causes and taking preventive measures (Ballard 2000; Hamzeh *et al.* 2015).

Combination of different planning methods

Recently, there have been several attempts to combine two planning methodologies, the CPM and the LPS. CPM observes the big picture and focuses on delivering a master schedule that helps management maintain control over large and complex projects. Complementary to this method, the LPS views tasks in granular detail, recognizing the most effective and collaborative way to complete each of them. Moreover, the industry has developed various database-driven tools that allow the combination of these two methodologies. These tools can record all types of data, hence allowing a systematic analysis of metrics.

Additionally, the combination of the EVM approach and the LPS has been tackled in literature. Olano *et al.* (2009) tested certain LPS metrics that measure planning reliability, such as PPC (Percent Planned Complete) and PCR (Percent Constraint Removal), in correlation with schedule performance. Results showed statistical correlation between workflow reliability and schedule performance while emphasizing the need to manage both methodologies to ensure successful planning and control on projects. Alarcón *et al.* (2014) analyzed planning reliability metrics using schedule progress curves for 'successful' and 'unsuccessful' projects. Results indicated that for successful project completion, the focus should be shifted from not only achieving high values of planning metrics, but also controlling their variability.

However, even with such successful attempts at integrating several planning and control methods to improve the reliability of projects, a comprehensive approach for monitoring several aspects of construction planning is needed to present a holistic perspective for planners. In this regard, another aspect of planning that is often overlooked is capacity planning. In what follows, a thorough discussion of this planning area is presented to show how critical capacity planning is for project success.

Capacity planning

Planning should not only be concerned with achieving proper workflow and adhering to set plans but should also consider the proper assignment of activities and tasks (the workload) to the available labor, equipment, and other resources (the available capacity). The balance between the workload and capacity is known as capacity planning. Workload is the quantity of work needed to be done in a specific time allotted by planners, and capacity is the quantity of work a crew can complete using their available resources (Ballard 2000). This concept differs from the 'capacity planning' concept provided by other engineering fields where the focus is on long-term expansion decisions. In contrary, the capacity planning concept in lean construction focuses on the short-term production planning. Nevertheless, matching load to capacity is a dynamic process that if not promptly managed would lead to schedule delays, increased costs, and wasted resources (Shehata and El Gohary 2011). This is due to the fact that, in practice, companies tend to either overload or underutilize their resources causing an imbalance that leads to the aforementioned detrimental consequences (González *et al.* 2010).

Several attempts have been made to address the challenging issue of matching load to capacity. In this regard, Kim *et al.* (2008) have developed a workforce information database to help resolve this conflict. The database system, which includes information on the workforce such as skill level and history of accidents, allows the user to develop proper strategies based on the workforce capacity and the available needed information during production planning (Kim *et al.* 2008). From a utopian perspective, this attempt can be easily achieved without any difficulties; however, two impediments stand in the planners' way from reaching accurate capacity planning. The first hurdle relates to predicting the workload itself. This is due to the inherent variability that results in an imprecise forecasting of what activities would be ready for execution since prerequisite tasks would not be completed as assigned (Tommelein *et al.* 1999). Although the Last Planner® System supports reliable commitment, it can be undermined by the second hurdle which is improper resource estimation and utilization.

Uncertainties on construction projects cause perpetual changes which in turn result in both inaccurate resource estimation and inaccurate workload valuation. Resource variations cause unfeasible and inefficient work in addition to incurring more costs during construction (El-Rayes and Jun 2009; Koulinas and Anagnostopoulos 2013). The fact that resources are limited and rarely sufficient on projects forces planners to prioritize their allocation strategies. Khattab and Søyland (1996) presented the priority rule method to address the issue of limited resources on projects or resource-critical projects. In this case, and unlike the CPM-based rule, the project schedule is established by allocating limited resources in a way that keeps the exceeded duration to a minimum. Additionally, the resource leveling problem has been tackled where efficient project sequencing efforts can aid in resolving the leveling difficulty (Damci *et al.* 2013). Damci and Polat (2014) considered different objective functions to optimize resource utilization for the purpose of measuring the efficiency of the construction sequence; yet, no metrics were suggested to reflect the actual variations between resource levels and tasks to be completed. Additionally, to resolve the problem of resource fluctuations, Ponz-Tienda *et al.* (2017) suggested an innovative algorithm and tested it with the Project Scheduling Problem Library for four metrics.

Research has addressed aspects of the variability present in both the capacity and workload sides. A reliable commitment model was suggested to improve short term prediction of work progress employing factors related to labor, buffers, and work plans. The model was first presented by Mundaca (2006) and Bustamante (2007) and later augmented by González *et al.* (2008) and González *et al.* (2010). However this model, as well as earlier research, do not introduce a method for monitoring and managing capacity planning.

Although previous research efforts aimed at improving production planning, no comprehensive method, metrics, or holistic measurement frameworks were presented to manage capacity planning within the bounds of the Last Planner® System. Nevertheless, in the following section, a list of currently available metrics in connection with the Last Planner® System is presented as a foundation for developing advanced construction planning metrics.

Summary of available LPS metrics

Metrics are used on construction projects for monitoring and measuring the progress of work to help planners formulate important decisions in regard to time and cost improvements. While many metrics are currently in use for real-time control on construction projects, researchers along with practitioners are continuously developing metrics to better understand the project performance when implementing the LPS.

The available metrics have been developed to address the different stages of the LPS, covering both long-term and short-term planning (Ballard 2000; Chitla and Abdelhamid 2003; Mitropoulos 2005; Jang and Kim 2007; González *et al.* 2008; Hamzeh *et al.* 2008; Hamzeh, Saab *et al.* 2015; Hamzeh, Zankoul *et al.* 2015; Emdanat and Azambuja 2016). The Percent Planned Complete (PPC) metric is the most commonly used metric under the LPS to reflect planning reliability on projects. PPC is obtained by calculating the number of completed tasks from the total tasks promised to be completed during the planning horizon, which is normally a week. However, perceiving the PPC metric in isolation will lead to an incomplete understanding of the overall planning process. Therefore, recent research expressed the need to complement PPC with other metrics to measure performance. Accompanying metrics, such as the Tasks Anticipated

(TA) and Tasks Made Ready (TMR), measure the success of anticipating tasks and removing constraints at the lookahead planning level. TA represents the total anticipated tasks out of the total tasks found on the Weekly Work Plan (WWP). TMR represents the ratio of executed tasks that were made ready by removing all their constraints and that continued from the lookahead week. Researchers continue to investigate the planning practices and study the extent to which available metrics are sufficient to successfully plan and control construction projects. Table 16.1 presents an overview of the different metrics discussed in the

Table 16.1 Description of available metrics – Summary of LPS metrics

Equation	*Metric*	*Formula*
(1)	Percent Planned Complete (PPC) (Ballard 2000)	$\frac{Did}{Will}$
(2)	Tasks Made Ready (TMR) (Ballard 2000; Hamzeh, Zankoul *et al.* 2015)	$\frac{Did}{Can}$
(3)	Tasks Anticipated (TA) (Ballard 2000; Hamzeh, Saab *et al.* 2015)	$\frac{Anticipated\,Tasks}{Total\,Tasks\,on\,WWP}$
(4)	Planned Work Ready (PWR) (Mitropoulos 2005)	$\frac{Work\ Expected\ to\ be\ Perfomed\ in\ Lookahead}{Work\ that\ Should\ be\ Performed}$
(5)	Delta 1 (Δ_1) (Mitropoulos 2005)	$\frac{Constraints\ Promised\ to\ be\ Removed}{Constraints\ Identified}$
(6)	Delta 2 (Δ_2) (Mitropoulos 2005)	$\frac{Constraints\ Removed}{Constraints\ Promised\ to\ be\ Removed}$
(7)	Delta 3 (Δ_3) (Mitropoulos 2005)	$\frac{New\ Constraints}{Constraints\ Identified}$
(8)	Percent Constraint Removal (PCR) (Jang and Kim 2007)	$\frac{Ready}{Can}$
(9)	Performance Factor (PF) (Ballard and Howell 1998)	$\frac{Actual\ Labor\ Hours}{Earned\ Labor\ Hours}$
(10)	Labor Utilization Factor (LUF) (Chitla and Abdelhamid 2003)	$\frac{Effective\,Work + \frac{1}{4}\,Essential\ Work}{Effective + Essential + Not\ Useful}$
(11)	Project Productivity Index (PPI) (González *et al.* 2008)	$\frac{\Sigma(Activity\ Pr\,ouctivity\ Indexes)}{N} \times 100$
(12)	Planning Reliability Index (PRI) (González *et al.* 2008)	$\frac{Actual\ Progress}{Planned\ Progress} \times 100$
(13)	Construction Flow Index (CFI) (Sacks *et al.* 2017)	$10\sum_{i=1}^{7} W_i P_i^{x_i}$
(14)	Commitment Level (CL) (Emdanat and Azambuja 2016)	$\frac{Required\ Will}{Should}$
(15)	Percent Required Complete and Ongoing (PRCO) (Emdanat and Azambuja 2016)	$\frac{Required\ Completed + Required\ Ongoing\ on\ Track}{Required\ Will}$
(16)	Capacity to Load Ratio (CLR) (Rizk *et al.* 2017)	$\frac{Total\ Completed}{WWP}$
(17)	Required Capacity Ratio (RCR) (Rizk *et al.* 2017)	$\frac{Required\ Completed}{Total\ Completed}$
(18)	Required Percent Complete (RPC) (Rizk *et al.* 2017)	$\frac{Required\ Completed}{Total\ Required}$

Source: Retrieved from Hamzeh, El Samad et al. (2019)

literature in relation to the LPS. These metrics quantify productivity, assess workflow reliability, and/or realize the progress on projects. In what follows is a brief explanation of each metric that is listed in the table.

The Planned Work Ready (PWR) metric was introduced by Mitropoulos (2005) to assess the quality of the lookahead planning process. PWR measures the percentage of work expected to be ready as planned in the lookahead horizon, thus acting as a forecast for the schedule performance. Moreover, Mitropoulos (2005) suggested three deltas to measure the make ready process: Delta 1 measures the constraints that are expected to be removed from the total constraints identified, Delta 2 measures the actual removed constraints, while Delta 3 looks into new identified constraints.

Additionally, the Percent of Constraint Removal (PCR) is a metric introduced to indicate workflow predictability and is considered a leading indicator which correlates well with PPC (Jang and Kim 2007). PCR is calculated by comparing the number of ready tasks when scheduling the WWP (Ready) to the total number of planned tasks found in the lookahead plan (Can).

On the other hand, some metrics were introduced to relate workflow reliability to productivity. The Performance Factor (PF) was introduced by Ballard and Howell (1998) to indirectly assess the impact of the LPS on productivity by calculating the ratio of actual labor hours to earned productivity (represented in labor hours). Likewise, Chitla and Abdelhamid (2003) investigated the effective and non-effective utilization of labors through the Labor Utilization Factor (LUF). Chitla and Abdelhamid (2003) found that an increase in PPC causes an increase in LUF; however, an increase in LUF does not result in an increase in PPC. Therefore, they recommended that project managers should focus mainly on PPC instead of focusing on the local production represented by LUF.

Another two LPS metrics were suggested by González *et al.* (2008) as performance indicators at the activity level. The first metric is the Project Productivity Index (PPI) calculated by finding the average of the Activity Productivity Indexes (API), with API being the ratio of the average labor productivity to the maximum labor productivity. The second metric is the Process Reliability index (PRI) concerned with the planning reliability at the activity level. PRI is measured by calculating the ratio of actual to planned weekly progress of a certain activity, thus overcoming the shortcoming of PPC for analyzing LPS at this level. Similarly, PPI overcomes the limitation of the PF metric by computing the maximum productivity on-site instead of the expected productivity.

Sacks *et al.* (2017) introduced the Construction Flow Index (CFI) which measures the quality of construction flow in repetitive construction projects. CFI enables project managers to improve production flow on projects by tracking and sharing this metric with trade crews.

Another attempt to align short and long-term planning was performed by Emdanat and Azambuja (2016) by introducing three metrics to complement PPC, TA, and TMR. The Commitment Level (CL) metric measures the percentage of the committed required activities (critical activities) to the total required activities in a given work plan cycle. Tracking CL fluctuations and relating them to the project status revealed the correlation between CL variations and the decrease in floats as a result of the increase in late paths. Another developed metric is the Percent Required Completed or Ongoing (PRCO). PRCO is the percentage of required/critical activities that are completed or expected to be completed (ongoing) on or before their promised completion dates with respect to the total required activities committed to on the work plan. The third metric is the Milestone Variance (MV) which represents the difference in days between the forecasted date to complete all remaining activities and the required date of the milestone. Emdanat and Azambuja (2016) demonstrated that teams who adjust plans to maintain high CL, PRCO, and PPC had a lower overall MV and were, therefore, more reliable.

Finally, Rizk *et al.* (2017) investigated the issue of matching load to capacity and proposed several metrics in this regard: Capacity to Load Ratio (CLR), Required Capacity Ratio (RCR), and Required Percent Complete (RPC). CLR measures the tasks completed based on the capacity available to the chosen load (tasks committed to be completed that week); the CLR can also be measured in terms of man-hrs. RCR gives the planners an indication of how many tasks were critical out of all those executed that week.

Lastly, RPC depicts the percentage of critical tasks completed on a certain week. Additionally, the Weekly Deviation (WD) is a metric, along with the three aforementioned metrics, that helps measure the extent of deviation from the WWP.

While these metrics equip planners with valuable input on certain planning aspects, each metric is solely concerned with a specific planning feature and disregards other facets. This leads to an incomplete understanding of the entire project performance and the needed planning intervention actions (Emdanat and Azambuja 2016; El Samad *et al.* 2017). Additionally, many of those metrics were either inconsistently used or required data that was too difficult and time consuming to collect (Emdanat and Azambuja 2016).

For proper management and communication among project stakeholders, the use of metrics is usually accompanied with the use of tools such as dashboards. Dashboards are substantial when multiple metrics are employed to measure and track progress on projects. Construction dashboards are thus a control tool used to facilitate communication on construction projects and to share the monitored metrics for control purposes. Such dashboards convert raw data into meaningful information regarding project performance by highlighting its weaknesses and strengths (Song *et al.* 2005; Suk *et al.* 2012). The conveyed information is used to aid planners in making responsible decisions on projects (Kerzner 2013). Furthermore, planners base such decisions on an evaluation process which considers the different metrics used and several accompanied factors. However, most metrics fall short when considering the activities' status. In fact, El Samad *et al.* (2017) demonstrated that many metrics assume that all activities are equally import when they are not. Therefore, in order to take precise decisions, planners need to have advanced metrics that reflect a deeper understanding of activities and their statuses.

The need for new metrics

The literature review section revealed the existence of numerous metrics pertaining to the Last Planner® System. However, the authors discussed several issues and gaps in connection with the available metrics, namely:

1- Capacity planning is not properly emphasized, and no clear measurement frameworks are found to assess capacity planning jointly with the LPS.
2- Studies employing the integration of metrics are scarce because most studies focus on certain aspects of planning rather than assembling and reflecting a holistic picture.
3- Research correlating the available metrics to overall schedule performance has been both challenging and time consuming; collecting the required data was found to be overwhelming. Additionally, metrics were found to be inconsistently used.
4- Few metrics consider activity status: whether the activity is critical/required, a backlog activity, or a newly emerging activity.

In an attempt to address these issues, this study proposes a set of new metrics to complement commonly used metrics and offers an LPS control framework to guide production planning processes. Additionally, the study utilizes the power of CPM, LPS, and database-driven tools to develop a project management and control dashboard as a merging tool of the different metrics for the ease of their use and interpretation. A detailed calculation method to enhance the planning efforts on construction projects is also provided. Furthermore, complimentary capacity planning metrics are offered to attain more reliability on projects. The derived metrics correlate capacity planning to project performance while examining resource allocation patterns. Tracking the attempts and performance of the different teams and planners in satisfying the required matching between load and capacity, would first help in identifying hidden problems and waste, then improve the planning processes accordingly. The framework will help in setting the foundation for a comprehensive understanding of capacity planning.

Consequently, the contributions of this chapter to the state-of-knowledge comprise the following: (a) introducing new planning metrics within a comprehensive framework to enhance planners' understanding of the project performance on both project and process levels; (b) helping planners in tracking and measuring various aspects of capacity planning in connection with the LPS; (c) formulating an understanding of the underlying factors shaping capacity planning; and (d) helping teams focus on critical/required activities.

Research methodology

The research methodology adopted in this research is Design Science Research (DSR). DSR is a multi-step process employed when there is a need to create and evaluate an innovative artefact intended to solve a certain identified problem (Hevner *et al.* 2004). The artefact makes a research contribution after undertaking several steps while focusing on three research cycles: the relevance cycle, the design cycle, and the rigor cycle (Hevner 2007). The relevance cycle includes field testing and a searching process to identify the opportunities and problems to be addressed. This cycle bridges the contextual environment with the research activities in the design cycle. The design cycle thereafter, includes building and evaluating the artefact. Afterwards, the rigor cycle connects the design science research activities to the knowledge base where the outcomes shall be clearly communicated (Hevner *et al.* 2004; Hevner 2007). Therefore, a good DSR process shall be effective in identifying the problem, investigating contribution to the knowledge base, and understanding the development of the solution (Weber 2018). In the construction management research field, the DSR method is suitable when researchers develop artefacts that can be implemented in construction to solve an existing problem (Rocha *et al.* 2012).

In this research, the newly introduced metrics and the framework guiding their use represent the artefact under study that need to be assessed and tested. Accordingly, the research was divided into three stages as per the DSR guidelines proposed by Hevner *et al.* (2004) and as shown in Figure 16.1. The first stage involves examining the available metrics in connection with the LPS and discussing their practicability and probable deficiencies. Hence, areas of need were underlined representing opportunities for novel methods to be developed. Based on the needs identified, the second stage comprises developing new metrics as well as introducing the framework and the accompanied dashboard to guide the metrics' use. This process included multiple iterations of refinements and adjustments to address the objectives of the study.

Finally, stage three includes testing the framework and the dashboard by employing an empirical study comprising two projects in the United States. Assessing the metrics on two case studies helps in cross analysis and comparing the results for a valid argument on their feasibility. However, the authors will not discuss the results of testing in this chapter, but will rather present an overview of the summary of findings. Further details on the experimental testing is explained thoroughly in Hamzeh, El Samad *et al.* (2019) and Hamzeh, Al Hattab *et al.* (2019).

The research group was divided into two teams to conduct the study; the first team recommended the metrics while the second team ensured the knowledge about the project and the day-to-day operations. As

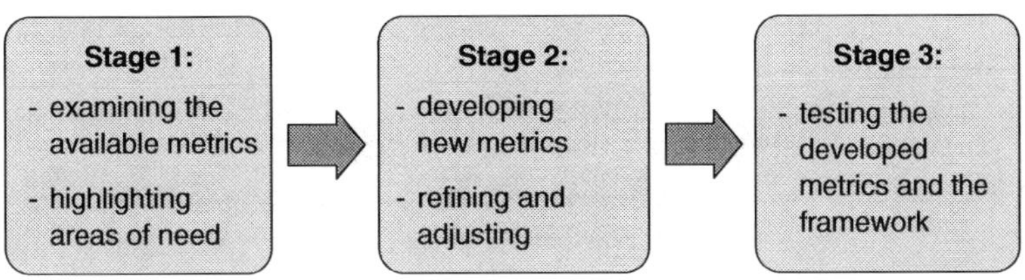

Figure 16.1 Research methodology

such, this grouping helps in testing the developed metrics while avoiding researcher bias. This reflects the actual project dynamics and tests whether the proposed metrics are good indicators.

Empirical study background

As previously stated, two large construction projects were selected for the application of metrics. Both projects were selected and deemed appropriate to the research due to the detailed implementation of the Last Planner® System which was based on the general guidelines and principles explained by Ballard (2000) and Ballard and Tommelein (2016). These principles include: maintaining plans visible to all parties all times; establishing milestones in master schedules then planning in greater details as the execution of the activities approaches; planning collaboratively with those executing the work; working as a team to remove constraints for planned activities to maintain workflow; committing to execute only activities that are well defined, sound, sequenced, and sized; generating reliable promises; performing root cause analysis to learn from failures; and maintaining workable backlog of ready-tasks to buffer against capacity and time loss.

Detailed data was collected from both projects over a two-year period during the execution phase. The research team examined the data acquired from the master schedule, phase schedule, lookahead stage, and weekly work plan. The projects used an integrated database driven tool called vPlanner to plan and control the production. The software helped in keeping track of milestone dates and in analyzing the data for evaluating the proposed metrics.

Each project included multiple teams who were involved in the planning process during various phases. Each team consisted of last planners who performed phase scheduling at the outset of each phase. The last planners, who were engineers, superintendents, and foremen from various disciplines – including the owner, the contractor, and subcontractors, met on weekly bases for collaboration purposes on planning endeavors. As such, lookahead planning was applied to identify the activities that need to be done and consider removing their constraints to make them ready. The last planners then committed to the activities that could be done based on the evaluation process of removing the constraints. Additionally, the teams considered the activities' statuses, focusing on the critical activities as a priority to work while considering the upstream constraints. Lastly, several metrics were tracked to measure the project performance including: PPC, Percent Required Complete or Ongoing (PRCO), Commitment Level (CL), and Milestone Variance (MV).

Along the project execution, the research team performed multiple actions to enhance the information gathering process and the sources of evidence used in this research. These actions included: (a) real time observations through attending planning meetings; (b) analysis of metrics applied to data collected from the project database; and (c) soliciting explanations from on-site planners and managers. This guaranteed data triangulation and bias avoidance.

The list of metrics developed is discussed next, then the overall framework for employing the combined metrics in a newly devised dashboard is presented afterwards.

Description of new metrics and sample calculation

The metrics identified in this research aim at complementing the existing LPS metrics to have a more comprehensive planning exercise. The new metrics assess the performance of the weekly work planning efforts. The first problem addressed here is concerned with the drawback of the PPC metric and other metrics that ignore activity characteristics and consider them of equal importance. Consequently, the suggested metrics differentiate between activities and consider their status as being either Critical/Required, Not Critical/Not Required, Backlog, or New. Required activities are critical activities that have a priority over other activities since their delay will lead to delaying the whole schedule or the phase milestone. Backlog activities are excess activities that are ready for execution but are Not Critical/Not Required. Whenever capacity becomes available, these activities could be completed. New activities, however, are activities that occur during the execution of the work without being previously accounted for, or were not allocated correctly within the schedule timeframe. These activities should be completed on the week of their emergence. The emergence of new activities can be attributed to several possible reasons which are related to either planning issues, ongoing construction issues, or uncertainties accompanied with construction (Rouhana and Hamzeh 2016).

Advanced metrics developed

REQUIRED LEVEL (RL)

RL calculates the number of Required activities, or otherwise known as critical activities, to the total number of activities committed to on the WWP. The RL metric complements the CL and PRCO metrics described earlier in Table 1.

$$RL = \frac{Required\ Will}{Will} \tag{1}$$

RL helps planners in identifying the criticality level of the committed activities; a high RL indicates that most of the activities on the WWP are critical. Therefore, calculating RL before the start of the week of execution is important for planners to understand the seriousness of the situation. A high RL shall be accompanied with serious attempts by planners to complete all activities and to achieve a high PPC.

COMPLETED UNCOMMITTED (CU)

The CU metric measures the number of activities completed but were not committed to on the WWP, out of the total completed work. Note here that the total completed work is different from the completed or "Did" in the PPC metric, since the Backlog and New tasks are included in the CU metric but disregarded in PPC. Hence, the purpose of the CU metric is to address this limitation of the PPC metric where PPC neglects New activities and Backlog activities.

$$CU = \frac{Total\ Completed - Completed\ from\ Will}{Total\ Completed} \tag{2}$$

$$CU = \frac{Completed\ from\ Backlog}{Total\ Completed} \tag{3}$$

Notably, a high CU indicates that teams are completing uncommitted work. An increase in CU thus reflects a tendency to complete activities that were not committed to by teams, a matter that needs to be addressed to understand the reasons behind it.

PERCENT COMPLETE NEW (PCN)

PCN measures the number of New tasks completed out of the total number of New tasks committed. This metric focuses on new emerging tasks that were not anticipated in the lookahead scheduling process as a result of mistakenly allocating them in a different time window. This indicates that these activities are not new in scope, but emerged during a certain week and need to be completed (Rouhana and Hamzeh 2016).

$$PCN = \frac{Completed\ New}{Committed\ New} \tag{4}$$

PCN helps teams in recording completed New tasks. These tasks, unexpected yet important, could be a burden on planning reliability. A high PCN indicates a high ability to complete New tasks on time. A low PCN, in contrast, reflects the inability of teams to complete the New emerged activities which can delay the schedule if these activities are critical, which is usually the case.

CAPACITY PLANNING METRICS

Existing production planning metrics, as revealed earlier, remain short in capturing a comprehensive assessment of capacity planning and its effect on workflow performance. Therefore, advanced metrics are proposed in this research to help in evaluating mechanisms of capacity planning.

Capacity planning balances the available resources with the workload demand. The workload could be of different types of activities. In their research, Hamzeh, Al Hattab *et al.* (2019) distinguished between three types of activity clusters, as represented in Figure 16.2. Within each cluster, critical and non-critical activities are found. The normal or non-critical activities are designated by green pebbles, while the critical or required activities are represented as red pebbles. As per the traditional definition of critical activities, those activities when delayed would lead to delaying the whole work-stream.

The first cluster is the WWP cluster which consists of critical and normal activities that planners committed to complete for a certain week. The second cluster is the New cluster involving all emerging activities not previously accounted for in the WWP. However, as explained earlier, these New activities which might be critical and/or normal, need to be completed on the week of their emergence (Rouhana and Hamzeh 2016). Another cluster is the Backlog cluster which includes excess activities that are ready

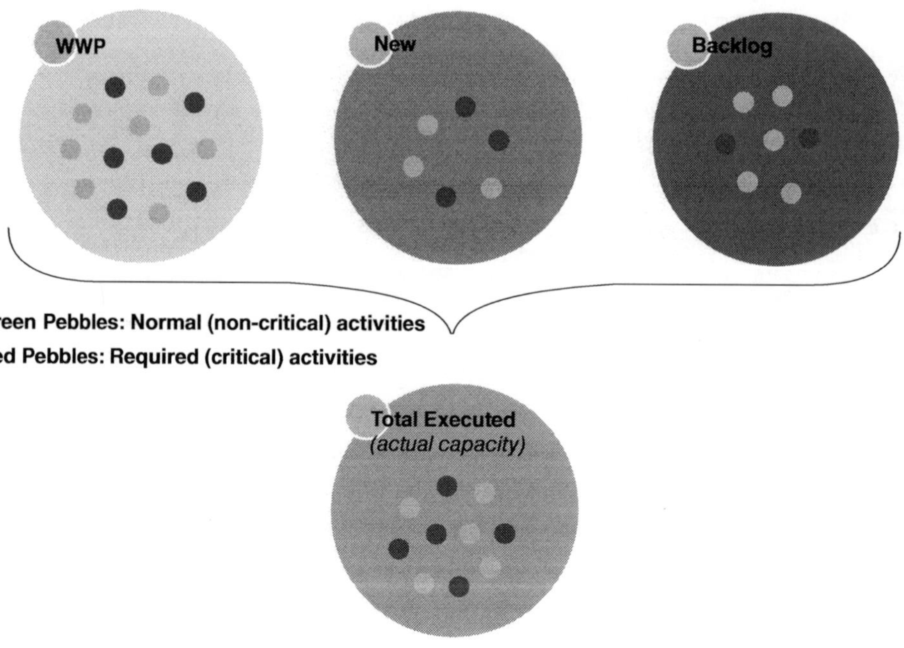

Figure 16.2 Activity clusters

Source: Retrieved from Hamzeh, Al Hattab *et al.* (2019)

for execution yet are not a priority and can be completed whenever capacity is available. In such cases, if resources are idle or pending other works, backlog activities could be completed. The fourth cluster is the Total Executed which includes the actual activities that were completed during the assigned week of execution and represent the actual capacity. The activities that have been completed in each of the three aforementioned clusters contribute to the Total Completed cluster.

After clarifying the different activity clusters, four new derived primary metrics and three secondary ones are presented hereafter. These metrics should be measured on a weekly basis, along with the PPC metric, when applying the LPS. Table 2 defines the parameters included in calculating the derived metrics.

(1) Capacity to Load Ratio (CLR) The CLR metric measures the total tasks that were completed representing the actual capacity (Total Completed cluster) to the tasks committed to on a given week (WWP cluster) representing the chosen load. CLR is calculated using Equation (1).

$$CLR = \frac{Total\ \ Completed}{WWP} \tag{1}$$

The CLR metric is a retrospective metric which reflects on how precise was the team in employing resources. Note that this research is not concerned with the number of workers assigned to an activity, it is mainly addressing the ability of planners to do their job based on the different circumstances.

(2) Required Capacity Ratio (RCR) The RCR metric measures the Required activities, or otherwise the critical activities, out of the total completed activities (Total Completed Cluster) as shown in Equation (2):

$$RCR = \frac{Required\ \ Completed}{Total\ \ Completed} \tag{2}$$

(3) Required Percent Complete (RPC) The RPC, which is the third primary metric developed, focuses on the Required activities or critical ones. RPC measures the total number of Required activities completed out of the total Required activities during a given week. It is calculated as shown in Equation (3):

$$RPC = \frac{Required\ \ Completed}{Total\ \ Required} \tag{3}$$

(4) Weekly Deviation (WD) From its name, the Weekly Deviation (WD) metric assess the team deviation from the WWP they prepared. WD is calculated by subtracting the total number of committed activities (WWP cluster) from the number of total Completed activities (Total Completed Cluster) as shown in Equation (4):

$$WD = Total\ \ Completed - WWP \tag{4}$$

If the deviation is less than zero (WD<0), then the team is overloading its resources. Otherwise, if WD>0, then the team is under-loading its resources. If WD=0, then the team has matched the load to capacity.

Additional to the above described primary metrics, three secondary metrics are established to reveal more nuances of the resource allocation issues taking place during the project execution.

(5) Glut The RCR metric described earlier is concerned with critical activities completed out of the total completed; 1-RCR thus represents the non-critical activities that are completed in that given week.

Table 16.2 Definitions of parameters included in the calculation of the derived metrics

Parameter	*Description*
WWP	Represents all the activities on the weekly work plan that have been committed to be completed for the given week
WWP Completed	Represents all activities on the weekly work plan that have been completed for the given week
Total Completed	Represents all the activities from the WWP, New, and Backlog Clusters that were actually Completed on the given week (indicated by all the pebbles in the Total Completed Cluster – Fig. 16.2)
Total Required	Represents all the critical activities that are in the WWP, the Backlog, and the New Clusters for the given week (indicated by all red pebbles in the WWP, Backlog, and New Clusters – Fig. 16.2)
Required Completed	Represents all the critical activities that were Completed on the given week (indicated by all red pebbles in the Total Completed Cluster – Fig. 16.2)

Source: Extracted from Hamzeh, El Samad *et al.* (2019)

Glut measures the number of non-critical activities completed and consequently taking available resources from critical tasks. Glut is calculated according to Equation (5):

$$Glut = Total\ \ Completed - Required\ \ Completed \tag{5}$$

(6) Starvation Likewise, if RPC represents the fraction of critical activities that have been Completed, then 1-RPC represents the fraction of required (critical) activities that have been missed or not completed. As such, the Starvation metrics actually represents the critical activities that are starving for resources. It is measured based on Equation (6):

$$Starvation = Total\ \ Required - Required\ \ Completed \tag{6}$$

(7) Misallocation Factor (MF) The Misallocation Factor (MF) is the sum of the Glut and the Starvation metrics, as shown in Equation (7). This metric reflects the overall waste in capacity planning.

$$MF = Glut + Starvation \tag{7}$$

These metrics need to be calculated together with two metrics developed in earlier research, one of which is the PPC metric as explained earlier. The second metric that needs to be tracked is the Milestone Variance derived by Emdanat and Azambuja (2016). MV is the difference in days between the forecasted date to complete all remaining activities and the required date of the milestone (Emdanat and Azambuja 2016). Hamzeh, El Samad *et al.* (2019) explained that MV can relate the developed capacity metrics to the project or workflow stream schedule performance and ensure that the remaining work is in alignment with the milestone targets.

Framework for employing the developed metrics

The literature review revealed a lack of a comprehensive framework combining different metrics to enhance their systematic use and reap the full potential benefits of improving the process. Accordingly, a Diagnosis/Control Framework is proposed covering all planning stages of the LPS, as represented in Figure 16.3. The

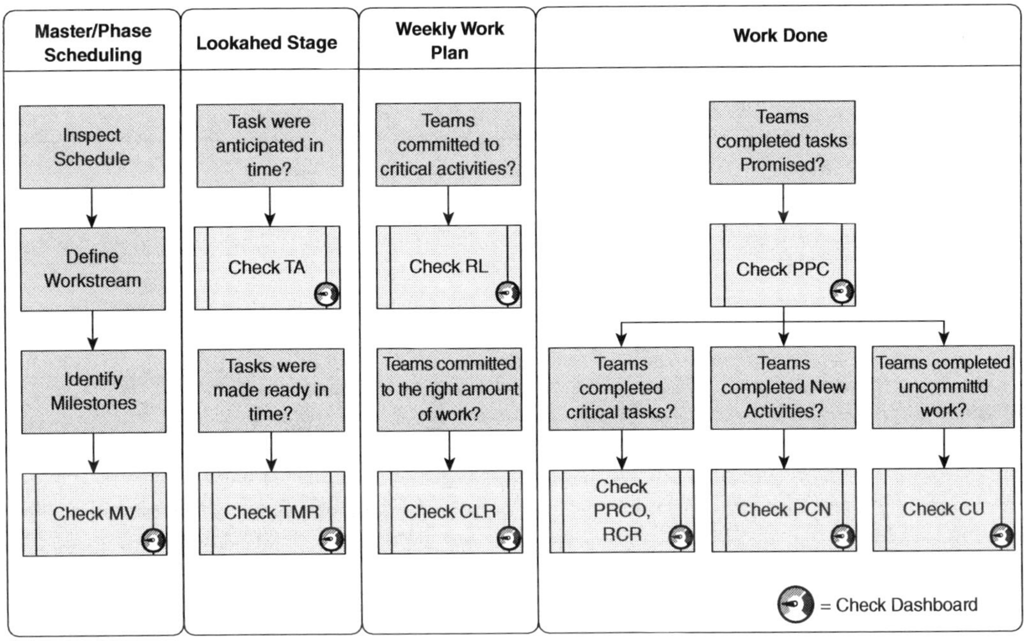

Figure 16.3 LPS diagnosis/control framework

Source: Retrieved from Hamzeh, El Samad *et al.* (2019)

suggested framework identifies at what stage and under which occasion the presented metrics shall be used. The framework reflects the ability of teams in anticipating tasks and making them ready. It also helps in recognizing the teams' commitment level and their responsiveness to critical activities as well as to the capacity planning problems.

The first step in the presented framework is inspecting the schedule to define the work stream. This is followed by identifying milestones from the master schedule to be able to track regularly using the Milestone Variance (MV) metric. The MV represents the difference between the expected date to complete the tasks and the assigned milestone date based on the initial master schedule. Second, in the lookahead stage, the planners need to check whether tasks are being anticipated and made ready in time. This is done through checking the TA and TMR metrics. The third phase represents checking the weekly work plan through emphasizing on the ability of teams to complete the critical activities on one hand, and to commit to the adequate amount of work capacity on the other. This is achieved through calculating the RL and CLR metrics. The final stage is performed after completing the work, where metrics are employed to examine the executed tasks whether Critical/Required, New, or Backlog. This step includes the calculation of PRCO, PCN, and CU metrics.

It is important to perform the aforementioned steps within the proposed framework on both the project and team levels. To have a complete understanding of the project progress and performance, there is a need to understand the big picture first through applying this framework on the project level. Then, to reveal more details, the exercise shall be repeated within the different teams of last planners who control workflows. The resulting trends from this exercise shall be analyzed and compared to the project key performance indicators to be able to take actions based on the analysis of metrics' performance and on what is revealed on site.

Sample calculation of metrics

The sample calculations in this section are in reference to the calculations made in Hamzeh, El Samad *et al.* (2019). Data collected from the empirical study was used to calculate the proposed metrics and a sample calculation is shown in Figure 16.4. The figure shows the number of activities anticipated two weeks ahead of execution, the number of tasks committed (WWP), as well as the tasks done while considering whether the task is critical, not critical, new, or backlog.

First, TA was calculated by dividing the number of tasks on the WWP that were anticipated in the lookahead stage (11) by the total number of tasks on the WWP (13), which resulted in a TA value of 84.6%. TMR was then calculated by dividing the anticipated tasks completed (8) by the total number of tasks in the lookahead (14), where the resulting TMR was 57%. Next, RL was calculated by dividing the number of critical activities on the WWP (10) by the total number of tasks on the WWP (13), resulting in an RL value of 77%. CLR is the ratio of completed tasks (11) to the number of tasks on the WWP (13), the calculated CLR was equal to 0.85. Finally, metrics assessing the work done were calculated. Accordingly, the resulting PPC value of 69% was calculated by dividing the number of tasks completed that were on the WWP (9) by the total tasks on the WWP (13); the PRCO value of 80% was obtained by dividing the number of critical tasks completed (8) by the number of critical tasks on the WWP (10); the RCR metric was calculated by dividing the critical tasks completed (8) by all tasks completed (11), resulting in an RCR value of 0.73; the CU metric was obtained by dividing the completed backlog tasks (2) by the total completed (11), resulting

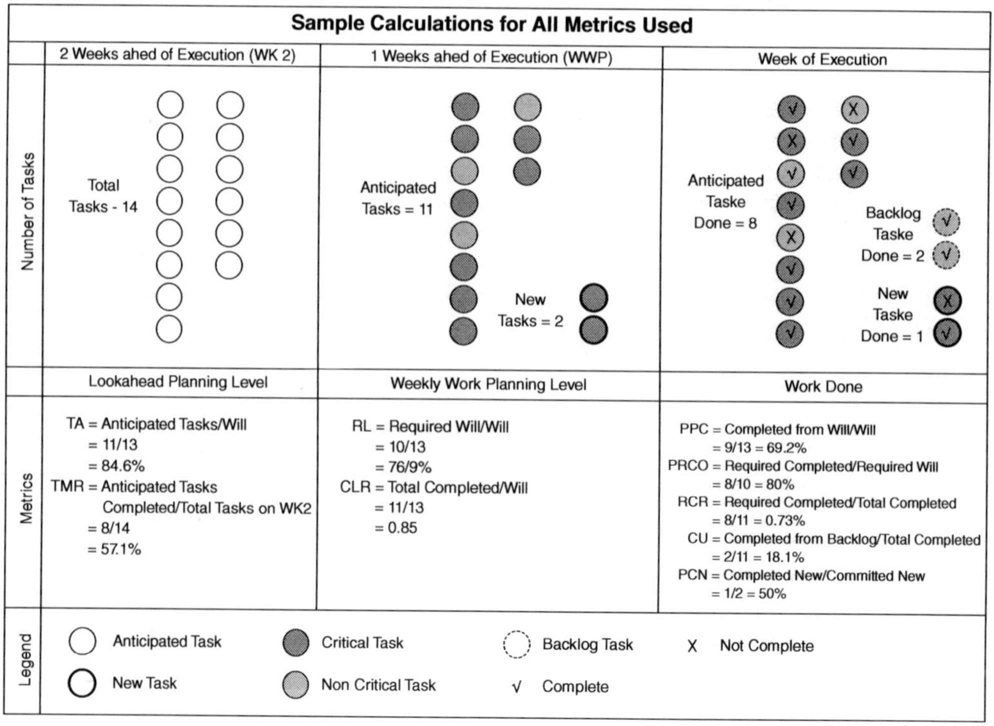

Figure 16.4 Sample calculations of metrics used

Source: Retrieved from Hamzeh, El Samad *et al.* (2019)

in a CU value of 18%; finally, the PCN value of 50% was measured by dividing the new tasks completed (1) by the new tasks on the WWP (2).

Planner's role in project planning

Project performance with respect to time, cost, and quality is often affected by external and internal factors that occur during the course of construction. Usually, external factors are beyond the control of the management team. However, internal factors, related to the design and performance of the product, can be controlled and should be properly managed to improve the ultimate performance of the project. In this regard, planning methodologies represent a major internal factor critical to the project's success, specifically when realized during the early phases (Dvir, Raz and Shenhar 2003). In the previous sections, the authors discussed existing and newly proposed metrics for enhancing project performance. Additionally, the authors highlighted that capacity planning presents a main challenge to planners because it is greatly affected by variability. However, an important yet overlooked feature critical to the capacity planning exercise is the planners' role and their decisions. In this section, the authors target the planner's role in the lookahead planning process. The research discusses the effects of the planners' capacity planning strategies on the cost, time, and quality of construction projects.

This study examines three aspects of variability that impact the balance between load and capacity: (1) variability in the load predicted for the weekly work plan; (2) variability in the capacity of resources decided by the planner, and (3) variability in the actual tasks ready during a given week. These three aspects, along with their effects on project performance, are not fully studied in the literature. Additionally, this study differentiates between two categories of planners, Informed and Un-Informed planners. Four different planning methods are associated with both types of planners as shown in Table 16.3.

A simulation model was developed based on nine different project profiles, and variability was embedded with different intensities. Further elaboration of the model is presented in Abou Ibrahim *et al.* (2019).

Table 16.3 Planners and planning methods

Category	*Planning Method*	*Description*	*CAP Calculated*
Un-informed Planners	Constant CAP	Planners are assumed to choose a CAP of 1.0 every week regardless of project performance.	$CAP_n = 1.0$
	Variable CAP	Planners are assumed to variably select weekly CAPs following a normal distribution of mean 1 and standard deviation of 0.15.	CAP_n = Normal(1,0.15)
Informed Planners	Average CAP	Planners choose weekly CAPs according to the average of previous optimum CAP values.	CAP_n = avg (CAP_{opt})
	Metrics' Average	Planners choose weekly CAPs according to the average of previous RR, NR, NBR, and N.	CAP = {($Ready_n$*RR_{avg}+NotQuiteReady$_n$*NR_{avg}+ All InitialActivities (excluding those falling back from the previous week)*NBR_{avg}*N_{avg})} /WWP

Source: Retrieved from Abou Ibrahim *et al.* (2019)

Under the proposed circumstances, different capacity planning methods were investigated with respect to time, cost, and quality parameters. The following sections present the major discussions and findings of the research.

Impact of planner's decisions on project's cost and time

The planner's decision is vital in balancing capacity to load during the lookahead phase. However, the planner's decision is also considered a cause of variability during production. The planner needs to estimate the number of the following activities: those that will actually be ready (RR) from the pool of (Ready) tasks, the Not ready activities but that can be made ready during the week tasks (Not Ready CMR), and the probable (New) emerging tasks within the execution week. As such, Equation (1) was developed to represent the planner's estimation for the probable load and the needed capacity. Note that CAP is a capacity factor assigned by the planner and is multiplied by the assessed number of activities.

$$Capacity = (Ready\ Activities + NotReadyCMR\ Activities) \times CAP \quad (8)$$

According to Equation (8), the planner strives to achieve an optimum CAP. The optimum CAP is the number upon which any guess under this value would result in an under allocation of resources and schedule delays, and any guess above this value would result in the over allocation of resources and a subsequent increase in cost. The output of the simulation exercise is based on this concept in addition to other developed equations and parameters. The developed parameters are concerned with cost, time, and quality attributes. After conducting the simulation experiments, the cost and time performances were measured against the different planning methods adopted by planners.

As summarized in Table 3, the planners are applying four different methods to estimate CAP: (1) a Constant CAP where planners are assumed to choose a CAP of 1.0 every week regardless of project performance; (2) a Variable CAP where planners are assumed to variably select weekly CAPs following a normal distribution of mean 1 and standard deviation of 0.15; (3) an Average CAP where planners choose weekly CAPs according to the average of previous optimum CAP values; and (4) a Metrics' Average where planners choose weekly CAPs according to the average of previous weekly LPS metrics as proposed by Hamzeh, Zankoul *et al.* (2015).

Results reveal that variability caused by fluctuation of available resources or even by planner's decision negatively affects the project's schedule and budget. However, informed planners who (a) use advanced planning metrics, (b) are aware of their project's characteristics, and (c) assess the type of variability they are facing, are able to attain better results in balancing load and capacity measures, thus minimizing the hidden risks associated with such uncertainties.

Results also show that the "Metric Average" method was performing the best with respect to cost and is followed by the "Average CAP". As for time performance, the "Variable CAP", represented by the random selection of capacity, leads to underestimating resources. This in turn results in activities starving on resources and, consequently, schedule delays. Surprisingly, the "Constant CAP" had the best time performance. Yet, this advantage is often associated with extra incurred costs. Accordingly, Informed Planners seem to have the best balance between cost and time performances.

Moreover, informed capacity planning, where planners use advanced metrics, helps in optimizing the use of buffers and in decreasing their sizes. This leads to minimizing waste in production and maximizing the buffers benefits as advocated by Park and Pena-Mora (2004). Planners who are continuously adding capacity buffers or extra resources eventually cause resources to be sub-optimally used in addition to incurring an increase in cost.'

Impact of planner's decisions on quality

Quality of capacity planning is another measure studied in this research in connection with capacity planning and the planners' decisions. The metric developed for measuring quality is represented in Equation (9).

$$CPQ = \sqrt{\sum_{i=1}^{i=n} \frac{\left(CLR_i - 1\right)^2}{n}} \tag{9}$$

CPQ is the deviation of the capacity to load ratio (CLR) from the desired value of one, where planned capacity is equal to the executed load (the optimum case).

Results from testing nine project profiles over the four different planning methods reveal that the "Variable CAP" method results in the highest CPQ for all profiles. Therefore, a planner's decision that adds variability to the process will negatively affect the quality of capacity planning irrespective of project characteristics. In summary, planners play an important role in construction projects and have direct or indirect effects on the project performance with respect to time, cost, and quality. Informed planners are more likely to achieve a good balance between load and capacity, and thus shield the project from the unfavorable effects of variability.

Summary of research findings

The framework and the metrics explained were tested on two case studies as clarified in the methodology section. The recorded results and the detailed explanations found in Hamzeh, El Samad *et al.* (2019) and Hamzeh, Al Hattab *et al.* (2019) reveal that the proposed metrics are suspenseful in showing aspects of planning not covered in current LPS metrics. Accordingly, a set of planning issues were revealed and are summarized as follows:

(1) Studying project performance at both the project and team levels is of great importance to attain the required benefits;
(2) Metrics shall not be selectively used, instead it is important to interpret metrics collectively since planning involves multiple independent yet interrelated aspects including: capacity planning, anticipating tasks, reliability, and prioritizing critical tasks;
(3) Metrics are established for tracking, measuring, and evaluating performance; therefore, they shall be always related to key performance indicators;
(4) Metrics should be studied on a weekly basis to assess weekly performance and an interpretation of their general trends helps show a more global perspective;
(5) Given that resources are limited, teams seem to struggle in prioritizing the allocation of resources to critical tasks, a matter that highlights implicit issues with capacity planning;
(6) New tasks are often expected to emerge as a result of uncertainty and interrelation of activities. Therefore, their impact could be alleviated through proper planning using a collective analysis of different metrics;
(7) The proposed framework for metrics use in a dashboard setting brings visibility earlier so teams could engage corrective measures to improve reliability of plans and reduce delays;
(8) Since planners play an important role in construction projects and since results show that informed planners are more likely to achieve a good balance between load and capacity, applying the new proposed metrics and training planners on using the proposed framework will enhance the planning process and project performance.

Conclusions

This research is concerned with improving planning and control methods on construction projects. The LPS is employed on projects to enhance reliability, workflow, and planning efforts. The existing metrics within the LPS system were discussed revealing their positive outcomes. However, certain aspects of planning are not covered where each set of available metrics only measures one facet of planning without reflecting overall project performance. Accordingly, newly developed metrics were proposed to help overcome the deficiencies and to complement existing metrics in better measuring, tracking, and evaluating different aspects of planning including capacity planning.

To improve the weekly work planning process, additional metrics are proposed. Required Level (RL), Completed Uncommitted (CU), and Percent Complete New (PCN) should be used to complement PPC. To acquire a more thorough analysis of planning and enhance capacity planning, four primary metrics (CLR, RCR, RPC, and WD) and three secondary metrics (Glut, Starvation, and Misallocation Factor) were derived to complement existing ones. A framework was accordingly proposed to organize and employ the various metrics. The presented framework encourages planners to track the LPS performance at both project and team levels. The metrics involved within the framework assess the team's agility in anticipating tasks, in making them ready for execution, in committing to required tasks, in completing critical/required, not required, and new tasks, and finally, in achieving an adequate capacity planning exercise.

The framework was tested on two projects in the United States to validate its importance. Results reveal several important issues that concern planners on any construction project. First, to be able to evaluate performance at the lookahead level, planners need to employ the PCN metric together with the previously established TMR and TA metrics. Second, since planners must prioritize critical activities that directly affect the project schedule and milestones, measuring the set of RL, PRCO, and PPC metrics will help planners in tracking performance and predicting any probable milestone variance. This, in turn, can guide planners to act and take the essential control measures. Finally, to improve capacity planning, planners should measure the CLR and CU metrics while considering the activity's status by calculating the RCR, RPC, and Misallocation Factor metrics. Assessing if the work completed is based on the system status is of utmost importance for planners to achieve an optimal performance.

This research is expected to improve future implementation of the LPS and better guide planning practices through employing the proposed framework and metrics. Having the advantage of tracking performance through multiple interconnected metrics will help planners perform a global evaluation of the issues at hand and then implement counter measures to steer the project towards its required goals.

Questions for discussion

1 Why is it important to employ and calculate metrics on construction projects?
2 What are some of the limitations that necessitated the development of new metrics to complement the existing metrics pertaining to the Last Planner® System?
3 How can teams cope with the emergence of new tasks and what are the metrics that deal with activities' statuses?
4 On what levels should the planners focus? On the project level or the team level?
5 How can planners and project managers apply the different established metrics on construction projects?
6 What role does the planner have in considering capacity planning, and why is it important to emphasize this role?

Notes

* Dr. Farook Hamzeh is an Associate Professor in Civil and Environmental Engineering at the University of Alberta, Canada. He earned his Ph.D. at the University of California, Berkeley. As an expert in Lean Construction and Production Planning and Control, his theoretical and applied research in the US, Canada, and the MENA region aim at improving the delivery of design and construction. (Corresponding author).

Salam Khalife, PhD Student, Dept. of Civil and Env. Eng., American University of Beirut, Beirut, Lebanon.

Ghali El Samad, Masters Graduate, Dept. of Civil and Env. Eng., American University of Beirut, Beirut, Lebanon.

Lynn Rizk, Masters Graduate, Dept. of Civil and Env. Eng., American University of Beirut, Beirut, Lebanon.

Hisham Abou-Ibrahim, PhD Candidate, Dept. of Civil and Env. Eng., American University of Beirut, Beirut, Lebanon.

Malak Al Hattab, Research Associate, PhD, Dept. of Civil and Env. Eng, American University of Beirut, Beirut, Lebanon,

Samir Emdanat, Director of Management Services, Ghafari Associates, LLC, Michigan, USA.

References

Abou Ibrahim, H., F. R. Hamzeh, E. Zankoul, S. Lindhard, and L. Rizk. 2019. Understanding the planner's role in lookahead production planning. *Production Planning and Control* 30(4): 271–284. DOI: 10.1080/09537287.2018.1524163, https://www.tandfonline.com/doi/full/10.1080/09537287.2018.1524163

Alarcón, L. F., J. L. Salvatierra, and J. A. Letelier. 2014. Using last planner indicators to identify early signs of project performance. *Proceedings of the 22nd annual conference of the international group for lean construction.* Oslo, Norway, 547–558.

Aziz, R., and S. Hafez. 2013. Applying lean thinking in construction and performance improvement. *Alexandria Engineering Journal* 52(4): 679–695.

Ballard, G. 2000. *The Last Planner® System of production control.* PhD Dissertation. University of Birmingham, UK.

Ballard, G., and G. Howell. 1998. Shielding production: Essential step in production control. *Journal of Construction Engineering and Management* 124(1): 11–17.

Ballard, G., and I. Tommelein. 2016. Current process benchmark for the last planner® System. *Lean Construction Journal*, 57–89.

Birrell, G. S. 1987. Discussion of 'Criticism of CPM for project planning analysis' by Ali Jaafari (June 1984, Vol. 110, No. 2). *Journal of Construction Engineering and Management* 10.1061/(ASCE) 0733–9364(1987)113: 2(343).

Bustamante, J. A. 2007. *Implementing of a rational commitment model for improving planning reliability in construction projects.* Masters Thesis, Faculty of Engineering, Pontificia Universidad Católica de Chile, Chile.

Cândido, L. F., L. F. M., Heineck, and J. D. P. B. Neto. 2014. Critical analysis on earned value management (EVM) technique in building construction. *Proceedings of the 22nd annual conference of the international group for lean construction.* Oslo, Norway, 159–170.

Chan, A. P. C., D. Scott, and A. P. L. Chan. 2004. Factors affecting the success of a construction project. *Journal of Construction Engineering and Management* 1(153): 130.

Chitla, V. R., and T. S. Abdelhamid. 2003. Comparing process improvement initiatives based on percent plan complete and labor utilization factors. *Proceedings of the 11th annual conference of the international group for lean construction.* Virginia, USA.

Damci, A., D. Arditi, and G. Polat. 2013. Resource leveling in line-of-balance scheduling. *Computer-Aided Civil and Infrastructure Engineering* 28(9): 679–692.

Damci, A., and G. Polat. 2014. Impacts of different objective functions on resource leveling in construction projects: A case study. *Journal of Civil Engineering and Management* 20(4): 537–547.

Demachkieh, F., and M-A. Abdul-Malak. 2018. Synthesis of improvements to EVMS key parameters representation. *Proceedings of the construction research congress*, 399–408. New Orleans, Louisiana, April 2–4, 2018.

Dvir, D., T. Raz, and A. J. Shenhar. 2003. An empirical analysis of the relationship between project planning and project success. *International Journal of Project Management* 21: 89–95.

El-Rayes, K., and D. Jun. 2009. Optimizing resource leveling in construction projects. *Journal of Construction Engineering and Management* 135(11): 1172–1180.

El Samad, G., F. Hamzeh, and S. Emdanat. 2017. Last planner system – the need for new metrics, *Proceedings of the 25th annual conference of the international group for lean construction.* Heraklion, Greece, 637–644.

Emdanat, S., and M. Azambuja. 2016. Aligning near and long term planning for LPS implementations: A review of existing and new metrics, *Proceedings of the 24th annual conference of the international group for lean construction.* Boston, MA, 103–112.

González, V., L. F. Alarcón, S. Maturana, F. Mundaca, and J. Bustamante. 2010. Improving planning reliability and project performance using the reliable commitment model. *Journal of Construction Engineering and Management* 136(10): 1129–1139.

González, V., L. F. Alarcon, and F. Mundaca. 2008. Investigating the relationship between planning reliability and project performance. *Production Planning and Control* 19(5): 461–474.

Hamzeh, F. R., M. Al Hattab, L. Rizk, G. El Samad, and S. Emdanat. 2019. Developing new metrics to evaluate the performance of capacity planning towards sustainable construction. *Journal of Cleaner Production* 225: 868–882, Elsevier.

Hamzeh, F. R., G. Ballard, and I. D. Tommelein. 2008. Improving construction workflow-the connective role of lookahead planning. *Proceedings of the 16th annual conference of the international group for lean construction*. Manchester, UK, 635–646.

Hamzeh, F. R., G. El Samad, and S. Emdanat. 2019. Advanced metrics for construction planning. *Journal of Construction Engineering and Management, ASCE* 145(11): 1–16. https://doi.org/10.1061/(ASCE)CO.1943-7862.0001702

Hamzeh, F. R., and B. Langerud. 2011. Using simulation to study the impact of improving lookahead planning on the reliability of production planning. *The 2011 winter simulation conference*, Phoenix, AZ, 3436–3447.

Hamzeh, F. R., I. Saab, I. D. Tommelein, and G. Ballard. 2015. Understanding the role of "tasks anticipated" in lookahead planning through simulation. *Automation in Construction* 49: 18–26.

Hamzeh, F. R., E. Zankoul, and C. Rouhana. 2015. How can 'Tasks Made Ready' during lookahead planning impact reliable workflow and project duration? *Construction Management and Economics* 33(4): 243–258.

Hevner, A. R. 2007. A three cycle view of design science research. *Scandinavian Journal of Information Systems* 19(2): 4.

Hevner, A. R., S. T. March, J. Park, and S. Ram. 2004. Design science in information systems research. *MIS Quarterly – Research Essay* 28(1): 75–105.

Howell, G. A., H. G. Ballard, and I. D. Tommelein. 2010. Construction engineering: Reinvigorating the discipline. *Journal of Construction Engineering and Management* 137(10): 740–744.

Jang, J. W., and Y. W. Kim. 2007. Use of percent of constraint removal to measure the make-ready process. *Proceedings of the 15th annual conference of the international group for lean construction*. Michigan, USA.

Junnonen, J., and O. Seppanen. 2004. Task planning as a part of production control. *Proceedings of the 12th annual conference of the international group for lean construction*. Helsingor, Denmark, 1–11.

Kerzner, H. 1998. *Project management, a systems approach to planning, scheduling and controlling*, 6th ed. New York: Wiley.

Kerzner, H. 2013. *Project management metrics, KPIs, and dashboards – A guide to measuring and monitoring project performance*, 2nd ed. Hoboken, NJ: John Wiley & Sons.

Khattab, M., and K. Søyland. 1996. Limited-resource allocation in construction projects. *Computers and Industrial Engineering* 31(2): 229–232.

Kim, S., Y. W. Kim, and C. Park. 2008. Introduction of workforce database system for effective production planning. *Proceedings of the 16th annual conference of the international group for lean construction*. Manchester, UK, 833–838.

Kim, Y. W., and G. Ballard. 2000. Is the earned-value method an enemy of work flow? *Proceedings of the 8th annual conference of the international group for lean construction*. Brighton, UK.

Koulinas, G., and K. Anagnostopoulos. 2013. A new tabu search-based hyper-heuristic algorithm for solving construction leveling problems with limited resource availabilities. *Automation in Construction* 31: 169–175.

Laufer, A., and D. Cohenca. 1990. Factors affecting construction-planning outcomes. *Journal of Construction Engineering and Management* 116(1): 135.

Laufer, A., and R. L. Tucker. 1987. Is construction project planning really doing its job? A critical examination of focus, role and process. *Construction Management and Economics* 5(3): 243–266. doi:10.1080/01446198700000023.

Mitropoulos, P. T. 2005. Planned work ready: A proactive metric for project control. *Proceedings of the 13th annual conference of the international group for lean construction*. Sydney, Australia, 235–242.

Mundaca, F. 2006. *Variability analysis and planning reliability improvement in construction projects*. Masters Thesis, Department of Civil Engineering, Universidad Técnica Federico Santa María, Chile.

Olano, R. M., L. F. Alarcon, and C. Razuri. 2009. Understanding the relationship between planning reliability and schedule performance: A case study. *Proceedings of the 17th annual conference of the international group for lean construction*. Taipei, Taiwan, 139–153.

Park, M., and F. Pena-Mora. 2004. Reliability buffering for construction projects. *Journal of Construction Engineering and Management* 130(5): 626–637.

Ponz-Tienda, J., A. Salcedo-Bernal, E. Pellicer, and J. Benlloch-Marco. 2017. Improved adaptive harmony search algorithm for the resource leveling problem with minimal lags. *Automation in Construction* 77: 82–92.

Rizk, L., F. Hamzeh, and S. Emdanat. 2017. Introducing new capacity planning metrics in production planning. *Proceedings of the 25th annual conference of the international group for lean construction*. Heraklion, Greece.

Rocha, C. G., C. T. da, Formoso, P. Tzortzopoulos, L. Koskela, and A. Tezel. 2012. Design science research in lean construction: Process and outcomes. *Proceedings of the 20th annual conference of the international group for lean construction*. San Diego, CA.

Rouhana, C., and F. Hamzeh. 2016. An abc approach to modelling the emergence of 'new tasks' in weekly construction planning. *Lean Construction Journal*, 35–56.

Sacks, R., O. Seppänen, V. Priven, and J. Savosnick. 2017. Construction flow index: A metric of production flow quality in construction. *Construction Management and Economics* 35(1–2): 45–63.
Shehata, M. E., and K. M. El-Gohary. 2011. Towards improving construction labor productivity and projects' performance. *Alexandria Engineering Journal* 50(4): 321–330.
Song, K., S. Pollalis, and F. Pena-Mora. 2005. Project dashboard: Concurrent visual representation method of project metrics on 3D building models. *Proceedings, computing in civil engineering*. Cancun, Mexico.
Suk, S.-J., B.-G. Hwang, J. Dai, H. Caldas Carlos, and P. Mulva Stephen. 2012. Performance dashboard for a pharmaceutical project benchmarking program. *Journal of Construction Engineering and Management* 138(7): 864–876.
Tommelein, I., D. Riley, and G. Howell. 1999. Parade game: Impact of work flow variability on trade performance. *Journal of Construction Management* 125(5): 304–310.
Weber, R. 2018. Ch. 11. Design-science research. In *Research methods*, eds. K. Williamson, and G. Johanson, 2nd ed. Burwood: Chandos Publishing, 267–288.

Bibliography

Hamzeh, F. R., G. Ballard, and I. D. Tommelein. 2012. Rethinking lookahead planning to optimize construction workflow. *Lean Construction Journal*, 15–34.

GLOSSARY OF LEAN TERMS

A3 report The A3 report is a disciplined and rigorous system for implementing PDCA management. As used by Toyota, the report fits on a single sheet of A3 paper, measuring approximately 11 × 17 and documents problems and the available information about them in a manner that helps users to understand quickly. It focuses users on improving processes and solving problems.

Activity definition model The ADM represents design tasks or construction processes and enables planners to examine scheduled tasks in detail and determine their readiness for execution. These tasks are released only if all constraints are removed, thereby improving performance.

Allowable cost The maximum amount the owner is prepared to spend for a facility asset. This cost is an important component of Target Value Design (TVD).

Autonomation Lean construction applies the principle of autonomation that is an important ingredient of the Toyota Production System (TPS); the people closest to the work are empowered to stop production if they determine that the upstream production is defective. The last planners and their crews fulfill this role when using The Last Planner® System.

Basis of design The basis of design (BOD) reflects the owner's needs and wants that must be satisfied by the design of a project. This may include the use of spaces, their sizes, finishes, and activities to be performed within them.

Benchmarking This involves comparing actual or planned project practices to those of other projects to generate ideas for improvement and to provide a standard by which to measure performance.

Buffer The means (capacity, inventory, time) used to cushion against the shock of variation in a process. Two types of buffers may be used to shield downstream construction processes from flow variation. Plan buffers are inventories of workable assignments. Schedule buffers are materials, tools, equipment, manpower, time, and so on.

Building information modeling Building information modeling (BIM) is the process of generating and managing building data during its life cycle. It is a model-based technology linked with a database of project information. A BIM carries all information related to a facility, including its physical and functional characteristics and project life cycle information, in a series of smart objects.

Chase UP Syndrome © represents the significant loss of time spent chasing people who have made promises but are tardy in meeting them. It also includes the additional loss of time in being chased by others. Ed Anderson, President and CEO of Lean Implementation Services of Florida, has identified and copyrighted this waste. In asking over 1,500 people in many countries during a 16 year period

Ed states that the answer to these two questions has never been less than 60% of work hours lost due to this waste.

Choosing by advantages (CBA) A tested and effective system for decision-making that was developed by Jim Suhr. It determines the best decision by looking at the advantages of each option. CBA has five phases of decision-making: – 1) Establishing the purpose for the decision 2). Innovation – formulating an adequate set of alternatives 3) Decision-making – choosing the alternative with the greatest total importance of advantages, 4) Reconsideration – changing the decision if it can be improved on 5). Implementing the decision, adjusting as needed.

Commissioning From the design phase to a minimum of 1 year after construction, this is a systematic process of assurance by verification and documentation that all facility systems perform efficiently and in accordance with reconsider the design documentation and intent and meet the owner's operational needs.

Commissioning agent The CxA (or Authority) is a professional that serves as an objective advocate for the owner and is responsible for directing the commissioning team and process; coordinating, overseeing, and/or performing the commissioning testing; and reviewing the results of the system performance verification.

Commitment planning Commitment planning is a method for defining criteria that lead to the selection of quality assignments. It results in commitments to deliver that other production teams can rely on by following a prerequisite that only sound assignments should be made or accepted.

Commitment reliability Commitment reliability is made by producers (contractors, designers, etc.) in the design/construction supply chain in order to meet schedules. Commitment reliability is calculated by comparing completed work (DID) with planned work (WILL).

Conditions of Satisfaction A set of criteria or expectations that specify how the success of an outcome will be determined.

Consensus Docs 300 This is a standard form of contract for the construction industry that is based on an Integrated Form of Agreement (IFOA). Twenty-two leading construction associations came together and published a consensus set of contract documents in 2007 that they felt was fair to all parties.

Constraint A constraint is an obstacle that inhibits the execution of a task that is required in the lookahead plan. It should also be beyond the control of the last planner.

Constraint analysis Foremen, project managers, schedulers, and other staff communicate to identify constraints and the reasons for their occurrence. After analysis, they are able to solve and remove the constraint and assign work tasks.

Continuous flow Producing and moving one item at a time (or a small and consistent batch of items) through a series of steps as continuously as possible, with each step making just what is requested by the next step. Also called one-piece flow or single-piece flow.

Continuous improvement Small improvements to reduce costs and ensure consistency in performance of products and services.

Control chart A graphic display in the progression of a process. It is used to determine if the process is under control and progressing on time.

Cycle time How often a part or product is completed by a process, as timed by observation. This time includes operating time plus the time required to prepare, load, and unload. The appropriate calculation of cycle time may depend upon context. For example, if a paint process completes a batch of 22 parts every five minutes, the cycle time for the batch is 5 minutes. However, the cycle time for an individual part is 13.6 seconds (5 minutes × 60 seconds = 300 seconds, divided by 22 parts = 13.6 seconds).

Design assist A project arrangement in which architects prepare a design and trade contractors estimate the cost based on the information provided to them.

Design-bid-build Traditional contracting method where the architect and contractor secure separate contracts with the owner to provide specified services.

Design-build A project delivery method in which one entity provides design and construction services under as single responsibility contract to an owner.

Design criteria Design criteria for buildings define the owner's requirements, including those that may exceed applicable codes. They primarily serve as a performance-based guide to designers, to ensure that ensuing designs meet the functional needs and preferences of the owner. Design criteria list applicable codes and standards such as weather-related and earthquake codes as are relevant to the location. A criteria document typically describes the building systems and materials to be used. It may also define project-specific requirements such as building orientation, space descriptions, and adjacencies to meet owners' operating needs.

Design of experiments A statistical method that identifies those factors that might influence specific variables.

Energy Star Energy Star is a reference system developed by the US Environmental Protection Agency that promotes efficient energy usage in facilities and the systems and appliances that are used in them. Buildings that have earned the ENERGY STAR label use an average of almost 40% less energy than average buildings and emit 35% less carbon.

Evidence-based design (EBD) The conscientious and judicious use of current best evidence, and its critical interpretation, to make significant design decisions for each unique project. These design decisions should be based on sound hypotheses related to measurable outcomes.

First Run studies These studies involve trying out new construction ideas on a pilot basis and applying the PDCA methodology to determine their success. First run studies also identify the best means, methods, and sequencing for a specific activity.

Fishbone diagram Also called the Ishikawa diagram or cause-and-effect diagram. The diagram illustrates how various causes and subcauses relate to creating potential problems. It may be used to determine the root of observed problems.

Five Ss One of the guiding principles of Lean meant to achieve standardization and stability in the workplace through visual management. (1) **Sort**: Remove everything from the workplace that is not needed for current production. (2) **Set in order**: Arrange items so they are easy to find and return and locate items to minimize motion waste. (3) **Shine**: Keep everything clean and in top condition so that it is ready to be used. (4) **Standardize**: On-going maintenance of sort, set in order, and shine. (5) **Sustain**: Create the conditions or structures that will help sustain commitment.

Five Whys A method of root cause analysis whereby the investigator repeatedly probes operators and other employees with the simple question "why?" until the problem is uncovered.

Flowcharting A flow chart is any diagram that shows how various elements of a system relate (cause and effect diagrams, system or process flow charts).

Gemba A Japanese term that means the place where things are actually happening. *Gembashugi*: shop floor oriented (Shugi = orientation or philosophy)

Hand-off the act of releasing an item or activity, e.g., a room that has had drywall installed being handed off to a paint crew for completion.

Hansei A Japanese term that means critical self-reflection.

Heijunka Creating basic stability in processes. To produce goods at a constant and predictable rate.

Hoshin Kanri Japanese term for policy deployment. Its components are Hoshin, a compass, course, and policy; Kanri, management control.

Information and communication technology The ICT encompasses computer hardware, software, and communications devices that allow the sharing and access of information between many users. In effect, it creates an "information village."

Integrated Form of Agreement The IFOA is a legal agreement that seeks to align the commercial relationships of a construction project's design and construction participants that are assembled as a temporary production system. It also requires collaborating through planning, design, construction, and managing the project as a network of commitments. As a result it optimizes the project as a whole, rather than any particular piece. The agreement also calls for a combination of learning with action and promoting continuous improvement throughout the life of the project.

Integrated project delivery The IPD is a relational contracting approach that aligns project objectives with the interests of key participants. It creates an organization able to apply the principles and practices of The Lean Project Delivery System™. The fundamental principle of IPD is the close collaboration of a team that is focused on optimizing the entire project as opposed to seeking the self-interest of their respective organizations.

IPD team performance contingency A provision in integrated project delivery for the sharing of costs and benefits based on the performance of the IPD team.

Jidoka Providing machines and operators the ability to detect when an abnormal condition has occurred and immediately stop work. This enables operations to build in quality at each process and to separate men and machines for more efficient work. Jidoka is one of the two pillars of the Toyota Production System along with Just-In-Time. Jidoka is sometimes called autonomation, meaning automation with human intelligence.

Just-in-time production A system of production that makes and delivers just what is needed, just when it is needed, and just in the amount needed. JIT and Jidoka are the two pillars of the Toyota Production System.

Kaizen Continuous improvement of an entire value stream or an individual process to create more value with less waste. There are two levels of kaizen: (1) system or flow Kaizen focuses on the overall value stream and (2) process Kaizen focuses on individual processes.

Kanban A signaling device that gives authorization and instructions for the production or withdrawal (conveyance) of items in a pull system. The term is Japanese for sign or signboard.

Last Planner® System The LPS is an important subset of The Lean Project Delivery System™ and is critical to its effective deployment. The LPS accommodates project variability and smooths workflow so that labor and material resources can be maximally productive. It uses lean methods to provide improved project control.

Last responsible moment In considering alternatives, the last responsible moment for one alternative is the time when, if that alternative is not selected and pursued, that alternative is no longer viable.

Lean operations Lean operations enable a facility owner/user to maximize the use of the built environment. Lean design should facilitate lean operations in a completed facility. A poorly laid out facility could be relatively expensive to operate – excessive travel by users from one point to another is one of the seven wastes documented in the Toyota Method.

Lean production A business system for organizing and managing product development, operations, suppliers, and customer relations that requires less human effort, less space, less capital, and less time to make products with fewer defects to precise customer desires, compared with the previous system of mass production. It was pioneered by Toyota after World War II.

Lean Project Delivery System™ The LPDS is a lean technique that integrates phases to facilitate the design and delivery of construction projects. These phases are: (1) project definition, (2) lean design, (3) lean supply, (4) lean assembly, and (5) use. The LPDS is based on a close collaboration between the members of the project delivery team. They are bound by codes of conduct – both written and unwritten – that focus on the success of the overall project rather than their individual success.

LEED Leadership in Energy and Environmental Design was developed by the US Green Building Council. It is a voluntary, consensus-based national standard for developing high-performance, sustainable buildings. The LEED® certification program is a feature-oriented rating system that awards buildings points for satisfying specified green building criteria.

Life cycle cost The basic premise of the (LCCA) method is that to an investor or decision maker can be aware of all costs arising from an investment decision, including future as well as present costs. Applied to buildings or building systems, the LCC encompasses all relevant costs over a designated study period, including the costs of designing, purchasing/leasing, constructing/installing, operating, maintaining, repairing, replacing, and disposing of a particular building design or system.

Linguistic action The theory of linguistic action describes the purposeful ways in which people communicate – the extent to which they mean what they say. In the lean construction environment it relates to the conversational nature of design, planning, and coordination. The theory influences the securing of reliable promises that are critical to the success ofThe Last Planner® System and lean construction in general.

Load leveling A procedure where demand is rearranged so it is distributed evenly during a specified time period.

Look ahead plan A plan that covers work for a designated period – typically 6 to 8 weeks for construction, or 2 to 4 weeks for design. It is based on work assignments pulled from the master schedule through the process of phase pull planning. It represents work that can be done.

Muda Seven classes of waste: over production, delays, transportation, processing, excess inventory, wasted motion, and defective parts.

Mura Japanese for unevenness in the production system relating to man power or materials. Includes day-to-day variation in customer demand.

Mura (variation) Mura internal to the enterprise does include day-to-day variation in customer demand as long as it is not part of a long-term trend. It also includes gyrations in orders and operational performance progressing up a value stream – through production facilities and suppliers – that are caused by the internal dynamics of the process. These variations are the norm in modern production systems, leading to firefighting and muri (overburden of employees and technologies.)

Muri Overburden of employees and technologies having a direct impact on morale in a negative manner.

Obeya In Japanese means simply "big room." At Toyota it has become a major project management tool, used especially in product development, to enhance effective and timely communication. Similar in concept to traditional "war rooms," an Obeya will usually contain highly visual charts and graphs depicting program timing, milestones and progress to date, and countermeasures to existing timing or technical problems.

Owner's project requirements (OPR) The needs established by an owner for a project (See Design Criteria)

Pacemaker process Any process along a value stream that sets the pace for the entire stream. (The pacemaker process should not be confused with a bottleneck process, which necessarily constrains downstream processes due to a lack of capacity.)

Pareto chart A diagram that is used to display the relative importance in a number of factors. In evaluating problems, it shows the types of problems and the frequency of their occurrence or cost impact based on the so-called 80/20 rule. The chart enables users to readily prioritize needed corrective action.

Percent plans complete In lean construction, work accomplishment is recorded as a graphical plot of PPC; it shows the percentage of the assigned plans (i.e., commitments that were completed, fractional completion is not considered). Some lean construction practitioners refer to PPC as commitment reliability.

Plan, do, check, act An improvement cycle based on the scientific method of proposing a change in a process, implementing the change, measuring the results, and taking appropriate action. The PDCA cycle has four stages: Plan, Do, Check, and Act.

Poka Yoke Japanese term that means mistake (or error) proofing. A poka yoke device is one that prevents incorrect parts from being made or assembled, or easily identifies a flaw or error. Error-proofing is a manufacturing technique of preventing errors by designing the manufacturing process, equipment, and tools so that an operation literally cannot be performed incorrectly.

Prevention over inspection A well-known mantra emphasized by proponents of quality. Prior to the emergence of the quality movement, the main focus of quality was on inspection. Both research and experience pointed to the fact that the net cost of inspecting is so high that it is better to spend money on preventing problems. "Quality must be planned in not inspected in."

Production lead time Also called **throughput time** and **total product cycle time**. The time required for a product to move all the way through a process from start to finish. At the plant level this is often termed door-to-door time. The concept can also be applied to the time required for a design to progress from start to finish in product development or for a product to proceed from raw materials all the way to the customer.

Productivity Productivity is the measure of how well resources are brought together and utilized for accomplishing a set of goals. Productivity is measured as the ratio of outputs to inputs. In the construction environment it may be represented as the constant-in-place value divided by inputs such as the dollar value of material and labor.

Quality assurance Planned and systematic actions to help assure that project components are being designed and constructed in accordance with applicable standards and contract documents.

Quality control The review of project services, construction work, management, and documentation for compliance with contractual and regulatory obligations and accepted industry practices.

Quality function deployment QFD is defined as "a technique to deploy customer requirements into design characteristics and deploy them into subsystems, components, materials, and production processes." QFD may be applied to the design and planning of construction projects; the built facilities meet users' needs more closely than is derived through practitioners' experience alone.

Reasons for noncompletion In conjunction with lean-based construction projects, incomplete work plans are studied each week to determine the root causes or RNC. At each weekly meeting time is devoted to learning why certain tasks were not completed, in order to improve the effectiveness of future work plans.

Relational contracting Relational contracting is a transaction or contracting mechanism that apportions responsibilities and benefits of the contract fairly and transparently, based on trust and partnership between the parties. It provides a more efficient and effective system for construction delivery in projects that require close collaboration for execution. The relationship between the parties transcends the exchange of goods and services and displays the attributes of a community with shared values and trust-based interaction.

Reliability The extent to which a commitment is fulfilled.

Reliable promise A promise made by a performer that others can depend on. 1. The promissor must be competent – having access to skills and/or resources 2. has estimated the time needed 3.has blocked the time needed for the task, 4 has no doubt of the capability to carry out the promise 5. is prepared to accept any upset that may result from not meeting the promise.

Reverse phase scheduling The RPS involves working backward from the deliverables of the completed project, determining what needs to be done at each stage to complete the preceding activity. It is a detailed work plan that specifies the hand offs between trades for each project phase (i.e., what SHOULD be done at each phase). In lean construction, the reverse phase schedule is developed through conversations with the primary subcontractors, including their last planners, typically with sticky notes on a display board.

Sensei A Japanese title used to refer to or address teachers, professors, professionals such as lawyers and doctors, politicians, clergymen, and other figures of authority. The word is also used to show respect to someone who has achieved a certain level of mastery in an art form or some other skill.

Set-based design A design management approach that defers design decisions to the "last responsible moment" to allow for the evaluation of alternatives that improve constructability. Toyota has used this technique to design new models in a much shorter time than the industry standard. They have learned that the best solutions are hybrids of original design options.

Six S Six S is an enhancement of the 5S methodology to include an emphasis on safety. Like 5S it also promotes Lean through visual management to achieve standardization and stability in the workplace.

Statistical sampling Involves choosing part of a population of interest for inspection.

Supply chain The term encompasses all the activities that lead to having an end user provided with a product or service – the chain is comparable to a network that provides a conduit for flows in both directions, such as materials, information, funds, paper, and people.

Sustainability The ability of a society to operate indefinitely into the future without depleting its resources. Sustainability includes concepts of green building design and construction, reuse and recycling of materials, reduced use of material and energy resources for building construction and operation, water conservation, and responsible stewardship of the surrounding environment.

Takt time The available production time divided by customer demand. For example, if a widget factory operates 480 minutes per day and customers demand 240 widgets per day, takt time is 2 minutes. Similarly, if customers want two new products per month, takt time is 2 weeks. The purpose of takt time is to precisely match production with demand. It provides the heartbeat of a lean production system.

Target value design The TVD involves designing to a specific estimate instead of estimating based on a detailed design. It seeks to address the problem that affects many projects – various design disciplines work from a common schematic design to do design development in their areas of expertise.

Team dynamics Influences problem solving, communications skills, conflict resolution.

Total quality management A philosophy that encourages companies and their employees to focus on finding ways to continuously improve the quality of their business practices and products.

Toyota production system The production system developed by Toyota Motor Corporation to provide best quality, lowest cost, and shortest lead time through the elimination of waste. TPS is comprised of two pillars, JIT production and Jidoka. TPS is maintained and improved through iterations of standardized work and Kaizen, following the scientific method of the plan-do-check-act cycle.

Trend analysis Involves using mathematical techniques to forecast future outcomes based on results. It is often used to monitor technical and cost/schedule performance.

US Green Building Council This is a non-profit membership organization whose vision is "a sustainable built environment within a generation," Its membership includes corporations, builders, universities, government agencies, and other non-profit organizations.

Value-added and non-value-added work activities Value-added activities provide desirable outcomes for a customer; that is, they meet the customer's value proposition. Conversely, non-value-added activities do not contribute to meeting a customer's expectations. For example, searching for a tool to perform a task does not get the task done and does not provide value.

Value stream All of the actions, both value-creating and non-value-creating, required to bring a product from concept to launch and from order to delivery. These include actions to process information from the customer and actions to transform the product on its way to the customer.

Value Stream mapping A simple diagram of every step involved in the material and information flows needed to bring a product from order to delivery. A current-state map follows a product's path from order to delivery to determine the current conditions. A future-state map shows the opportunities for improvement identified in the current-state map to achieve a higher level of performance at some future point.

Variation The divergence of a process from an intended plan. The lean philosophy seeks to reduce process variation in order to obtain reliable flow through a process.

Visual management Using visual displays for better communication. It works in conjunction with empowerment so that anyone, not just a supervisor, can detect process anomalies and take prompt corrective action.

Waste Any activity that consumes resources but creates no value for the customer. There are seven wastes as identified in Lean: overproduction, excessive inventory, unnecessary conveyance, overprocessing, excessive motion, waiting, and corrections (of defects).

Work package A compilation of information regarding the availability of material, equipment, resources, and other pertinent information that is necessary to perform a specific scope of work. The work package is coordinated with other interdependent work packages and requirements.

Work structuring The process of defining and organizing activities into logical, practical groups of work. It is a process of subdividing work so that the pieces are different from one production unit to the next to promote flow and throughput, and to have work organized and executed to benefit the project as a whole.

Workable backlog Is used to describe assignments that have met all quality criteria, but may need other prerequisite work to be done before they can be started. Workable backlog enables crews to continue working productively if there is a constraint that prevents the completion of an item on the weekly work plan.

INDEX

Note: Page numbers in *italic* indicate a figure and page numbers in **bold** indicate a table on the corresponding page.